FOURTH EDITION

UNDERSTANDING
GLOBAL
CULTURES

Martin Gannon dedicates this book, with fond memories, to the late Mr. Francis X. Long, distinguished faculty member at the Scranton Preparatory School, Scranton, Pennsylvania.

Rajnandini Pillai dedicates this book to her family and friends. Thank you all for your support.

FOURTH EDITION

UNDERSTANDING GLOBAL CULTURES

Metaphorical Journeys Through 29 Nations,
Clusters of Nations, Continents, and Diversity

Martin J. Gannon

Professor, California State University, San Marcos
Professor Emeritus, Robert H. Smith School of Business, University of Maryland

Rajnandini Pillai

Professor, California State University, San Marcos

Los Angeles | London | New Delhi
Singapore | Washington DC

For information:

SAGE Publications, Inc.
2455 Teller Road
Thousand Oaks, California 91320
E-mail: order@sagepub.com

SAGE Publications Ltd.
1 Oliver's Yard
55 City Road
London EC1Y 1SP
United Kingdom

SAGE Publications India Pvt. Ltd.
B 1/I 1 Mohan Cooperative Industrial Area
Mathura Road, New Delhi 110 044
India

SAGE Publications Asia-Pacific Pte. Ltd.
33 Pekin Street #02-01
Far East Square
Singapore 048763

Printed in the United States of America.

Library of Congress Cataloging-in-Publication Data

Gannon, Martin J.
Understanding global cultures : metaphorical journeys through 29 nations, clusters of nations, continents, and diversity / Martin J. Gannon, Rajnandini Pillai. — 4th ed.
    p. cm.
Includes bibliographical references and index.
ISBN 978-1-4129-5789-2 (pbk. : acid-free paper)
   1. Cross-cultural studies. 2. Cross-cultural orientation. 3. National characteristics.
I. Pillai, Rajnandini. II. Title.

GN345.7.G36 2010
306—dc22   2008049628

This book is printed on acid-free paper.

10 11 12 13 10 9 8 7 6 5 4 3

Acquisitions Editor:	Lisa Cuevas Shaw
Editorial Assistant:	MaryAnn Vail
Production Editor:	Catherine M. Chilton
Copy Editor:	Jacqueline Tasch
Typesetter:	C&M Digitals (P) Ltd.
Proofreader:	Annette R. Van Deusen
Indexer:	Molly Hall
Cover Designer:	Gail Buschman
Marketing Manager:	Jennifer Reed Banando

Brief Contents

Detailed Contents

Preface

Understanding Cultures in Depth

If we are right in suggesting that our conceptual system is largely metaphorical, then the way we think, what we experience, and what we do every day is very much a matter of metaphor.

—Lakoff and Johnson (1980, p. 1)

The emphasis in this book is on an in-depth understanding of a culture through the use of the cultural metaphor, which is any activity, phenomenon, or institution that members of a given culture consider important and with which they identify emotionally and/or cognitively, for example, the Turkish coffeehouse and the Chinese family altar. As such, the metaphor represents the underlying values expressive of the culture itself.

Frequently outsiders have a difficult time relating to and/or understanding a specific cultural metaphor such as American football and the underlying values of a culture that it expresses, and this book is designed to address this difficulty. Culture allows us to "fill in the blanks," often either semiconsciously or unconsciously, when action is required, and cultural metaphors help us to see the values leading to action. This is probably the most interesting feature of culture (see Brislin, 1993; Triandis, 2002). Cultural metaphors, the basis of this book, represent an effective way to profile and to learn about: ethnic cultures, national cultures, the diffusion of a base culture such as the Chinese culture and the English culture across nations, continental cultures, and even diversity within a nation, as the description of kaleidoscopic diversity in India in Chapter 29 demonstrates.

In Chapter 1 we describe how cultural metaphors are developed and how we integrate other approaches to cultures. For example, there are prominent bipolar or dimensional profiles of many national cultures that allow a researcher to rank-order them on various scales, such as power distance or the degree to which there is an unequal distribution of power and rewards (see Hofstede, 2001, and the GLOBE study, House, Hanges, Javidan, Dorfman, & Gupta, 2004). In every chapter we use these other approaches to gain additional knowledge and perspectives and integrate them within our discussion. Cultural anthropologists such as Kluckholn, Strodtbeck, and Hall developed some of the approaches described in Chapter 1. The bipolar or dimensional approach comes from cross-cultural psychology and cross-cultural management.

The first edition of this book in 1994 had 17 chapters describing 17 cultural metaphors and no parts into which the chapters logically fitted. This fourth edition includes 12 major parts and 34 chapters; three new parts were added for this edition. As such, the book has evolved and become richer over time both in terms of new perspectives and depth of material.

We have focused on Scandinavian egalitarian cultures separately from other egalitarian cultures such as Canada, for they are unique in several ways. Hence there are now two parts devoted to egalitarian cultures rather than only one part. We have also included a final integrative chapter in a new part; this chapter links directly to Chapter 1 and the framework we have employed throughout the book. Similarly we have a new separate part of the book focusing on India. As such, there are three chapters on Chinese culture, as in the 3rd edition—covering China at home, Chinese around the world, and Singapore—and two chapters devoted to India in this edition.

An example of the inclusion of new perspectives in this fourth edition is Chapter 29, "India: A Kaleidoscope of Diversity." Most other nations can be assessed in a similar fashion, although the diversity in India is startling. Given the importance of diversity today, we feel that this chapter has enriched the book with a new perspective. We have also updated all chapters and have significantly rewritten several of them, including Chapter 16, "American Football"; Chapter 24, "The Russian Ballet"; and Chapter 20, "The Israeli Kibbutzim and Moshavim." In addition to the new chapter on India, Chapter 10, "The Finnish Sauna," and Chapter 34, "An Integrative Summary," are new in this edition. We have also strengthened the emphasis in the previous editions on a nation's history, religions, geography, and demographics as they relate to and influence a nation's culture.

When Martin Gannon began work on cultural metaphors in 1988, there was a concern that there would be an automatic negative reaction to the concept, because cultural metaphors are related to stereotypes. From the 1950s until about 1990 the typical reaction to any generalization about groups was "That is a stereotype," and the clear meaning was that such a generalization was totally biased. As explained in Chapter 1, however, humans use such generalizations automatically, and some stereotypes are legitimate whereas others are not. Clearly, one illegitimate stereotype is a generalization that allows no exceptions. Cultural metaphors, on the other hand, are probabilistic statements that apply to a group but not to every individual within it. And, as the introductory quote from Lakoff and Johnson points out, humans tend to learn through the use of metaphors.

We would also like to add a short explanation about the many statistics presented in this book and their relationships to the cultural metaphors and the numerous examples used to illustrate them. Frequently the statistics are presented in a comparative fashion, for example, the number of citizens per square kilometer in the United States (31.7) and Singapore (6729.3). These and similar statistics are taken from The Economist's *Pocket World in Figures* (2007) and the World Bank's *The Little Data Book* (2007). We have also used many other sources and have duly noted them. In some instances the statistics are presented without a citation simply to facilitate the ease of reading. When doing so, we have checked with at least two sources. In most if not all such instances, the statistics are generally well known. As suggested, we have used statistics not only to enrich understanding but also to facilitate ease of reading.

There is some work in cultural anthropology that bears directly on the concept of cultural metaphors, particularly Clifford Geertz's (1973) well-known description of male Balinese society in terms of the cockfight. Each of the metaphors in this book uses three to six characteristics of each metaphor to describe the culture of the nation being assessed. Through the use of each metaphor and its characteristics we can begin to see the society in a new and different way and, we hope, in the same manner as its members do. We can also compare societies through the use of these metaphors and their characteristics. There is empirical evidence that (1) cultural metaphors do exist; (2) individuals, both within their own national culture and from other national cultures, tend to think of national cultures in this fashion; and (3) they are important for understanding cross-cultural differences (Gannon, Gupta, Audia, & Kristof-Brown, 2005–2006).

Although each metaphor is a guide or map, it is only a starting point against which we can compare our own experiences and through which we can start to understand the seeming contradictions pervasive in most if not all societies. Thus a cultural metaphor provides an in-depth understanding but one that must be enriched periodically if not constantly by new information and experiences.

Also, while we are describing a dominant, and perhaps the dominant, metaphor for each society, other metaphors may also be suitable. In this book we do not address the issue of suitable alternatives or supplementary metaphors. Furthermore, our descriptions explicitly recognize and focus on the regional, racial, and ethnic differences within each nation.

Still, the unit of analysis for most of the chapters in this book is the nation, as a good amount of evidence suggests that there are commonalities across regional, racial, and ethnic groups within each nation that can be effectively captured by cultural metaphors. As pointed out in "The Malaysian *Balik Kampung*" (Chapter 18), if citizens in a nation cannot identify at least one activity, phenomenon, or institution expressive of their values, the probability is high that the nation will experience difficulty and may even be rent asunder. Furthermore, as indicated above, we can employ cultural metaphors to profile a major base culture such as that of China and its evolution across borders; continents; and even diversity within one nation.

This book contains 29 metaphors for specific nations. There are just 220 nations in the world. Realistically, then, the approach could be used for all or most nations, thus providing a starting point for understanding commonalities across nations and differences between them. We are currently working on several additional cultural metaphors, but they are not ready to see the light of day.

Chapter 1 contains a section describing how to read and use this book. Also, faculty who adopt the fourth edition will receive an instructor's CD containing sample tests, discussion questions, new exercises, short case studies, and PowerPoint slides that can be shown in class.

Acknowledgments

Work on this book began in 1988 when Martin Gannon conducted a seminar for MBA and doctoral students at the University of Maryland. The task proved to be very difficult, and only a small number of papers reached a stage where they could be included in the first edition in 1994. As indicated above, only 17 chapters describing cultural metaphors for specific nations were in that edition.

Clearly this book could not have been written without the help of individuals who are intimately familiar with the specific cultures for which we sought to construct metaphors. They served as coauthors of the various chapters. Our basic approach was for each coauthor to use the methods for constructing metaphors as outlined in Chapter 1 when writing his or her chapter, preceded and followed by intensive discussions both among the coauthors and with other knowledgeable individuals. Each chapter was rewritten in light of the suggestions offered by these individuals.

For the first edition, Martin Gannon wrote the Preface, Chapter 1, and several other chapters and was responsible for restructuring and editing the remaining chapters. The coauthors of the remaining chapters in this and previous editions were Diana Liebscher and Eileen Fagen, Britain; Stefania Amodio and Lynne Levy, Italy; Douglas O'Bannon and Julie Kromkoski, Germany; Peter Brown and Sharon Ribas, France; Ana Hedin and Michelle Allison, Sweden; Paul Forward, Canada; Amy Levitt, Russia; Stacey Hostetler and Sydney Swainston, Belgium; Katherine Feffer Noonan and Manuel Bacerra, Spain; Daniel Cronin and Cormac MacFhionnlaoich, Ireland; Michele Lopez, Mexico; Amy Levitt, Turkey; Efrat Elron, Israel; Isaac Agboola, Nigeria; Diane Terry, Japan; Amit Gupta and Jeffrey Thomas, India; Hakam Kanafani, the United States; Pino Audia, Chinese Family Altar; Takako Sugiyama, Singapore; Jennnifer Lynn Roney, Poland; Christin Cooper and Maria Masatroianni, Brazil; Rozhan Othman, Malaysia; Louise Warberg, Saudi Arabia; Carlos Cantarilho, Portugal; and Paul Dowling, Africa.

Several of the doctoral students who were both participants in this cross-cultural management seminar and eventual coauthors of individual chapters have now become faculty members at major universities. They include Pino Audia, Dartmouth;

Amit Gupta, Indian Institute of Management, Bangalore; Douglas O'Bannon, Webster University; Cormac MacFhionnlaoich, University College–Dublin; Efrat Elron, University of Jerusalem; and Manuel Bacerra, Instituto de Empresa, Spain. Others with university affiliations are Jennifer Lynn Roney, Southern Methodist University; and Roshan Othman, Universiti Kebangsaan Malaysia.

In addition to the authors, many others contributed to all editions of the book. Most if not all coauthors interviewed several people, and each chapter was read and critiqued by several citizens or residents of the countries being described. Martin Gannon would like to thank his former colleagues at the Robert H. Smith School of Business, University of Maryland at College Park, for their advice and suggestions, including Michael Agar, Mercy Coogan, Stephen Carroll, Edwin Locke, Sabrina Salam, Allyson Downs, and Guenther Weinrach. He also wants to thank the following for providing a supportive environment: Rudolph Lamone, former dean of the Robert H. Smith School of Business; Dean Howard Frank; Maryann Waikart, former director of the MBA Program; and Mark Wellman, former director of the MBA Program. Both Rajnandini Pillai and Martin Gannon want to thank Dean Dennis Guseman and Professors Kathleen Watson and Gary Oddou of the California State University, San Marcos, for helping to facilitate the final stages of writing.

Furthermore, given both the scope and depth of this project, we were almost overwhelmed, not only with its complexity but also with its details. We have painstakingly attempted to eliminate any errors, however small, that would serve to detract from the general focus of the book. We accept responsibility for any inadvertent errors that might have occurred. If you encounter even a minor error, we hope that you will bring it and any suggestions for improving the book to our attention. Our e-mail addresses are listed at the end of the Preface.

For this fourth edition, the two authors have updated all chapters and have shared the work equally. Rajnandini rewrote the original chapter on India focusing on the base Indian culture as represented by the Dance of Shiva and Hinduism. She is also the author of the new chapter devoted to diversity in India. Rajnandini would like to thank Prashant Valanju, Shama Gamkhar, and Shakuntala Desai for their suggestions on Chapter 28, "India: The Dance of Shiva." She would also like to thank Arisa Isayama for her suggestions on Chapter 3, "The Japanese Garden," and Durgesh Rajsingh for his suggestions on Chapter 17, "The Traditional British House." Martin Gannon rewrote the chapter on American football and the final integrative chapter, and significantly updated both the chapter on the Israeli kibbutz and moshav and the chapter on Russian ballet. He would like to thank his son, Reid, who provided invaluable input to the chapter on American football; Reid is much more knowledgeable than his father about the intricacies of this game, and the many discussions they had while watching televised football games over the years led to many new insights that have been incorporated into the revised chapter. Professor June Poon of the Universiti Kebangsaan Malaysia provided suggestions for Chapter 18, "The Malaysian *Balik Kampung*," as did her graduate students: Kok Wai, Chow Sook Woon, Syamala Nairand, and Kalaimar Vadivel.

We are very pleased that Michael Berry, a U.S. American who went to Finland as a Fulbright scholar and remained as Senior Lecturer at the Turku School of Economics and Adjunct Professor of Intercultural Relations and History in three Finnish universities, has written Chapter 10, "The Finnish Sauna," for this fourth edition. Michael, who is married to a Finnish Senior Lecturer of English and Communication, was a Visiting Scholar at the University of Maryland in 1990–1991 and a very involved member of the seminar in which the original ideas for this book evolved. His contributions to the area of cross-cultural understanding are well-recognized, and they include a well-received approach for facilitating communication across cultures and several insightful articles on the differences between Finnish and U.S. American communication styles. One of his courses, Culture and International Management, was designated the exceptional Finnish intercultural business course for 2007 (by the Finnish Association of Graduates in Economics and Business Administration).

As usual, the staff at Sage Publications performed its work admirably, and we are especially appreciative of the help that our editors, Al Bruckner and Lisa Shaw, provided. We would also like to thank the following at Sage Publications: MaryAnn Vail, editorial assistant; Catherine M. Chilton, production editor; and Jacqueline Tasch, copy editor.

—Martin J. Gannon
California State University, San Marcos
E-mail: mgannon@csusm.edu

—Rajnandini Pillai
California State University, San Marcos
E-mail: rpillai@csusm.edu

PART I

Introduction

Chapter 1 provides a description of a cultural metaphor, that is, any major phenomenon, activity, or institution with which its members closely identify cognitively and/or emotionally. The primary focus of this book is the cultural metaphor for a specific nation, but it also offers cultural metaphors for ethnic groups crossing into and settling permanently in other national cultures (e.g., the Chinese) and even cultural measures for continents. Likewise, the book highlights how nations close to each other tend to have similar cultural metaphors and features, as in the case of the Scandinavian nations. In the final section of the chapter, there is a description of when culture does and does not matter.

Understanding Cultural Metaphors

The great pedagogic value of figurative uses of language is to be found in their potential to transfer learning and understanding from what is known to what is less well-known and to do so in a very vivid manner. . . . Metaphors are necessary because they allow the transfer of coherent chunks of characteristics—perceptual, cognitive, emotional, and experiential—from a vehicle which is known to a topic which is less so.

—Andrew Ortony (1975, p. 53)

There are many good and obvious reasons for studying cross-cultural differences, including a conservative estimate that somewhere between 25% and 50% of our basic values stem from culture (for such estimates, see Haire, Ghiselli, & Porter, 1966; Hofstede, 2001). Other aspects of workforce diversity, such as age and socioeconomic status, also account for significant variances in our values and attitudes, but clearly culture is critical.

Failures in cross-cultural communication and negotiation have very tangible bottom-line results. For example, Disney negotiated a very favorable contract to establish Euro-Disney outside of Paris in the 1980s, only to face near disaster because of such issues as the corporation's ban on the sale of wine, which usually accompanies eating in France. The French stayed home or went elsewhere rather than endure such inconveniences. More recently, Daimler-Chrysler Corporation was dissolved in 2004 at least in part because of the cultural conflicts between the German and U.S. executives, including the latter's lack of understanding of the German system of corporate governance. Large German corporations are required to have two boards of directors, the first of which appoints and has some jurisdiction over the second corporate board. For example, the first German board—composed of representatives of trade unions, employees, large-scale investors such as banks, and government—appoints the second board. A U.S. board, because there is only one, can unilaterally authorize such decisions as large-scale employee layoffs and changes in reward systems. In Germany, the second board must receive authorization from the first board before proceeding with such major changes. Because of the U.S. penchant for downgrading the importance of two boards and the German penchant for having many more formal meetings than the U.S. executives before decisions can be enacted, change was extremely difficult to implement. David Ricks (2006) provides numerous additional examples demonstrating how cross-cultural differences create major and sometimes fatal difficulties in many areas of business.

An even more fundamental reason for studying culture is that our globalized world demands cross-cultural expertise if we are to survive. Amy Chua (2003) clearly supports this point of view:

> After the fall of the Berlin Wall a common political and economic consensus emerged, not only in the West but to a considerable extent around the world. Markets and democracy, working hand in hand, would transform the world into a community of modernized, peace-loving nations. In the process, ethnic hatred, extremist fundamentalism, and other "backward" aspects of underdevelopment would be swept away. The consensus could not have been more mistaken. Since 1989, the world has seen the proliferation of ethnic conflict, the rise of militant Islam, the intensification of group hatred and nationalism, expulsions, massacres, confiscations, calls for renationalization, and two genocides unprecedented since the Nazi Holocaust. (p. 123)

As argued elsewhere, there are as many reasons supporting the view that global disintegration rather than global integration may be our fate (see Carroll & Gannon, 1997). We are just as likely to become a global battlefield as we are to become a global village or a global community.

To highlight how difficult but critical it is to understand the importance of culture, it is useful to examine a short case study developed by Martin Gannon. By way of background, he attended a 10-day cross-cultural training program in 1990 led by Professor Richard Brislin at the East-West Center, University of Hawaii at Manoa. All 35 attendees were professionals involved in cultural studies in some way, and they included professors from a diverse range of disciplines and immigration officials from several nations. During the course of the program, a well-known cross-cultural training exercise, The Albatross (Gotchenour, 1977), was conducted, which produced a number of insights. Gannon has used this case study more than 200 times in a variety of settings involving students and managers. Readers are invited to read Case Study 1.1 and answer the questions after it before we provide any additional details.

CASE STUDY 1.1

I recently participated in a cross-cultural training session at the East-West Center, Hawaii. There were six male volunteers (including me) and six female volunteers. We walked into a room where a man was dressed in Eastern or Asian garb but in a somewhat indistinguishable manner; he could have been a king or a Buddhist monk. A woman sat beside him, and she was dressed in a similarly indistinguishable fashion.

There was no talking whatsoever in this training session, which lasted for about 10 minutes. The "king" beckoned the males to sit on chairs, after which he indicated that the females should sit at their feet. He then greeted each male silently and in standing position; he clasped each male by the arms and then gently rubbed his hands on the male's sides. The males did as the king instructed, but there was some nervousness and laughter, although no talking. The king then bowed to each female.

Next, the king presented a large vase of water to each male and he invited them to drink from it. He then did the same thing with each female.

The king and queen then walked before the volunteers, peering intently at the females. After a minute or two, the king put on a satisfied look and made a noise as if satisfied. He then looked at the queen, who nodded in agreement. The queen then took the hand of one female to lead her to a sitting position on the ground between the king and queen. Next, the king and queen tried to push the female's head toward the ground as she sat on the ground between them (they were on chairs), but she resisted. They tried once again, but she still resisted. The training session then ended.

Instructions. Each small group should appoint a recorder/secretary to report back to the larger group. Time limit is 10 minutes. Please answer the following questions:

1. What kind of a culture is this? Please describe.

2. How would you interpret the differential treatment of males and females in this culture?

In each of the 200 sessions Gannon has held on this case study, there have been about six subgroups, so he has received about 1,200 interpretations of the story. In 9 of 10 instances, the subgroup describes the culture in the following manner: A male-dominated traditional culture, probably Asian or African or Mideastern, ritualistic, and conservative. Sometimes a subgroup tries to identify the religion involved, and frequently Buddhism or Islam is cited. And, although almost all subgroups feel that females are in a subordinate role, a few believe that females have high status, clearly separate from that of the dominant males.

In fact, this is an Earth-worshipping culture in which males are clearly subordinate to females, and the only way to integrate all of the information provided is to use this framework. For example, the male leader was not being friendly when he patted the males; rather, he was checking for weapons, as males tend to have too much testosterone and too strong a tendency to engage in immature fighting. Similarly the females were seated in the place of honor (nearest to the ground), and the males were relegated to the bleachers. The males drank first to test for poison, thus ensuring the safety of the females. Even the "king," whose ambiguous position is highlighted by the quotation marks, must ask permission of the female leader before selecting a favored female, who was placed nearest the ground for the ritual in an honored position between the two leaders. Frequently Gannon asks why a particular female was chosen, and rarely does anyone guess the reason: A visual inspection indicated that she had the largest feet, an obvious sign of importance in an Earth-worshipping culture. In many cultures the number three is used, and it was being used in this ritual until the favored female resisted.

This exercise is usually sufficient to make the point that having a framework is very useful in understanding any culture. If the trainees had been told that the culture was Earth-worshiping, they could have integrated the various stimuli that were overwhelming them. Furthermore, the feedback session after the original training proved to be insightful, as the young woman selected for the ritual was asked why she resisted. This young woman was a very accomplished cultural anthropologist who has devoted her career to the study of village life around the world. She was in her mid-thirties, well published, and tenured at a good university, attractive, and divorced but without children. Her response focused on the maltreatment that she had experienced at the hands of various men in her life and on her resolve never to allow such maltreatment to occur again. Thus she had interpreted the ritual as a form of subservience to men, as the "king" was pushing her head toward the ground; so was the "queen," but she did not mention this fact. The pedagogical point is that this young woman, given her educational training and work experiences in different villages, was as knowledgeable as or more knowledgeable than any professional in the room, but her perspective— warped by unpleasant experiences with men—had led her to react emotionally, even to the extent that she was not able to think about an alternative framework such as an Earth-worshipping culture. Gannon also points out that he felt overwhelmed and had no idea what was going on.

The case study also highlights other critical aspects of culture, which operate subtly, often on the unconscious or semiconscious level. Culture has been aptly compared to a computer program, which once activated by a few commands or stimuli begins to operate automatically and seemingly in an independent manner (Fisher, 1988; Hall, 1966; Hofstede, 1991, 2001). Clearly such automaticity occurred, but unfortunately the stimuli were not properly matched to the cultural framework because of the negative relationships with males that this young woman had experienced.

Frequently, when foreigners violate a key cultural value, they are not even aware of the violation, and no one brings the matter to their attention. The foreigners are then isolated and begin to experience negative feelings. As one U.S. businessman in Asia aptly pointed out, one of the central problems doing business cross-culturally is that once a visitor makes a major cultural mistake, it is frequently impossible to rectify it, and it may well take several months to realize that polite rejections signify isolation and banishment. Sometimes foreigners make such a mistake and eventually leave the country without even realizing or identifying what they have done.

Even genuinely small cultural mistakes can have enormous consequences. Many older and even some younger Germans, for instance, do not like to converse too much during meals. They will ordinarily begin the meal by taking a sip of beer or soda, then pick up the knife and fork and hold them throughout the meal, putting them down only when they have finished eating. For many Germans eating is a serious business, not to be disturbed by trivial comments and animated conversation. Many Italians, on the other hand, tend to talk constantly during meals and wave their hands repeatedly. As a result, a German and an Italian dining with one another may feel aggrieved by each other's behavior, and much time is wasted negotiating acceptable rules of behavior that could otherwise be spent on substantive issues, including the development of trust.

Furthermore, while technological and societal changes have been rapid in recent decades, many key aspects of culture tend to change only slowly, frequently at a snail's pace, and the influence of culture persists for centuries even after mass immigrations take place. The U.S. Irish have the "gift of the gab," befitting a cultural heritage that has a strong oral tradition, and they are disproportionately represented in fields such as trial law and politics, where this gift is an asset. English and French residents of Canada think and feel differently in large part because of their respective cultural heritages, and these differences have threatened the very existence of that country.

Language Barriers

Individuals from English-speaking countries are at a particular disadvantage culturally because the people of many non-English-speaking countries use both English and their own native languages. It is common for English-speaking visitors to a non-English-speaking country to assume cultural similarity when dissimilarity is really the norm. English has become the language favored in international business and mixtures of languages such as Chinglish and Spanglish have become prominent, thus creating both opportunities and pitfalls for natives of English-speaking countries.

However, it should be noted that knowing a country's language, while clearly helpful, is no guarantee of understanding its cultural mind-set, and some of the most difficult problems have been created by individuals who have a high level of fluency but a low level of cultural understanding. Glen Fisher (1988), a former Foreign Service officer, describes a situation in Latin America in which a U.S. team's efforts were seriously hampered because of the condescending attitude of one member, whose fluency in Spanish was excellent. Fortunately another member of the team helped to save the day because she showed a genuine interest in the culture and its people, even though she was just beginning to learn how to speak Spanish. Moreover, members of a culture tend to assume that highly fluent visitors know the customs and rules of behavior, and they judge those visitors severely when violations occur.

U.S. Americans are at a particular disadvantage in trying to understand the mind-sets of other cultures. U.S. businesspeople and travelers tend to follow frantic schedules, sometimes visiting Hong Kong, Thailand, Japan, and Taiwan within the space of 2 weeks. To expect these U.S. travelers to understand these cultures in such a short period of time is unrealistic. Even fewer U.S. Americans spend any time residing in foreign countries and, when doing so, tend to isolate themselves from the natives in their "golden ghettoes." By contrast, Europeans speak two or more languages, including English, and they experience great cultural diversity simply by traveling a few hundred miles from one country to another. Many Asians, because of their knowledge of the English language and education in Europe and the United States, are similar to these Europeans in terms of cultural sophistication.

Using Cultural Metaphors

This book describes an innovative method, the cultural metaphor, for understanding easily and quickly the cultural mind-set of a nation and comparing it to those of other nations. In essence, the method involves identifying some phenomenon, activity, or institution of a nation's culture that all or most of its members consider to be very important and with which they identify cognitively and/or emotionally. The characteristics of the metaphor then become the basis for describing and understanding the essential features of the society.

For example, the Italians invented the opera and love it passionately. Five key characteristics of the opera are the overture, spectacle and pageantry, voice, externalization, and the interaction between the lead singers and the chorus (see Chapter 21). We use these features to describe Italy and its cultural mind-set. Thus, the metaphor is a guide, map, or beacon that helps foreigners understand quickly what members of a society consider to be very important. This knowledge should help them to be comfortable in the society and to avoid making cultural mistakes. The cultural metaphor is, however, only a starting point and subject to change as the individual's first-hand knowledge increases.

Cultural metaphors can be used to profile ethnic groups, nations, clusters of nations, and even continents. We have taken this approach in this book for nations, the base culture and its evolution across national borders, and even for two continents.

Constructing Cultural Metaphors

Countless social scientists, particularly cross-cultural psychologists and cultural anthropologists, have devoted their lives to the study of culture. Our cultural metaphors are based partially on the work of cross-cultural psychologists and cultural anthropologists, who emphasize a small number of factors or dimensions such as time and space when comparing one society to another.

Six Age-Old Dimensions

The first of these dimensional approaches was described by two anthropologists, Florence Kluckhohn and Fred Strodtbeck (1961). They compare cultures across six dimensions, pointing out that philosophers, social scientists, and commentators interested in understanding cultural differences have focused attention on these dimensions for hundreds of years. These six dimensions are:

1. What do members of a society assume about the nature of people, that is, are people good, bad, or a mixture?
2. What do members of a society assume about the relationship between a person and nature, that is, should we live in harmony with it or subjugate it?
3. What do members of a society assume about the relationship between people, that is, should a person act in a self-serving manner or consider the group before taking action (individualism versus groupism or collectivism in terms of such issues as making decisions, conformity, and so forth)?
4. What is the primary mode of activity in a given society, that is, *being* or accepting the status quo, enjoying the current situation, and going with the flow of things; or *doing*, that is, changing things to make them better, setting specific goals, and accomplishing them within specific schedules?
5. What is the conception of space in a given society, that is, is it considered *private* in that meetings are held in private, people do not get too close to one another physically, and so on; or *public,* that is, having everyone participate in meetings and decision making, allowing emotions to be expressed publicly, and having people stand in close proximity to one another?
6. What is the society's dominant temporal orientation: past, present, or future?

Kluckholn and Strodtbeck note that each society has a dominant cultural orientation that can be described in terms of these six dimensions but that other weaker orientations may also exist simultaneously in its different geographical regions and racial and ethnic groups.

Hall on Communication Patterns

Another well-known anthropologist, Edward T. Hall, has spent more than 40 years developing and writing about a similar dimensional classification system (for a good summary of it, see Hall & Hall, 1990). He basically focuses on the communication patterns found within cultures, and he emphasizes four dimensions along which societies can be compared:

1. Context, or the amount of information that must be explicitly stated if a message or communication is to be successful

2. Space, or the ways of communicating through specific handling of personal space; for example, North Americans tend to keep more space between them while communicating than do South Americans

3. Time, which is either *monochronic* (scheduling and completing one activity at a time) or *polychronic* (not distinguishing between activities and completing them simultaneously)

4. Information flow, which is the structure and speed of messages between individuals and/or organizations

Hall then arrays societies along an overarching high-context/low-context dimension. In a high-context society, time tends to be polychronic, and there is a heavy investment in socializing members so that information does not need to be explicitly stated for it to be understood. Members of such a culture have known one another for long periods of time, and there is strong agreement as to what is expected and not expected. In the high-context Japanese society, there is even an aphorism that expressly addresses this issue: He who knows does not speak; he who speaks does not know (see Chapter 3). Hence, verbal communication is frequently not necessary and may well impede the transmission of the message. Also, members of high-context societies tend to have less physical space between them when communicating than those in low-context societies.

As Hall notes, high-context societies tend to require a strong leader to whom everyone else expresses submission or at least great respect. In the Arabic countries, such a leader will sit in his office surrounded by people seeking his help and advice. He will not address the issues and people sequentially, as would tend to happen in monochronic countries such as the United States and Germany. Rather, he will deal with several issues and people as conditions seem to warrant, going from one group to the other in a seemingly haphazard fashion that takes into consideration their sensitivities and need to save face or avoid embarrassment.

Hall tends to array societies he has studied in the following way, going from high-context to low-context: Japan, the Arab countries, France (approximately in the middle of the continuum), the United States, and Germany. Hall has a bias against low-context societies, even though he recognizes that it is much easier to interface with a low-context society because information about rules and permissible behaviors is explicitly stated. To him, such societies tend to be too mechanical and lack sensitivity to the needs of individuals. However, he does not critically analyze some of the problems found in high-context societies, particularly the overwhelming power of the leader, which can be used indiscriminately, or the in-group bias that hinders relations with anyone outside of the culture.

Hall's system begins to break down when he talks about the low-context way that the Japanese interact with foreigners but the high-context way in which they interact among themselves. Thus he seems to be describing the classic in-group/out-group phenomenon rather than an overarching dimension along which societies can be arrayed. Triandis, Brislin, and Hui (1988) have argued that the major dimension separating societies is individualism-collectivism, in which the in-group and out-group distinction is critical, and this seems to be the dimension that Hall is describing. Furthermore, as described in the various chapters of this book, there are many specific kinds of individualism and collectivism.

Still, Hall's work has been significant and insightful, particularly his treatment of time and space. Throughout this book we will use some of his basic concepts, especially the monochronic-polychronic distinction and that between high-context and low-context communication.

Establishing Country Profiles

The third major dimensional approach was developed by Geert Hofstede (1991, 2001) and has been independently confirmed and in some instances refined by other researchers such as the Chinese Culture Connection (1987), a group consisting of Michael Bond, a long-time professor of cross-cultural psychology at the University of Hong Kong, and 25 Chinese associates. (For examples of frameworks similar to Hofstede's, see Schwartz, 1994; Trompenaars & Hampden-Turner, 1998.) Hofstede is a prominent organizational psychologist whose research is primarily based on a large questionnaire survey of IBM employees and managers working in 53 different countries, completed in the period 1967 through 1973. Hofstede's work is especially significant because the type of organization is held constant; his is the only large-scale cross-cultural study in which the respondents all worked for a multinational corporation that had uniform personnel policies. He develops empirical profiles of these 53 countries across five dimensions of basic cultural values:

1. Power distance or the degree to which members of a society automatically accept a hierarchical or unequal distribution of power in organizations and the society

2. Uncertainty avoidance (acceptance of risk) or the degree to which members of a given society deal with the uncertainty and risk of everyday life and prefer to work with long-term acquaintances and friends rather than with strangers

3. Individualism or the degree to which individuals perceive themselves to be separate from a group and to be free from group pressure to conform

4. Masculinity/femininity or the degree to which a society looks favorably on aggressive and materialistic behavior and clearly separates male from female roles

5. Time horizon (short-term to long-term) or the degree to which members of a culture are willing to defer present gratification to achieve long-term goals

A fourth dimensional approach was developed by Robert House and a team of 162 researchers in 62 national societies in the landmark Global Leadership and Organizational Effectiveness (GLOBE) study published in 2004. We will refer to this study throughout the book as the GLOBE study. GLOBE used the terms *societies* and *societal culture* instead of *country* or *nation* to indicate the complexity of the culture concept; in a few instances the researchers sampled two subcultures from a single nation, for example, black and white citizens of South Africa. Based on previous work and their own findings, the GLOBE researchers (House, Hanges, Javidan, Dorfman, & Gupta, 2004) identified nine cultural dimensions: uncertainty avoidance, power distance, institutional collectivism, in-group collectivism, gender egalitarianism, assertiveness, future orientation, humane orientation, and performance orientation. They also developed separate measures for the value or importance and the actual practice of each of these nine dimensions, for example, valuing equal treatment for all and actually instituting practices that reinforce this value.

The GLOBE Study

Uncertainty avoidance/acceptance of risk and power distance are similar to their counterparts in Hofstede's framework. With regard to collectivism, they distinguish between institutional and in- group collectivism. In-group collectivism represents the extent to which people are loyal to their organizations and/or families. In many cultures organizations tend to hire only those with whom they are very familiar, that is, family members, members of the larger family or kinship group, and individuals trusted by family members. The employees are typically guaranteed employment, even during difficult economic times, and in return they are expected to work hard for the company, sometimes even putting the company's needs ahead of their own. In many instances non-family members will work loyally for a family business for several years, and in return the family provides the capital that allows them to establish their own businesses. Another classic example of in-group collectivism is Toyota, which retrains workers during economic recessions rather than terminating them.

Institutional collectivism describes the extent to which members of a culture identify with broader societal interests. As a culture increasingly values its institutional collectivism, the emphasis on both the value and practice of in-group collectivism also rises significantly, as we might expect, according to the GLOBE study (House et al., 2004, p. 467). Gender egalitarianism refers to the degree to which male and female roles are distinct from one another, while assertiveness refers to the extent to which a culture encourages individuals to be tough, forceful, and aggressive as opposed to being timid and submissive in social relationships. Thus the GLOBE researchers have separated Hofstede's masculinity-femininity dimension into two dimensions: gender egalitarianism and assertiveness. Future orientation refers to the extent to which people of a culture engage in planning and investing in the future. It is very similar to the short-term/long-term time horizon dimension in Hofstede's framework. Humane orientation describes the extent to which a culture rewards people for being fair, altruistic, generous, caring, and kind to others. Finally, performance orientation describes the extent to which an organization or society rewards people for setting and meeting challenging goals and improving performance.

The GLOBE researchers divided the data from the 62 societies into regional clusters based on prior research, common language, geography, religion, and historical accounts. They arrived at 10 different clusters. These clusters are very similar to those derived by other researchers: Anglo, Latin Europe, Nordic Europe, Germanic Europe, Eastern Europe, Latin America, Middle East, Sub-Saharan Africa, Southern Asia, and Confucian Asia. The GLOBE analysis indicated that scores of respondents correlated within a cluster but were unrelated to the scores of respondents in different clusters.

Societies in the Anglo cluster (Canada, the United States, Australia, Ireland, England, South Africa [white sample] and New Zealand) were high in performance orientation and low in in-group collectivism.

The Latin Europe (France, Portugal, Spain, French-speaking Switzerland, Italy, and Israel) cluster had fewer high scores on any of the cultural dimensions and the scores were quite moderate, but the cluster scored low on humane orientation and institutional collectivism.

The Nordic European cluster of countries, which include Denmark, Sweden, and Finland, scored high on future orientation, gender egalitarianism, institutional collectivism, and uncertainty avoidance/acceptance of risk and low on assertiveness, in-group collectivism, and power distance.

The Germanic Europe (Austria, The Netherlands, Switzerland, and Germany) cluster of countries scored high in performance orientation, assertiveness, and future orientation and low in humane orientation and both forms of collectivism.

The Eastern Europe cluster included Greece, Hungary, Albania, Slovenia, Poland, Russia, Georgia, and Kazakhstan. These countries scored high on in-group collectivism, assertiveness, and gender egalitarianism. They scored low on performance orientation, future orientation, and uncertainty avoidance/acceptance of risk.

The Latin America (Ecuador, El Salvador, Colombia, Bolivia, Brazil, Guatemala, Argentina, Costa Rica, Venezuela, and Mexico) cluster of countries predictably scored high on in-group collectivism and low on performance orientation, future orientation, and uncertainty avoidance/risk acceptance.

The Middle East cluster, which included Qatar, Morocco, Egypt, Kuwait, and Turkey, had high scores for in-group collectivism and low scores on gender egalitarianism, future orientation, and uncertainty avoidance/acceptance of risk.

Sub-Saharan Africa included Zimbabwe, Namibia, Zambia, Nigeria, and South Africa (black sample); these countries had high scores on humane orientation and had average scores on most other dimensions.

The Southern Asia (Philippines, Indonesia, Malaysia, India, Thailand, and Iran) cluster of countries had high scores on in-group collectivism and humane orientation and average scores on most other dimensions.

Finally, the Confucian Asian cluster, which includes Singapore, Hong Kong, Taiwan, China, South Korea, and Japan, had high scores for performance orientation and high scores for both forms of collectivism (institutional and in-group). They too had average scores on most other dimensions.

The various dimensional approaches developed by Kluckholn and Strodtbeck, Hall, Hofstede, and the GLOBE researchers, along with similar works of others, have become enormously influential. Relying on a small number of dimensions so that profiles of various societies can be constructed, they by necessity leave out many features of the cultural mind-sets that are activated in daily cultural activities, and they neglect the institutions molding these mind-sets. These dimensional approaches are helpful for understanding cultures and providing an overall perspective on cultural differences, but an individual will experience great difficulty in applying these approaches to daily interactions. In effect, these dimensions are instructive but somewhat lifeless and narrow in that they leave out many facets of behavior.

Using Metaphor

The metaphoric method highlighted in this book supplements and enriches the four major dimensional approaches so that a visitor can understand and, most important, begin to deal effectively with the flesh and blood of a culture. While the metaphor itself cannot encompass all of the reality that is found within each society, it is a good starting point for understanding and interacting effectively with it. At the same time the various chapters of the book are linked together through the use of the four dimensional approaches.

Throughout this book we have attempted to identify metaphors that members of given societies view as very important if not critical. However, we needed to identify metaphors that would be relatively complex so that we could make several direct comparisons between the metaphor and the nation being represented by it. Also, we wanted to have a metaphor for each society that would have several suitable features that we could then use to describe it. In addition, we sought to include numerous factors or variables such as religion and small-group behavior when using the metaphor to describe the society, recognizing that some of these factors are important in some societies but not others. For each society we used the dimensions of the four dimensional approaches described above. In addition, we focused on all of the following:

- religion
- early socialization and family structure
- small-group behavior
- public behavior
- leisure pursuits and interests
- total lifestyle: work/leisure/home and time allocations to each
- aural space or the degree to which members of a society react negatively to high noise levels
- roles and status of different members of a society
- holidays and ceremonies
- greeting behavior
- humor
- language: oral and written communication
- non-oral communication such as body language
- sports as a reflection of cultural values
- political structure of a society
- the educational system of a society
- traditions and the degree to which the established order is emphasized

- history of a society, but only as it reflects cultural mind-sets or the manner in which its members think, feel, and act; not a detailed history
- food and eating behavior
- social class structure
- rate of technological and cultural change
- organization of and perspective on work, such as a society's commitment to the work ethic, superior-subordinate relationships, and so on
- any other categories that were appropriate

Using all of these categories initially, we studied each society in depth and interviewed several of its natives. After an initial draft of a chapter was written, it was presented at seminars and reviewed by natives and long-term residents of the society being described. The chapter was then rewritten in light of the suggestions that were offered, and additional comments were solicited. This iterative process typically led to rewriting a chapter five or six times, and sometimes nine or ten times.

Reading and Using This Book

To understand the book thoroughly and use it most effectively, readers should understand the rationale behind each of the nine major parts of the book. Many nations can fit into more than one part of the book. We have placed chapters in various parts because they fit comfortably there and are representative of the themes for the parts themselves, as our discussion suggests. The book itself can be used flexibly and selectively, as it is not necessary to read the chapters sequentially or to read all of them. Ideally the reader should understand the material in Chapter 1. The four dimensional frameworks are used to integrate the material and to link the chapters to one another.

Cultural metaphors can be employed to profile an ethnic group, a specific nation, a base nation from which a cluster of nations sharing similar values and attitudes emerges, the cluster of nations itself, and even a continent. In this book, the unit of analysis for each chapter is the nation and its national culture. However, three chapters on the Chinese in Part VIII illustrate the importance of each root culture, not only the spread of this culture but also its transformation as its members move permanently to other nations, and the implications for one specific nation, Singapore. We have also developed cultural metaphors for two continents, Australia and Africa, particularly Sub-Saharan Africa, in Part XI. Furthermore, there are two chapters on India in Part IX of the book, one explaining how the dance of Shiva helps to illuminate how the dominant religion of India, Hinduism, must be understood before trying to interact with members of this culture. The second chapter then looks at the amazing diversity in India and describes various non-Hindu religious and ethnic groups and their importance in understanding not only India but almost all nations, if not every nation. Fewer than 10% of the approximately

220 nations in the world are monocultural, and these two chapters on India are reflected in the numerous comparable multiethnic situations found elsewhere.

Two parts of the book are based on the distinction that Samuel Huntington (1996) proposed: torn national cultures, that is, cultures such as Russia and Mexico that have experienced dramatic changes in values at one or a few periods in history; and cleft cultures, or those in which it is difficult to integrate the national culture due to the existence of sharply different subcultures. We have expanded on Huntington's concept of cleft cultures to include those that experience difficulty integrating into a national culture because of geographic differences, that is, the north and south of Italy.

Three parts of the book highlight a framework linking culture and economics more closely. This framework was independently developed by Harry Triandis (2002) and Alan Fiske (1991b). There are minor differences between the two frameworks, and sometimes Fiske does not clearly demarcate the cultural level from the individual level. For our purposes, however, the two frameworks can be treated as identical. From this framework we can identify why we have chosen three major parts or topics of the book.

Individualism and Collectivism

Both authors seek to identify generic types of cultures. They begin their analyses with the cultural dimension of individualism-collectivism, which has been the dimension of most interest to researchers because of its obvious importance. Also, they emphasize the extent to which there is a large degree of inequality or power distance in the culture. Thus there is a four-cell typology of cultures emphasizing individualism-collectivism and power distance. There are two generic types of collectivism (horizontal and vertical) and two generic types of individualism (horizontal and vertical). Horizontal collectivism reflects community sharing, in which members of the in-group share all of their goods, as in a small village, even to the extent that there is no such phenomenon as theft. There is not much differentiation between individuals, and ethics are based on group membership: in-group or out-group. In essence, members of out-groups are viewed as nonpersons.

Vertical collectivism or authority ranking, found in large parts of Asia, Africa, and Latin America, involves a psychological relationship between the leader or leaders and all others in the culture. Frequently such a culture is symbolized not by the handshake, which reflects equality, but by different forms of bowing. Only a relatively few U.S. Americans and Europeans have experienced such a culture in depth because the relationship between superior and subordinate in most U.S. firms is instrumental and focuses only on work-related goals. In contrast, there is a dynamic two-way relationship between subordinates and leaders in authority ranking cultures: While the leaders receive more rewards, they are responsible for safeguarding the livelihoods of subordinates, even to the extent of finding them new positions when bankruptcy occurs. In turn, the subordinates are expected to be committed to the leader and the organization. Terminating employees to save money is anathema. Ethics is still determined in large part by group membership (in-group and out-group), and status as signified by family background, position at work, and so on is also critical.

It is important to realize, however, that there are at least two major types of authority ranking cultures. One type is paternalistic and the image of the leader is that of a kind father. The second is authoritarian and the image is that of a leader who is a harsh father. It is sometimes difficult to separate paternalistic and authoritarian authority ranking cultures, but we attempt to do so in our descriptions.

Horizontal individualism or equality matching is dominant in Scandinavian nations such as Sweden and Norway. All individuals are considered equal, even when some are taxed heavily, and it is expected that those who cannot make individual contributions to the common good will do so at a later time if possible. Finally, vertical individualism or market pricing is found in the United States and other market-dominated nations. Although individualism is emphasized, so too is the free market, and inequality resulting from its operation is deemed acceptable. There is, in principle, equality of opportunity and a level playing field, but not equality of outcomes. Ethics revolves around the operation of such a market.

A Scaling Perspective

Fiske, in particular, relates these concepts to the four types of statistical scales: Nominal, ordinal, interval, and ratio. His argument is that individuals have difficulty making decisions and use these scales as rough approximations for determining how to interact with others. Thus community sharing represents nominal scaling, as names are given only to entities (in-group versus out-group). In an authority ranking culture, Individual A may be more important than Individual B, and Individual C may be more important than Individual B, but there is no common unit of measurement. The scale is ordinal in nature. Hence we cannot say C is twice as important as A. In equality matching, the culture has a common unit of measurement, but it does not make value judgments about individual worth, as there are too many dimensions along which individuals can be measured. In this sense the scaling is interval. Finally, in market pricing, there is a common unit of measurement and a true zero point (zero money), which allows members of the culture to transform every other dimension and compare them monetarily. In this case the scaling is ratio.

Fiske provides an insightful example of these four types of culture in his discussion of a small town's decision about the purchase of an expensive fire truck. The issue becomes: Who should receive the new fire protection? The reader may want to stop at this point to consider the matter. The answers are: (a) community sharing, only members of the in-group; (b) authority ranking, all members of the in-group, but the leaders receive more attention and monitoring of their homes; (c) equality matching, everyone is protected; and (d) market pricing, only those who can pay the taxes are protected. This example is not far-fetched. In the United States there have been several recorded instances when fire trucks did not respond, sometimes because a home was just outside of the fire department's district and sometimes because the owners of the homes did not contribute monetarily to the fire department's upkeep.

In Hofstede's (1980b) original analysis of 40 nations, he divided them at the median score both on individualism-collectivism and power distance. No nation is in the community-sharing quadrant, probably because this form of collectivism is not appropriate for such large entities as nations. Interested readers can view the dispersion of the 40 nations into the other three quadrants by consulting the original work. As the Table of Contents indicates, we have devoted separate parts of the book to authority ranking cultures, equality-matching cultures, and market-pricing cultures.

Part X of the book focuses on the same metaphor (bullfighting) but the different meanings associated with the manner in which bullfighting is practiced in two national cultures, Spain and Portugal.

When Culture Does, and Does Not, Matter

The last important issue that we address in this chapter is: When does culture matter? There are times when culture is not important and other times when it is critically important. In this book we emphasize culture, but we do caution the reader to consider other factors.

Limitations on Culture

Frequently occupational similarities neutralize culture. For instance, when two medical doctors are jointly working on a problem, their medical backgrounds can help them to work together smoothly regardless of cultural backgrounds. Also, similarity of social class can diminish the importance of culture. For example, throughout the world middle-class families tend to use positive reinforcement in raising their children and provide them with opportunities to develop skills and a strong sense of self-esteem. These families may provide their children with music lessons and ask them to perform in front of guests, who respond enthusiastically. Conversely, blue-collar families throughout the world tend to emphasize negative reinforcement and punishment, which negatively influence skill development, opportunities to function in a public or leadership role, and feelings of self-esteem (see Kagitcibasi, 1990).

However, sometimes powerful groups will exclude others from opportunities and then stereotype them negatively, thus consigning them to permanent inferior status. This clearly happened in Ireland and India when the English ruled those nations for centuries. Apartheid, now outlawed in South Africa, began as a reaction to scarcity of jobs and led to the stigmatization of native Africans for nearly a century (see Olson, 1982). If the playing field is level, as is more probable when markets are genuinely competitive, this outcome is mitigated.

At times social class or occupational similarity and culture become confused in the minds of visitors. Some U.S. Americans, for example, complain about the rudeness of Parisian shopkeepers, whereas other U.S. Americans describe wonderful relationships with their occupational peers in France. Presumably what is occurring reflects social class or occupational similarity as much as if not more than culture.

Sometimes the nature of the problem minimizes the importance of cultural differences. For example, when companies from two or more nations are working together on a joint project that their top managements strongly support, organizational members are more likely to forget cultural differences, especially when ample rewards for goal attainment are available, along with punishments for failure.

When trust is present, culture decreases in importance. Jarvenpaa, Knoll, and Leidner (1998) studied 75 virtual work teams throughout the world, which were integrated via the Internet. The major finding was that, if "quick trust" can be established, culture is not a major issue. However, developing trust is frequently neither easy nor quick. For example, a virtual Internet team with members from several different nations must adjust to differences in time zones and the lack of commitment that may accompany not actually knowing the other team members on a firsthand basis.

The Impact of Technology

One of the most controversial issues is the degree to which technological changes such as the Internet influence culture. As Patricia Wallace (1999) points out, the Internet has not resulted in a global village, as the case is so often stated. Rather, individuals with similar interests—including crime and terrorists—seek one another out on the Internet. As such, the Internet has led to as much differentiation as integration and possibly more differentiation, as Amy Chua's (2003) book, *World on Fire*, persuasively demonstrates. Still, the world is becoming more flat as globalization has spread and individuals from different cultures are learning more about one another due to increased travel, the Internet, more business contact with those of other cultures, and so on (Friedman, 2005).

But any indirect form of communication such as e-mail presents special difficulties. For example, a high-context and high-level manager in Indonesia became angry when he received a terse message from his U.S. counterpart, not because of the content but because of the manner in which the message was phrased. That is, instead of starting the e-mail with "Dear Mr. XXX," the writer just started putting his ideas forth concisely and succinctly, without following the protocol associated with authority ranking cultures. As a general rule, technological and economic changes do matter, especially when disruptive of cultural patterns. Still, problems are minimized when changes are introduced gradually and are not directly injurious to deep-seated values. And, as the comparison of northern and southern Italy indicates, the cultural patterns found in the 11th century have persisted into the present, leading to concrete economic differences between these two major regions (Chapter 21).

When Culture Matters

As suggested in the Preface, perhaps the most interesting feature of culture is that it triggers unconscious values leading to action. Thus it is not surprising that culture is important when individuals must communicate directly. If individuals expect that outsiders will follow their cultural rules and are unwilling to facilitate the relationship by developing new rules acceptable to all, communication is likely to break down. Conversely, a Texan businesswoman was having difficulty in China until she started wearing red dresses, a favorite color among the Chinese signifying good fortune. Business improved dramatically.

As shown in numerous research studies, culture is particularly important in cross-cultural negotiations. Understanding both the similarities and differences of the cultures represented by the negotiators is a good way to facilitate interaction and goal attainment. U.S. Americans, for example, have a reputation for being direct and low-context when communicating information and this becomes a problem when the communication is phrased in terms of "take it or leave it." Billion-dollar deals have died on the table because of such behavior.

Culture is also important when individuals move to another nation or culture for an extended period of time. The well-known phenomenon of culture shock does occur and, if not handled properly, can lead to major problems. In this regard, it is not surprising that many managers from a company's headquarters who are sent to work in a subsidiary for an extended period cling to the values and ways of behaving found in their base culture, even to the extent of isolating themselves in "golden ghettoes."

The Impact of Stereotypes

Culture is also relevant if distorted stereotypes are present. There is some confusion surrounding the definition of a stereotype, but at a minimum it represents a distorted view or mental picture of groups and their supposed characteristics, on the basis of which we tend to evaluate individuals from each group. Stereotypes can be erroneous and can lead to unwarranted conclusions, particularly if no exceptions are allowed; for example, if we assume that all Italians are emotional and talk constantly while moving their hands and arms. In this sense a stereotype is a universal syllogism. In the case of Italians the stereotype is clearly erroneous in many instances, as anyone who has transacted business in northern Italy will confirm; there, the level of formality is frequently high in business relationships, especially in terms of dressing properly, showing up for meetings on time, being prepared, and so on. Ironically, in-group members frequently use universal stereotyping as a form of humor but react negatively if out-group members employ it.

However, all human beings use stereotypes, as they are a shorthand and easy way of classifying the multitude of stimuli to which we are exposed. The issue is not stereotyping itself but whether the stereotypes are accurate.

Most of us take an extremely negative position on stereotyping. It can be very embarrassing to be accused of stereotyping, especially since it is frequently so difficult to refute the charge. In today's world the accusation is frequently raised, and as a result, it has become difficult to discuss genuine differences. However, many social psychologists now take the position that there are real differences between groups and societies and that the negative connotations associated with stereotyping have led us to deemphasize these legitimate differences. From this perspective a stereotype represents only a starting point that is to be rigorously evaluated and changed as experience with groups warrants. Nancy Adler (2007) argues persuasively that it is legitimate and helpful to use stereotypes if they are descriptive rather than evaluative, the first best guess, based on data and observation, and subject to change when new information merits it.

Metaphors are not stereotypes. Rather, they rely on the features of one critical phenomenon in a society to describe the entire society. There is, however, a danger that metaphors will include some inaccurate stereotyping, and we have attempted to guard against this possibility by having the various chapters of this book reviewed by natives or long-term residents of the societies being described. In some instances we were unable to construct a metaphor that satisfied natives, residents, or ourselves. Hence this book includes only metaphors about which there is a consensus.

Admittedly, it is very difficult to test the validity of these metaphors empirically, at least at this point in time. However, we have completed one six-nation study of cultural metaphors and confirmed that they are critical (Gannon, Gupta, Audia, & Kristof-Brown, 2005–2006). In addition, we employ two tests in deciding whether a particular phenomenon can be considered as a valid cultural metaphor: whether there is consensus and whether a metaphor other than the one we have selected increases our understanding of a particular society. Also, we have noted in many instances that not all members of a society adhere to the behavioral patterns suggested by the metaphor by using such phrases as "Some Germans," "Many Italians," and "The Irish tend to." In effect, we are highlighting patterns of thought, emotion, and behavior that a society manifests and that are clearly and concisely portrayed by means of a simple and easily remembered metaphor. In this way visitors can use the metaphor as an initial guide, map, or beacon to avoid cultural mistakes and to enrich cross-cultural communications and interactions. In addition, instructors using the previous editions of this book have received very positive reactions from students and experienced international managers to the descriptions of each national culture in the book. Also, numerous book reviews have been uniformly very positive. In sum, as experience accumulates, the visitor can modify and enlarge the frame of reference, but each metaphor provides an initial starting point and basis for discussion.

In sum, culture frequently does not matter, but at other times, it is very influential. We have described only some of the instances when culture does and does not matter. Culture probably counts the most when there are feelings of inequity (perceived or real) and a scarcity of resources and opportunities. It is comforting to cluster with others similar to ourselves, especially when we are rejected by dominant groups. As Huntington (1996) persuasively argues, the major threat to world security is the increase in ethnic wars that has accompanied globalization and privatization, and this tension has remained strong since the publication of his thesis (Ferguson, 2006). Cultural differences are especially exacerbated when accompanied by extreme religious and ideological viewpoints. Many cultural problems are solved in the long run through intermarriages and increased social and business contacts, all of which are hindered by religious and ideological differences.

All of the factors described above, and some others not described because of space limitations, are important for evaluating when culture does or does not matter. The position taken in this book is that culture is important and is of critical significance in many situations but not in all of them. Culture also interacts with political, social, and economic forces and is, in that sense, a fuzzy concept. But clearly it is possible to understand cultures and use this understanding to enhance relationships between individuals and groups. Throughout the remainder of the book, we will be employing the methodology described in this chapter to demonstrate how cultural metaphors can strengthen understanding and to show how those metaphors are related to the core values, attitudes, and behaviors of various nations.

PART II

Authority Ranking Cultures

In this part of the book we focus on authority ranking cultures, in which there is a high degree of collectivism but also a high degree of power distance. Unlike market-pricing cultures, in which the relationship between the superior and subordinate is primarily one way, these cultures emphasize that superiors and subordinates have obligations toward one another that transcend job descriptions. This is particularly true of nations such as Japan and Thailand and to a lesser extent nations such as Poland and Brazil. Ordinal statistical scaling dominates relationships; that is, Person A is more important than Person B, and Person C is more important than Person B, but there is no common unit of measurement. Thus, it is impossible to say, for example, that C is twice as important as A.

There are at least two major types of authority ranking cultures: paternalistic and autocratic. Some cultures are a more or less pure form of one type or the other, but in other cases, the two types may overlap or the culture may transition from paternalistic to autocratic. Such transitions occur when a leader is initially viewed as helpful but paternalistic, but he or she later demands not only allegiance but complete subjection of subordinates. In this book the focus is primarily, although not exclusively, on paternalistic authority ranking cultures.

CHAPTER 2

The Thai Kingdom

Bhumibol, the world's longest-serving monarch, enjoys a reverence seldom seen in modern times. Regarded by some as semi-divine, he is known for his lifelong dedication to helping the country's needy. Bhumibol is a constitutional king with no formal political role, but has repeatedly brought calm in times of turbulence and is considered the country's moral authority.

—"Thais Tickled Pink" (2007)

One of the shabbiest but most popular tricks in Thai politics is to accuse your critics of disloyalty to revered King Bhumibol. Anyone can file a police complaint of lèse-majesté on the king's behalf and the penalty is up to 15 years in jail.

—"No Disrespect" (2008)

Thailand, with its population of 64 million and a land mass about equal to that of France or the states of New York and California in combination, is a nation that is at the crossroads of Southeast Asia. Various peoples from China and India and the bordering nations of Laos, Burma, Malaysia, and Cambodia originally populated it. All of these peoples have influenced the development of Thai culture in one way or another.

The nation itself is divided into several distinct regions. Low-lying Bangkok, one of the largest cities in the world, represents one such region. Built on a delta, Bangkok is sinking at the rate of 2 inches a year and may be completely inundated by 2050. Southern Thailand borders Malaysia, whereas northern Thailand borders Laos and Burma. Northern Thailand includes the city of Chiang Mai, Thailand's second-largest city, and is home to the hill tribes, some of which move periodically from mountainous area to mountainous area. The Thai government is trying to stop this practice, as these tribes frequently cut down trees that are hundreds of years old for firewood. It is also trying to put curbs on the logging companies in the northeast for the same reason. Excessive cutting of trees has led to erosion and sometimes mudslides that kill the villagers living below the mountains.

About 45% of the Thai workforce is still involved in agriculture, and Thailand is one of only seven or eight food-exporting nations in the world. While the food is abundant and the Thai cuisine world famous, agriculture contributes only 9.9% to Thailand's Gross National Product, industry 44%, and services 46%. There are wide disparities in income, both within and between regions (such as Bangkok and the northeast). Thailand's economic growth rate was extremely high for about 20 years, averaging 7% until 1997, when it was negatively affected by the Southeast Asian Financial Crisis of 1997. The economy is recovering, but slowly. Between 1994 and 2004 the annual growth in gross domestic product averaged 3.2%.

U.S. Americans and Thais share many similarities, but they also are quite different from one another. When Americans visit Thailand, they are frequently entranced by the friendliness of the people, the many Buddhist temples or *wats*, and the contrast between rural and urban life, particularly the urban life found in Bangkok, the nation's capital. Thailand has a parliamentary system heavily influenced by Britain, for Thais were often educated in England rather than the United States until recent years.

In 1932 Thailand abolished the absolute monarchy under which citizens could not touch members of the royal family, which has resulted in such bizarre situations; one royal, who could not swim, drowned because numerous onlookers felt they were barred from offering a helping hand. Since 1932 the country has experienced 17 coups, 55 governments, and 16 constitutions, with the most progressive constitution taking effect in 1997. Given such rapid change, it is little wonder that the most distinctive symbol in Thailand is King Bhumibol, who has been in power since 1947. The Thai kingdom was in decline until he became king when his older brother, King Mahidol, was killed in 1946.

Since 1947 the king has created what is arguably the most powerful and workable kingdom in a world that increasingly views kings as anachronistic. He is a talented and sophisticated man known for his love of the arts and is also a noted jazz composer. Most important, he symbolizes critical features of Thai culture and, it can be reasonably argued, is the glue that is holding the nation together. In this chapter we will not speculate on the issue of succession to his rule. Rather, the focus is on the Thai kingdom and why it is consonant with the underlying core values of all or most Thais. The characteristics of this cultural metaphor are loose vertical hierarchy, freedom and equality, and the Thai smile.

Loose Vertical Hierarchy

Thailand is an authority ranking culture in which vertical collectivism is emphasized. However, most probably because it is at the crossroads of many competing cultures, the Thais follow far fewer rules than most authority ranking cultures. As Harry Triandis and Michele Gelfand (1998) have shown empirically, the Thais are at the extreme end of the cultural dimension of looseness-tightness of rules. This means that there are probably more contradictions and tensions in Thai culture than in most other cultures.

Visitors to Thailand are immediately confronted with some of these contradictions. About 95% of the population is Buddhist, and there are numerous Buddhist *wats* or temples in Thailand, but most Thais do not frequent them except for scattered ceremonial days. Still, Buddhist values are ingrained in the Thais, as discussed below, both in the culture in general and in the schools.

During the Vietnam War the U.S Americans used Thailand as a recreational area for military personnel on leave, which helped to promote the well-known and widespread prostitution found in Thailand. This has had some very unfortunate consequences, including the fact that Thailand experienced an abnormally high rate of HIV infection, a crisis that has decreased significantly in recent years. However, upper middle-class Thai women can be Victorian in their attitudes. It is not unusual for a young man to visit a girlfriend's home on dates for many months and, when they are outside the home, for them to be escorted by an older sister or aunt.

The King's Role

The king has been particularly effective in using vertical hierarchical rules to manage the nation. While the king does not involve himself in daily governmental affairs, he becomes active during times of crisis and in the past has requested that prime ministers and generals voluntarily vacate their positions of power. Nothing more needs to be said to accomplish their removal, and there is no additional discussion about it.

Similarly, until recent years, the king or members of his family handed out individual diplomas at all university graduations. Each proud graduate wanted to have his or her picture taken with the king, and so two cameras were used for each photo in the event that one of them failed to record this moment. The king also appears at prominent *wats* for Buddhist ceremonies.

As these activities suggest, the king is very involved in the lives of Thais, and they revere him. Thais can become infuriated when the king is insulted in any way. One unfortunate visitor to Thailand became angry with a waiter in a restaurant, threw his cash payment for the meal on the floor, and stomped on it. He failed to realize that the king's picture is found on Thai money, and an enraged Thai attacked him mercilessly because of the perceived desecration. Another unfortunate visitor ripped up his Thai money as a sign of dissatisfaction at the airport, only to be arrested and jailed for 6 months. At movie theaters every film is preceded by a special royal anthem during which all are expected to stand. The government banned YouTube in 2007 because of a 44-second video that satirized the king and Paul Handley's (2006) critical biography of the king received similar treatment.

The essence of a paternalistic authority ranking culture is that there is a dynamic and two-way relationship between a superior and others. *Kreng cai,* or taking the other person's feelings into account, is a key concept in Thailand. All of the actions of the king reflect this orientation and both feelings and obligations are two-way or circular in nature. *Kreng cai* is quite similar to the Japanese concept of *amae* or looking to others for security and assurance. Thais are very sensitive to feelings and clearly recognize nuances of behavior that the typical U.S. American does not see. Similarly, Thais are adept at sorting themselves out in a reception line by status, even though they may not have met previously.

As might be expected in an authority ranking culture, members of the military are also important players. Many generals have business interests. There is a good amount of corruption in Thailand, at least in part because of the extremely complex social class structure involving the competing interests of businesspeople, the military, the politicians, the king, and others.

Personal and Family Interactions

Rather than shake hands, although the handshake is becoming more common, particularly in international business, Thais tend to *wai* to one another: They hold both their hands together as if in prayer and bow their heads when greeting one another. Bowing lower than the other person signifies lower social status, and bowing at the same level signifies equality in social class ranking. Sometimes a superior will just nod his head, and at other times, he will complete an attenuated *wai* that is scarcely noticeable. Supposedly, this pattern of the *wai* and its relation to social class structure emerged when someone conquered in battle would show the victor that he was totally subservient; his bow to the victor exposed his head to any blow that the victor wanted to administer. If a blow was not administered, the victor was signaling that a two-way relationship of fealty had been born, but one in which there was a clear superior and a clear inferior. Thus it is not surprising that at public ceremonies such as university graduations, anyone who wants to leave before the king must do so in an unobtrusive manner, making sure that he is lower than the king at all times.

In relationships within families this vertical ranking also prevails. Many Chinese Thai families live in a compound in which there are several homes, one for the father and mother and one each for married family members. There is a common area in the center where they meet at night and other times. Typically, at least one night per week, there is a large family gathering and dinner. The oldest son usually is the major decision maker in the family business if the father has retired, and the younger sons serve as vice presidents. If one of the sons is not very effective, he will retain the title but be assisted by either an in-law or outsider who will be the real decision maker in that part of the business. The family frequently rewards such important outsiders after several years by subsidizing the creation of their own businesses, thus strengthening the family's business network.

Furthermore, each family emphasizes this hierarchical ranking. One young woman who received her MBA in the United States did not want to return to Thailand and marry a man she had known since childhood, but she did so when her mother said, "If you don't return and marry him, I never want to see you again." However, the young woman and her mother did follow the dictates of loose authority ranking, agreeing that she would not have children for 5 years; if the marriage did not live up to expectations, she could divorce her husband and remarry, which in fact she did. In another situation a Chinese Thai family met to discuss an emergency problem: A younger female member of the family was very unhappy in the United States at a university in which she had just registered. In this instance a family member was immediately dispatched to bring her home.

An Educational Tradition

This form of vertical ranking is exemplified in the Flower Ceremony held in universities once a year. For decades Buddhist monks were the educators in Thailand, and the ceremony originated in their schools. A U.S. Fulbright professor was so startled by this ceremony that he wrote the following description:

> Today, students paid homage to their professors—a symbolic celebration of rather common significance to them. I found it an astonishing phenomenon.
>
> In a large auditorium, representatives from each department within the Faculty crawled up, in the manner of Asian supplication, and gave beautiful floral offerings to their "Aacaan" (professors). Their choral chants asked for blessing and showed gratitude. Their speeches asked for forgiveness for any disrespect or non-fulfillment of expectation. They promised to work diligently.
>
> In a moment of paradox, I remembered I must not forget to pay the premium on my professional liability insurance this year. (George, 1987, p. 5)

Still, Thais implicitly recognize both the humor in this ceremony and the loose nature of authority ranking. While many students make such declarations, they have difficulty honoring them.

Education in Thailand, as in other authority ranking cultures such as Japan, has historically stressed memorization and the taking of copious notes in lectures. Discussion is not emphasized. If a student asks a professor a question and the professor does not know the answer, the professor may well give an incorrect answer that the student will dutifully record, even though the student is aware that the answer is incorrect. In this way face is saved for both people. A similar pattern of behavior involving managers and employees can be found in some traditional Thai firms, but it is less common in the multinationals and larger firms now operating in Thailand. Similarly, an MBA education in Thailand tends to stress Western-style case discussion and interaction.

Ethnic Relations

One positive feature of the looseness of rules involves the ethnic groups in Thailand, particularly the Chinese and the ethnic Thais. About 80% of the population is composed of ethnic Thais and they are powerful in politics and the military. Another 10% of the population is composed of the ethnic Chinese, who generally believe in Confucianism and, to some extent, Taoism (for an explanation of Confucianism and Taoism, see Chapter 25). The ethnic Chinese tend to own prosperous family-run firms, most of which are relatively small in size. There is a good amount of intermarriage between these two groups, and some Thais will argue that they are more liberal and accepting than their counterparts in other nations, even those in Southeast Asia.

One reason for the high rate of intermarriage is that these two groups are compatible in terms of the overlapping religious and ethical perspectives of Buddhism, Confucianism, and Taoism. However, this is not always the case; discrimination and hostility exist, but seemingly to a much smaller extent than in most nations.

In short, the king's actions are quite consistent with the core values of the authority ranking Thais. He does, however, respect the loose nature of the rules, as evidenced by his hands-off policy on daily governmental affairs and related issues.

Freedom and Equality

Thailand means "land of freedom," and the name is apt, as it is the only nation in Southeast Asia—and one of the very few in the world—that has never been conquered. In the 1700s the Thai army was in grave danger during a war with Burma, but its leader, Taksin, reorganized the remaining 500 soldiers and led a brilliant counterattack. Taksin became king and established the capital in Bangkok. Later, he went insane and was replaced by a king from whom the current king traces his lineage directly.

A Military Tradition

With Thailand bordering several nations, the Thais have always been concerned about their military strength, and Thai soldiers are well trained, many of them in America and Australia. The military is involved in governmental affairs, even to the extent of overthrowing elected governments. However, the king's word is followed by all, including the military, even to the point of relinquishing power to citizens when the king requests it. It would be difficult for Thailand if the military were weak, given the complex region in which Thailand resides, and the issue is the balance between civilian and military rule.

The king has always been physically fit and exercises regularly and some of the rapport he enjoys with the military may stem from this fact. His son, in fact, served in the military for many years, although the Thais are not fond of him, probably because of the carefree lifestyle he chose rather than the ideal lifestyle exemplified by his father.

Westerners are frequently perplexed by the contradictions of Thai behavior. On the one hand it reflects authority ranking. But it also reflects pride stemming from this tradition of freedom from foreign domination. Even during World War II Japan did not invade Thailand because of a diplomatic agreement motivated in large part by the Japanese fear of a drawn-out and costly confrontation with the fabled Thai military.

Also, Thais see themselves as equal to Westerners because of this tradition and expect to be treated accordingly. Thus both freedom and equality are key components of Thai culture. As indicated previously, in the relationship between superiors and subordinates, there are obligations on both sides, and Thai subordinates expect to be treated with respect.

Thai Boxing

Thai fighting, particularly Thai boxing, is a unique manifestation and expression of freedom, equality, military prowess, and even the relationships between males and females. The roots of Thai boxing can be found in the experiences of the Chinese, who immigrated to Thailand 800 years ago and were forced to fight marauding tribes with a formless type of combat using head, teeth, fists, knees, ankles, and elbows (*Muay Thai*). This hand-to-hand combat is still part of the military training.

Modern Thai boxing as practiced in the two major stadiums in Bangkok is something to behold. Light gloves are worn and kicking is permitted. There are typically 10 fights per night, each of which lasts five rounds of 3 minutes each. Each stadium is small, holding only about 1,000 customers, almost all of whom are males, but many other Thais stand outside and follow the action intently by radio and, in some cases, TV. Before a fight each fighter performs a small ritual indicating the school or philosophy of fighting in which he has been trained. An Asian band plays throughout each round, but slowly in the first two rounds. In the third round, however, the band increases its sound and speed, thus exciting the crowd, and many fans rise excitedly and begin to place bets through the use of hand movements. In some ways this behavior is similar to that surrounding the Balinese cockfight (see Geertz, 1973).

About 50% or more of the fights do not go the full five rounds, as the fighting is bone chilling; knockouts are common. While some people are repelled by Thai boxing, the Thais have excelled at it and have produced many famous champions. As an expression of military prowess, Thai boxing is outstanding. It also symbolizes how far Thais are willing to go to preserve freedom and equality. The separation of sexes, which can be Victorian at times, also occurs during Thai boxing matches.

The Thai Smile

Anyone who has visited Thailand for even a few days can be dazzled by the friendliness of the Thais and the Thai smile is legendary. Some of this demeanor comes directly from the Thai practice of Buddhism. There are two general streams of Buddhism, one of which, Theravada Buddhism, emphasizes an internal focus and meditation more than the other (see Smith, 1991). Thailand is representative of Theravada Buddhism.

The Impact of Buddhism

Thai Buddhism, given its inward focus, stresses the key concept of the Middle Way, which emphasizes keeping emotions and even body movements under control. The Thais believe that anger and emotionality lead to more anger and more emotionality, which restricts freedom due to the fact that individuals engage in activities that they would otherwise avoid. Thais tend to be sophisticated diplomats and negotiators because of these beliefs. Similarly, the king is reserved during his public appearances. He does not use emotional approaches and arguments as many of his counterparts throughout the world do.

The Thais, as Buddhists, believe in karma or the concept that one's behavior leads to consequences. Thus Thais believe that behavior in this life determines the life form an individual will assume in the next life and that there will be many life cycles. Behaving inappropriately toward others, including subordinates or superiors, will help to determine the next life form. For example, Buddhist monks frequently rescue dogs, as many Buddhists believe that dogs were formerly humans who were assigned to a lower life form due to misbehavior in a past life.

Furthermore, as might be expected in a face-saving culture, Thais hate to say "no" directly. They use statements such as "We'll need to think it over" and "That may be a problem" as a proxy for no. Many Westerners spend months negotiating a particular issue only to realize belatedly that a slight movement of the shoulder is the equivalent of saying no. When negotiating with Westerners, Thais prefer to emphasize personal relationships and to make the contract of secondary importance, even to the extent of not using a contract at all. In one instance negotiators from Northrop Grumman negotiated effectively with their Thai counterparts until they indicated that company lawyers would be joining the discussions, at which time the Thais—both politicians and businesspeople—became unavailable through any form of communication.

Smiles in Context

No matter what happens, however, the Thai will keep smiling. Thus a smile should not be interpreted as conveying deep friendship; rather, it is a mechanism for making life pleasant and avoiding difficulties that might lead to the dreaded expression of negative emotions. Thais, in fact, genuinely dislike complainers, including demanding Western tourists, and will avoid them. In contrast, Israelis complain about many things at parties that have come to be called "griping parties," but they still have a good time (see Chapter 20).

However, the Thai smile can be genuine and, as is common in collectivistic cultures, must be evaluated in terms of the context or situation. For example, Thais love to have *sanuk* or fun, and they punctuate the workday with periods of group activity stressing it. If work is boring and monotonous, the Thais may quit, especially if the periods of *sanuk* are denied. Japanese firms, in contrast, sometimes pay Thai workers less than the U.S. firms do and work them longer hours, but they ensure that the workday is broken by such periods of respite.

As this discussion suggests, the Thais have a different conception of time than Westerners. Being Buddhist, they do not make sharp distinctions between the past, present, and future. In Buddhism time is only one circle, not three. In his first MBA class at Thammasat University, Martin Gannon arrived at 5:50 p.m., one half hour before the class period, to make sure everything was prepared, but the first Thai student did not appear until 7 p.m. and the last student to arrive did so at 8 p.m. Given his U.S. orientation, Gannon complained about this behavior, but one smiling Thai student confided, "We love to start class on Thai time and quit on American time." It was difficult to respond to this comment.

Accepting Things as They Are

There is another related Thai belief that is captured in the virtually untranslatable phrase, *mai pen rai.* Essentially, it means that humans have little if any control over nature, technology, and many other forces. Carol Hollinger (1967/1977), a U.S. high school teacher in Thailand, fell in love with Thai culture, particularly this aspect of it, and titled her book *Mai Pen Rai Means Never Mind.* This phrase does not signify the acceptance of a fatalistic view of life, sometimes found among conservative Muslims or Christians. Rather, it is an acceptance of things as they are and the willingness to make life as pleasant as possible regardless of life's circumstances. When Thais use this phrase, they tend to flash their distinctive smiles, and even that small action immediately lessens the magnitude of the problem, whatever it is.

The royal family, and particularly the king, follows similar patterns of behavior. Although serious, the king enjoys parties and performs his own jazz compositions at them. He clearly is happy interacting with all types of people and will smile through boring ceremonies that other international leaders would not even think of attending. As indicated previously, throughout his life he has maintained excellent physical conditioning, although his health has been declining in recent years. Given the many onerous duties that he must perform while smiling, such conditioning has probably been critical.

John Fieg (1976; Fieg & Mortlock, 1989) has captured the essence of Thai culture and the three characteristics highlighted in this chapter (loose authority ranking, freedom and equality, and the Thai smile) in his classic study comparing Thais and U.S. Americans. In both cultures there is a love of freedom, a dislike of pomposity, and a pragmatic outlook. But the differences are significant and Fieg uses the image of a rubber band to highlight Thai values, attitudes, and behaviors. When the rubber band is held loosely between the fingers, its looseness is comparable to the manner in which most Thais interact with one another during the day. However, as suggested previously, Thais also have a complex status system in which relationships are vertical and hierarchical. Once this status system is activated in any way—for example, a superior giving a direct order to a subordinate—Thais tend to respond immediately to its dictates, and so the rubber band tightens. As soon as the demands of the status system have been met, Thais can return to looser and more liberated behavioral patterns.

To describe U.S. Americans, in contrast, Fieg uses a string held tightly between the fingers for most of the day. Periodically the string is loosened, but U.S. Americans do not enjoy the same degree of behavioral freedom that Thais experience; that is, the string can never be as loose as the rubber band. In the United States many internal and external controls motivate individuals in this achievement-oriented society. Examples of external control include the numerous legal, accounting, and governmental forms that U.S. Americans routinely fill out. Imbuing children with the desire to work hard and to respond enthusiastically to the demands of the Protestant work ethic reflects internal control. Such internal and external control is present in Thailand to a much lower degree than in the United States.

As indicated previously, Thailand experienced severe economic difficulties starting in 1997 when the value of its currency was halved. The new constitution in 1997 has addressed the need for reform, focusing on such issues as democratic election reforms and the elimination of bribes. There are frequently more than 1,000 protests against the government each year, prompting *The Economist* to portray Thailand as a "land of frowns" (McBride, 2002). Such activity is to be expected as a nation's educational level and wealth increase and the nation's citizens express a desire for popular elections rather than military-influenced dominance. Still, at the personal level, the Thai smile is one cultural characteristic that remains unchanged.

In sum, Thailand is a fascinating nation and its core values overlap with those of the United States. But it is clearly different from the United States, and the Thai kingdom is an apt metaphor for capturing the essential features of this land of freedom.

CHAPTER 3

The Japanese Garden

Lacking wide-open spaces, and living close together as they do, the Japanese learned to make the most of small spaces. They were particularly ingenious in stretching visual space by exaggerating kinesthetic involvement. Not only are their gardens designed to be viewed with the eyes, but more than the usual number of muscular sensations are built into the experience of walking through a Japanese garden. The visitor is periodically forced to watch his step as he picks his way along irregularly spaced stepstones set in a pool. . . . Even the neck muscles are brought into play.

—Edward T. Hall, 1966, pp. 52–53

Japan is an island nation of 127.8 million citizens. There are 338.4 people per square kilometer, making Japan one of the most densely populated nations in the world. Compared to the United States it is relatively small in size: 377,727 square kilometers versus 9,372,610. Japan ranks first among nations in terms of its population's median age, with 26.3% over 60. Its economy is the second largest, right behind the United States. However, it is also the land of "disappearing children" with the number of children declining for 27 consecutive years; it is on track to lose 70% of its workforce by 2050, according to a report by the nonprofit Japan Center for Economic Research in 2008 (Harden, 2008). In comparison, children in India today exceed the population of the entire United States.

In the early 1980s almost all of the U.S. writing on Japan praised this country wholeheartedly and tended to highlight only its positive features, for example, William Ouchi (1981) in his best-selling book, *Theory Z*. Ezra Vogel (1979) even titled his best-selling book, *Japan as Number One*. By 1990 many U.S. writers and citizens had radically altered their view of Japan from one of uncritical admiration to one of "fear and loathing," as a *Fortune* article so aptly expressed the viewpoint (Smith, 1990). Some reasons for this shift in perspective were the economic successes of Japan at the time and the difficulty U.S. firms had in penetrating Japan's markets. Starting in 1990 Japan experienced a continuous and deleterious recession but it has steadily come out of it since 2002 with what *The Economist* (Standage, 2007) calls a hybrid model of capitalism that incorporates some dynamic elements from the U.S. system and melds it with the old Japanese model, somewhat akin to the hugely successful Prius, the best-selling hybrid car from Toyota. However, rising international fuel and commodity costs and the global financial crisis of 2008 threaten the well-being of the economy. The latest forecast by *The Economist* indicates that the gross domestic product is likely to grow by 1% in 2008 and 1.2% in 2009.

Japanese companies declaring bankruptcy frequently attempt to place their employees with other companies, and the shame associated with failure has led to many suicides among managers. It is hard to imagine comparable action by managers in the United States, who supposedly are more motivated by guilt than shame. Suicide is very much part of the Japanese cultural tradition, going back to the days of the samurais, who viewed it as noble and possibly preferable to being captured and tortured. Nobuo Shibata, the 48-year-old president of a small sheet metal company, and his brother took their own lives and left this plaintive note: "We apologize to all our employees for the slump in our business" (Sugawara, 1998, p. G3). In general, the total number of suicides in 2007 was 33,093, which represented the second-highest tally on record, according to the National Police Agency. In recent years, there has also been an alarming increase in the number of suicides among elderly people, largely because of health and economic troubles. Another alarming trend is the use of Internet suicide pacts by strangers. The government has earmarked more than $200 million to help with mental health problems and combat the stubbornly high suicide rate.

Corporate Cultures

Japanese companies have now begun to accept some Western-style management practices, including a direct link between individual performance and compensation, vast differences in individual compensation, and downsizing of the workforce. However, many Western companies that have introduced such practices into their Japanese subsidiaries without tailoring them to cultural norms have experienced great difficulty and low performance, including the Ford Motor Company, which owns a large share of Mazda. Once Mazda proudly advertised its cars under the motto "Mazda Engineering," but no longer, and its sales have decreased significantly. By contrast, Carlos Ghosn became CEO of the embattled Nissan and turned the company around by introducing drastic changes such as downsizing and plant relocations while being sensitive to cultural norms (referred to sometimes as the "Carlos Ghosn effect"). For example, he downsized the workforce by offering a generous buyout to current employees; he opened new factories that were typically 50 miles from the old factory;

and although he helped workers to relocate, the buyouts simultaneously averted involuntary layoffs. Ghosn has become a symbol of enlightened management in Japan, among both managers and employees. He soon became a hero to ordinary Japanese, and salarymen (the commonly used Japanese term for white-collar business workers) mimicked his business wardrobe. At one time, he popped up on short lists for the post of prime minister, even though he was ineligible. Sony's Welsh-born CEO, Howard Stringer, is another example of a foreign CEO of an iconic Japanese company, signifying a more globalized Japan.

However, the reality of the Japanese situation and culture seems more complex than U.S. Americans can imagine. The concept that Japan is a collectivistic nation needs to be qualified. In Geert Hofstede's (2001) classic study of 53 nations, Japan ranked 22 to 23 on individualism, which indicates that many nations are more collectivistic. Still, Japan is a classic authority ranking nation, and to expect it to become a market-pricing or vertically individualistic nation overnight is unrealistic. The interesting balance between individualism and collectivism in Japanese culture is illustrated by the more recent GLOBE study (House et al., 2004) in which the Japanese were found to score high on societal collectivism practices but in the medium range on in-group collectivism. Apparently large Japanese multinational corporations such as Mitsui & Co. and Mitsubishi Corp. have noticed this gap, as they have returned to the practice of having dormitories for single men and single women (separated by gender) so that they can get to know one another outside of work, thus facilitating their in-group collectivism and effectiveness at work.

To gain some perspective on Japanese culture, it is useful to compare the basic metaphors or mind-sets through which U.S. Americans and the Japanese see the world. Whereas football reflects the U.S. perspective, most if not all Japanese would immediately identify the Japanese landscape or wet garden as an appropriate metaphor for understanding the Japanese culture. In fact, the *Sakuteiki* or "Records of Garden Making," written almost 1,000 years ago by a Japanese court noble, is considered to be possibly the oldest book on gardening ever written.

Wet and Dry Gardens

There are two general types of Japanese gardens: the wet or landscape garden (*Tsukiyama*) and the dry Zen Buddhist or religious garden (*Karesansui*). The major difference between them is that the dry garden does not have water or a pond but, rather, raked pebbles to simulate water. Hence the wet and dry Japanese gardens are functionally equivalent to one another, and both are designed to create both a sense of integration between the observer and nature and an atmosphere in which meditation can take place. A third style of garden called *Chaniwa* gardens are built for the Japanese tea ceremony. In this chapter the focus is on a wet Japanese garden.

Like the water flowing through a Japanese garden, Japanese society is fluid, changing yet retaining its essential character. Alone, each droplet has little force, yet when combined with many others, enough force is produced to form a waterfall, which cascades into a small pond filled with carp. The pond appears calm, yet beneath the surface a pump constantly recirculates water back to the top of the waterfall. A rock tossed into the pond causes a ripple, then sinks to the bottom, ever to remain separate. Undoubtedly the scene is designed to replicate nature and to capture its true essence, and the sound of the gently flowing water amid this small creation of beauty evokes a feeling of harmony, oneness with nature, and perhaps even timelessness.

The garden itself serves as a reminder of the centrality of nature to the development of Japanese society, religion, art, and aesthetics: not just nature, but the way in which the Japanese perceive nature. As the garden is part of nature, the Japanese tend to view themselves as also integral with it. In contrast to the Western view of nature as something to be confronted and subdued, the Japanese generally see it as something to be accepted.

A Love of Nature

The Japanese, an agricultural people, developed a passionate love of nature and a keen awareness of the beauties inherent in it. Japan's temperate climate, abundant rainfall, and proximity to the sea on all sides contribute to a view of nature as a friendly blessing and source of growth and fertility. Thus it is not surprising that the earliest Japanese literature manifests a deep appreciation of the beauties of the sea, mountains, and wooded glens.

The three basic elements of the Japanese garden include stone, plants, and water. Usually these are combined to form larger elements common to Japanese landscape gardens, such as flowing water, ponds, and groupings of stone, trees, and shrubs, each as natural looking as possible to evoke the feeling of artlessness.

Japan, like the water in a Japanese garden, is seldom still. It is a complex, dynamic society that has undergone enormous change in the past 150 years, transforming itself from a feudal state into a modern industrialized nation that has fought two world wars, the latter of which resulted in utter defeat. Since then the country has risen to become a major economic power. In the process the Japanese have absorbed Western technology, science, education, and politics while retaining their unique cultural identity.

Still, underlying this profound change is the perspective that is mirrored in the Japanese garden. To see how the elements of such a garden are manifested in Japanese society, we focus on four topics: *Wa* (harmony) and *shikata* (the proper way of doing things), with emphasis on the form and order of the process; the combination of droplets or the energies of individuals in group activities; *seishin* or "spirit" training; and aesthetics.

Wa and *Shikata*

To understand Japan it is important to know its history, which can be viewed as a continual search for *wa* or harmony in interpersonal relations or, in the Japanese garden, the harmonious relations among elements so as to create a feeling of the effective interaction between human beings and nature. The Japanese are a homogeneous people, who have lived in relative isolation for centuries, although the geographical features of the country do make for diversity in the population with regard to dress, food, customs, language, and so on. Prior to the end of the last ice age about 11,000 years ago, Japan was connected by land to the rest of Asia. Because those people who wandered into it could go no further, they remained and mixed with those who came later. Historical records show that a considerable number of people flowed into Japan from the Korean peninsula until the 8th century CE. By that time the mixing was nearly complete, and for the past 1,000 years, immigration has been infinitesimal.

Borrowing From the Chinese

Virtually devoid of natural resources such as iron and oil, the Japanese borrowed the technique of rice farming from the Chinese sometime around the year 1000 BCE. Each of its villages became self-sufficient, but the complexity and interrelatedness of the tasks associated with rice farming led the Japanese to emphasize the importance of group activities and group harmony without which there would be starvation.

In the 7th century BCE, this emphasis on harmony was expressed in the country's first "constitution," written by Prince Shotoku. The first of the 17 articles made harmony the foundation for all of the others. This is in marked contrast to the Western conception of the importance of life, liberty, and the pursuit of individual happiness.

Furthermore, the Japanese have always had a distinct awareness of the difference between things foreign and native, and early on, they recognized the value of borrowing from others while maintaining their Japanese-ness. After 552 CE, when Buddhism was formally introduced to the Yamato court from China, the Japanese increasingly became conscious of the superior continental civilization of the Chinese. From the 7th to the 9th century, they made a conscious and vigorous effort to borrow Chinese technology and institutions; they studied the philosophy, science, literature, arts, and music of China in depth, and these, in turn, deeply influenced Japanese thought, culture, and habits. The Japanese even adopted the Chinese writing system, which had a unique character for each of the thousands of Chinese words. However, even though the Japanese wrote in Chinese, they spoke only Japanese, which is as different from Chinese as it is from English. Japan's island location was far enough from China to escape invasion but close enough to benefit from interacting with this nation, which at that time was the most technologically advanced and powerful in the world. From the 9th to the 12th centuries, the Japanese blended these new elements with their own culture to form a new synthesis.

The Shogun Era

In the 12th century, Yoritomo Minamoto first used the title of shogun or generalissimo of Japan and established his headquarters in Kamakura, which is one hour's drive from Tokyo, or Edo as it was called at that time. Although the imperial family living in Kyoto was still allowed to perform ceremonial activities, the real power rested with the shogun, and clan lords reported directly to him. This shogun era lasted until 1868, at which time the imperial family's power was restored (Meiji Restoration). In 1868, the shogun was deposed, the shogunate ended, and the country's capital and Imperial Residence were moved from Kyoto to Tokyo.

During the shogun era the emphasis on harmony continued, but in a distinctly different manner. Surprisingly, the shogun system was similar to European feudalism. About 10% of the population consisted of samurai or warriors, who pledged absolute loyalty to the shogun. Thus this system is quite different from the Chinese Confucian system, in which loyalty to the family is the overriding value. The subjugation of the samurai to the shogun paved the way for the transference of loyalty to the ultimate family, the nation. Also, unlike Europe, Japan did not have a cult of chivalry. In addition, the samurai were also men of learning, and they prided themselves on their fine calligraphy and poetic skills.

However, most of the population had only limited rights, and a samurai could kill commoners on the spot if they failed to abide by the many rules and practices pervasive throughout the society. Under such conditions it is little wonder that the Japanese sought to achieve harmony and to protect face, which is an unwritten set of rules by which people in a society cooperate to avoid unduly damaging each other's prestige and self-respect. While all cultures emphasize the protection of face, the Asian concept includes "giving face." For example, something is always left on the table for the defeated adversary in business negotiations while baseball games may end in ties rather than wins. Even today the Japanese tend to apologize in advance before making a critical comment on another person's project so as to avoid the loss of face for either party; many of them begin a formal speech in a comparable manner.

Japan's sense of isolation was heightened during the reign of Shogun Ieyasu in the 16th century when he banished Jesuit missionaries, cut off the ears of Christians, and mandated that only coastal vessels not fit for ocean travel could be built. With the exception of a small Dutch trading area in the Kyushu port of Nagasaki, relations with the outside world were cut off.

Out of Seclusion

In 1853 Japan was forced to end its self-imposed seclusion when Commodore Perry and a quarter of the U.S. Navy arrived off the shores of Edo. The Japanese government had no choice but to sign unequal trade treaties with the United States and principal European powers. These developments led to a weakening of the shogun system and its replacement by the imperial family in 1868.

Furthermore, it is extremely difficult for a foreigner or *gaijin* (or *gaikokujin,* which has become the more acceptable term in recent years) to be fully accepted in Japan, as this would upset harmonious relations between and among groups that have taken years to develop. A foreigner is like the rock tossed into the garden's pond; it momentarily disturbs the harmony but quickly disappears from sight. However, many visiting U.S. Americans misperceive the extreme care and attention given to them as genuine friendship or at least genuine acceptance. The reason for this leads us naturally to *shikata*, which is necessary if *wa* is to be maintained.

Because of the subsistence-level economy that developed in the rice-growing villages and the emphasis on nature, the Japanese came to believe that there is an inner order (the individual heart) and a natural order (the cosmos) and that these two are linked together by form (see De Mente, 1990). *Shikata* is the way of doing things, with special emphasis on the form and order of the process; in compound forms, the term is *kata*. Thus there is a *kata* or several *kata* for eating properly, for using the telephone, for treating foreigners, and so on. To the Japanese, the process or form for completing an activity is just as important as completing the activity successfully, whereas U.S. Americans have historically been concerned with the final result rather than the process of achieving it. The Japanese emphasis on Total Quality Management (TQM) and *kaizen* (continual improvement) is consistent with *shikata*, as the Japanese tend to feel that doing something in the proper manner will ultimately lead to doing it in the most successful manner.

The Proper Way

Kata were developed within this hierarchical society because it was assumed that everyone has specifically categorized and defined life roles *(bun)* in which obligations are spelled out in detail. Edwin Reischauer (1988) has termed the Japanese the most punctilious people on Earth because of such explication, if not the most polite. A person's outward conduct, as manifested in the obedience to rules, is a reflection of inner character. There are even *kata* associated with Japanese baseball, and visiting U.S. baseball players frequently find it difficult to adjust because they must hold the bat in a particular way and observe other rules that they consider irrelevant.

The importance of *kata* cannot be overemphasized, and there are specific *kata* for activities that U.S. Americans tend to handle in many different ways, such as greeting visitors and exchanging business cards. When a *kata* for a new activity does not exist, the Japanese may have difficulty completing the task. Thus some Japanese officers who wanted to discontinue fighting in World War II bemoaned the fact that no *kata* for surrender had been developed simply because it was so unthinkable. As they stated, "We don't know how to surrender" (De Mente, 1990, p. 68).

Seishin Training

To live comfortably in such a rule-ordered society, people need a great deal of self-control, and they are socialized from birth to acquire it. In the Japanese garden the carp symbolize masculinity, valor, endurance, and tenacity. Analogously the concept *seishin* or spirit, which is an integral part of the Japanese philosophy of life, stresses the importance of self-discipline and devotion to duty. Through discipline and adversity, people achieve self-development and, most important, self-mastery. *Seishin* training has been most commonly applied to the study of martial arts, flower arrangements, and the tea ceremony. In the case of the martial arts, physical training is seen as a means to the spiritual end of attaining inner balance and harmony. Mastery of skills requires self-control and self-discipline; this leads to the development of inner strength. In the past, a number of Japanese companies also offered *Seishin* training to new recruits, although in some quarters this was considered to be a relic of Japan's prewar educational system, which was considered to be nationalistic and militaristic.

Buddhist Influence

Seishin training tends to be associated with Zen Buddhism, which is a unique form of Buddhism that has enjoyed a resurgence of interest in Japan since the 1970s. While Hinduism incorporates the belief in many personal gods, Buddhism rejects them. It does, however, accept the Hindu emphasis on meditation as a major vehicle for achieving nirvana or salvation. Buddha emphasized four Noble Truths: Suffering is inevitable; it occurs because of selfish or self-centered desires; such desires can be overcome; and the method of overcoming them is to follow the Eight-Fold Path, which is the Buddhist version of the Ten Commandments.

Zen Buddhism proposed that people can free themselves from their self-centered world by perceiving the timelessness and oneness of the universe. Likewise, in a Japanese garden, the whole of nature can be seen as well as the individual parts. Also, each part of the garden can capture the essence of nature in its own way. The garden depicts a freed world, which has been physically reduced and spiritually enlarged to suggest the size and grandeur of nature. Japanese, who recognize that they are part of nature and hence mortal, essentially stop time by allowing it to have its own way. In the same manner, although the seasons may change the appearance of the Japanese garden, the stones, trees, and water in it are always visible and unchanged.

Perhaps the best way to understand Japanese behavior is to take part in such martial arts as karate and sword play. Karate (literally, "the way of the empty hand") originated in Okinawa, Japan, and is a combination of preexisting Okinawan martial arts called *te* and Chinese martial arts. It is both a form of self-defense against marauders and a form of spiritual training. Once again, forms or *kata* constitute much of the training that karate students receive, and they practice these forms by themselves for hours before engaging in any kind of combat. Eventually karate students are taught to "think through" an obstacle, and one of the most difficult challenges facing a parent is to see a nine-year-old son or daughter trying to "think through" or break a one-inch-thick plywood board with an openhanded blow. Sometimes the karate instructor will have unsuccessful students repeat the exercise three or four times so as to exhibit self-control and learn to handle adversity, and some young students have broken their hands while engaged in this activity.

Going to "Hell Camp"

Although theoretically the ultimate result of *seishin* is an improved state of personal spiritual growth and freedom, many Japanese use it to attain practical ends, such as better performance in school or work. Some large Japanese firms appeal to this element of Japanese thought in their training programs, which have been called "hell camp" because the trainees are allowed to sleep only a few hours per day and undertake physically demanding activities designed to instill a sense of self-discipline and devotion to the company; the trainees are given several "badges of shame" that they can remove from their clothing only when they successfully complete the various hurdles and examinations. In many modern transformations of the philosophy, the central emphasis is on the development of a positive inner attitude by means of which external constraints can be overcome and difficulties resolved. Japanese tend to believe that any obstacle can be overcome provided one tries hard enough. Conversely, U.S. Americans emphasize ability as much as or more than effort.

Like the Chinese, the Japanese believe strongly in the principles that Sun Tzu, the ancient Chinese general, put forth in *The Art of War* (see McNeilly, 1996, whose book also includes the translation by Samuel B. Griffith, 1963). They see the marketplace as a battlefield but accept Sun Tzu's first principle, namely to avoid war or confrontation if at all possible simply because both sides deplete critical resources. Still, the Japanese revere Miyamoto Musashi, a legendary samurai who combined kendo and Zen while killing 37 opponents; then he retired to a cave, where he died. Musashi's (1974) *A Book of Five Rings: The Classic Guide to Strategy* describes the set of rules that many Japanese managers follow when doing business. This paradoxical and simultaneous acceptance of inner reflection and merciless activity against business opponents is similar to the Chinese paradox of being both sincere and deceptive (see Chapter 25). A popular 14-hour film was made of Musashi's life and book.

Combining Droplets or Energies

Achieving harmony through *kata* and *seishen* training leads naturally to the topic of combining energies of individuals in groups or, analogously, combining droplets of water in the garden to form a waterfall. This emphasis on the group begins at birth, for mothers tend to shower an excessive amount of love and attention on their children, especially boys. In Japan infants and children are constantly in contact with their mothers, are treated permissively, are seldom left alone, and often sleep with their parents until the age of 7 or 8. For the first 2 years of life, children are swaddled and carried by their mother in a special device attached to her body. All of these practices result in a high degree of dependence on the mother and encourage an attitude defined as *amae* in Japanese, which means to look to others for affection. As children grow up, this dependence on their mothers as the source of gratification is transferred to the group, particularly the mentor or *senpai*. The system of *senpai* and *kohai* is a complex relationship with obligations on both sides. The Japanese often meet their first *senpai* in school when they join a club or organization and this relationship can last for a long time. The closest Western concept is the mentor-apprentice relationship but the Japanese tie is much more formalized. The *kohai* are trained like soldiers or sportsmen and they are expected to respect and obey their *senpai*. They also use formal and polite language, which is very different from the language used with their friends or peers. The *senpai* has to earn the respect of the *kohai* and vice versa (Hogg, 2006).

An Elder Adviser

During the first few years at a business firm, employees may be assigned to an older individual or *senpai*, who will show them how to perform their job and help them adjust to their new environment. More commonly, *senpai-kohai* relationships develop informally. The *senpai* offers advice and encouragement and may play an important role in the socialization of the youngest employees. This relationship allows the *kohai* to express views informally without fear of censure from a larger group.

The Japanese emphasis on the group is quite apparent and permeates practically every aspect of Japanese life. Some people argue that Japan's low crime rate can be attributed to the importance of the group in Japanese society and the strong desire not to bring shame on the group. It can be seen in the educational system, the structure of business organization and work, and the political system, among other social institutions.

At school, children are identified by their class. Japanese children enter the first grade of elementary school in the April after their sixth birthday and spend about six years in elementary school, three years in middle school, three years in high school, and four years at university. There are 30 to 40 students in a typical elementary class and they are divided into teams for several activities. They learn the traditional subjects that children all over the world learn in their elementary schools but they also learn traditional Japanese arts like haiku (a form of poetry that was developed in Japan more than 400 years ago). Many schools offer English classes and use informational technology extensively.

There is intense competition to get into schools. If attending public school, children will go to an elementary school and middle school located in the same district in which they live. They will then take an exam to determine which high school they will attend, because public high schools are rank-ordered based on test results. If attending a private school, students must pass an entrance exam for the appropriate level (elementary, middle, high), but once admitted, they can normally have access to a university, because most private schools are affiliated with universities.

Student Attitudes

A well-known Japanese proverb, "The protruding nail will be hammered," aptly explains the behavior of individual students. Adhering to this proverb does not pose a problem for most students, who have learned that security, acceptance, and love flow from the group. In fact, most students display a sense of belonging that would be envied by many of their counterparts in the West, who often feel alienated and alone.

On the other hand, life can be intolerable for the rare student who does not fit comfortably in a group. Fellow students relentlessly bully, pick on, tease, and persecute such individuals. This phenomenon, *ijime*, usually translated as bullying, leads to several deaths every year, and the number of reported incidents has been increasing recently. If such a student has any friends, they quickly desert him or her for fear of being excluded from the group, which is the worst fate that can befall a Japanese child or adult.

Students achieve group identification not only with their class but also with their school, particularly at the college level. The university they attend frequently determines their future career prospects, as the top businesses hire primarily if not exclusively from the top universities. Ties made in college days are important in Japanese life, and throughout their lives Japanese will identify themselves with the university they attended.

Managers and Workers

In Japan a job means identification with a larger entity, through which people gain pride and the feeling of being part of something significant. An individual's prestige is tied directly to the prestige of his or her employer. The company is typically viewed not as an entity trying to take advantage of its employees to make profits but as a provider of individual security and welfare. When the Japanese are asked what they do, they usually respond with the name of the company for which they work and not the job they perform. By contrast U.S. Americans will normally respond to the question of "what they do" by mentioning the occupation or job first, and they may not even divulge the name of the company.

Japanese workers or managers usually perceive their company and other institutions with a strong sense of *us* versus *them*. According to Akira Saito of Chuo University, as quoted in *The Economist*, this is a vestige of the feudal system of Bushido, which governed a samurai warrior's allegiance to his shogun or overlord and a master's duty to take care of his subjects. However, the 1990s were characterized by a slumping Japanese economy and this brought about a number of changes in the workforce, including a rise in the temporary workforce from 20% to 35%. An interesting outcome of this was the institution of a law to protect whistleblowers in 2006, after a rise in tip-offs about improper practices.

In recent years, there has been much discussion about the future of lifetime employment in Japan. Matsushita's decision to cut 8,000 jobs in 2002 was widely seen as a turning point in industrial relations in Japan. In the 2000s, the practice of layoffs and downsizings has continued but according to labor experts, Japanese lifetime employment is too deeply embedded into its institutions and human resources practices to be completely abolished (Moriguchi & Ono, 2006). According to Akira Kawamoto, a director at RIETI, a government think tank, the system is slowly crumbling but only at the edges. Most of the salarymen are expected to stay inside the system until they retire (*The Economist*, 2007). Another distinctive structure of the business firm in Japan is supposedly that it fosters group identification. Prior to World War II much of Japanese industry was organized into six huge *zaibatsu* or family-owned companies, each of which consisted of about 300 smaller companies and their suppliers. Each *zaibatsu* combined the activities of many subcontractors with whom it had long-term contracts, a manufacturing organization, a major financial institution, and an export-import organization. Such a form of organization is outlawed in the United States and other developed countries and was actually forbidden by law in Japan after World War II. However, a nonfamily variant of the *zaibatsu*, the *keiretsu*, has emerged and become prominent.

U.S. and European managers frequently complain bitterly about the operations of the *keiretsu*, as these organizations have a great amount of power over many activities in the marketplace and they can persuade Japanese distributors not to carry the products of foreign companies. Ironically, however, the most dynamic and prosperous companies in the Japanese economy tend to be those that are not members of a *keiretsu* (Tasker, 1987).

Since 1990 there has been an increase in entrepreneurial activities among younger managers, some of whom have started their own firms rather than accepting employment in established firms. Such entrepreneurial efforts will increase as the world economy becomes more global, but identification with the company will probably remain much stronger in Japan than in the United States.

The Importance of the Group

Furthermore, the organizational structure and work allocation in Japanese firms also emphasize the group. Work is often assigned to various office groups and is viewed essentially as a group effort. Frequently there are no formal job descriptions existing separate from work groups. A company will sometimes reward the office group and not the individual for work well done. Even the physical arrangement of the office emphasizes the group mentality: The manager sits in front of the workers in a classroom-style setting, and they work in subgroups with their own supervisors. If the work group is small, everyone will sit around a table, with the most senior members closer to the manager. As a general rule the Japanese do not allow managers to have their own offices and, even if the work requires that they be given one, managers tend not to prefer this arrangement. This sometimes frustrates U.S. employees of Japanese companies that set up shop in the United States. The U.S. TV show, *The Office*, which parodies bureaucratic office culture and dimwit bosses and is based on the popular British show of the same name, is considered to be an unlikely candidate for a Japanese version because the office culture in Japan is so sacrosanct. Still, this did not prevent the show's creator, Ricky Gervais, from creating a short Japanese satirical version on *Saturday Night Live*.

Japanese groups abound in society—women's associations, youth groups, PTA, and hobby groups, to name just a few. Political parties and ministerial bureaucracies often divide into opposing factions. When sightseeing, the Japanese tendency to perform activities in a group is particularly evident, as they tend to wear the same type of clothes and behave as if they were one person.

Decision making in Japan also reflects this emphasis on the group. For example, in a Japanese company a business proposal is usually initiated at the middle or lower levels of management. The written proposal, called *ringisho*, is distributed laterally and then upward. Each person who reviews it must provide a personal seal of approval. Making decisions requires a great deal of time, largely because a good amount of informal discussion has preceded the drawing up of the *ringisho*. In spite of its time-consuming nature the consensus approach to decision making has important merits. Once a decision is made, it can be implemented quickly and with force because it has the backing of everyone in the department. By contrast, U.S. managers frequently make decisions quickly, but a great amount of time and effort is required for implementation, often because only a few key people have actually been involved in the decision-making process itself.

A particularly telling contrast occurred when Travelers Group and Nikko Securities were contemplating a joint venture. After Sandy Weill of Travelers and his Japanese counterparts agreed on the basic issues, Weill placed a conference call to New York to obtain board approval, which he secured in 30 minutes. Conversely, the Nikko board required 12 days of briefings and discussion to approve a decision that had been resolved in its favor. However, decision-making practices on both sides of the Pacific have evolved over the years as Japanese and U.S. companies have borrowed best practices from each other.

Responsibility to the Group

Just like the water droplet, the individual is significant only insofar as he or she represents the group. If individuals disagree with one another, the overall interest of the group comes before their needs. Cooperativeness, reasonableness, and understanding of others are the virtues most admired in an individual. Harmony is sought and conflict is avoided if at all possible.

The extent to which the individual is responsible to the group, and the group responsible for the actions of the individual, is illustrated by the following incident. A U.S. teacher was accompanying an eighth-grade Japanese class on its annual field trip to Kyoto when one of the Japanese teachers kindly informed her that the students would be required to sit in the school auditorium for an hour on returning from Kyoto. When the U.S. teacher inquired why, he responded that one of the students had been 10 minutes late to the school meeting earlier that morning; therefore, all would be punished.

Like the water that flows in the garden, the Japanese people prefer to flow with the tide. Observers have noted the relative absence among the Japanese of abstract principles, moral absolutes, and definitive judgments based on universal standards, in contrast to the West. The Japanese think more in terms of concrete situations and complex human relations; therefore, while to Westerners the Japanese may seem to lack principle, to the Japanese, Westerners may seem harsh and self-righteous in their judgments and lacking in human feeling.

Similarities and Contrasts

As Ellen Frost (1987) points out, many Japanese refer to this difference by saying that the Japanese are "wet" people whereas Westerners are "dry." She elaborates:

By "dry" they mean that Westerners attach more importance to abstract principles, logic and rationality than to human feeling. Thus, Westerners are said to view all social relations, including marriage, as formal contracts which, once they no longer satisfy individual needs, can be terminated. Their ideas of morality are generalized and absolute, with little regard for the particular human context. . . . By contrast, the "wet" Japanese are said to attach great importance to the emotional realities of the particular human circumstances. They avoid absolutes, rely on subtlety and intuition, and consider sensitivity to human feelings all-important. They notice small signs of insult or disfavor and take them deeply to heart. They harbor feelings of loyalty for years, perhaps for life, and for that reason are believed to be more trustworthy. (p. 85)

Some see the Japanese identification with nature as the explanation for the origin of the relativism of Japanese attitudes. Others point to the influence of Chinese thought, with its situational approach to applying principles (see Chapter 25). Whereas in the West the division is between good and evil, in China the division is between yin and yang, two complementary life forces. Thus the Japanese tend to see situations more in terms of tones of the color gray, whereas U.S. Americans see them as black and white. However, others suggest that this relativism stems from the child-rearing techniques discussed previously. Most probably all of these explanations are interrelated and have some validity.

The emphasis on situational ethics is illustrated in the Japanese legal system. If people who are convicted show that they are genuinely repentant for what they have done, then it is highly likely that they will receive a more lenient sentence. Furthermore, the laws passed by the Japanese Diet or central government are loosely structured so that the courts can interpret them in ways that different situations demand. By contrast, the U.S. Congress attempts to write laws in such a way that all possible issues are spelled out carefully, and as a result, judges must operate within relatively strict limits. Furthermore, the Japanese orientation toward harmony and group loyalty means that conflict resolution is valued, not the individualistic assertion of legal rights. Consequently, the legal system and Japanese values reinforce the emphasis on group consensus or solutions prior to coming to court. There is also a strong cultural aversion to litigation. The opposite situation exists in the United States, where individuals involved in legal disputes are aware that judges must interpret laws that are already finely delineated.

This practice does not mean that the Japanese do not have a sense of right and wrong. It is simply that they place greater emphasis on the particular situation and human intentions than Westerners do. The major issue is whether an action harms others and has a disruptive effect on the group and community. Many experts assert that the Japanese are motivated primarily by a sense of shame to correct a problem when such disruption occurs, whereas individualistic Westerners are motivated mostly by a sense of guilt at their failure to fulfill responsibilities. Such an orientation is particularly evident in attitudes toward sex and drinking.

Private Lives

Traditionally marriages in Japan were arranged, and even today, both families and companies act as intermediaries although this practice is being seen as outdated. Women are increasingly passing up marriage for their careers. Japanese government figures indicate that the percentage of women ages 25 to 39 who have never married has surged from 40% to 54%. This is comparable to 40% in this age group for U.S. women according to U.S. census data. A book about single career women by Junko Sakai titled, *Howl of the Loser Dogs,* sold more than 300,000 copies. The typical Japanese male expects the wife either to give up her job or to manage both the career and the children while he keeps long hours at work, and modern Japanese women are rebelling against these expectations. They are increasingly looking for a partner who will give them a lot of autonomy and freedom in the relationship. All of this makes for a low birthrate of 1.29 per woman (compared to 2.13 for the United States) and has grave implications for labor policy.

Drinking and even drunkenness are good-naturedly tolerated and even encouraged. In fact, almost anything people do while drunk, except driving, is considered forgivable. Intoxication allows the Japanese to express themselves freely without fear of repercussion, and it is normal for members of a Japanese work group to spend several hours after work drinking and eating as a group before catching a late bus home. However, Western women sometimes find this behavior difficult to accept, even when a Japanese man is very polite when not drinking. In one celebrated instance a British female high school teacher finally lost patience with a Japanese coworker who had repeatedly harassed her while he was drinking, and she "decked him" at an office party. However, while Japanese may indulge in unruly behavior while drinking, they rarely engage in displays of hostility and violence. Indeed, with the Japanese emphasis on politeness, it is quite common to see tipsy "salarymen" bowing to each other as they bid goodbye after a get-together!

Like the natural order of the Japanese garden, the Japanese believe in a natural order in society. Therefore, as reflected in the history of hereditary power and aristocratic rule in feudal Japan and in modern Japanese organizations, different ranks and status are considered natural. Within an organization, a person's rank is usually more important than his or her name. For example, the principal of a school is often addressed simply by the Japanese word for principal, *kocho sensei*. This emphasis on title establishes an individual's rank within an organization while serving to reinforce group identity.

Establishing Status

The Japanese are great believers in establishing a person's status as quickly as possible so that the proper interaction and communication can take place. When meeting for the first time, Japanese businessmen follow the *kata* or form of immediately exchanging business cards, *meishi*, largely to establish each person's specific position and group affiliation. This exchange is not to be taken lightly, and the recipient of such a card is expected to read the card carefully and even repeat the person's name and title before putting the card away. To put the card away immediately, as many U.S. Americans do, would be insulting to the giver.

Language is also used to reinforce the natural order of rank and status. Various endings are added to words of introduction that subtly indicate the status of a person, particularly in the case of a foreigner or *gaikokujin*. The Japanese language is famous for such subtlety. Even at work, honorifics are used to address higher status managers, but they are frequently dropped during the later stages of a working day, as these managers indicate by their language and behavior that everything is going smoothly.

Even slight differences in age or in length of time at a company establish rigid differences in status and seniority that employees must acknowledge. Historically, junior employees were expected to use an honorific term when addressing a senior employee. However, this practice is declining; today, 59% of companies with more than 3,000 employees have a policy of not using such honorifics, compared to only 34% in 1995.

A good part of personal self-identification derives from status. Furthermore, people are expected to act according to their status. To do otherwise would be unbecoming. As they serve as role models and are addressed by the honorable title *sensei*, teachers in Japan must pay particular attention to their behavior outside of school. For example, if caught driving while under the influence of alcohol, a teacher would lose face and probably have to resign his or her position.

A hierarchy exists not just within the group but between different groups. Most Japanese have a clear idea of the informal rank among universities, business firms, and even countries. This latter ranking helps to explain Japanese attitudes toward people from different countries. For example, those from developed countries, such as the United States, England, France, and Canada, are considered to possess higher status than those from developing Asian or African countries. Even within Japan certain groups suffer discrimination, particularly the 600,000 Koreans. At one time in history the Japanese had a caste system similar to that of India. Although this system is outlawed, the outcasts or *burakumin* are still considered inferior by many Japanese.

Because of their long historical isolation from foreign influences, the Japanese have a complex relationship to outsiders. Foreigners in Japan report experiencing indignities like being passed by a taxi driver who chooses to pick up a Japanese passenger or being denied admittance to some restaurants and night clubs. National Public Radio's correspondent in Japan, Eric Weiner, reported that even a U.S. American who went through the convoluted process of becoming a Japanese citizen was considered to be a "*gaikokujin* who carries a Japanese passport" and was barred from "Japanese only" establishments. In 2008, many Japanese chefs looked down their noses at the new Michelin guide to Tokyo's restaurants because they could not imagine how a "bunch of foreigners" could rate Japanese food. At the same time, the Japanese love French luxury goods and Japan is the single largest market for Louis Vuitton goods.

Race and Class

One of the great dangers of overemphasizing the uniqueness of being Japanese and group harmony is that a destructive groupthink may occur. Some prominent Japanese have openly criticized minority groups in the United States in ways that many U.S. Americans consider racist, and Dr. Tsunoda of Tokyo Medical and Dental University has argued that the Japanese are unique because they have an unusual brain structure. According to Dr. Tsunoda, the Japanese brain mixes rational and emotional responses in the left hemisphere while the Western brain is divided into the rational left and emotional right side.

As indicated previously, Japanese political parties are not parties per se but representatives of powerful interest groups, and political scandals that would not be countenanced in other developed countries have occurred in Japan at least in part because of their existence. These examples should suffice to demonstrate that there are some downside risks associated with the emphasis on Japanese uniqueness and group conformity.

Contrary to expectation, this emphasis on status and rank coincides with a lack of class consciousness. About 90% of the Japanese consider themselves to belong to the middle class. In Japanese society ties are vertical; loyalties overwhelmingly tend to lie with the immediate group. For example, a Japanese factory worker for Toyota identifies himself as a Toyota man and has little solidarity or identification with a factory worker for Honda.

Promotion is based significantly on seniority in Japan, although this is not the only consideration, and today firms are emphasizing individual performance. People may not jump a rank and must serve a prerequisite number of years in the previous rank to be eligible for promotion. However, only a limited number of positions are available for a much larger number of eligible employees. Therefore the company promotes a certain percentage of eligible employees from each group until people reach a specific age, at which time it is unlikely they will be promoted further.

It is important to note that hierarchy in Japan is not directly associated with clear authority and clearly delineated roles as it is in the West. For example, senior but less competent employees may be given relatively high titles that entail minimal responsibility, and extremely able young people may be given more responsibility than their title indicates. Also, while the Japanese emphasize group consensus and harmony, in fact Japanese managers tend to share information with their subordinates and delegate decision-making authority to them less than U.S. managers. As these examples suggest, appearance does not always reflect reality in Japan, just as the water in the garden's pond appears to be quite calm, even still. Yet the architect of this garden knows that beneath the surface, the water is continually being sucked into a horizontal pipe and pumped back to the top of the waterfall.

Room for Competition

Like the pond, from the outside, the Japanese appear to be an extremely harmonious people. However, this appearance does not take into account the intense competition that exists in Japanese society. Firms compete fiercely for market share, rival political parties compete for power, and individuals compete ferociously to enter the top universities. On the personal level conflict occurs between family members, between friends, and between coworkers. However, individuals often suppress their true feelings and, if open conflict erupts, it is kept within the confines of the group.

In general, Japanese usually express the group opinion, *tatemae*, although it may differ from what they really think, *honne*. They associate this duality with the legitimate needs of the group and the reinforcing role of each person within it. Frost (1987) explains this dichotomy as follows:

> Tatemae . . . is used in connection with a view made from an accepted or objective standpoint. *Honne* has the meaning of one's true feelings or intentions. . . . This doesn't mean anyone is lying. There is a delicate but important difference between *tatemae* and *honne* which comes up in any situation where a person must consider more than one's own feelings on the matter. (p. 92)

To Westerners this lack of frankness smacks of insincerity. However, to many Japanese the failure to observe *honne* and *tatemae* shows insensitivity and selfishness. Japanese tend to appreciate foreigners who outwardly conform to certain rules of behavior. Whether they actually respect these rules is not relevant; observance of them is a sign of appropriate and hence sincere conduct on their part.

However, this suppression of the self may have some negative effects. Dean Barnlund (1989) conducted a comparative study of U.S. Americans and Japanese on the issue of the private and public self that was well received by both Japanese and U.S. experts. He basically argues that, because U.S. Americans tend to expose their private selves much more than the Japanese, they have a much better understanding of their private selves and are able to cope in an active manner with threatening interpersonal experiences. He concludes with the following:

> The Japanese appear to be more socially vulnerable and to cultivate greater reserve, are more formal and cautious in expressing themselves, and communicate less openly and freely. Americans, in contrast, appear more self-assertive and less responsive to social context, are more informal and spontaneous in expressing themselves, and reveal relatively more of their inner experience. (p. 64)

A study of shyness among several nations confirms this portrait. Bernardo Carducci developed a questionnaire measuring the degree to which people feel shy in the following nations: the United States, Germany, Mexico, Israel, Japan, and France. Japan proved to be the shyest nation: 6 of 10 believe they fall into this category. In contrast, the least shy nation was Israel, at 31% (see Morin, 1998).

Some Japanese respond that U.S. Americans say and do things that they later regret. The aphorism, "He who speaks does not know; he who knows does not speak," touches directly on this issue, for frequently the most powerful member of a group is the last to speak if he deigns to speak at all. What matters, however, is that the Japanese emphasis on the group is protected by the use of *honne* when they interact among themselves and with foreigners.

Nature-Based Religion

The Japanese aesthetic sense is well developed and distinctive, and in fact, it may be this culture's most important contribution to the world. As with most if not all of the activities discussed previously, this aesthetic sense is heavily based on the interaction between human beings and nature.

Japan has only one indigenous religion, Shintoism, which is nature based and animistic; even the word for God is translated as "up the mountain" in Japanese. Joseph Campbell (1962) captures the aesthetic sense of this religion in the following passage:

> Such a place of worship is without images, simple in form, wonderfully roofed, and often painted a nice clear red. The priests, immaculate in white vesture, black headdress, and large black wooden shoes, move about in files with stately mien. An eerie music rises, reedy, curiously spirit-like, punctuated by controlled heavy

and light drumbeats and great gongs; threaded with the plucked, harp-like sounds of a spirit-summoning *kot*. And then noble, imposing, heavily garbed dancers silently appear, either masked or unmasked, male or female. These move in slow, somewhat dreamlike or trancelike, shamanizing measure; stay for a time before the eyes, and retire, while utterances are intoned. One is thrown back two thousand years. The pines, rocks, forests, mountains, air and sea of Japan awake and send out spirits on those sounds. They can be heard and felt all about. And when the dancers have retired and the music has stopped the ritual is done. One turns and looks again at the rocks, the pines, the air and the sea, and they are as silent as before. Only now they are inhabited, and one is aware anew of the wonder of the universe. (p. 475)

The importance of nature can be seen not only in Shintoism but also in Japanese painting, literature, and language. Long before it was considered acceptable in the West, landscape painting was a major theme of the *sansuiga* painting introduced from China at the end of the Kamakura period (1185–1333). Nature is often the subject of the seasonal references in *waka* and *haiku* poetry and in the seasonal introductions that the Japanese customarily use in correspondence. The Japanese language is full of references to nature; it even has a word to describe the sound of cherry blossoms falling.

But the Japanese are acutely aware that nature can wreak havoc. In 1923 the great Tokyo earthquake resulted in the loss of 140,000 lives, and there is a one in five chance that a similar occurrence will happen in the next few years ("The Flowers of Kobe," 1995). In 1995 the 7.3 magnitude Kobe earthquake resulted in more than 6,400 deaths. While the Japanese are trying to make their buildings earthquake resistant, they realize that their efforts may prove futile because Japan is situated in one of the Earth's most seismically active areas.

Aesthetic Theory

Aesthetically the water in the Japanese garden and the garden itself try to represent nature as it is, not to order and impose a visible form on it as a Western garden may, but to capture its intrinsic being. Gardeners do not create the beauty but merely allow it to express itself in louder and plainer terms. In doing so they not only consider the placement of the rocks and trees but the effect that time, namely the change of the seasons, will have on the garden. They observe the laws of *mujo* or mutability and *sisei ruten* or perpetual change of the universe. For example, the garden may be designed in such a way that in winter the bare branches have their own particular appeal. Given this emphasis on the natural impermanence of things, it is not surprising that the three great nature-watching rituals in Japan are snow gazing in February or thereabouts, cherry blossom viewing, and viewing the ninth moon of the year or the harvest moon, all of which activities preferably occur in a Japanese garden.

Gardeners also observe another Japanese aesthetic theory that developed from the Japanese acceptance of nature: uniqueness. No two trees are alike. Therefore gardeners do not search for the perfect rock, for such a rock cannot exist in a world characterized by uniqueness, but look for a tree or rock that expresses its own individuality. Gardeners also do not seek to perfectly order a garden, which is made of imperfect and unique objects. To insist on a harmony other than the underlying, naturally revealed one would be unnatural.

Zen Buddhism's influence on the development of Japan and the Japanese garden, not just the religious garden but the landscape garden, has been enormous, as suggested previously.

In Japan, the Buddhist concept of transience has been integrated into the indigenous concept of nature as an extension of oneself. This integration can be seen in the oneness with nature and the Buddhist feeling that worldly display counts for nothing. For example, Zen teachings gave rise to the principles of *wabi* and *sabi* inherent in the tea ceremony, Japanese garden, and flower arrangements. *Wabi* is the aesthetic feeling of discovering richness and serenity in simplicity; it is an emotional appreciation of the essence of things, including the ephemeral quality of life. *Sabi* refers to a feeling of quiet grandeur enjoyed in solitude, and it normally involves the beauty that comes from the natural aging of things. To the Japanese, something old is to be respected, even its imperfections. Clearly, the Japanese garden is an ideal setting for experiencing *wabi* and *sabi*.

Tranquility and Impermanence

In *The Global Business*, Ronnie Lessem (1987) discusses two principles that explain the Japanese aesthetic, *shibui* and *mononoaware*. *Shibui* refers to a quality of beauty that has a tranquil effect on the viewer; the object, whether natural or manmade, clearly reveals its essence through perfection of form, naturalness, simplicity, and subdued tone. *Mononoaware* refers to the merging of one's identity with that of an object or mood, especially one tinged with recognition of the impermanence of things. Lessem suggests that this aesthetic awareness leads to a sensitivity to man and nature that is rare in Europe or the United States.

The Japanese may stroll in the garden, or perhaps just sit on its edge, listening to the water and soaking in its beauty. The garden provides a place of escape from society, both physically and spiritually, just as many Japanese seek refuge through their hobbies.

This refuge usually takes the form of some kind of identification with nature, as in the case of cultivating one's own tiny landscape garden or arranging flowers. Millions of Japanese express themselves through the traditional arts, dancing, music, and literature. In developing their individual skills, they practice self-control and self-discipline. Pursuit of a hobby, called *shumi*, which means tastes, is important to self-identity and even self-respect in Japan.

Except for the sound of the flowing water, all is quiet near the Japanese garden. Likewise, silence is an important value to the Japanese. As mentioned previously, a common Japanese proverb is, "Those who know do not speak; those who speak do not know." Conversation is often punctuated by long stretches of silence. The Japanese say that it is during these silent intervals that real communication takes place and that they can sense others' feelings and thoughts. Westerners should not feel compelled to fill in the gaps with conversation, which may only annoy the Japanese, who already perceive Westerners as talking too much.

Similarity and Contrast

To return to Hofstede's (2001) research on the five dimensions of culture, the Japanese were portrayed as tending to avoid uncertainty and seeking comfort in familiar situations. The more recent GLOBE study (House et al., 2004) finds the Japanese not much different from U.S. Americans in this regard, being somewhere in the middle range. In addition, Japan ranked first on masculinity in the Hofstede study, thus indicating that its people are aggressive in the pursuit of worldly success and material possessions. On the comparable dimension in the GLOBE study, gender egalitarianism, Japan ranked somewhere in the middle. However, according to the United Nations "gender empowerment measure," which is an index of female participation in a nation's economy and politics, Japan ranks as the most unequal of the world's richest countries, placing 42nd among 75 countries and ranking just above Macedonia (Fackler, 2007). In 1985, women held 6.6% of all management jobs and by 2005 that number had risen to only 10.1%. Japanese work culture makes it practically impossible for a woman to have a career and a family and as former cabinet minister in charge of gender equality, Kuniko Inoguchi, put it, "Japan is losing half of its brainpower as it faces a labor shortage." As indicated previously, the carp in the Japanese garden symbolically represents this aggressiveness or masculinity. Unlike the Chinese, who are sometimes hampered by their excessive devotion to the past and ancestor worship, the Japanese have historically been great borrowers from other cultures and are quick to change when conditions warrant such action. In this sense, the Japanese are similar to the individualistic U.S. Americans.

But the Japanese are clearly different from U.S. Americans, as our metaphor of the Japanese garden indicates. Still, worried Japanese elders are concerned that the water in the garden will not continue to flow smoothly because the "new humans" may have lost their appreciation of things Japanese. Iwao (1990) argues that there have been three major changes in the Japanese character since 1980: a tendency toward diversity and individuality, a need for swift results and instant gratification, and a desire for stability and maintenance of the status quo. It is true that young Japanese seek more leisure time and are more likely to devote themselves to personal goals than their elders did. As Iwao suggests, consumer behavior reflects a movement toward diversity and individuality. Female roles, as well as attitudes toward marriage and work, are also changing as indicated earlier.

In 2008, concerned by a rapidly aging society's burgeoning health care costs, the government enacted a law that required companies and local governments to measure the waistlines of Japanese people between the ages of 40 and 74 as part of their annual checkups, setting limits of 33.5 inches for men and 35.4 inches for women (Onishi, 2008). There is a campaign against becoming a *metabo* (a more acceptable term than obesity and a shorthand for metabolic syndrome, which is a condition characterized by high blood pressure, abdominal obesity, etc.). As this chapter indicates, Japan has suffered economically since 1990 and has been buffeted by many winds of change, including the imposition of Western-style individualistic management practices. While individualism is clearly increasing, millions of Japanese love to watch a TV show, *Project X: Challengers*, which showcases the passion and single-minded efforts of Japanese workers who have helped Japan become such a powerful economic force since World War II (see Kashiwagi, 2002). Such nostalgic longing for traditional values highlights the tensions that this group-based culture confronts. However, the past few years have also witnessed the emergence of the *hodo-hodo zoku* or "so-so folks," a generation of young people who are not very interested in promotions and advancement. They would much rather have less money and prestige if it allowed them to have more time for their families and other pursuits.

The water in the Japanese garden is flowing quickly and with force. It may even cause the shape of the pond to alter somewhat. The change of the seasons will also bring changes to the garden. Yet the basic design and elements of the Japanese garden will always remain, for this island nation of relatively homogeneous people relies heavily on its groups to maintain stability and ensure change and progress within the framework of a high-context culture emphasizing the natural ordering of individuals, groups, and activities.

CHAPTER 4

Bedouin Jewelry and Saudi Arabia

This is a very, very conservative society. They have moved in about 50 years from, frankly, living in the desert in mud huts to now skyscrapers and super-highways and luxury cars. But the roots of this society go back centuries and centuries as a very insular, inward-looking, family-oriented and tribal-oriented society.

—Ambassador Robert W. Jordan,
speaking to *Frontline* in 2004

"This is the twist in this country," said a wealthy Saudi businessman who is close to the royal family. "The government is pushing toward moderniza-tion, and the culture is going backward."

—Lancaster (1996, p. A26)

Westerners, and the United States in particular, could easily relate to the sentiment above on September 11, 2001, when 19 Muslim fanatics, of whom 15 were Saudi by birth, bombed the World Trade Center and part of the Pentagon, an act claiming thousands of lives. According to then-U.S. Ambassador to Saudi Arabia Robert W. Jordan, the Saudis were taken by surprise and appalled and embarrassed that their sons and members of their tribes could be involved in such a horrific act. This caused them to go into a period of self-reflection and introspection. There are more than 1 billion Muslims in the world, and their numbers are rapidly increasing, but it is critical to realize that the extremist groups among them are in a minority. It is also important to realize that Saudi Arabia, the largest of the Arab gulf states in population and size, is quite different from the others in many ways. For example, Saudi Arabia restricts what women can do, and alcohol is forbidden, while quite a different situation exists in Oman.

Like many nations, Saudi Arabia is confronting the tension between traditional and modern beliefs, values, and lifestyles. The Saudi royal family is the official guardian of the Muslim holy shrines of Mecca and Medina, but it is supporting modernization, at least partially because the economy is too dependent on oil, and there is only a limited supply of it. For example, the royal family is encouraging tourism, even to the extent of permitting the building of a ski resort with artificial snow. But, although the Saudi family erected a modern concert hall completed in 1989, no symphonies or operas have been performed because of the conservative opposition's insistence on strict Islamic standards. Mutaween, or the religious police, routinely punish women they deem immodestly dressed or merchants who do not close their shops during the five daily prayer periods. No religion other than Islam may be practiced in public; churches and synagogues are illegal; and proselytizing for any non-Muslim faith is illegal. Similarly the government executes prisoners by ritual decapitation with a sword. Even foreign workers, constituting 8.2 million of the 25 million population, are not exempt, and some have been seized without formal legal process and tortured. The Saudi Labor Ministry estimates that about 67% of the workforce are foreigners and 90% to 95% hold private sector jobs. Many of these individuals were attracted by the promise of plentiful jobs and largesse but ended up feeling dehumanized and exploited. They complain of labor laws being nothing more than "ink on paper."

This tension between modernization and conservatism is likely to increase in all the gulf states. More than half of their population is under age 20. In Saudi Arabia, 43% of the 25 million residents are under 15. By contrast, in neighboring Kuwait only 26% of the population is in the same age range. The tension was acute in Saudi Arabia during the early to mid-1990s because of the 1990 Gulf War with Iraq, in which Saudi Arabia and other Arab nations were allies of the Western powers; 24 U.S. Americans were killed in Saudi Arabia. After September 11, 2001, tensions between the West and Saudi Arabia increased. More recently, the Islamic traditionalists have muted their criticisms and seem to accept the need for modernization or at least some aspects of it. The gross domestic product per capita is more than $15,210, according to Economist.com, and the rise in oil prices over the last few years has generated a sizable petrodollar economy. According to the Institute of International Finance, Persian Gulf countries earned $1.5 trillion in oil revenue from 2002 to 2006, which is twice as much as in the previous 5 years. In 2007, the government recorded a budget surplus of $48 billion but oil has not been the only engine of growth. The private sector now accounts for 45% of the economy (Mouawad, 2008), which is fortunate given the fluctuation in oil prices following the global financial meltdown of 2008. However, the country has one of the world's highest birth rates and the population is expected to reach 40 million by 2025. As suggested by the figures on population growth, it is easy to imagine a return to traditions and a rejection of modernization. Given this tension between tradition and modernization, Bedouin jewelry is an appropriate cultural metaphor for Saudi Arabia.

History and Geography

Saudi Arabia became a nation in 1932, when Ibn Saud united disparate tribes into one nation and named himself king. Oil was discovered 10 years later, but the country remained fairly isolated until the 1970s. Then, suddenly, Saudi Arabia was slaking the world's thirst for oil, and the country was yanked into the 20th century; it possesses 24% of the world's known oil reserves. This nation is still stumbling, struggling to reconcile deeply held traditional values with disconcerting new realities. By examining the ancient art form of Bedouin jewelry, we can address the values and beliefs of Saudi culture. Who the Saudis are today began with the Bedouin of the Arabian desert.

Saudi Arabia stretches across three fourths of the hot, dry, rugged Arabian peninsula. Its land mass is about 24% of that of the United States. The country's borders are: the Red Sea on the west; Jordan, Iraq, and Kuwait on the north; the Arabian (Persian) Gulf, Qatar, and the United Arab Emirates on the east; and Oman and Yemen on the south. Small, jagged mountains rise in western Saudi Arabia near the Red Sea. From there, the land gradually slopes down to the Arabian Gulf in the east. Three major deserts lie in between. The largest is the Rub al-Khali, the Empty Quarter. It is the greatest continuous sand area on Earth. Although several oases spring out of this harsh landscape, the country has no natural rivers or lakes. But Saudi Arabia does have one enormous consolation. Beneath this austere terrain are the largest reserves of oil in the world.

Several million years ago this area was covered by shallow seas filled with mollusks, brachiopods, and plankton. When these creatures died, they sank to the floor of the sea. Eventually the waters receded, and the sea floor compressed at high temperatures under layers of sedimentary rock. The organisms were transformed into oil and natural gas.

Much later, the Arabian peninsula was a grassy savanna. Until about 18,000 years ago, ample rainfall supported vegetation and wildlife. But the last Ice Age ended in a prolonged drought. Rivers evaporated, leaving dry valleys called *wadis*. The fertile land turned into desert. Now, rainfall averages 0 to 4 inches per year. When a rare downpour does occur, the wadis are briefly filled with water once again before the water soaks into the ground or evaporates away. During much of the year, temperatures are 90°F to 120°F. However, strong winds and cooler temperatures can make winter days quite cold. These raging winds also generate the stinging sandstorms common to the Arabian desert. Surviving in this environment takes skill, courage, and fortitude.

The Desert Bedouins

For centuries the Bedouins of Saudi Arabia have made the desert their home. Their way of life is a highly sophisticated adaptation to an extremely harsh environment. The fact that they have been successful at it for so long is a source of fierce pride. Many Bedouins feel far superior to townspeople, and they display a certain arrogance for having braved the deserts and won. Surely theirs is a purer, worthier life, as they see it.

The word *Bedouin* comes from the Arabic *badawi,* which means desert dweller. Bedouins live in extended family groups, moving when they need new pasture or water. The only animals that can comfortably live in the desert are goats, sheep, and—most of all—camels. Goats supply milk and hair for tents. Sheep give meat and wool for carpets and blankets. Camels provide milk, meat, hair for tents, dung for fuel, urine for shiny hair, and transportation, all of which almost makes up for their nasty tempers. In addition, camels can go many days without food or water. Raising and selling camels to townspeople has always been a source of income for these desert dwellers. According to Gardner (2003), the truck, which has become the foundation of contemporary Bedouin livelihoods, is increasingly changing the communication networks of Bedouins by moving them from the tent to the marketplaces of towns and villages. Trucks enable Bedouins to scout for distant pastures, visit friends and family, carry animals to the market, and transport crops back to camp (Gardner, 2003).

Because the desert provides scant food for livestock, Bedouin camps must move often, usually every couple of weeks. Camps consist of 2 to 30 tents, with most camps having fewer than 10. Each tent houses a nuclear family plus perhaps an unmarried uncle, mother, sister, or brother. A Bedouin tent is a long, low, black structure made of camel and goat hair, with a wide front opening. The shape is ideal for protection against wind and sandstorms. Each tent is divided into two sections: one side is for the mother and daughters, the other for the father and sons. The tents and all possessions can be packed up and the family can be on its way within 2 hours.

Bedouins pay little attention to national borders. Saudi Arabian borders are poorly defined, but the Bedouin need for pasture and water takes precedence over any arbitrary lines, anyway. The families camp on the edge of the desert, occasionally visiting towns for supplies. Their habit is to live in the higher areas during the hottest part of the year so they can escape the worst heat and sandstorms. Later, they move back to the lowlands to avoid cold temperatures and winds. Bedouins are continuously on the alert for chance thunderstorms, which mean lush pastures are sprouting, flowers are blooming, and wadis are filling.

Camel raiding has been a traditional pursuit of Bedouin men. Although it is now outlawed, the raiding rules are still important for illustrating the strict code of Bedouin honor. First, the hostility must be explicitly declared; using the element of surprise is shameful. Therefore raiding is not done at night. Second, no women or children may be harmed. Third, sunrise raids are fairer, so that victims have more chance to track and retrieve their camels before nightfall. Breaking these rules brings shame to the perpetrators, severely tarnishing their honor.

Personal Characteristics

Bedouins are known throughout the world for their great generosity. A guest, even an unexpected stranger, will be fed and housed for 3 days. In fact, if those raiders arrived at the opposing camp, they would receive the same welcome, and they would not be harmed during their stay. While an ordinary meal consists of milk and milk products, dates, flat bread, and perhaps rice, the guest's feast also includes fresh meat and pine nuts. Pale green coffee with cardamom, as well as tea, is served. Afterward, poets and dancers (always male) provide entertainment. Poets, or storytellers, are important and valued in Bedouin culture. Because few Bedouins can read or write, poets function as historians and educators as well as entertainers.

Bedouin loyalty goes first to the extended family. No one can survive the hard desert life alone. Other people are essential for shared chores and protection. But the meager desert offerings cannot support whole clans, let alone tribes, in the same spot at once. The family unit is large enough, yet small enough, for safety and survival in the desert. Intense feelings of loyalty and dependence are fostered and preserved in this unique family group setting. Home is viewed as people, not place.

One way of strengthening the Bedouin extended family is by consanguineous marriage (marriage to a close relative). A Bedouin girl marries young, at about 14 years of age. For several reasons her ideal spouse is a first cousin: She is already familiar with her cousins, having grown up in the same camps; a husband cannot mistreat her with fathers and brothers nearby; she can remain near her mother and sisters; her future children will enhance and perpetuate the group; and the dowry she brings to the marriage will remain within the family. For centuries intermarriage has bound Bedouin families to one another and to their culture. (In fact, the very conservative Wabhabi sect and the Saudi royal family intermarried generations ago, and some writers have suggested that the religious conservatism of this theocratic nation was strengthened by this union.) Thus it is not surprising that Saudis are much more collectivistic and more likely than U.S. Americans to denigrate out-group members (Al-Zahrani & Kaplowitz, 1993).

The Forces of Progress

Bedouin life remained constant for many hundreds of years. Until oil. The discovery of the treasure trove beneath the desert has affected every Saudi citizen, and that includes even these inscrutable, self-sufficient nomads. The wandering Bedouin lifestyle, which withstood the severity of the desert, is succumbing to the forces of progress.

What are these forces? First, camels were once the main source of Bedouin income. But Saudi preferences have changed as beef has become more available through international trade, modern domestic transportation systems, and improved refrigeration methods. Camels are now an expensive hobby rather than a primary source of food. Second, more Bedouins are taking permanent jobs in towns for a steady income, requiring them to settle nearby. Third, they like the conveniences that make life easier: canned tomato paste for rice, flashlights, portable radios, insulated coolers, sewing machines, and trucks. These possessions are cumbersome to haul to a new camp every few weeks. Fourth, some of their locales have been breached by oil pipelines, highways, storage tanks, and industrial sites. Fifth, the Saudi government wants them settled.

The government considers the Bedouin a disruptive element. They do not pay taxes, their trucks are unlicensed, their women do not always wear veils and often drive, they can disappear at will into the desert, and their loyalties are to the tribe rather than to the government. Basically the Bedouin refuse to be controlled, which is threatening for an absolute monarchy like the House of Saud.

Along these lines, it is interesting to note that many Egyptians and Kuwaitis, when discussing Saudi Arabia, frequently employ a mild form of putdown by saying: Oh, those Bedouins! Once again, such statements, while stereotypic, suggest that Saudi Arabia does indeed have a distinctive national culture.

Some governmental efforts to settle the Bedouin have failed. Housing for the Bedouin has been built in major population centers, but most of it does not meet their needs. It remains empty. The desert dwellers feel cramped in the stationary rooms, and they are baffled by appliances. They prefer carpets spread on sand rather than on hard concrete floors. Most of the housing units are small and designed for nuclear families, with high privacy walls between houses. Because there is no place for their animals, sometimes Bedouin families will camp outside their new walls and keep the animals inside the house.

A more appropriate governmental attempt to encourage settlement is to bring necessities to the desert. Water trucks and nurses visit the more permanent camps. Sometimes cash grants are distributed. Special hospitals that have campsites right outside the rooms treat Bedouin patients. The government recruits young men for Bedouin National Guard units, where their strength, agility, and desert skills are highly valued. These transitional measures are somewhat successful, and increasing numbers of Bedouin are settling permanently.

Wandering Bedouin probably now make up 5% to 10% of the Saudi population (Mackey, 1992, p. 22; Wilson & Graham, 1994, p. 12). They have never made up more than half of the population (Alotaibi, 1989, p. 36).

Most Saudis feel ambivalent about the Bedouin. The nomads are envisioned as simplistic, dirty, and crafty—in general, uncivilized. They are both ridiculed and feared. Yet they are admired for their virtues of generosity, boldness, and courage. Bedouin are idealized, much as knights and cowboys are in other places. But no matter how far Saudis are removed from the desert, the Bedouin ethos is the bedrock of their culture.

Jewelry as Wealth

Silver Bedouin jewelry is one of the few art forms of the desert dwellers, made all the more beautiful by their spartan surroundings. But it is more than adornment; it is also a woman's personal wealth. Usually a woman acquires her first jewelry as part of her dowry. Later on family savings are invested in either jewelry or livestock. The designs are similar from one woman to the next, although the cost varies depending on the silver content. It is impossible for an amateur to determine the amount of silver in a particular piece. In this way all women can own similar types and amounts of jewelry, regardless of financial status. A woman will choose to wear certain pieces every day and then add other pieces for special occasions.

Traditional Bedouin jewelry is purchased from either settled artisans or itinerant silversmiths, the vast majority being men. Sometimes a craftsman will live among a tribe, if it is a large one. His workmanship ranges from exquisite to crude, but all of it is eye catching. Necklaces, bracelets, rings, belts, earrings, nose rings—the kinds of jewelry worn by women everywhere are worn by Bedouins, too. The artisan incorporates multiple silver beads, chains, bells, and cylindrical pendants into his basic designs. Then, he adds filigree, heavy granulation, and gemstones. The stones can be amber, coral, carnelian, garnet, or turquoise; sometimes, they are only colored glass or plastic. The stones are often associated with specific beliefs and values, with red being the most popular: Coral is associated with wisdom and garnet is associated with the alleviation of some illnesses. Turquoise, which is said to glow when the wearer is happy, along with coral and amber, have been among the most common stones in the adornment of Bedouin silver jewelry. Many pieces have tiny objects placed inside of bells and pendants that rattle melodiously with movement. Others have Koranic verses permanently sealed within cylindrical charm cases. Craftsmen add one item that is unique to Bedouin jewelry: coins.

Progress now threatens the popularity of Bedouin silver jewelry, just as it does the Bedouin lifestyle itself. Townspeople prefer gold jewelry, and as the Bedouin people leave the desert, they are also beginning to choose to work in gold. In addition, many of the silversmiths are retiring, and younger people are not interested in pursuing the trade. Westerners living in Saudi Arabia who are interested in handcrafted items are largely credited with keeping the craft alive. Still, traditional silver jewelry can be studied as a metaphor for Saudi culture. Salient features of Bedouin jewelry are its bold form, handcrafted appearance, traditional design, and female ownership.

Bold Form

To an outsider, Bedouin silver jewelry is overwhelmingly ostentatious. Westerners would feel garish and conspicuous wearing the favorite Bedouin styles: multiple chains, pendants, bells, and coins, decorated with knobs of thick granulation and strands of filigree, inlaid with colored stones—all in the same piece. One, called a *kaff* (glove), has a ring for each finger, a section for the back of the hand, and a bracelet, all connected with chains and bells. Sometimes a piece will have only one type of decoration, such as a ring with filigree, but the decoration will be used very liberally for an exotic effect.

Likewise the Saudi personality is lavish, too. Their generosity is legendary. When entertaining, they overwhelm their guests with abundance. Saudis are proud to serve two or three times the required amount of food, and they insist that guests consume more than their fill. The host typically does not eat along with his guests. He walks around checking on everyone's intake, urging them on and plying them with additional courses.

Good personal relationships are valued professionally and individually. Saudis will do almost anything for even a casual friend. They like to be helpful and endeavor to fulfill expectations. But an oral promise has its own value as a generous response, even if action does not follow. Results can be a separate thing altogether. For themselves, Saudis do not hesitate to ask for special favors, for rules to be bent, or for special treatment (Nydell, 1987, p. 17).

Opulent Language

Just as the silver jewelry of the Bedouin gives the owner an aura of opulence, the Arabic language surrounds Saudis with its mystique of extravagance and vitality. Saudi Arabia is the birthplace of Arabic, and Saudis are secure in the belief that their language is superior to all others (Khalid, 1979, p. 130). Arabic has phonetic beauty, rich synonyms, rhythmic cadences, and majesty. The structure of the language encourages repetition and exaggeration. But thoughts themselves are often vague, and sometimes only general meanings are inferred. The language overemphasizes the significance of words in themselves and underemphasizes their meaning. This unbounded medium has far-reaching effects.

In the first place, eloquence and verbosity are equated with knowledge. Like silver jewelry that appears expensive but has low silver content, an effective speaker can appear prudent and rational by using his verbal skills. Poets are held in high esteem, and they frequently mesmerize and motivate audiences to action. Advanced education and sophistication do not reduce an Arab's susceptibility to a good sermon or a political harangue. The grace and fluency of the words count more than the logic and veracity of the argument (Khalid, 1979, p. 131). Therefore oral testimony is considered superior to circumstantial evidence, businessmen often get bogged down in language as they trail off from the main point, and students are better at memorizing course work than using it.

Bedouin jewelry is bold in design, but the individual components of each piece are often flamboyant as well. Perhaps this is related to the Saudi's characteristic generosity and feeling of superiority. More must be better. Or possibly it is *proxemics*, which Edward T. Hall (1966) describes as "man's use of space as a specialized elaboration of culture" (p. 1). Certainly Saudis love to have ample space in their homes. Ideally ceilings should be high and views unobstructed. A sitting room may have 25 chairs arranged around the edges, with what seems like a cavernous speaking distance between them. But in public, Saudis do not mind being crowded.

Privacy in Public

In fact, Hall (1966) concluded that Saudis do not have any sense of privacy at all in a public place. They find it difficult to understand that privacy can exist while being surrounded by other people. No zone of privacy surrounds Saudi bodies in public. Consequently, they are not bothered by the jostling and noise of public places in the Middle East. Lack of public privacy zones explains their aggressive behavior at accidents, where Saudis surround the scene, yell advice, and shake their fists—even if they did not witness the incident. It is not unusual for them to join haggling sessions they happen to come across. They nonchalantly cut off other drivers, perhaps because they believe no one has a right to claim the public space into which they are about to travel.

Rarely do these situations lead to physical confrontation. Actual violence is almost unheard of here. Likewise, as a country, Saudi Arabia does not invade enemy territory, although it often engages in strong rhetoric. In fact, Saudis are premier mediators. They aim for consensus, although their initial position may be extreme. They willingly make concessions and expect others to reciprocate (Adler, 2007).

So, the idea that someone could have privacy in a public space confuses Saudis. There is not much concept of personal privacy, either. The Arabic word closest to *privacy* means loneliness. Saudis feel that friends should see each other often, and there is no compunction about dropping in unannounced. Greetings are effusive and extensive: "Praise be to Allah," "Welcome to you," "Hello, friend." Common inquiries center on the state of a person's health or the cost of a recent purchase, but never refer to female family members. Saudis are quick and generous with praise for others and like to highlight their own accomplishments as well (Samovar & Porter, 1994, p. 290).

Geert Hofstede (2001) concluded that Arab countries scored 53 out of 100 on the masculinity scale, giving them a rank of 23 out of 53 nations. This average score might seem surprising, but feminine societies stress solidarity, emotional displays, relationships among people, and resolution of conflicts by compromise. Masculine cultures stress achievement, competition, and resolution of conflicts by violence. Arab culture incorporates several of Hofstede's feminine characteristics. In the more recent GLOBE study (House et al., 2004), gender egalitarianism was defined as the way in which societies divide roles between men and women. Although Saudi Arabia was not included in the study, the Middle Eastern cluster, which included countries like Egypt, Qatar, and Kuwait, scored low on gender egalitarianism; this could be an indicator of Arab societal culture.

Handcrafted Appearance

The second feature of Bedouin jewelry is its handcrafted appearance. While some of the jewelry shows intricate and precise workmanship, other pieces are crude and casually made. This does not diminish or detract from their desirability, however. Both kinds are considered legitimate, just as the Saudi considers the Bedouin to be desert riffraff and the embodiment of deep cultural values at the same time. According to Gardner (2003), the Bedouin occupy a distinct and complex position in the Saudi cultural milieu: Although they are often disdained as uneducated, provincial, and irreverent, they fascinate their countrymen with their perceived resilience and fierce independence at the same time, thus inspiring pride and admiration. Cursory examination of the jewelry leaves no doubt that both types are handmade: Bells are often mismatched, pendants have irregular shapes, stones are slightly jagged, granulation is lumpy and uneven, fasteners are often simple hemp strings, and cheap cloth backings are attached to prevent the wearer from burning her skin. These raw aspects of the jewelry mirror the harshness of desert life.

Saudis believe in fate: that people have little control over their lives. Bedouin jewelry reflects this acceptance in several ways. The jewelry itself represents an adaptation to life in the desert. The cloth protects from the hot sun, the gemstones are found locally, and the coins are acquired from trade caravans and pilgrimage routes—all demonstrating improvisation, using what is at hand.

Saudis have traditionally felt "subjugated to nature," as defined by Florence Kluckholn and Fred Strodtbeck (1961, p. 13), living at the whims of their natural environment (and Allah). Fate decreed that they accept these natural forces, but economic advancement has meant leaping ahead to "mastery over nature." Today desalinization plants process drinking water, irrigation systems water crops, and drilling equipment extracts oil and natural gas. Creeping sand dunes are still a problem. Saudis are becoming expert at mastering nature, but their mind-set is still fatalistic.

"Being" is more important than "doing" in Saudi culture. Adler (2007) defines a being orientation as a condition in which people, events, and ideas flow spontaneously within the moment. The future is not nearly as important as the past and the present; present gratification is not delayed for future gain. You could say people with a being orientation live for today. Fate and Allah are in charge of the future.

Edward T. Hall (1983) suggests that time is not fixed or segmented for polychronic people. Time is considered loose and fluid rather than compartmentalized. Therefore occasions do not have definite beginnings and endings, and schedules are not as important as people. Saudis are polychronic and irreverent about time boundaries. Consequently, guests may be kept waiting, deadlines are often missed, and appointments are broken. Polychronic people are able to carry on many activities simultaneously. For example, several customers may be helped at the same time, or different business negotiations may be managed simultaneously in the same office. This behavior exemplifies the being orientation.

Traditional Design

The Bedouin favor traditional styles of jewelry that are decorative and valuable. Styles worn today are the same as the ancients wore, and they are always silver. New and unusual designs are seldom crafted. Individualism and creativity are discouraged in both Bedouin silversmithing and in Saudi culture. This country ranks 26th of 53 nations on Hofstede's (2001) scale for individualism, which places it on the collectivist side of the spectrum. Collectivism is having a "we" consciousness, belonging to a cohesive in-group beginning at birth, and taking the group's opinions as one's own. Private opinions and individual achievement are not highly valued in this setting. The more recent GLOBE study (House et al., 2004) placed the Middle East cluster of countries (Turkey, Egypt, Kuwait, Qatar, and Morocco) at the high end of the collectivism scale.

One of the strongest manifestations of Saudi collectivism is the family. The extended family is the norm, although with new prosperity, more nuclear families are living alone. Even then, other members are either close by or in the same walled compound, the Bedouin camp replicated in modern times. These families collectively form clans, tribes, and the nation of Saudi Arabia. Like a piece of Bedouin jewelry that has several different components, which are assembled together by silversmiths through the process of soldering, the country has distinct segments also.

For Saudis, family is paramount. It determines how they think, whom they marry, and where they work. Familial involvement is not viewed as interference but is welcomed as support. Loyalty to family takes precedence over work or friends, and it is rewarded with protection by the group. In the workplace relatives receive preferential consideration in hiring and promotion opportunities. Families never leave a member in need. There are no daycare centers or nursing homes. The oldest male member is honored as head of the family. Children are reared not only by their own parents but by all adults in the family. In fact, friends will feed, care for, and discipline others' children. This results in a remarkably homogeneous upbringing for the country's young. Like the desert nomads, people with this background cannot comfortably live alone. To be separated from the family is the ultimate punishment.

Honor and Shame

Honor and shame are major features of modern Saudi culture, just as they were in camel-raiding days. When people's honor is tarnished, they feel shame and lose face. So does their family, which by association is also shamed. Members of collectivist cultures place importance on fitting in harmoniously and saving face. Honor can be lost through being stingy, treating the old and weak poorly, fathering only daughters, being passed over for special favors, and yielding in traffic and through the immoral sexual conduct of female family members.

Understanding the importance of honor and shame in Saudi culture provides insight to certain behaviors. Mackey (1987) maintains that shame is a factor in Saudi generosity. She even argues that hospitality is provided to add to the giver's reputation, not necessarily to benefit the recipient (p. 116). Fear of making a mistake and suffering the resulting shame may explain excessive delays in the business world. Many Saudis drink or gamble when outside of their country, even though these acts are forbidden by Islam. In fact, Westerners flying into Saudi Arabia are often surprised by the Clark Kent–Superman behavior of returning Saudis, who quickly take off their Western attire and don traditional Arab clothes before departing from the airplane. Perhaps

Saudis feel honor is not lost if no one is there to bestow shame. It could be that Saudis rarely resort to violence because they feel the worst harm comes from the shame of grievous verbal insults.

Another cohesive force, besides family and honor, is marriage. The marriage model is the same for modern Saudis as it was for the nomads. Partners are chosen by elder family members. Marriage is such a momentous undertaking that young people welcome the group's opinion. Although it is rarely invoked, both the man and the woman have the right of refusal and can ask for another choice when selecting a mate. Marriage to first or second cousins is common. The benefits of keeping money and people in the family supersede any genetic concerns. Most marriages take place before age 20, and the new couple usually lives with the husband's family (Mackey, 1987, p. 150). However, larger dowries and college educations are pushing this age higher. New marriages are based not on romantic expectations but on a desire for companionship, children, and later, love. Although Muslims may have up to four wives, most men have only one at a time. An old proverb states, "A man between two women is like a lamb between two wolves" (Peters, 1980, p. 24).

The Role of Religion

Religion is like the hemp that draws all the parts of a Bedouin necklace together. All Saudis are Muslim. The king is not only their political leader but their religious leader as well, and the Koran is the framework for running the country. Piety is honorable for a Saudi, and religion is not separated from daily life. Fatalism is expressed by the frequent use of the qualifier *inshallah,* or God willing. A Muslim's greatest joy comes from performing the *haj* or pilgrimage to Mecca. The charm cases incorporated into silver jewelry contain verses that praise Allah and ask for daily protection. Permeating the cohesion of family, honor, and marriage are the inviolate religious beliefs of Saudi Muslims.

Saudi Arabia ranks 7th out of 53 nations on power distance (Hofstede, 2001), which is the extent to which less powerful citizens accept that power is distributed unequally. It is also a high-context culture, which means that only a small amount of information must be explicitly stated for understanding to occur because so much information is implicit. In short, Saudi Arabs adhere to many written and unwritten rules, tolerate hierarchies, follow formalities, practice obedience, and encourage conformity.

Female Ownership

Custom and tradition play prime roles in the acquiring and wearing of Bedouin jewelry. What is striking is that it is wholly female. Not many other joys of Saudi culture belong entirely to women. Their reasons for wearing adornment are universal: vanity, superstition, sentiment, and pleasure. Ownership bestows honor on her as a woman of property and guarantees her the security of having negotiable assets. Her wealth also adds beauty to an austere desert life. Bedouin jewelry is rarely old. Unlike Western jewelry, often handed down through generations, a Bedouin woman's jewelry is generally melted down upon her death. It would be unacceptable to a new bride (Ross, 1979). Sometimes, in times of dire need, it is sold to augment the family's finances.

Like the serendipitous nature of Bedouin jewelry, the role of Saudi women contains surprises, too. The world disdains the limitations under which Saudi women live, but perhaps outsiders are not qualified to pass judgment. What do Saudi women themselves think?

Saudi females do not necessarily see themselves as repressed. Many think of themselves as protected and are frequently shocked at the crimes such as rape and physical beatings that their Western female counterparts suffer. Neither men nor women want to subject females to the stress, temptations, and indignities of the outside world. Most women feel satisfied that the present system provides them with security and respect, and they pity the rest of the world's women. Besides, even though males head the family, women have enormous influence inside their own homes. Through their extended family they have wonderful support systems. Women do not defer to their husbands in private as they do in public. Like elsewhere in Saudi life, there is much discussion between a husband and wife when they have a difference of opinion. This, then, has been the world of Saudi females for centuries.

The long black *abaya* and veil, the separate entrances for men and women, women's exclusion from university education, and the prohibitions against their driving may seem a little less protective today and a lot more restrictive than they once did. The barriers to change are daunting, although change has occurred in other Arab countries. Some Saudi women have already started pressing for greater personal and social freedom, while at the same time realizing what they will be losing. Like the flamboyant desert jewelry that has adorned Bedouin women for centuries, the traditional life of Saudi women will slowly lose its luster. In February 2009, King Abdullah reshuffled his cabinet and chose the first ever woman minister, Norah al-Faiz.

Saudi Arabian culture is rooted deep in the desert, where the Bedouin people developed an incredible ability to survive under extreme conditions. Most Saudis feel they are of this desert, even if they have no Bedouin ancestry. For hundreds of years, Saudi values and beliefs were safe, sheltered from the outside world. For hundreds of years, Bedouin women proudly wore their distinctive jewelry, with its bold form, handcrafted appearance, traditional design, and female ownership. By using Bedouin jewelry, we can create a metaphor for modern Saudi Arabian culture.

The perception of the jewelry has undergone radical change in recent years. There may soon come a time when the country's values and culture are challenged at their core as well. The vision for the future is to turn the country into a major industrial power by 2020 and this is to be accomplished partly by building six new cities throughout the country with industrial centers which will double as housing and commercial hubs (Mouawad, 2008). Saudi Arabia has one of the largest economies in the Middle East and is an important political power in the region. Its future is inextricably intertwined with that of the world and it behooves us to understand the forces that shaped its rich past and those that are at play in the present that will shape its future.

CHAPTER 5

The Turkish Coffeehouse

Coffee should be black as hell, strong as death, and sweet as love.

—Turkish Proverb

Turkey is a remarkable land of contrasts. Joining the continents of Asia and Europe, it has elements of old and new, Islam and Christianity, mountains and plains, and modern cities and rural villages. Often referred to as the cradle of civilization, Turkey is filled with archeological remnants of the ancient Greeks and Romans, along with treasures of the sultans and caliphs of the Ottoman dynasty. Perhaps most interesting of all are the people of Turkey, rich in emotions, traditions, and hospitality. Turkish origins can be traced to the Mongols, Slavs, Greeks, Kurds, Armenians, and Arabs. Like the Thais, the Turks have never been conquered or colonized by other peoples in the modern era. This distinction, along with a geographic boundary straddling the two continents, makes Turkey unique. Turkish culture today is a marriage of Turkish traditions and Western ideologies.

Turkey is a mountainous peninsula slightly larger than Texas, and it shares borders with Bulgaria, Greece, Iran, Iraq, and Syria. Considered a strategic location due to its control of the straits linking the Black and Aegean Seas, Turkey is the only NATO country other than Norway to border the states of the former Soviet Union. The western portion of Turkey, Thrace, is located in Eastern Europe, while the larger eastern section, Anatolia, is part of Asia. Of 72.3 million people, 80% are Turkish and 20% Kurdish, and about 30% are less than 15 years of age. More than 10 million agricultural workers endure hot summers and cold Turkish winters. Adult literacy estimates at more than 97.7% are higher than ever, and the average life expectancy is about 70 years of age. Unlike most other Islamic societies, the Republic of Turkey was established in 1923 as a secular democratic nation. There is, however, a small but growing minority that would like to see Turkey become once again a theocratic nation. The Turks are proud of both their achievements as a modern state and their rich heritage.

An appropriate metaphor for understanding Turkish culture is the coffeehouse, a part of everyday life in Turkey. In the oft-cited famous words of the great Turkish 20th-century poet, Yahya Kemal, coffee has created its own "culture" in Turkey. What the Turkish coffeehouse represents is quite different from its counterparts in other countries. An emphasis on both Islam and secularity is the first of four characteristics of the coffeehouse that mirror Turkish culture. Coffeehouses also provide an important forum for recreation, communication, and community integration. Moreover, the customers who frequent coffeehouses reflect a male-dominated culture. Finally, the Turkish coffeehouse found in villages and towns is modest in comparison to exotic taverns, distinguished pubs, and chic cafes that are found in the larger cities.

A Unique History

Before we can understand how these four characteristics of the coffeehouse reflect Turkish culture, we need to know something about the unique history of Turkey. Ancient Anatolian civilization dates back to 6500 BCE. As early as 1900 BCE the Hittites occupied Turkey. About 700 years later the Phrygians and Lydians invaded the land. These early Anatolian groups ruled until the Persian Empire took hold in the 6th century BCE. Then the Greek-Hellenistic rule emerged, followed by the Romans in 100 BCE. Influences of each succeeding group left their mark on the people by contributing customs, language, and trade practices.

In the year 330 CE, Constantine the Great named Constantinople the capital of the Byzantine Empire. Arabs relocated westward, and by 700 CE the empire included Anatolia, Greece, Syria, Egypt, Sicily, the Balkans, most of Italy, and some areas of North Africa. Turkish tribes began to migrate from central Asia to Anatolia in the 1000s. The Oguz Turks, who embraced the new religion of Islam, occupied a vast part of Anatolia during the 11th century. After their setback in the Crusades, Turkish gazis, or Islamic warriors, fought against the Byzantine Empire and began to build the Ottoman dynasty. The Ottomans conquered the region and expanded their reign to the Christian Balkans. In 1453, the Ottomans conquered Constantinople, which was virtually the last stronghold under Byzantine rule. Constantinople, which is known today as Istanbul, was then rebuilt as the capital of the Ottoman Empire and the center of Sunni Islam.

The Ottoman Empire reached its peak of wealth and power under the rule of Suleiman the Magnificent in the 1500s. He controlled all or part of states including present-day Turkey, Iran, Iraq, Egypt, Israel, Syria, Kuwait, Jordan, Saudi Arabia, Albania, Algeria, Libya, Tunisia, Greece, Bulgaria, Sudan, Romania, Hungary, Yugoslavia, Czechoslovakia, Ethiopia, Somalia, and the Soviet Union. His vast empire reached into three continents and was noted for its system of justice and expansion of the arts, literature, architecture, and craftwork. The Europeans called him the

Magnificent but the Ottomans also referred to him as *Kanuni* or the Lawgiver. His legal influence by itself earned him an important place in Islamic history. For almost six centuries the Turks lived in or near Europe and interacted with Europeans, assimilating parts of European culture over time. Sultan Suleiman's death in 1556 marked the end of a wonderfully creative era. The empire began its decline with the loss of Hungary and the Crimea in the 1700s. In the next century the Ottomans lost control of Egypt and most of the Balkans, and this was followed by the loss of Serbia, Romania, Cyprus, Algeria, and Tunisia.

In 1908 a coalition called the Young Turks attempted to restore power to the empire. Instability in the region brought about drastic territorial changes and reduced the Ottoman holdings even further. The shrunken empire entered World War I on the side of the Central Powers in 1914. By the end of the war the Arab provinces were lost and a nationalist movement began. These events led to both the proclamation of a new nation and additional contributions to Turkish culture.

In 1923 the Treaty of Lausanne established the Republic of Turkey as we recognize it today. Mustafa Kemal, also known as Atatürk, renounced all prior conquests and introduced widespread reforms to guide the country. Atatürk literally means "father of the Turks." His deliberate reforms transformed the diverse empire into a nation while shaping daily lifestyles for every Turkish citizen. Along with the internal modernization of the country, the republic turned toward Western principles and conventions. Turkey became a member of NATO in 1952 and an associate member of the Common Market in 1963.

Turkish coffee was first introduced during the Ottoman Empire. In the 15th century, merchants from the Far East traveled along the Silk Road to trade their exotic spices and other wares in European markets. In an effort to encourage trade and offer hospitality, hostels for travelers called caravansaries were built. As they passed through ancient Turkey, the merchants bartered along the way and offered their products in exchange for hospitality. According to the strictest rule of the Koran, the sultans forbade the drinking of coffee because it is a drug. In spite of the restriction, drinking of coffee became so popular that the palace rescinded the rule and allowed its consumption. Today, Turkish coffee is as popular as ever.

Islam and Secularity

One of Atatürk's most important reforms was the adoption of a constitution that encouraged secularism. Although an overwhelming number of Turkish citizens are Muslims (about 99%), other religions, such as Greek Orthodoxy and Judaism, are tolerated in Turkish society. During the rule of the Ottoman Empire, religion as well as culture was totally integrated with government, and Islamic world power was concentrated in Turkey. Atatürk and his followers felt strongly that adherence to the caliphate was an obstacle to the country's attempts to Westernize. Major changes resulting from secularism included a shift from Islamic to European legal codes, closing of religious schools and lodges, and recognition of the Western calendar rather than a religious one. Despite these sweeping changes Islam still thrived, and in the 1950s religious education was reintroduced. Today the Turkish government still oversees a system in which children are exposed to a measure of religion in schools and the clergy receive salaries from the government.

A Secular Nation

The distinction between Turkey and the other Islamic nations is noteworthy. Unlike other Islamic countries, Turkey is a secular nation. Governments, schools, and businesses are operated independent of religious beliefs. Even traditional clothing, such as veils for women and the fez for men, were abandoned in the shift from Islam to secularity. Consequently, Islam does not affect daily Turkish life as much as it does in an Arab nation. Religious power over Turkish institutions is nonexistent, and this fact reflects a preference for association with other Western cultures.

However, as noted above, a small but growing political movement has evolved around the issue of making Turkey a theocracy comparable to other Arab nations. While Turkey has a Western-style system of government in which there are free elections, the military is extremely influential and will intervene if a political party seems to be too theocratic in orientation. In 1998 the Supreme Court, prodded by the military, forbade the operation of the Islamic Welfare Party. However, Islamic politicians regrouped in a new but similar political party, the Virtue Party. (Political parties frequently have colorful names in Turkey, for example, the Motherland Party.) Prosperity has witnessed the growth of a prosperous middle class, which has become a new center of power and influence, and it staged the first modern protest against military and political corruption in 1997. Hence there are three major centers of power and influence in Turkey: the military, the middle class, and the Islamic movement (Pope, 1997).

Although Islam does not influence state matters, its traditional practices, such as that of women wearing headscarves in public, has generated controversy. In 1999 Merve Kavakci, a Texas-educated politician in the Virtue Party, created an uproar by appearing in parliament in a headscarf. Almost immediately an appeal was filed with the Supreme Court to outlaw the Virtue Party for this and other assumed infractions. Furthermore, Mustafa Karaduman has become a successful businessman by establishing an Islamic-style clothing chain for women. One of his creations is the *nesrin,* an ankle-length skirt with a long jersey to which a hood is attached. Conservative Islamic women can easily take the hood off in places like parliament where it is banned (Zaman, 1999). The headscarf ban became one of the most emotional issues in Turkey, signifying a power struggle between an increasingly wealthy middle class of observant Turks on the one hand and the secular elite of the judiciary and the military on the other (Tavernise, 2008). Some people see the issue as a clash between tradition and modernity and Islam and democracy. Parliament voted to lift the ban on wearing headscarves in university settings in February 2008.

Islam is also evident in the five times for prayer spread throughout every day. Turkish Muslims do not necessarily converge on the mosque when the call to prayer is sounded, the way Catholics in such nations as Poland and Ireland flock to church for Sunday Mass at specified times. Islam is a religion in which individuals communicate directly to God without a need for intermediary spokesmen. Consequently those who pray at mosques do not need to coordinate their visits, although many arrange their visits to coincide with the five daily prayer periods. Afterward Turks may head to the coffeehouse for refreshments and socializing.

The Pillars of Islam

Five well-known pillars embody the essence of Islam. The first is acceptance of the creed, "There is no God but Allah and Muhammed is his prophet." The second pillar comes from the Koran, or Holy Book, and involves keeping life in its proper perspective. This can be done through prayer five times a day to submit oneself to God's will. Third is the observance of the holiday of Ramadan by fasting during daylight hours for a month as determined by the lunar calendar, and fourth is giving to charity at least 2.5% of one's income. Fasting underscores humanity's dependence on God, teaches self-discipline, encourages compassion, and forces one to think. The final pillar is a once-in-a-lifetime pilgrimage to Mecca for those who are able. It is understood that those who reach the holy city where God revealed himself have demonstrated their devotion.

In Turkish towns the two most important places for social gathering, the mosque and the coffeehouse, are located in the town square. The coffeehouse is often adjacent to the village mosque and may even provide a small operating fund or rent for the general upkeep of the mosque. Throughout the world inhabitants of small towns tend to be more conservative than city residents, and Turkey is no exception. As a result the Turkish townspeople tend to follow their faith more strictly than their urban counterparts. Also, eastern and central Turkey is more religious than western Turkey, due in part to the diversity of the people living in the western region. Although there has been a resurgence of religious practices since the 1950s, Islamic fundamentalism remains a minor if vocal influence today, and even moderate Muslims have tried to revive some traditions that were eliminated with the arrival of a secular nation. In fact, the Justice and Development Party (AKP), which came to power in 2002, has been pushing Turkey closer to the Muslim Middle East and further away from the West with Islamist ideology being very much part of its foreign policy calculus (Cagaptay, 2006). In 2008, Turkey's highest court struck down, by a narrow margin, an attempt to outlaw the governing party, averting a political crisis but highlighting the ongoing struggle between the secular establishment and the ruling party, which is seen as too Islamic.

In addition to rent, many of these coffeehouses occasionally collect donations from patrons to pay for the mosque's water and cleaning. As noted above, Muslims are called to prayer five times each day, although the frequency of prayer for contemporary

Muslims is a function of both the level of piety and the availability of time. Thus a visit to the coffeehouse can also be considered a ritual ironically linked with the daily ritual of prayer. Furthermore, cleanliness is an important related Islamic value, and the owner of a coffeehouse typically squirts a bottle of water on the floor from time to time to keep the dust from rising. Some of the original popularity of coffeehouses was due to the Islamic prohibition of alcohol and the use of coffee as a substitute and acceptable beverage.

Many other Turkish cultural values and beliefs are derived from the Islamic faith, some of which are not directly related to the operations of the coffeehouse. In particular, Muslims believe that the future will be better than the past. Village parents who struggle to support their children may rely on this optimism for decades. Also, the ideas that the soul lives forever and that people are responsible for their actions are important considerations for Muslims in accepting their lot in life.

Some of the beliefs and practices of the Islamic faith that are suitable for an agricultural people have become problematic in modern cities like Istanbul. For example, the Muslim Feast of Sacrifice—3 days of killing sheep, which are given in three equal portions to family members, friends, and charity—has horrified animal rights groups and created health issues in crowded cities. The Istanbul government has created several municipal sacrificial centers to decrease the health problems and make the sacrifices as painless as possible.

The Ottoman Empire governed the people by *Sharia*, or Islamic law. Turks also adhered to moral and social rules, such as respect for their father, on a daily basis. These ideas persist in modern Turkey. Unlike purely theocratic Islamic nations, however, Turkish legal decisions are not judged by one code alone. Likewise civil marriage ceremonies must be conducted whether or not there has been a religious wedding to obtain recognition under Turkish law. Civil code based on European models was introduced in 1926. Still, basic customs from the Ottoman days, just like the coffee-house, have been carried over into modern Turkish society. Some Turks still refuse to accept interest accrued from savings accounts because such profits are prohibited in the Koran. Widely used Turkish phrases derived from Islam include *salaam*, a friendly greeting that literally means "peace be upon you," and *masallah*, which means "God protect you from harm." Both phrases are used in everyday language, in much the same way that U.S. Americans respond with "God bless you" when someone sneezes. Other examples include abstinence from or moderation in drinking alcohol, fulfilling oral promises, and using the catch-all phrase "*Bismillahir-rahman-irrahim*." This translates to "I am starting this in the name of merciful Allah" and is said when starting a task—be it a journey, a test, a wedding, or a meal.

Family Values

In Turkish tradition the household extends beyond the nuclear family and individuals are loyal to the entire family or kinship group. Women especially are expected to offer hospitality to all guests. If visitors knock on the door before or during mealtime, they are almost always invited to join the family in the dining room or at the kitchen table. As in mosques, Turkish guests offer to remove their shoes as a gesture of cleanliness on entering someone's home. Marriage decisions are influenced by families in rural areas, and in some instances dating is not permitted. Of course, these are merely generalizations about Turkish customs and do not apply to all Turkish citizens, particularly those living in large cities. Turkish compassion and hospitality come from the ethical aspects of the Islamic faith, as do racial equality and religious tolerance. Islam acknowledges the legitimacy of the other Semitic religions that preceded it but recognizes both Judaism and Christianity as incomplete.

Part of the greatness of the Ottoman dynasty is attributable to Islam, which brought together different peoples and united them through spiritual means. When Muslim warriors conquered an area, they did not kill the people, and they allowed residents to practice their religion, although as second-class citizens. Still, this was an advance over the Christians in the Middle Ages, who frequently gave conquered people a choice of conversion to Christianity or death. Bernard Lewis (1995) points out that the Ottoman dynasty was much more socially advanced than Europe until the beginning of the 18th century, in large part because Europe practiced hereditary feudalism. While land was given for exceptional service in the Ottoman Empire, it provided only rent and space in which to live to the grantee; at his death the land was not given to his heirs but returned to the empire.

The concept that every event in people's lives is predetermined has its roots in the religion of Islam. Like many other religions, Islam attributes circumstances beyond one's control to a supreme being. Muslims believe that people will act according to their own decisions under a given set of circumstances. For example, much of the younger population is migrating toward cities to obtain jobs and better schooling. This trend does not reflect anyone's fate in particular, but people who decide to change their lifestyle by leaving the village subject themselves to *kismet,* or chance. In other words, God meant for that person to move on and start a new way of life. If someone eventually relocates back to the village, that, too, is *kismet.* This belief allows an individual's shortcomings to be more easily accepted.

God's Will

The Turkish orientation toward time goes hand in hand with the belief in destiny. It is easy to say that most Turks are not very time conscious, and one of the attractions of the coffeehouse is that customers can linger there if they so choose. Hosts and hostesses are concerned if their guests do not arrive promptly at the time given for an invitation to a social event, but they are very understanding. Delays at airports and train stations are not a cause for alarm. Polychronism best describes the Turkish ability to concentrate on different things simultaneously, whether at work, at home, or in the coffeehouse. Although schedules and appointments are important, especially in

Western-oriented firms, people and relationships are also valued. To the polychronic Turks time is intangible and supports the development and maintenance of long-term relationships. Elements of the past, present, and future tend to merge, and they permeate Turkish values and norms.

If Turks were asked to ponder the ideal position of their country in international dealings, they would probably refer to the halcyon days of the Ottoman Empire. The history of the country is crucial to understanding the Turkish mind-set, and a comparison of today and yesterday is absolute. Conversely Turks place much emphasis on what future generations will accomplish. This explains why it is important that children get optimal educational opportunities, irrespective of costs and degree of parental sacrifice. *Kismet* and time orientation are inextricably linked together in Turkish culture.

Understanding the idea of destiny helps us to comprehend why time is relatively unimportant. In the United States, if a business project falls behind schedule, a firm risks embarrassment and scrutiny by consumers, investors, shareholders, and the government. In Turkey the same type of delay is attributed to God's will and is more easily forgiven. Thus the phrase, *Inshallah,* "if it is God's will," is heard quite often. A famous Turkish joke illustrates the degree of urgency in daily business:

> One day, a fellow named Hodja brought some material to a tailor and requested that he measure him and make a shirt from the material as soon as possible. The tailor measured him and said, "I am very busy but your shirt will be ready on Friday, *Inshallah.*" Hodja returned on Friday, but the tailor apologized that the shirt was not ready and said, "Come back Monday and, *Inshallah*, it will be ready." On Monday, Hodja went back to the tailor only to find out that the shirt was still not finished, and was told, "Try again Thursday, *Inshallah*, and I promise it will be ready." This time Hodja wisely answered, "How long will it take if we leave *Inshallah* out of it?"

Recreation, Communication, and Community Integration

Coffeehouses became popular during the 16th century. The coffeehouse has always been a source of information, especially when illiteracy rates were high. In those days one person would read the newspaper to an eager audience. Throughout the reign of the Ottoman Empire, defense strategies were discussed and developed there. Coffeehouses had battery-operated radios before they were introduced in homes, and the same was true when television arrived. Today patrons enjoy camaraderie when viewing a soccer match at the coffeehouse instead of watching it on television by themselves at home, and this pattern fits the group-oriented lifestyle of the Turks.

Turkish men go to the modern-day coffeehouse to become part of a group. They feel very comfortable when surrounded by friends and family and prefer the stability of belonging to an organization to individualism. And, as an old Turkish proverb states, one cup of coffee is worth 40 years of friendship. According to Bisbee (1951),

> The prime ingredients of the Turks' idea of fun and amusement seem to be relaxation, imagination, sociability and humor. Sitting is almost, if not quite, the most popular recreation of all. Turks sit at windows, in gardens, at coffee-houses . . . anywhere and everywhere they can see a pleasing view and relax in conversation. (p. 145)

In Turkey there is a relationship between collectivism and accomplishing the goals of the group. In small villages the coffeehouse is often the setting in which important village decisions are made. Prior debate of the issue also takes place at the coffeehouse. A custom, *imece*, describes a Turkish social gathering at which everyone pitches in to help a neighbor undertake a large task, such as building a new home. In many instances of *imece,* the initial request for assistance and plans for the event are proposed at the coffeehouse. Turks prefer to approach problems in a logical but personalized manner, and they will discuss practical solutions at length in the coffeehouse. In small Turkish towns the mayor can frequently be found at his second "unofficial" office, the coffee-house. It is not uncommon to find community notices tacked onto a coffeehouse bulletin board. Typical examples are government announcements that crops will be sprayed with pesticides on a particular date or that a public hearing to discuss a new water project will be held. Messages are left there for others who will eventually pass through; the coffeehouse is the information center in which most communication takes place, including political discussions and gossip. Being a "member" of the coffeehouse is considered important, regardless of the topic under discussion.

Women also have a need for affiliation; they get together at weekly coffees or teas at the homes of friends. The weekly coffee presents an opportunity to show some hospitality and become involved with others. Traditional Turkish hospitality extends to strangers, particularly in rural areas. The group itself is highly valued, and individual identity is determined on the basis of group membership. Conformity to group norms and traditions is expected; trust and reliance within the group is important (Dindi & Gazur, 1989, p. 17). In Geert Hofstede's (1991) 53-nation study of cultural values, it is not surprising that Turkey clusters with those nations emphasizing collectivism rather than individualism. In the more recent GLOBE study (House et al., 2004), which distinguishes between societal institutional collectivism and in-group collectivism, Turkey scores in the same band as the United States with respect to institutional collectivism but much higher along with India on in-group collectivism practices. This may very well reflect the unique position of Turkey as a geographic and cultural balancing act between the East and the West.

Collectivism at Work

Even at work, collectivistic values are expressed in unusual ways. At McDonald's the "employee of the month" is not selected for outstanding work. Rather, the selection is made on a rotational basis. Furthermore, because Turkey also clusters with high power-distance nations in Hofstede's 53-nation study (and this remains unchanged in the recent GLOBE study findings), there is little if any two-way discussion or participation with subordinates in many firms. However, Turkish managers link collectivism and power distance—and by extension a distinctive concept of employee participation—by telling employees that they are an essential part of the organization and that, as long as loyalty is shown, they will not be fired, even if performance is substandard.

As might be expected in a collectivistic culture, communication tends to be high-context in nature. For example, if someone says "it is very hot," people assume that the person wants a ride in their car. This indirect form of communication is common, even among very close family members.

Family gatherings are very important, as are deep friendships. Turks are nostalgic when it comes to family traditions and special occasions. Just as a man can stop by the coffeehouse whenever he wants, so too it is not considered impolite to drop in at a friend's home without an invitation. This is true regardless of economic, social, or educational status. Fostering good relationships requires time, and Turks feel that such time is always well spent.

Coffeehouses open early in the day, and for many Turks, they are the place to stop in before beginning the day's work. In the late afternoon, workers stop in prior to going home to their families. After dinner various men visit the coffeehouse to catch up on the latest news. In the summertime customers sit outside on sidewalks or patios to escape the oppressive Turkish heat and catch an occasional cool breeze. In the winter everyone gathers around the warmth of a stove or furnace in the coffeehouse. Just as the Italian piazza is the place to stroll to see others and be seen, the village coffeehouse provides a place for social gathering.

People Who Need People

In Turkish culture, involvement and interdependence are the ideal. Turks care about others, and they show their concern by stopping to help perfect strangers. People are important in Turkish culture, and most Turks will sympathize with those who are less fortunate. Above all, Turks pride themselves on hospitality and being helpful to strangers. The response to a traffic accident should illustrate this aspect of Turkish culture. Within moments of the accident, everyone who witnessed the collision, along with those who did not, is on the scene. No one present would dream of minding his or her own business. Strangers then offer help, comments, and interpretations of the accident, and chaos results. If a vehicle breaks down at night, Turkish passers-by generally do not hesitate to stop and offer assistance to the driver. To ignore someone in distress would be considered indecent. By contrast many U.S. Americans would not dare to stop for fear of their own personal safety.

However, these customs are decreasing in importance in urban areas, particularly Istanbul, which suffer from all of the problems associated with rapid migration from rural areas. As in other countries, residents of rural areas are drawn to the cities because of opportunities for work and diversity of experiences. Pockets of extreme poverty, many times in and around downtown areas, have arisen. Militant Islam, which was previously strongest in the rural areas, is on the increase in urban areas, at least in part because of such conditions.

Whether sharing small talk at the coffeehouse or within the hospitality of their home, the Turkish people tend to be curious by nature. Consequently, very little in people's lives is a real secret. Communication is the essence of Turkish existence. When one Turkish stranger meets another, one typically asks the other where he or she is from. Of even more interest is who the stranger's parents are and where they are from. The next few minutes are spent quizzing each other with questions like, "Oh, my brother-in-law's uncle is from there, do you know Ahmet?" or "Surely you must know Ali." "Turkish geography" is really a game played to discover to which clan the stranger belongs, as happens in other countries, such as Israel, India, and Ireland.

A Male Domain

A Turkish coffeehouse is a place where men assemble throughout the day and throughout the year. Regardless of age, men gather at the coffeehouse to play backgammon, share gossip, or just enjoy company. Young boys accompany their fathers and play alongside the tables until they are old enough to join discussions. Old men sit on wooden chairs and sip coffee while reminiscing, offering words of wisdom, and solving the world's problems.

A visit to the Turkish coffeehouse is a favorite pastime for male Turks of all ages. As in bygone eras, the village coffeehouse remains a center for everyone, regardless of income, education, or social status. But this is less true in the larger cities, and the traditional coffeehouses are becoming associated with unemployed workers and those from the lower socioeconomic rungs of society. There is even sometimes a small stigma associated with frequenting such a coffeehouse, especially among adults who are supposedly working full-time or going home to a family dinner.

A correlation can be drawn between the prominence of coffeehouses and cultural values in cities versus rural areas. In the larger cities of Istanbul, Ankara, and Izmir, life is hectic, and people have little time to sit and chat. However, Turkey is still an agrarian society in large measure; 34% of the workforce is in agriculture. People working outside industrialized areas are often limited by weather conditions and daylight. With extra time on his hands a farmer may visit the coffeehouse, and it is a welcome diversion for a man out of work. Coffeehouses remain popular in rural areas where Turkish traditions and male roles are strong. The popularity of the city coffeehouse has declined as a new generation places less emphasis on traditional values and a greater emphasis on earning a living. Within the city older citizens tend to be more devout and frequent the coffeehouse more than younger Turks, who often spend most of their leisure time on other activities such as cinema, theater, and concerts, as alternatives to the coffeehouse.

As cities become more cosmopolitan, urban coffeehouses are less likely to attract clientele in comparison with coffeehouses in less populated locales. The coffeehouse remains a cherished institution outside of the cities, where lower- and middle-class customers enjoy good coffee and good conversation, and city dwellers recognize its historical importance and speak of the coffeehouse fondly even when they don't frequent one.

Male Dominance

There is no question that Turkey has a male-dominated culture, and the nature of such domination is ever present at the coffeehouse. According to the findings of the GLOBE study (House et al., 2004), Turkey had low scores on the dimension of gender egalitarianism practices, which indicates that the society generally believes that biological sex should determine the roles of members in homes, business organizations, and communities. While women are welcome, the coffeehouse has always been considered a male domain. This is true of other public places in Turkey, even today. The Islamic origins of the population still influence the role of women. For example, most women pray at home rather than at the mosque. The killing of women and girls by relatives who believe that the females have brought shame to the family (honor killings) has plagued Turkey and other Islamic countries. In 2007, however, the Turkish government took unprecedented steps to campaign against this horrific practice by soliciting the help of imams, pop music stars, and soccer celebrities (Wilkinson, 2007c).

Even in Western-oriented Istanbul, which accounts for about 40% of Turkey's GNP, the separation of sexes is evident. During the day, there always seem to be eight men to every two women in public places, and at night, the number of women decreases significantly.

A long tradition of men as breadwinners, leaders of the family unit, and decision makers in local politics and matters of national concern exists in Turkey. As heads of the families, men in rural areas decide which crops will be planted and at which market crops will be sold. Village coffeehouses often become a market where brokers trade farm commodities. Within the clan structure, it is understood that sons will be in charge of the family and daughters will marry into their husband's family. As in other Western cultures, sons carry on their family name. Within the traditional family a father's request is always obeyed, and Turkish fathers are especially protective of their daughters.

A Modest Environment

The simplicity of the coffeehouse itself is noteworthy. The furniture is basic, and the chairs are uncomfortable. A typical coffeehouse in a small village may consist of one small room, a kitchen, and several small wooden or aluminum tables and chairs. A smoky atmosphere from cigarettes or *nargile*, the traditional waterpipe, is unmistakable. Although each coffeehouse differs, the sounds are always familiar; steady chatter, heated discussion, and hearty laughter can be heard above the clatter of cups and saucers, and Turkish music comes from a radio or TV. Likewise, the Turkish people are concerned more with substance than form. The true draw of the coffeehouse is the opportunity for release and self-indulgence. The release is only temporary, allowing escape from the stress or monotony of the moment, and the indulgence reflects the element of Turkish culture that says "experience life."

As noted earlier, Turkish life in the city is very different from that in the village. In the cities neighborhood coffeehouses abound. However, the middle and upper classes tend to spend time at various elite establishments, such as bars or cafes, instead of being loyal to a particular coffeehouse. Apartment living is very popular in Turkish cities. Turks are less transient than U.S. Americans and tend to live in the same building for years, where they are surrounded by family and friends. Simply put, people are valued in Turkey. Respect for others is important; however, like other Mediterranean peoples, Turks have little fear of consequences for failing to respect laws. Along with liberal interpretations of traffic signals, patience is not a Turkish strong suit. A prime example is the early morning scene when workers try to catch a bus. It's "every man for himself" as people struggle on the downtown streets of Ankara to get on the crowded buses. In most other situations, however, Turks have considerable respect for the rights of others.

Coffeehouse Operations

Coffeehouses are family-run establishments. Often the owner's son takes orders and delivers teas and coffees to the table. People should not expect to be served a meal there; however, patrons are more than welcome to bring in food. Many hungry patrons will first stop at a local carry-out for a freshly baked snack. *Simit*, a round flat bread topped with sesame seeds, is a familiar sight at coffeehouses. A coffeehouse owner does not mind food being brought into the coffeehouse because he knows that the more food eaten, the more drinks will be ordered. Unlike Turkish restaurants, coffeehouses invite customers to remain as long as they wish. The coffeehouse is not a high-profit enterprise. By the same token, it is rare that a coffeehouse folds. The reason is that coffeehouses provide an incredible service to the community as a meeting place and information center.

Turkish coffee is always made in a special way. First the grains are ground into a fine powder, then water and sugar are blended with the powder, and finally the mixture is boiled over a flame. The three possible versions of Turkish coffee are *sade, orta*, and *sekerli*, which mean no sugar, some sugar, and more sugar, respectively. At the coffeehouse the waiter will call out the orders to the kitchen so fresh pots can be made. When he returns, the waiter never seems to remember who ordered what. A running joke is that one day Atatürk, who is greatly respected by the Turks, went to a village coffeehouse with his entourage. The waiter took everyone's order and brought the coffees to the table. Atatürk was amazed that the waiter had gotten the order correct and told him so. When asked by Atatürk how he remembered, the waiter responded, "Yours is the only order that counts!"

Life Outside the Coffeehouse

Many activities occur outside of the coffeehouse, and we need to understand them and the manner in which they reflect cultural values. For example, observance of Islam in Turkey today has led to popular misconceptions about the role of women. Since the adoption of this religion more than 1,000 years ago, conservative ideas have evolved, including the seclusion and submission of women to men as a form of protection for them. At one time, women were expected to retreat to the harem out of deference to men. However, even in the harem, women had some degree of power in terms of training daughters and young sons in household management, arts and crafts, and religion. The harem was highly political; concubines were selected for sultans by their mothers, and royal marriages were arranged in the harem in those days. In 1925 the wearing of veils by Turkish women was officially discouraged. In the following year Atatürk, the founder and first president of modern Turkey, introduced civil marriage codes that abolished polygamy and established divorce and child custody rights for both men and women.

Women's Role

As in earlier times, Turkish women have always been capable of subtly influencing their husbands' way of thinking. In Turkey women are brought up to show respect for their husbands, and they do not openly flaunt any power they may hold over their men. Turkish women rarely make firm denials of a husband's decision, yet they may quietly oppose decisions with which they are unhappy. In either case the wife recognizes the needs of her husband's ego and is typically shrewd enough to offer small suggestions rather than demand compliance with her wishes. Many Turkish women possess an emotional inner strength and maturity that developed from the burdens of maintaining a stable family life under harsh economic conditions. This is particularly true for those who live in remote areas. In rural villages women are responsible for meals, housework, and child care. Turkish kitchens are busy all day long as women prepare delicious meals of roasted lamb, rice pilaf, or other specialties. Throughout Turkey women take pride in a cooking style that intermingles Arabic, Greek, and European influences. An important development for women in the villages is that they have become responsible for working outside the home, too. The unemployment rate tends to be high, averaging 9.6% between 1995 and 2004, and many men have difficulty finding work. According to *The Economist's* Intelligence Unit forecast, the rate of inflation rose to more than 10% in 2008, and gross domestic product is expected to slow down to about 4% in 2009; this will affect the availability of jobs. Lack of employment opportunities for men in small towns and villages has sustained the appeal of the coffeehouse.

External employment is often a necessity for women, who will work outside of their homes on local farms and in factories or will sell handmade items to support their families. Beautiful woven carpets, ceramic pottery, mosaics, and other items handcrafted by Turkish women are appreciated throughout the world. The women's determination in taking care of daily life affords them some measure of power over their husbands' behavior. Women who work in the fields cook, eat, clean, and even sing together.

Urban wives, who may be on a more equal educational footing with their husbands, may also affect their behavior and take part in decisions as trivial as recommending what clothes a husband should wear or as monumental as deciding to risk a new business venture. Finally, career-oriented women reflect modern Turkish societal norms, which recognize joint decision making and cooperative patterns of husband-wife authority, along with traditional respect for the husband. It should be noted that families still supervise closely the activities of unmarried women, and strong attitudes prevail about the proper conduct for women.

Turkey was one of the first countries to give women the vote more than 70 years ago and although most Turkish party leaders are men, Turkish men interviewed at the coffeehouses where they usually retreat to escape from their wives are in support of women leaders in politics and business. In 2007, as a protest against the lack of representation in Parliament, Turkish women launched a unique protest that featured famous women pop stars and business leaders sporting fake mustaches on TV and in ad campaigns sending the message that you had to be a man to be in parliament. Although the numbers are still quite small (72 years ago, Turkey was in the top five and is now 127th in women's representation in parliament), the percentage of women increased from 4.36 to 9.1 (50 women) following the elections.

The Extended Family

Furthermore, respect for parents is absolute. In turn, parents are expected to be concerned about their children's future. The extended Turkish family is proof of the respect afforded to elders. Grandparents, aunts, uncles, and cousins often live in the same dwellings or nearby as core family units. The notion of a home for elders is nonexistent. Likewise, hired babysitters are unusual because family members care for children. Parents are willing to stretch the family budget to provide tutors or send children to better schools. Most parents feel obligated to give their children a moral education as well. Children are a continuation of a family's reputation, and famous children are an even greater source of pride for doting parents.

Boys and girls are brought up similarly with a few exceptions. Prior to reaching 7 years of age, boys are circumcised in a religious ceremony that symbolizes their ascent to manhood. Upon graduation from high school there is a mandatory 18-month period of military service. Men who attend a university have the option of deferring their service until they reach 32 years of age. There is no compulsory service for women, although they are allowed to enlist for noncombatant positions if they desire. Traditionally new brides became part of their husband's family only after they had produced an heir. Except in rare circumstances, this is no longer true. Today, Turkish daughters-in-law are immediately accepted into the family household. The family structure is important, and the divorce rate is relatively low. In Turkey, it is unusual for people to live on their own, unless they are students living in dormitories.

Education and Class

Another significant reform, the right to education for the masses, took place in 1924. Atatürk believed that education was an essential element of Westernization. Today education is seen as the legitimate discriminator for the social classes in Turkey. Elementary school is mandatory for all children. As in the United States, the school year begins in September and ends in June. In keeping with national policy, secularity is practiced. Although *laik*, or separation of church and state, is understood, school vacations coincide with the Islamic holidays Ramadan and Kurban. In addition to monitoring regular schools, the Turkish Ministry of Education also controls religious schools to limit their influence. For those schools with limited resources, the school day is taught in half-day shifts. Wearing uniforms is required for the young children. A dress code also requires neat grooming such as conservative haircuts for boys and hair ribbons for girls with long hair. Children learn respect for teachers at an early age. In the classroom students normally stand whenever a teacher enters the room. It is considered disrespectful to disagree with the teacher.

The curriculum includes mathematics, science, history, Turkish language, art, music, physical education, and morals. After fifth grade, as in Germany, children study for a placement exam that determines their future path of study. In larger towns and cities tutors are often hired to complement formal school instruction. For those who score well, the next year is spent at a private secondary school to learn a second language, most often English, French, or German. The language study is followed by 2 years of preparation for high school. Students attend high school for 3 years and on graduation, those who wish to continue their studies at a university take a difficult entrance examination. Students who do not score as well on the initial placement exam attend a general secondary school, followed by entry into a vocational or technical school to learn a trade. Those students interested in entering a university are also invited to take the entrance exam.

The competition to attend universities is keen due to the limited number of universities in the country. Each year, about one fifth of the students who take the entrance exam are accepted at the universities. There is an obvious inequity in this system. Those who are privileged enough to have access to better quality academic resources have a better opportunity for admission to a university. The selection process is highly competitive; within the universities in Turkey, many disciplines have a limited number of student spaces available. Thus university students may aspire to a profession from which they are effectively barred, unless they are able to study for that profession abroad. While selection to an institution is competitive, interaction among students is cooperative. It is common practice for students to work together on homework and studying. Students tend to take studying seriously and to appreciate their learning opportunities.

The Use of Humor

Furthermore, humor is an important mechanism in Turkish culture and is a good way to make friends and maintain friendships, both in the coffeehouse and outside of it. Turkish humor is distinctive; Turks do not mind making fun of themselves. When things become too serious, a joke is welcomed to strike a balance and provide relief. The coffeehouse itself provides a similar relief from life's predicaments. Whether in political satire, television shows, or within the coffeehouse, humor is used frequently. For example, Turkish coffee is served in a small cup and has a thick, almost muddy, consistency; it is usually served with a glass of water. A popular joke involves differentiating between Turks and foreigners. The difference? Turks drink the water first to clean their palates; foreigners take the water after the coffee to wash it down.

Humor is taught early on, and folk heroes such as Nasreddin Hodja are loved by children and grown-ups alike. As a boy Hodja always had something clever and humorous to say in every situation. Hodja stories are well known and relate to everyday life in Turkey. A favorite story goes as follows:

Hodja was walking down the street when he noticed something glittering in the gutter. He ran over to pick it up and noticed it was a small metal mirror. He looked in it and said to himself, "No wonder they threw this thing away. I wouldn't keep something as ugly as this either!"

What makes Hodja special is that he always expresses a note of encouragement and tries to live a happy life of honesty and simplicity. His jokes run the gamut from religion to mothers-in-law to sultans and even to his own hearty appetite. Not only are his jokes retold at the coffeehouse, they often begin with "One day when Hodja was at the coffeehouse. . . ." The proof of this character's popularity is that many people in Turkey like to claim that Hodja was born in their village.

Humor is an effective communication technique and can even serve as a way to avoid confrontation and conflict. Turks are generally quite modest about their accomplishments. This makes sense because besides being impolite, boasting would tend to separate people from their groups. Instead, it is perfectly acceptable for good friends or family members to brag about others. Modesty complements the Turkish values of tactfulness and diplomacy. Turks prefer to avoid confrontation and to handle disputes indirectly whenever possible. Being direct is considered rude and insulting; directness goes against the respect Turks feel for other people (Dindi & Gazur, 1989, p. 19). Respect for authority is highly regarded and unconditional in many instances. In his 53-nation study, Hofstede (1991) showed that uncertainty avoidance epitomizes Turkish culture. However, this seems to have changed over the last several years because the GLOBE study (House et al. 2004) indicated a score of 3.63 for Turkey for society practices of uncertainty avoidance as compared to 4.15 for the United States and 5.37 for Switzerland, which had the highest score. For example, the strong need for consensus, the undesirability of conflict and competition, and the strong belief in experts and their expertise are understandable when examining the face-saving nature of Turkish culture.

Looking Ahead

Turkey faces considerable challenges in the near future. High unemployment and inflation have left an unstable economy, and past government corruption and military coups cast doubt over true Turkish democracy. The question of Kurdish independence still needs to be addressed for the estimated 12 million to 15 million Kurds who live in Turkey; the army waged a fierce war with them during the 1990s. In 2008, Turkish forces attacked the guerillas of the Kurdistan Workers Party (PKK) from land and air over Iraqi sovereign territory. In general, however, the ruling AK party has had better relations in the last few years with the Kurds both within Turkey and the two Kurdish parties that jointly administer the autonomous region in northern Iraq that shares a border with Turkey. Although the Turks prefer further integration with Western Europe, Turkey has not yet been offered full membership in the European Union (EU) as "Turko-skepticism" continues over the AK party's more openly Islamist foreign policy ("Coming Apart," 2006). At the same time, Islamic fundamentalists point to a growing economic gap between rich and poor in their attempt to discourage Westernization. Disputes over Cyprus continue to strain Greek-Turkish relations, terrorism exists within Turkish borders, and Turkish minorities in the Balkan states are persecuted.

On the other hand, the future holds promise for Turkey. The end of the Cold War, growing support for full integration of Turkey into the EU throughout the world, and important Turkish energy resources provide hope for increased stability in the region. Water projects on the Tigris and Euphrates rivers will help provide better irrigation of Turkish farmland when completed. Trade with many of the EU countries and Russia now totals hundreds of billions of dollars each year. Despite inflationary problems, economic reforms of the 1980s have caused a steady average growth of 6.9% per year in Turkey's GDP from 2003 through 2007 although it is expected to slow down to about 4% in 2009. In a World Bank study evaluating the ease of doing business, Turkey moved up from 65th in 2007 to 57th in 2008. PricewaterhouseCoopers predicts that Turkey's economy will equal Italy's by 2050. Foreign direct investment reached a record $22 billion in 2007. Together, these events demonstrate that Turkey possesses a good amount of strength and resilience.

This, then, is our portrait of Turkey, a traditional agrarian nation that is seeking to modernize rapidly. The coffeehouse reflects many of the important values and behaviors of the Turks, particularly the relationship between Islam and secularity, recreation, communication, and community integration, male dominance, and the importance of substance over form within such a modest environment. Although Turkey is changing, its citizens still recognize the historical and cultural importance of the coffeehouse, even when they frequent it only periodically. Thus, there will always be a place for the coffeehouse in Turkey, especially in the rural areas and small towns that still dominate life in this country.

The Brazilian Samba

It is probably unwise to draw too many conclusions about Brazilian perceptions on the grandeur of the country and of the grand endeavors of its people, but it would not be excessive to assume that they naturally regard Brazil as an extraordinary place, which would imply that other countries are somewhat less so. In a country with so many monumental things, both natural and man-made, any comparisons with its immediate neighbors make the neighbors look pretty puny.

—Jon Vincent (2003, p. 15)

Brazil, with a population of 186 million, is extremely diverse and spread across several distinct regions and 26 states. In land area, Brazil is almost equal to the entire United States: 8,511,965 square kilometers versus 9,372,610. It lays claim to many aspects of bigness and grandeur, including the Amazon, the world's most voluminous river; the largest oxygenating forest, the Amazon forest; the world's largest wetland; the world largest soccer stadium; and the world's largest samba dance stadium. There are 12 Brazilian cities with populations of more than a million each, and São Paulo with its population of 18.3 million is tied with Mumbai for being the fourth-largest city in the world. (Tokyo has 35.3 million, followed by Mexico City at 19.3 million.) Goldman Sachs, the investment bank, has designated Brazil as one of the BRIC countries, along with China, Russia, and India. Its research suggests that these four developing nations will be the dominant world economies by 2050.

Brazil is also a land of contrasts. The industrialized south with São Paulo as its center is far wealthier than the north. Average income in the northeast is only two fifths of that in the southeast. However, the rate of poverty declined from about 35%

to 23% between 1992 and 2005 and is still falling. As Brooke Unger (2007) emphasizes, many issues require attention, particularly a high rate of taxation, excessive government bureaucracy, and the destruction of the Amazon forest, of which about 18% has disappeared since the 1960s. However, the future looks bright and efforts to address these issues are ongoing.

Brazilians have tremendous spirit in the face of adversity, particularly economic problems, which plagued the country for decades and still do, but to a lesser extent. In fact, in recent years Brazil's economy has improved significantly. The well of this spirit is continually replenished by the Brazilian's passion for life. Metaphorically the samba encapsulates this passion for life and serves to replenish the well in both good times and bad.

The samba has truly become a national symbol. For example, at a beauty pageant for Italian-Brazilian young women, held in Brazil, all contestants were asked what they first thought of when hearing the word *Brazil*. Immediately all of the young women began to samba. The samba's technical definition is "a binary, percussive rhythm in which the first beat is never sounded, causing a continual, hesitant urgency" (Krich, 1993, p. 73). This idea parallels the Brazilian's polychronic nature and seemingly constant movement, although the business-centered south tends to be more monochronic. For example, for a business meeting in the south, participants are expected to arrive on time, and doing otherwise is viewed negatively. Brazilians are naturally uncertain about their future when the first indication of the direction in which they are going cannot be seen nor heard. One long-standing joke is that Brazil is the country of the future and always will be. However, since 1988 the nation has undergone significant changes. Inflation is now manageable at less than 10% per year, and frequently less than 5% per year, and both the government and private businesses are capitalizing on the vast resources that Brazil possesses to strengthen the political system, the economic system, and institutions such as the educational system and legal system.

There is some truth that Brazil's potential far outstripped its achievements until recent years. It is not that Brazil is the country of the future and always will be. Rather, Brazil did not emphasize the future until recently, as the three founding groups—the conquering and rapacious Portuguese, the indigenous Brazilians, and the slaves—had limited reasons for doing so. Today, this generalization is no longer valid (Unger, 2007).

Evolution of the Samba

The very existence of the samba is owed to the musical talent of the plantation workers. It was on the plantations that the first whispers of what is now known as the samba were heard. Decades later the samba's predecessors appeared in civil society in the form of elegant tangos, as blacks rose in social status in Brazil's bigger cities. The initial characterization of this dance as the samba occurred at free slaves' dance parties in Bahia in the 1870s. Bahia was and is a center of the vibrant and rich black culture in Brazil. As the story goes, a woman from Bahia named Tia Amelia took the samba further south to the slums of Rio de Janeiro. Her son, Donga, composed the first recorded samba in 1916. As early as 1923, samba schools were born. These schools flourish today and symbolize the spirit of the samba to most of the outside world, especially during Carnival, the annual pre-Lenten celebration. Although part of the spirit of the samba is captured in the glorious dancers and floats in the samba school parades during Carnival, the samba is much more complex. Its physical characteristics and musical qualities weave an intricate web around Brazilians and influence many facets of life.

The samba takes on many forms including "samba *pagode*, samba *raiado*, samba *de partido alto*, samba *do morro*, samba *de terreiro*, samba *canção*, samba *enredo*, samba *choro*, samba *do breque*, *sambalero*, *sambalanço*, and *sambā*" (Krich, 1993, p. 73). These various sambas are appropriate for specific situations and social levels. For instance, the samba *enredo* is composed and used in the samba schools. The samba *de terreiro* is less sophisticated and likely to be danced in more rural areas. Other sambas are simply musical innovations. For instance, the samba *do breque* incorporates a rhythmic spoken part, much like rap, in the middle of the traditional samba music.

The lyrics of the various types of sambas are varied as well, although all have a common theme, for the words center around people's trials and tribulations across history and in daily life. Popular subjects include corruption, poverty, historical events, and local heroes. The number of samba songs that have been composed is enormous: On the average samba schools produce 2,000 new songs a year for the Carnival.

Five characteristics of the samba, when explored in detail, delineate the culture of this country with respect to its people, their behavior, and their manner of conducting business. We have chosen to discuss primarily the physical attributes of this dance as opposed to the musical and lyrical aspects for the sake of brevity, although there is an obvious overlap. These characteristics are small-step circularity, physical touch, undulation, spontaneous escape, and the paradox of dancers.

Small-Step Circularity

When dancing the samba, people move in small, somewhat controlled steps in a circular pattern, while holding the upper torso still. This notion of small steps and circularity has been present in Brazil since its discovery in 1500. Pedro Alvares Cabral, a Portuguese trying to find an oceanic passage to Asia, accidentally and literally ran into it. Columbus ran into land along the same parallel, and Cabral was presumably hoping to do the same. But earlier, in 1494, Spain and Portugal had divided the New World in the Treaty of Tordesilhas, and undiscovered Brazil was given to Portugal, even though no one was aware of its vast size. In other words, this land became part of the Portuguese empire largely through default and not through competition or war.

Historical Evolution

In general, "Brazil's development as a nation has been essentially evolutionary with few sharp breaks or drastic discontinuities" (Schneider, 1996, p. 35). Brazil grew amid small steps that were often circular, for example, from military to civilian rule and back. In the early 1800s the Portuguese emperor, Dom João VI, moved to Brazil because he feared Napoleon's power in Europe. Under his rule growth truly began. After the situation in Europe stabilized, Dom João returned to Portugal and left his son, Dom Pedro I, to rule Brazil. The latter realized Brazil's need to separate itself from Portugal officially; on September 7, 1822, he declared Brazil's independence.

Several years later, he, too, returned to Portugal, but he left his son, Dom Pedro II, in charge of Brazil. While Dom Pedro II was away on business, his daughter signed a document freeing the slaves. Soon thereafter two army generals took the initiative to oust Dom Pedro II from Brazil. Hence, in 1889, the monarchy was dead and the Republic of Brazil was formed. This transition from monarchy to a republic demonstrates how Brazilians move in small steps. It took them 67 years after their country became independent to remove the Portuguese monarchy from power.

The existence of several different forms of government over time, as well as the vacillation between these forms, also supports the notion of small steps and their circularity. It is significant that the republic was proclaimed by two generals, as military dictatorship has been prevalent in Brazil for most of the nation's history. Many presidents have been military officers, thus further confounding military and democratic rule. In 1985 civilian rule and elections supplanted military rule, although the generals are still key players.

There are currently 21 political parties, in part because of the system of proportional representation in the legislature based on the total percentage that each party receives in the national elections. In comparison to the U.S. practice of having two major parties and one or two independent parties in any election, the Brazilian system creates the conditions for its complex multiparty system of government. This frequently results in weak governments ruled by coalitions of political parties and deals that political parties make with one another. It also results in excessive and inefficient bureaucracy; recently Brazil was ranked 121st out of 125 nations in terms of ease of doing business. According to Unger (2007), Brazil holds the world's record for the number of hours that the average firm requires to prepare its taxes: 2,600. On average, opening a new business necessitates 17 procedures and 152 days, 115th out of the 125 nations. There are attempts to change such practices, but they contribute to the probability of small-step circularity.

Admittedly there is much circularity, overlay, and repetitiveness in the ideologies and activities of these various parties. But this phenomenon also shows how people are given significant freedom in expressing their beliefs and forming organized groups representing these beliefs. Again, this notion dates back to Brazil's founding, when different groups were not forced to fully assimilate to the Portuguese culture.

This liberal attitude extended to the practice of interracial unions, resulting in many mixed marriages and mixed children. All types of Brazilians integrated themselves into a unified society through the small steps of increased racial integration. In fact, Krich (1993, p. 82) believes that "a new generation of blacks is finding the samba one of the handier means of advancement" in Brazilian society. Although practiced more devoutly in particular regions, the samba has been pervasive throughout the country, and knowing how to samba is part of being Brazilian. Hence this dance is a vehicle through which all of the diverse peoples and regions of this vast country can seek some degree of unity.

Economy and Education

The monetary and educational systems also demonstrate evolution and small-step circularity. Since 1940 Brazil has used five different currencies including the *mil reis*, the *cruzeiro*, the *cruzado*, the new *cruzeiro*, and currently the *real*. Hesitant to bring about sweeping change, the Brazilians attempted to keep the *cruzeiro* by devaluing it many times. As indicated previously, the yearly rate of inflation reached a mind-boggling 2500% in the early 1990s, but it has since leveled off to less than 10%, and it is frequently lower than 5%. When it was finally apparent that the situation was hopeless, a move was made to introduce the *cruzado*. However, when this did not work, a move back to the *cruzeiro* was made, and the currency was declared to be the new *cruzeiro*; again, the notion of circularity is evident.

The educational system in Brazil is quite interesting, as it is both lax and controlled simultaneously. As with most educational systems, there are levels of education such as grade school, high school, and college. To enter a university, people must take an exam called the *vestibular*. Entrance to the university is based solely on performance on this exam; high school grades and extracurricular activities are not important. Thus there tends to be a more relaxed attitude toward school work among some students in the early stages of high school. However, the high school curriculum is very structured and demanding; for example, everyone must take 3 years of mathematics, physics, chemistry, biology, and so on. It is impossible to go through high school without taking consecutive years of physics and chemistry, and failing one of them requires staying in the same grade for another year.

Before the *vestibular* there is a big push to pass this exam. Of significant interest is the fact that students take this exam in a particular subject area, such as history or physical education. Hence, at the age of 18, they must already commit to a field of study in college. Once they are accepted into a university for this field, they study only subjects related to it. This system is more similar to the German system than to the U.S. system, although grades in precollege classes are given more weight in the German system than in Brazil.

A typical university campus has separate buildings for each subject, for example, those studying history spend all of their time in the history building. Students are allowed to take only a few electives. If they decide that they do not want to continue studying history, they may not simply transfer into another department; they must take the *vestibular* over again in a new subject. Hence students lose valuable time. In this manner the educational system is rather rigid, as symbolized in the tight control of the torso and movement found in the samba. However, it is also clear that many steps must be taken to reach a degree in the person's preferred field, thus confirming the notion of small-step circularity.

Education is an area in which the conflicts and contradictions in Brazilian culture become apparent. In recent years Brazil has improved significantly, achieving near-universal primary schooling and large increases in secondary and college enrollments. Still, Brazil came in last in mathematics and fourth from the bottom in reading in one study of standardized performance among students in 40 nations; half of 10-year-olds are functionally illiterate (see Unger, 2007, p. 13). Many of the schools are decrepit and underfunded, thus putting Brazil at a disadvantage in a globalized economy in which an educated workforce is critical. Sometimes the conflicts and contradictions are startling.

For example, the Bulhões family in the early 1990s exerted incredible power and influence over all aspects of life in the state of Alagoas, where the people suffered from the highest infant mortality rate in the region, the lowest child vaccination rate, lowest rate of prenatal care, and one of the lowest rates of schooling, 1.8 years per adult. Yet, by contrast, the local elite looked forward to the inauguration of cellular telephone service in January 1994. Wearing designer clothes, Mrs. Bulhões, the governor's wife, roamed among the shantytown dwellers and peasants proclaiming: "Poor people have just the same right to see me pretty as people in society."

About that time, she chartered an executive jet three times to fly to São Paulo and back, with the state treasury picking up the $24,000 tab. Meanwhile the state's public schools were closed for 7 months because teachers were protesting salaries of $80 a month. The person taking the hard line with the teachers was the secretary of education, also the governor's brother. This family represents a pointed example of the political oligarchy existing in some of the poorer states, which blurs lines between public and private consumption and responsibilities. Fortunately this is an extreme example from one of the most corrupt and poor states in the northeast of Brazil. Many other regions can offer contrasting examples.

Personal Lives

Small steps and circularity are characteristics that are found in the personal lives of Brazilians as well. Although it is quite easy to make acquaintances or *colegas* in Brazil, it takes some time to cultivate deep and lasting friendships. A typical pattern for foreign students is to establish a network of *colegas* and to socialize outside of school and create friendships with a few individuals who become more than *colegas*. Later, such students may be invited to the homes of their friends and eventually may be treated as members of the family. Similarly business is typically conducted in restaurants, and it is a sign of honor and trust to be invited to a business colleague's home.

The transition of Brazilians from their parents' home to their own home also consists of many small steps. Family is the most important institution and, hence, greatly influences the individual. For instance, the dating process is regulated by the family, although much less so than in previous generations and in the urban areas. Even though times are changing, many young adults leave their parents' home only when they marry.

As in many cultures, relationships between men and women at work are complex, and what would be considered offensive behavior, even as the basis for a lawsuit in the United States, takes on a different meaning in Brazil and other Latin and non-Latin cultures. It is still common for a man to compliment a woman in a way that would be problematic in the United States and common—but much less so—for men to make remarks to and about women that only they consider humorous. Frequently, however, the tables are turned, and the man's worst fear is to have a woman make an effective retort, particularly in a group setting. Still, while these patterns of communication are prevalent, they are counterbalanced by the conservative Brazilian perspective on marriage and family relationships, as noted previously.

The notion of extended courtship is found in business relationships as well. When conducting business in Brazil, people usually first establish a solid friendship. Only after a committed relationship is in place can a deal be struck. A typical piece of advice is "Be prepared to commit long-term resources of time and money to establishing strong relationships in Brazil. Without such commitments, there is no point in attempting to do business there at all" (Morrison, Conaway, & Douress, 1995, p. 44). Hence, it is clear that patience and the willingness to cultivate solid personal ties through small steps are crucial to the success of business negotiations in this country.

Brazil is generally a patriarchal society where the father of the family is the final decision maker for many important decisions. As they age, older members of the family also are accorded respect and are well taken care of by younger family members. Hence it is not surprising that in Geert Hofstede's study (2001) of 53 countries, Brazil ranks 14th in power distance. However, the more recent GLOBE study of 62 national societies (House et al., 2004) ranked Brazil low on the value placed on power distance (54th), even though it ranked 24th on its practice. In terms of business dealings, there are many layers of authority, and foreigners often must take many steps to find the person who can make things happen.

Also, compared to U.S. Americans and Japanese, Brazilians are much more likely to use commands and say no in negotiations (Adler with Gundersen, 2007). The use of commands can be traced back to the extensive military rule experienced by the people of this country. This behavior is, thus, a small step away from the historical political system. The excessive tendency to say no is also a way of keeping negotiations from forging ahead too quickly, behavior that again demonstrates the need to proceed in small steps.

Physical Touch

Through the on-again, off-again military rule and its continued but declining influence, there is a definite feeling that the elite hold control over the general population. In fact, modern society contains strong vestiges of colonial times, when the elite group controlled nearly all of the wealth. Brazil has one of the widest disparities of wealth worldwide: The richest 10% control more than 50% of all wealth and the poorest 10% less than 1% but, as indicated previously, poverty has declined significantly since 1992 and there are vast regional differences.

Generalizations about national cultures must take into account regional differences, and Brazil is no exception. Many of the behaviors such as small steps and circularity apply less well to some areas, such as São Paulo, where business is conducted in typically Western fashion, and people are so organized that they will even line up to get onto buses. Still, São Paulo suffers from wide differences in wealth, even to the extent that wealthy businessmen use helicopters to travel between their places of business and their gated communities to minimize the chances of becoming crime victims.

The uncertainty about the future brought about by the small steps and circularity of this country's history also places stress on Brazilians. Brazil has been termed a big emerging market for quite some time and now is known as a BRIC nation. The long wait for the awakening of this sleeping giant has instilled a sense of tension in the population, which has developed means to deal with this oppression and uncertainty on a daily basis. A dominant means is the reliance on personal relationships and human warmth. We explore this idea further by analyzing the samba with regard to its characteristic of physical touch.

The samba is normally danced in close proximity to other people, and the movement is strenuous and continuous to keep up with the samba beat. The very history of the samba is indicative of the importance of physical touch. Reputedly the samba is the descendant of an Angolan fertility rite that features the "violent bouncing together of belly buttons" (Krich, 1993, p. 73). The *umbigada* ritual is thought by some to be another influence on the modern samba. This ritual is conducted in a circle, with people given the chance to dance by a thrust of the pelvic region. The word *umbigada* is derived from the word *umbigo*, meaning belly button. Hence the connection between this ritual and the former is clear. Also, the Ngangela word *kusamba* can be defined as "to skip, to gambol, to express uninhibited joy" (Krich, 1993, p. 73). Thus the importance of physical touch and human warmth is well captured in the metaphor of the samba.

Family Ties

This need for human contact is best exemplified through the Brazilian family. As noted earlier, family ties are of paramount importance in this country. Because children often do not leave their parents' home until they are married and then frequently establish their new homes close by, family members are usually found within a defined geographic area. This living arrangement allows family members to communicate with each other in person on a regular basis. The most common way to reaffirm family ties is the Sunday meal. Normally the head of the clan hosts this meal, and all of his children and grandchildren congregate at his house to share conversation and good food. This idea is conveyed through the following quote, "We can't forget that with the Portuguese cooking we inherited the taste for a plentiful table, for varied foods, and the pleasure of eating with company, friends and family—the basis of Brazilian hospitality" (História do Brasil, 1971, p. 121).

A natural progression of the strong family unity is the extensive use of nepotism in the business world. Family members take special care to make sure that all relatives are provided for. Unfortunately many employment opportunities are given to family members regardless of their abilities to perform the job correctly.

If their family structure is weak or nonexistent, Brazilians are at a great disadvantage, given the importance of social networks based largely on family ties. An extreme example of being outside of such social networks that can protect individuals is the killing of orphans and sale of their body parts, fortunately a relatively rare occurrence. Similarly, the police sometimes act in a corrupt fashion, even to the point of killing civilians with little or no justification. A few years ago there was public outrage when a judge did not punish two upper-class young men who poured kerosene over a lower-class Indian who was demonstrating on behalf of his people. The judge argued that the young men were sensitive to others, as shown by the treatment of their dogs.

Still, there are some mechanisms, and increasing support for them, to increase social mobility, including the *vestibular*, although children of wealthy parents can afford tutors and better schools, thus enhancing their chances of getting into college. Brazil recently passed its first equal employment opportunity legislation, similar to the law in the United States. Currently, 53% classify themselves as white, 6.2% as black, 38% as "mixed," and 2.8% as either indigenous or no designation.

The Self and Collectivism

Still, it is important to remember that individual Brazilians have a sense of equality and the freedom to pursue their dreams. Throughout history people from different races were allowed some degree of religious freedom and had the right to procreate with whomever they chose. In business negotiations, for example, Brazilians are much less likely than Japanese or U.S. Americans to use a normative appeal in which people use societal norms to prove their point (Adler with Gundersen, 2007). Hence, Brazilians have a definite sense of self within the close-knit group. This notion of collectivism in the face of the general liberties granted to the individual is well represented by Hofstede's (2001) rating of Brazil along this dimension. Because Brazil is precisely in the middle of the individualism-collectivism continuum among the 53 nations, it is not surprising to see such a contrast in ideals. Similarly research has demonstrated that Brazilians expressed much more concern for the outcomes of others than did U.S. Americans (Pearson & Stephens, 1998). While the GLOBE study (House et al., 2004) ranked Brazil 52nd on institutional collectivism, it also ranked Brazil as second on the value placed on it.

When conducting business, it is important for U.S. Americans to remember that although their Brazilian contacts may seem to be similarly individualistic, they are linked to many groups and often place the needs of those groups, particularly their family, ahead of their personal goals. It is not uncommon for businesspeople to cancel a meeting or arrive late to a meeting because they needed to take care of a family issue.

Similarly, when negotiating with Brazilians, simply forging ahead to discuss the business at hand is not appropriate. Getting to know one another is very important to Brazilians. Because personal relationships are so important, asking after the well-being of family members is generally a good place to start. Like any other relationship where personal time and care are important, a carefully cultivated business relationship with a Brazilian can yield long-term benefits to both parties.

This notion of physical contact, closeness, and human warmth is also evident in more specific negotiating behaviors. Brazilians have a much greater tendency than U.S. Americans and Japanese to engage in facial gazing and touching (Adler with Gundersen, 2007). Facial gazing is probably due to the desire of Brazilians to read the expressions of their business partners and gain knowledge about their personality, and the use of physical touch is often simply a reaffirmation of the relationship and a transfer of concern for the other person. "A normal conversation between two Brazilians generally takes place somewhere between 6 inches and 12 inches apart" (Morrison et al., 1995, p. 34). In the United States the typical space between two people is 2 feet or more.

Undulation

The characteristic of the samba described as undulation, a distinctive moving of the hips, also provides insight into the Brazilian culture. Undulation is an extension of the idea of circularity described earlier. Where circularity may result in either traveling a lengthy path to a solution or reaching a dead end, undulation most often results in taking a shortcut to a solution.

In Brazil it is common to get things done by circumventing rules and obstacles. Accomplishing a task in the face of the huge amount of bureaucracy found in this country is literally done through the art of dancing around things, as the metaphor of the samba suggests. This art can be dated back to the founding of the country. As Krich (1993, p. 169) points out, many of the original Portuguese were underdogs or bandits who were suddenly given land and titles. The behavior of these people, now in high places, remained entrenched in what they knew best, how to survive on the streets. More often than not, their survival depended on personal connections and navigating their way around societal norms.

It is particularly interesting to note a popular myth in Brazilian culture that exemplifies this ideal of ignoring rules. The protagonist of this myth is the Saci, a short black man with one leg who smokes a pipe and dons a red cap. There are many stories about this Saci, all with the same central theme of mischief. This character generally plays pranks on people and gets away with it. For example, in one story, he goes to a farm and, because of his presence, the milk sours. According to Solomon (1996), a myth is "a story containing symbolic elements that expresses the shared emotions and ideals of a culture. In this way, a myth reduces anxiety because it provides consumers with guidelines about their world" (p. 543). Hence it is not surprising that in many social circles in Brazil, an individual who succeeds through mischief is seen as a hero or as a worthy successor to Saci.

The Brazilian penchant for breaking rules and reaching goals outside of proper procedures is evident at traffic lights, which are frequently disregarded. However, stopping at a red traffic light can be dangerous, especially at night, for drivers are frequently accosted.

The Impact of Status

Another way to find a way around the rules is through using status. Because Brazil has a hierarchical structure, respect must be paid to those of a higher status, as evidenced by a standard response, "Do you know to whom you are speaking?" This response is intended to point out the status the individual wishes to invoke as justification for contending that the law was not intended for such a special person (Schneider, 1996, p. 193). Again, the speaker demonstrates the importance of personal relationships and who you know. An effective personal network can frequently get people much farther in accomplishing their goal than following the law. Even the 1988 constitution, which restored democracy after two decades of military rule, did not abolish the culture of *cordialidade*, that is, the primacy of personal bonds over rules (Unger, 2007, p. 4).

This type of personal bonding, when not invoked by rank, is called a *jeitinho*. To give a *jeitinho* is loosely translated as enlisting a fellow human being to help one circumvent the rules due to one's personal circumstance. In general, this notion is closely connected to the Brazilian belief in the equality of individuals and in preserving the human relationship in the face of the extensive bureaucracy. "From this perspective, the *jeitinho* is a flexible way of dealing with the surprises of daily life, a way of humanizing the rules that takes into account the moral equality and social inequalities of persons in the society" (Hess, 1995, p. 40). In this instance pulling rank when asking for a *jeitinho* will surely backfire on the individual asking for a special exception.

For instance, in dealing with the Brazilian embassy, travelers can ask the clerk if they could make a *jeitinho* and process the paperwork sooner. Travelers might say that they need to return to Brazil as soon as possible to be with a relative who is seriously ill. This appeal to the clerk's compassion and strong family values will tend to facilitate the process. But to accompany this appeal with any display of superiority will likely result in the denial of the request. Engaging in the *jeitinho* ritual allows Brazilians both to bypass impersonal rules and to exert a certain amount of control over their daily lives.

Navigating the Bureaucracy

In purely business transactions, asking for a *jeitinho* may not be enough. In this case business persons might hire a *despachante* or agent to aid in the navigation of Brazil's numerous and complicated bureaucratic rules. Twenty five years ago the extensive bureaucracy inherited from the Portuguese was simplified to a certain extent, and the need for such a service has diminished somewhat. There are, in fact, whole offices composed of *despachantes* whose work is devoted to wading through regulations for their clients. Furthermore, Brazilians have such little faith in the written word, in particular in personal signatures, that every conceivable document has to be notarized. Thus the usefulness of the *despachante* becomes evident. Ironically, it is easy to file income taxes; Brazil had a Web-based electronic system in place by 1995, far ahead of many nations.

The structure of the *despachante* business is somewhat complex. There is often a head *despachante* who takes leadership over a person's individual request. However, the head *despachante* has several layers of subordinates who perform the more tedious tasks. One such task involves waiting in line for hours to get a particular document signed by the appropriate person. Hence the very system that has evolved to aid an ordinary citizen in navigating the Brazilian bureaucracy has become quite cumbersome in its organization as well. However, wealthy U.S. Americans and business firms hire temporary workers for the same purpose.

The notion of circumventing rules can also be found in issues that affect international relationships. In the arena of intellectual property rights (IPR) Brazil has a very poor record, even though it is a signatory to the Paris, Berne, and Universal Copyright conventions on IPR and is in agreement with the GATT Trade-Related Aspects of Intellectual Property accord. In fact, in terms of software piracy, for every legal copy sold in Brazil, about four illegal copies are made. This is typical of an underdeveloped nation with an inefficient bureaucracy. Here, again, the degree of collectivism in the Brazilian culture, coupled with the importance of personal relationships, encourages this behavior. While some outsiders may view Brazil's emphasis on humanity in a positive light, others think otherwise. Former French President de Gaulle, for example, said of Brazil, "This is not a serious country." It is evident that Brazil must alter its behavior so that it can be seen as a legitimate business partner in the global environment. Dancing around the obstacles that numerous and complex regulations create has become so ingrained in this culture that seemingly only the need to conduct international business will result in Brazilians conforming to more universal norms.

Spontaneous Escape

Spontaneous escape represents the fifth characteristic of the Brazilian samba. As might be expected, Brazilians tend to get lost in this notion of escape and lose track of time. Frequently they do not strive to avoid the uncertainty of their lives, but too much uncertainty motivates them to escape tension through the music and dance of the samba. For Brazilians the samba represents this notion of a spontaneous escape from the everyday reality of unemployment, low standard of living, and poverty that affects many Brazilians, as noted previously.

Yet, on a closer look, the country is still smiling. Why? Perhaps it's because they can dance the samba and, for a brief moment, can be the center of attention in a world full of freedom. In a country that was under military rule for many years and plagued with a distrust of government and widespread bribery and corruption, escape is something many Brazilians crave. The samba is an escape because its music and dance sweep people away, and they lose track of time and the environment. It is no wonder that Brazilians are described as being polychronic. Again, however, in such a land of diversity and contrasts, this generalization must be tempered. In the business-like south, for example, the expectation is that participants will arrive on time and even early for a business meeting, and southern Brazilians generally frown on not completing a work assignment on time. Such expectations represent monochronic behavior.

As noted earlier, the samba came to be seen as synonymous with carnivals, the most well-known of which occurs in Rio. Throughout the world most carnivals last one day, but in Rio, the Carnival extends over 4 days. Each of the 16 samba schools works steadily over the year preparing its float for various parades and competitions, and the final parade occurs in the Sambodromo, a one-of-a-kind parade stadium that can hold 60,000 people. Various theories are used to explain the uniqueness and popularity of the Brazilian Carnival. One prominent explanation emphasizes the escape from daily drudgery and the spontaneous and unexpected activities during which members of different social classes mingle easily with one another at parties and parades.

The spontaneity of the Brazilian culture can be seen at times within its institutions. One Brazilian remembered her college years in Brazil by describing situations when students would just spontaneously take whatever was in their reach and start invoking the samba beat. For example, a pencil may be tapped on a desk or a ruler slapped on a chair. Many Brazilians can describe their school days with stories in a similar manner. For this reason the idea of samba as a cultural representation of Brazil becomes even more evident. This idea of spontaneity and escaping reality through the dance and motions of samba is taught to all Brazilians from their early years of education through their university graduation.

Soccer or what most of the world calls football is yet another cultural aspect of Brazil that reflects the spontaneity of the samba. In fact the Brazilian style of play is called "samba soccer" and it is characterized by flashy, complicated, and high-risk offensive maneuvers. Brazil has won more world championships than any other nation, and arguably the greatest soccer player of all time is the Brazilian, Pele.

Rich and Poor

This need for escape among the people of Brazil is also evident within the economy. Despite significant displays of prosperity, Brazil's largest cities are overpopulated and home to some of the poorest people in the world. In fact, Brazil is the country with the most *favelas* or ghettos in Latin America, and escape at least for these Brazilians is essential to their endurance and survival.

This idea of escape is very much a part of the Brazilian culture and language. One of the most famous phrases that Brazilians use, which best sums up the Brazilian civilization, is *tudo bom*—everything about life turns out to be good in some way or another. This phrase is illustrative of the Brazilians' attitude toward life and the negative twists and turns that are encountered in daily life. Brazilians are a people who have endured much political and economic hardship, but they have frequently emerged from difficult situations dancing . . . the samba. One Brazilian demonstrated this phenomenon daily at the office when she would use the phrase *tudo bom* even when she was having a bad day. It was her belief, as with most Brazilians, that disappointment is a fact of life; only by maintaining a positive attitude will everything eventually work out.

In terms of business practices, U.S. managers trying to do business in Brazil should realize that meetings will not always start or finish on time, that agendas may not be followed, and that requests may not be granted right away. As mentioned earlier, time is less important to Brazilians than to Americans, and schedules are frequently viewed simply as a formality. Again, however, this is much less true in the south than in the north.

Conventional U.S. rules of behavior in business settings, for example, negotiations, are virtually nonexistent. One of the strongest differences between communication behavior in Brazil and in the United States is the number of silent periods and conversational overlaps (Adler with Gundersen, 2007). Brazil tends to be a more polychronic than monochronic society, and Brazilians tend to be a spontaneous people. For example, Brazilians typically feel that their opinions are most important, and they speak when they feel like doing so, even to the point of interrupting others abruptly.

Furthermore, when negotiating with Brazilians, being aware of the culture's relative rankings in terms of collectivism/individualism is important in understanding why Brazilians have many conversational overlaps in their negotiations. Brazilians tend to interrupt each other constantly, perhaps reflecting the need for individual expression for the good of the whole. Brazilians are very open to negotiations, and nothing is considered absolute. This concept reflects the need of Brazilians to be individualistic and not to conform.

Mixing Business and Social Life

In addition, given the collectivist orientation of Brazilian society, it is not uncommon to see a high correlation between people's personal and work lives. For example, major firms in Brazil help their employees out with their financial problems and are very accommodating to family illnesses and other personal needs.

Another aspect of conducting business in Brazil that relates to the spontaneous escape of the samba is the need Brazilians have to socialize at lunch with potential business partners. Specifically, as in most Latin cultures, Brazilians do not feel life is all work and no play. The lunch setting is the typical Brazilian businessman's escape from an often frustrating work environment. So when attending a business lunch with a Brazilian, one should accept small talk and non-business-related conversations and adapt to it. Managers should also be open to spontaneous questions about their hobbies and families. The business lunch in Brazil is an opportunity for potential business partners to get to know each other and not necessarily to talk about the bottom line. Brazilians take their meal times very seriously and cherish the chance to get away from business issues. Typical business lunches in Brazil can last anywhere from 1½ to 3 hours. For U.S. Americans hoping to form alliances in Brazil, it is helpful to recognize and respect this need to escape as part of the Brazilian culture.

Moreover, people hoping to conduct business in Brazil might be most comfortable if they realize and appreciate that Brazil is a country with few clear-cut rules. For example, in Brazil, behavior such as reneging on a contract is treated as ethical when the person is seeking to protect long-term relationships not covered by the contract. For many Westerners this type of behavior would be considered unethical and unacceptable. When doing business with Brazilians, it is useful to understand their way of going around things instead of confronting them . . . literally dancing around things. Thus, for most Brazilians, the samba represents a stable factor over time, amid changing currencies, inflation, and political parties.

Paradox of Dancers

It is not unusual for Brazilian factory workers to spend a month's salary on the costumes they will wear at the Carnival parade. With friends from their samba school, with whom they have been rehearsing the entire year, they will dance on top of a gigantic dragon during the parade and experience a short but heavenly sense of existence. Such experiences portray very well the influence and impact of samba in Brazil. Samba schools in Brazil started more than 70 years ago and are as much a part of regional identity as soccer teams. Such behaviors, however, best illustrate the paradox of the dancers of the samba.

As the story demonstrates, in Brazil the dancers of the samba are full of contrasts. When preparing to compete with other samba schools or to dance samba in Carnival, Brazilians escape to another world where they can be anything they want. For that one moment the samba is their ticket to paradise. For example, samba costumes can cost up to $4,000 (a year's savings for slum dwellers). This figure is probably at the higher end; most slum dwellers, while spending a good amount of money, do not spend their entire year's salary on a costume. For some, however, "their investment pays off by bringing to life a daydream appropriate to the ancestors of slaves . . . they for a day become a friend of the king" (Krich, 1993, p. 69). This paradoxical way of living is seen in many aspects of the Brazilian culture.

As Krich (1993, p. 80) points out, whole samba parades are constructed as a tribute to an obscure, bygone composer, a forgotten slave revolt, the history of rice cultivation, or even the glory of the banana. This contrast between pleasure and pain can be seen throughout the Brazilian countryside. Eugene Robinson (1997) identified the city of Salvador da Bahia as the pulse of Brazilian music. In this city the contrasting images that Brazilians carry within them become most evident because of the activities taking place in the central plaza. The plaza is a tourist mecca where travelers pose for pictures and dreadlocked entrepreneurs sell local crafts; years ago, this same plaza was the site of the public torture of slaves.

Contrast and Paradox

Historically Brazil has been a country of contrasts and paradoxes. As noted earlier, Brazilian people come from various social and economic backgrounds and races, and few nations can match its diversity. Although many experts believe people of European descent have made the greatest contribution to the physical appearance of Brazilians, they also believe that the extensive interracial marriages throughout the history of Brazil have created a population so genetically complex that Brazilians constitute a new and distinct people.

The people of Brazil can easily confuse outsiders. Although most Brazilians look "black," at least to North American eyes, only about 6% tell census takers they consider themselves black. This characteristic of the people of Brazil is also illustrated through an excess of common terminology describing *mestiços*, or mixed races. Brazilians use numerous and creative metaphorical names for describing the various skin types found in Brazil. For example, they refer to very dark African types as *cor do carvão* (color of coal) and to dark-skinned people occupying important positions in society as *branco da terra* (white of the earth). One household survey indicated that respondents employed an astounding 136 terms to describe individuals along the black-white continuum ("No Black," 2006, p. 38). These results may seem surprising because Brazilian society has been successful in combining many racial and ethnic elements in its art, food, and music. Although these differences in social status exist, there are times, such as when dancing the samba, that the overall feeling of the people is "We are all Brazilian." This slogan identifies the overarching feeling of Brazil as a culture and perhaps even an ethnic unit.

Still, the issue of race is extremely complicated in Brazil. There is a good amount of income inequality, and much poverty is concentrated among those with darker skin pigmentation, even though Brazilians of all types tend to interact more easily in daily life than their U.S. counterparts. Eugene Robinson (1995, 1999), an African American journalist, attributes the difference between Brazil and the United States to the "one-drop" theory prevalent in the United States, that is, one drop of African blood leads to a classification as black. As a result there is a wide "black versus white" gap in the United States, substantiated by demographics. In recent years, African Americans represent the only ethnic or racial group that has not had a significant increase in out-group marriage, which typically leads to cultural integration.

In Brazil the opposite theory was prevalent, that is, one drop of white blood lightens, making the person white. Hence it is popular to use the term "Brazilian rainbow" but not "American rainbow," even though gradation of skin color is a point of discussion among African Americans. Because of this indeterminacy of skin color, some observers argue that social class is much more important than race in Brazil, and even more important in Brazil than in class-conscious England (Bond & Smith, 1998). Ironically, scientists now argue that race is a useless and superficial classification, given the high rates of miscegenation throughout the world. To complicate matters, it should be noted that there are large numbers of other ethnic groups in Brazil. In fact, few nations have a population more diverse than Brazil, which includes people of Native American, European, African, and East Asian origin of both pure and mixed blood. Brazil has not only the largest concentration of blacks and their descendants outside of Africa but also the largest concentration of Japanese and their descendants outside of Japan. All or almost all of them identify primarily with Brazil. For example, 100,000-plus Japanese Brazilians work in low-level jobs in Japan because of the lack of jobs in Brazil, but virtually none speak the Japanese language, and they are shunned by the Japanese. Similarly, German Brazilians celebrate Oktoberfest in the traditional German manner until midnight, when Brazilian music like the samba transforms the event. Commenting on such phenomena, *The Economist* ("Eine Kleine," 1995) argues that "The Brazilian rainbow," even with its defects, should be studied in depth: "Brazilians are remarkable more for what brings them together than for what keeps them apart" (p. 49).

This ironic unity is also reflected in the Brazilian cuisine. Many traditional dishes, such as seafood stews, replete with palm oil and coconut milk, have been heavily influenced by the Africans. The arts and crafts, particularly the wood carvings, look as though they could have been produced in Senegal or Ghana. These examples are part of the everyday surroundings of Brazil.

The contrast between rich and poor appears in almost every aspect of Brazilian culture. For example, many cities in Brazil are a paradox of modern-day skyscrapers and industry and historical pathways as represented by colonial churches and villages. It is not uncommon to see luxury apartments and houses at the foot of slum hills, or *favelas*. The government also reflects this paradox of dancers in that its business is half privatized and half government-run, again placing Brazil in the middle of the individualism/collectivism continuum.

The Role of Religion

In addition, religion illustrates the paradox or contrasts of the Brazilian people. The pervasive religion in some parts of Brazil is a syncretism of Roman Catholicism and African cult worship, featuring a pantheon whose origins can be traced directly back to the continent of Africa. In Brazil there is great tolerance for contradiction. The many religions of Brazil reflect this paradox also in that numerous witchcraft and cult religions have borrowed symbols, such as the crucifix and various saints, from Catholicism. Hence there is an intermingling of religious components where conservative traditions of the Catholic faith are adapted to ideas of witchcraft and supernatural powers. In the census, 73% of Brazilians identify as Catholics, 15% as evangelicals, 5% as other religions, and 7% as of no religion.

In Latin America the estimated number of evangelical Protestants increased from 5 million in 1960 to 60 million in 1997. To counter this trend, Brazilian Catholic priests now rely on rap music, sambas, and related music as a means of attracting Catholics back to the fold (Faiola, 1997).

As a final statement on the paradox of Brazil and its people, we can look to the Brazilian flag, which has a large yellow diamond on a green background. The color green stands for the lush fields and forests of Brazil. The color yellow represents its wealth in gold, which is found in many areas of the country. In the center of the yellow diamond there is a blue sphere that symbolizes the usual navy blue sky that one finds in the tropical areas of the Earth. In that sky are stars that represent the capital of the country and its federal states. In the middle of the sphere is a white banner with a legend, *Ordem e Progresso,* which means "Order and Progress." This description of Brazil by the Brazilian people is a paradox in itself. Although Brazilians take pride in their abundant national resources, these resources such as the Amazon forest are actually being eroded by current business practices. Furthermore, the notion of "Order and Progress" is quite contrary to how the country has largely been governed in its colorful history.

The paradox of the dancers of the samba has very important implications for businesspeople wanting to work with Brazilians. As we have illustrated above, Brazilians are a melting pot of people with characteristics that are not easily categorized or, for that matter, understood. When trying to work with Brazilians, it is helpful to remember their rich history and cultural triumphs. Patience is the key virtue in forming business partnerships in this country, although Brazilians are very hospitable to visitors, going out of their way to ensure that they are comfortable. Still, sensitivity to class differences and various special-interest groups is necessary when forging political alliances and creating strategic advantages. Brazil is a country of many differences, yet it can be united through dancing the samba for a common cause. For this reason it is important to understand the units in the Brazilian population to become comfortable with its economy, society, and culture (people).

In sum, Brazil is a vast nation in which there is great diversity and contrasts. Still, as the opening quote for this chapter suggests, most if not all Brazilians regard the nation and its culture as special and extraordinary. Despite the significant changes that must be made in the Brazilian infrastructure and regulatory environment, even though there has already been significant progress in recent years in these areas, Brazil will continue to be diverse, complex, and mysterious, and the samba will continue to express these cultural themes in new and exciting ways.

CHAPTER 7

The Polish Village Church

Swept by converging streams of economic, political and social events, Poland has been affected by all. Yet she has remained stubbornly unique . . . she has absorbed much, resisted much, and remained very much herself. In many respects, Poland presents a remarkable example of continuity not often met within history.

—Sula Benet (1951, pp. 26–27)

Turn off the main road anywhere in Poland, drive for a bit, and you will find yourself in a different country. The roads become worse, sometimes abominably so. The one-storey log cabins and horse-drawn carts of Poland's rural past make an appearance. . . . Nearly 15 million people, over a third of the population, live in the countryside.

—Edward Lucas (2006, p. 12)

The first-time visitor to Warsaw, Poland, is frequently surprised by its modernity, as exemplified by the buildings in the central business district, many of which could easily be mistaken for buildings in U.S. cities. At the very modern Marriott hotel, businesspeople are constantly on their cell phones and Blackberries or in intense discussions making deals. A few older buildings are left over from the Communist era, but not many. Although Warsaw was largely destroyed in World War II, visitors would have a difficult time finding remnants of this destruction and of Poland's past.

Yet Poland, with its population of nearly 39 million, is still a largely traditional society consisting mainly of small towns and villages. About 19% of the population is engaged directly in agriculture (and it contributes only 3% to Gross National Product), at least a third of the population resides in rural areas, and others live in towns close to the major cities of Warsaw, Poznan, and Gdansk. Moreover, many of the Poles in the large cities come from traditional rural areas. Currently this central European nation is making the exhilarating but disquieting and rapid transition from traditional to modern society.

In 2004 Poland became a member of the European Union (EU). Its economy has since grown, but not as much as its Central European counterparts such as the Czech Republic and Estonia, which were also admitted to the EU. Corruption, excessive government bureaucracy, and high tax rates represent some of the reasons why this state of affairs exists. The employment rate is 51% of the working-age population in comparison to the 60% average for its Central European counterparts in the EU. This has motivated Polish workers to seek work in other more prosperous EU nations. Since 2004, it is estimated that between 600,000 and 2 million of Poland's citizens have migrated, and they send billions of dollars home. This migration is particularly troubling because Poland needs the workers' skills and abilities as it makes a full transition to a prosperous capitalist economy (Fleishman, 2007).

Given Poland's long and tortuous history, briefly described below, it is understandable that the Polish people are characterized by contradictions held together by a strong social system and a sense of shared adversity and experience. The Polish people are guarded in their manner of dealing with each other or outsiders and in their trust of institutions and people. Despite this guardedness they possess a richness and depth of spirit that has helped them persevere through many shared adversities. Poland, in the years following the collapse of communism in 1989, is successfully struggling to find a place in the world as a free country and a self-determined people. The path Poland is now taking is very much influenced by the nation's peasant roots, a history of foreign power domination (including communism), and Catholicism. These three forces shape the image the Polish people have of themselves, their behavior, and their institutions.

The village church, found in virtually every Polish town, stands as a symbol of these three forces and presents a particularly appropriate metaphor for Polish culture. In each village a church sits in a prominent location and serves as the focal point of town activity. The churches are the beneficiaries of much of the village money and attention. Some are as old as 500 years and others were built since the collapse of communism. Church events are considered compulsory and failure to attend is not viewed favorably. During holy periods, such as Christmas and Easter, residents attend the many church services held every day of the week. For services like the Easter Resurrection ceremony, held on the night before Easter, churches are packed, and people overflow into the street. Men often congregate outside the church as the women, children, and the elderly find seats inside. The men and older boys gather around the entrances to listen. Church personnel have adapted to these circumstances by sending a priest outside the church to administer communion at the appropriate time. Like the Malaysian *balik kampung* (Chapter 18), the village church is important as a focal point for Poles; pilgrimages back home for religious holidays are common. The church represents a safe haven from the forces of history that have shaped and at times attempted to destroy Poland.

Historical Background

Whenever Poles want to explain some aspect of their work, they start by talking about Polish history. They are sensitive about Polish history and, as Edward Lucas (2006) notes, "prickly to a fault" about it, especially those who are older. History has played an important role in shaping the Polish way of thinking. The history of Poland is largely the history of peasants who were dominated by exterior groups (e.g., wealthy landowners, foreign powers, and communist leaders).

From the medieval period (6th century) through the late 19th century, the people who farmed the land were predominantly serf labor. Emancipation and refeudalization occurred several times during this period, which witnessed a continuing struggle between peasants who sought rights and freedom and landowners who secured legislation to force the peasants to stay on their land so that they could exploit their labor for immense profits. The Polish bourgeoisie were perceived as a threat to these wealthy landowners. Landowners, therefore, encouraged Jews, Germans, and Armenians to immigrate to Poland and take technical jobs because the foreigners' economic prosperity did not pose a political threat.

As the bourgeoisie diminished in size, the population of Polish towns declined and the power and tax base of the king fell. Landowners during this period became increasingly tied to specific German and Dutch bankers for capital. This left the landowners relatively powerful while they developed an increasing dependence on the core of the European economic system (northwestern European states). Some peasants during this period owned the land they worked, but they remained primarily subsistence farmers. To a great extent the technology used in Polish agriculture remained primitive because of the ample supply of serf labor.

During the 16th century Poland was large and powerful; the Polish state during the 16th through 18th centuries was run by a parliament that consisted of members of the Polish gentry, who selected the king. These wealthy landowners successfully maintained their independence and wealth through their employment of the *liberum* veto, a procedure giving any member of parliament the power to defeat any legislation with a single vote. Needless to say, few meaningful actions occurred. This procedure helped to maintain a balance in Polish politics and defined the strong individualistic values of the Polish gentry. It did, however, prevent Poland from amassing strong collective behaviors required to fend off foreign invaders through political means.

This weakened state and dependence on foreign capital allowed Poland's relatively stronger neighbors—Russia, Prussia, and Austria—to partition Polish land beginning in 1772 and to assimilate it into their own borders. This partitioning of Poland ushered in an era of domination by these three great powers that lasted through the end of World War I. The result was the complete elimination of the Polish state. During this period Prussia, Russia, and Austria broke the power and wealth of the landowners. In addition, pressure throughout Eastern Europe forced the abolition of serfdom and redistributed land to foreign parties and to the peasants themselves.

Poland was formally reconstituted as a state in 1918. From then until 1939, Poland was a free country, and the distribution of land ownership to the peasants (land reform) was a primary political objective. A soaring population growth rate, coupled with a tradition of inheritance in which land was divided among all living children upon the father's death, led to the economic failure of land reform. The process resulted in lower production and lower profits from agriculture during the interwar period. Germany and Russia invaded Poland in 1939, marking another break in its independence as a nation.

The Communist Period

The aftermath of World War II brought a division of the spoils among the victors, namely the Western allies and Russia. Devastated by war, during which about 30% of its population died, Poland became a pawn in this process. Communism was adopted as the official state ideology in 1948, and despite its lack of general support among the Polish populace, Soviet government support for the Communist leadership made it impossible to sustain significant resistance. Economic rebuilding and a change in governmental priorities prompted the postwar Polish leaders to divert resources from traditional sectors such as agriculture to industry (manufacturing, mining, service, and construction). Inadequate and unsupported reforms, however, produced a worsening economy and growing social discontent.

Collectivization of farming, a key tenet of communism, took place in Poland in the years 1949 to 1956. At its height, however, only 8.6% of total cultivated land was collectivized. A powerful peasant commitment to the land and often to a specific plot of land proved to be the reason for the failure of collectivization.

For peasants, the farm was as much a social as an economic unit. Its output was produced for the satisfaction of family needs. Market pressures running counter to family consumption needs and demanding production increases were often perceived as oppression. The reaction to these pressures was often to reject increased production and to produce only enough for family subsistence. Government agricultural policies had succeeded in reinforcing the essential features of the peasantry. This fact gave the peasants power that could significantly disrupt the government systems, and they were seen as a threat.

As the state required more and more agricultural production, the peasant farmers tended to resist through delaying tactics, highlighting their power over the government system. The government responded with greater pressure to collectivize, thus creating a vicious downward spiral of lower levels of production. In the 1950s this process hit a crisis point when peasants lost their pride of ownership in their farms and became reluctant to improve or even to maintain them, resulting in low productivity and workforce apathy. Eventually the Polish government repealed its policy of collective farming in Poland.

Other policies were also met with strong resistance. The Pole's basic distrust of government systems produced an unwillingness to make sacrifices to turn the economy around. In addition, the socialist ideology on which the Polish social and political system was built altered the outlook of traditionally hard-working Poles, leading them to believe that it was the state's responsibility to provide for their households. When the state could not provide adequately for the Polish people, they became resentful. To compensate for the growing unrest, the government increased the workers' wages fivefold between 1979 and 1986. The Polish government was close to proclaiming state bankruptcy when it declared martial law in 1981 to quell the growing unrest and the calls for reforms by the Solidarity movement.

Solidarity and Independence

In 1989, spearheaded by the Solidarity movement and prompted by the weakening political and economic position of the Soviet Union, roundtable meetings in Warsaw resulted in the peaceful overthrow of the Communist state in Poland and the beginning of significant economic, political, and social reforms that continue today.

It is against this backdrop that we explore Polish culture. The village church resides in the center of most Polish villages, exemplifying the central place that Catholicism plays in Polish life. In addition, the church's austere exterior and plush interior represent the contrasting nature of Polish identity, partially produced by the many years of foreign power domination and the partitioning of the Polish state. Finally, the village church endured and flourished during a period when religious worship was banned in Poland. Poles are very proud of this fact and the village church stands as a symbol of the ability of the Polish people to survive during extreme adversity.

Central Place of the Catholic Church

The village church sits in the middle of the village, often in a place of particular prominence, such as the crest of a hill. This placement signifies the preeminence of the church in Polish life, both in the past and in the post-Communist period. Although the power and influence of the church are decreasing among some citizens, it is difficult to overstate the importance of the Catholic Church in the public and private lives of the Polish people. Hann (1985) goes as far as to say, "Polish has come to mean Catholic" (p. 100). Virtually every major holiday is religious in post-Communist Poland. Public processions of clergy and lay people carrying religious banners and relics are the primary rituals in many of these celebrations. In these rituals, neighborhoods, youth, and Easter baskets are blessed. These occasions feature full Masses, sometimes lasting as long as 3 hours. In addition, Polish youth and adults celebrate "name days" rather than birthdays. These name days are set aside to celebrate the saints after whom many Poles are named.

The Catholic Church has played an important role for rural citizens in modern Poland. Parishioners continue to be a particular focus for the church. During the period of peasant emancipation at the end of the 19th century, the church played an active role in acquiring freedom and land for the peasants, even though the Polish state was not an independent entity and was under foreign domination. This championing of the peasant cause positioned the church perfectly to play an active role in the creation of national sentiment and it bound the Poles even more closely to the church. The result was that the parish became the focus for a collective Polish peasant identity.

The collectivizing influence of the Catholic Church is naturally accompanied by moral codes of conduct, including the belief in equality and fair division of resources and the church's sense of social justice. Also, the church placed importance on collectivity over individuality and on suffering rather than short-term gratification of personal needs. Sacrifice is seen as the best means by which Catholics can find salvation. This belief has spread beyond religious life in Poland and has permeated Polish views of national identity. It has given solace to a people who have suffered long periods of political and economic domination for centuries, including the relatively recent horrors of World War II. The search for explanations as to why they have had to suffer so terribly has generated a natural connection between Poles and the Catholic Church.

After World War II, Poland was unique among Communist nations in that the Catholic Church and the symbols of Catholicism were not outwardly diminished by Communist attempts to condemn and destroy them. In fact, commitment to Catholicism was probably strengthened by the condemnation, as were other aspects of Polish life. Polish peasants continued to abstain from meat on Fridays, to name their children after saints, and to marry, baptize, and administer communion in religious ceremonies. The Catholic Church also played an active role in the political life of peasants. It openly sanctioned much of the opposition to government policies in Poland. In addition, the church provided the structure to aggregate and articulate peasant interests, as a result of which the devout Polish peasants were able to collectively oppose communism through their religious affiliation.

This peasant-church connection is further manifested in the almost spiritual link that the Polish people have with their land, as demonstrated by the All Saints Day commemoration celebrated every November 1. This commemoration is an amazing cultural demonstration of respect for the dead that has turned into an event. Its religious meaning serves to connect the Polish people both symbolically and literally to their roots and to the land, as demonstrated by the actions of many Poles who have moved to the cities within the last few generations. Many of them travel back to their villages for this commemoration, and some of them to the villages of both sides of the family. City dwellers who have these agrarian roots usually choose to be buried in the villages from which they came, further preserving the connection and the importance of family lineage and agrarian heritage. In addition, the family plots establish a permanent link of generation to generation in the village, a lineage that is literally "planted in the ground."

The Partitioning and Polish Identity

The exterior architecture of the old and new churches alike is simple. New churches are large with brick or painted stucco exteriors. The old churches are rustic, made from meticulously matched split logs with steep curved metal roofs. They are simply adorned with wood crosses and rosebud-shaped steeples. The inside of these old churches is like the inner light of the Polish people: filled with color, attention to detail, wealth of spirit and surprise. But the church exterior hides the riches of golden altars, antique wooden carvings, and hand-painted walls. Still, the interiors of these village churches are not gaudy but simply elegant and rich with years of protecting the unfulfilled souls of the peasant people.

Like the contrast between the austere exterior and the rich interior of the village churches, the public and private lives of the Polish people are remarkably different and separate. This separation is in great part a result of the partition of the Polish state in 1772 and the resulting lack of Polish control over the Poles' own official business. Communism also contributed to this lack of control. It promised social peace and security in exchange for social control and private ownership of property. Poles desiring certainty and security after the perils of years of foreign invasions and rule, as well as the harsh consequences of World War II, entered into a social contract with the Communist government. The Polish people, however, chose to focus on the welfare of close family and friends as opposed to a larger group, as the Communist system required. The result of this social contract was an abdication of responsibility for much of the state of affairs that touched them directly. One example of this mentality is the upkeep of public property. In the socialist state, public property was theoretically the responsibility of everyone, but in practice, it was badly neglected. In this system, responsibility simply disappeared.

This does not mean that Poles do not care for their property. They do, in fact, take great pride in their simple possessions. Most Polish homes and apartments have a room where visitors are welcomed and in which the residents display prized possessions, special glassware, and religious symbols in glass-enclosed display cases. But exterior spaces are not viewed with ownership or pride and thus do not elicit particular care.

Attention to Appearance

Poles value private ownership of property and many village homes are "works in progress." Young men will begin building a home whenever money is available. Many Polish citizens working in other European nations send money home to support the construction of their homes, where they hope to live eventually. Houses are built in steps, and when one section is habitable, families will move into this space. Other floors may remain as open shells until more money is available to complete them or the expanding family needs the space. Grandparents and even siblings are likely to work together on a house, sharing their money and efforts and eventually the finished home. Cohabitation by the extended family is common in Polish villages as well as in the cities. Despite the strong influence of the Catholic Church, Poles in cities such as Krakow and Warsaw tend to limit their families to one or two children, whereas village families are likely to have three or more offspring.

Poles are proud people who value appearances, particularly as they relate to their clothing. During the Communist period, Poles did not have a good selection of apparel and had to settle for drab colors and poor styles. Today the individuality of Poles is reflected in their choice of apparel and the pride they take in wearing nice clothes. Young men and women in the cities wear designer fashions. While they may own only a few such outfits, they will wear them frequently and care for them meticulously. Older women can be seen on city streets in winter wearing a drab Communist-era coat with a brightly colored (purple or pink) hat. Village women wear on their heads traditional scarves, which differ in design depending on the region of Poland. Such scarves are also frequently colorful.

Despite the attention to their appearance, the face that Poles present in public is guarded and private, with smiles reserved for more informal and private moments between friends and family. People tend to mind their own business. In the cities Poles are reluctant to get involved in the affairs of others. For example, if someone falls in the street, Poles are unlikely to intervene unless it appears that there is serious injury. Older women are the rare exception to this norm. Women who reach the status of grandmother are only too willing to render advice, particularly to younger women with children. Children are the public domain, and Poles easily give advice on appropriate attire and discipline, as well as offers of candy or fruit. In addition, in the villages, involvement in the private matters of neighbors is quite common, but it is highly resented by those who are the subject of the attention. Still, offers of assistance, when made without invasion of privacy, are welcomed and usually serve to generate a strong bond between the parties, which will be remembered and reciprocated.

Ironically, Poles identify strongly with Poland but feel that the Polish people tend to be lazy and quarrelsome (Bond & Smith, 1998). Poles' suspicions about their neighbors' intentions are based on a strong and pervasive distrust in Polish society, which is often connected with accusations of bribery or favoritism. Bribery is an ever-present phenomenon in Polish life, a source of jokes, folklore, and resentment, and it is often the explanation for the "way things are." Payments on the side are still necessary to receive adequate medical services and even entrance into universities. For example, to get into many universities in Poland, prospective students must pass a particular entrance exam. To pass these exams, they must obtain special tutoring from university professors because the contents of the exams are a closely guarded secret. Some Poles call the tutoring fees a bribe to the university professors, while others see them simply as payment for services.

Bribery Customs

Bribery permeates many aspects of Polish life, including the workplace. For example, Polish workers often use the expression, "we came here on a rabbit," when bemoaning their particular lack of benefits or poor working conditions. This expression refers to refusal—or probably inability—to make some form of payment to their supervisor. Those who are favored and receive benefits or privileges are described as arriving "on a calf," the calf being the bribe to the new employer.

Bribery is just one manifestation of a system of privileges, power dispersal, and political access that is called *kumoterstwo* in the Polish language. *Kumoterstwo* is translated into English as favoritism, but the Polish word tends to encompass a great deal more in the Polish experience. Through a system of networks, the Poles place a heavy reliance on social contacts for a number of official or unofficial services or products. Unlike U.S. networks, which often involve occupation or profession, these networks are based, first, on familial ties and, second, on economic or social standing. Such networks may be a necessary organizational tool for rural Poles, but they tend to generate resentment and suspicion among those who do not belong. Monetary payments for favors are often required by the networks associated with power. Poles tend to view this system not as right or good but rather as necessary or a fact of life. While the specifics are usually hidden under layers of denial, the general existence of this bribery-based system is fair game for discussion and ridicule.

This form of social organization is a response to the difficulties associated with the Communist state. At the end of World War II Poles were subjected to a government that was ineffective and illegitimate in their eyes. Poles needed to find a way around the system. They relied on their traditional individualistic values and their dogged determination to guide their approach to dealing with their environment.

In brief, Poles live a dual life, public and private. Behaviors that are considered appropriate to "get things done" in the public sphere, such as bribery or deception, are not acceptable within the small private spheres of Polish life. Poles have learned to expect different rules for different situations and are able to compartmentalize their lives into these different spheres. The village church stands as a symbol of the divided lives that Poles live and the attempts to pull these two worlds together. Similarly the internal and external appearance of these churches also represents compartmentalization, a survival mechanism that has assured the longevity of the Polish identity and the Polish spirit.

The Survivor Transition

The Catholic Church survived and even flourished during the Communist period in Poland but not in other Communist states. Like the Catholic Church, the Polish people have survived under great adversity. Today the changing economic and political context of democracy and a market system presents new challenges and opportunities for the Polish people. But this new world leaves Poles without a clear enemy and with a great deal of uncertainty.

Resistance to a common enemy is a key feature in Polish self-identity. During this long period of the partition of Poland, beginning in 1772, the governing powers (Russia, Prussia, and Austria) were viewed as enemies to be resisted. This resistance pulled Poles together for a common purpose and helped maintain their unique cultural heritage. During the post-World War II period, the enemy was communism and the Soviet influence over the political and economic system. Poles are proud of their history, which is marked by these dramatic periods of resistance to external enemies.

Most notably, Poles have resisted much larger and stronger armies with great valor. For example, Polish underground resistance during World War II was instrumental in providing the allied forces with critical information and the Poles were able to thwart many Nazi efforts through sabotage. Resistance to collective farming during the Communist era is anther key example. Finally, the efforts of the Solidarity Union and others were instrumental in hastening the collapse of communism. These collective efforts were encouraged by the Catholic Church, which supported resistance to communism and earlier had championed land reforms during the period of the partition of the Polish state. The Poles have always adapted and resisted, and it is precisely because of their ability to adapt well to their environment that their unique culture has survived.

Dark Humor

One way that Poles have adapted to adversity is through humor. Polish humor is dark humor that often pokes fun at the Polish condition. For example, Polish winters can be very harsh, with bleak gray skies and extremely cold temperatures. One joke goes:

When the Germans were about to invade Poland during World War II, one Polish farmer was bemoaning to his friend the integration of Poland into the German state. The friend said to the farmer, "Well I don't think it is all that bad." The farmer looked at his friend in amazement. "How can you find good in this situation?" His friend replied, "Well, at least we won't have any more long Polish winters."

A second joke speaks to the lack of support Poles give to each other:

Satan was showing a reporter around Hell. The reporter saw many deep pits with guards around the top of each. The reporter asked what they were. Satan said that there was a pit for each nationality of people. "Here are the Germans, over there are the Americans, and here are the Poles. . . ." The reporter noticed that the Polish pit did not have any guards around it. When he asked Satan why this was so, Satan replied, "Well, the guards are stationed at the mouth of each pit to keep people from crawling out. In the Polish pit, we don't have to worry about that, because whenever anyone tries to get out, a Pole from inside pulls him back in again."

These jokes are a common and a necessary form of social interaction in Poland, but it is not acceptable for foreigners to tell these same jokes. Rather, these jokes are reserved for Poles to turn the hardships of their lives into humor, which they can share.

As these examples imply, Polish people communicate in a high-context manner, especially when compared to U.S. Americans. A good example is the hugely successful commercial for Pollena 2000, a laundry detergent, made by the Polish company Pollena-Bydgoszcz (later acquired by Unilever). The commercial is based on *The Deluge*, a novel by the pre-eminent Polish writer Henryk Sienkiewicz, a trilogy that is still the most loved novel in Poland. The story takes place during the Swedish invasion of Poland in 1655, called the deluge. Two characters are twin brothers who were pretty good warriors, but only because of their physical strength and not because of their mental capabilities. Before getting involved in any fight, they would always ask their father: "*Ociec prac?*" meaning, "Father, should we fight?" The commercial takes advantage of the fact that *prac* also means to wash. So when the brothers ask their question in the commercial, the father answers, "Yes, but only with Pollena 2000!" In contrast, low-context German detergent commercials focus on the physical attributes of the product, which is typically discussed by two housewives who explicitly compare the quality of the wash; these are perceived by the majority of Polish TV viewers as dull and boring.

The collective efforts that so define the Polish people are usually followed by periods marked by extreme difficulties in managing and operating as a unified people. Factions split off and individual interests take precedence. Poles refocus quickly on the issues of their own families, leaving a disunity that often appears like anarchy. But it is also in these periods that the ingenuity of individual Poles is unleashed.

Market Opportunities

However, the opportunities of a market system have divided modern Poland along generational and geographical lines. Opportunities abound for young Poles, particularly in the cities. This generation had less exposure to Communist ideology and employers can more easily mold them. In fact, when Marriott was recruiting workers in Warsaw, one advertisement was tailored to this situation: "No experience wanted." Marriott did not want workers used to Communism but, rather, young workers who would respond to the capitalistic incentives associated with high employee productivity.

In addition, younger Poles tend to be willing to take the risks associated with entrepreneurial activities. Compared to their older colleagues, they have traveled internationally much more and are much more comfortable using the English language, now generally recognized as the dominant language in international business. But change is slow. For example, in 1989, Solidarity won the election after several years of activity, and researchers established that there was a slight but insignificant move from autocratic to participative leadership style among Polish managers from 1984 to 1989 (Jago, Maczynski, Reber, & Bőhnisch, 1996). Also, Polish and Czech managers tend to be more autocratic than French, German, Swiss, and American managers (Jago et al., 1993). This latter study was conducted in the early 1990s and may confirm the Russian influence on Polish and Czech managers, which over time has most probably decreased, especially in private companies that have adapted smoothly to capitalism.

Entrepreneurship is at the heart of the Polish spirit and has flourished since 1989. Poles can fix virtually anything by recycling parts from other items, often using their vast networks to procure or barter needed parts. Poles deal with their economic challenges by establishing small, sometimes one-man or one-woman businesses to sell products of every kind. The roads into Poland from Germany have become makeshift markets and Western Europeans travel along them to buy products of every kind from Polish entrepreneurs. Likewise, the grounds of the Communist-era Palace of Culture and Science in the heart of the Polish capital, Warsaw, have become a flea market with rows and rows of concession stands. Many Poles now dream of being their own bosses, just as their forefathers dreamed of owning a farm. However, as noted previously, many Polish citizens have immigrated to other nations due to the lack of incentives and jobs.

The Church's New Role

The changing role of the church in Poland today is both positive and negative in the eyes of many Poles. Former Communist leaders, particularly in the Polish villages, now attend church regularly, thus uniting the village people under the church's steeple after many years of division. In fact, until recently former Communists won federal elections, and many of them loudly proclaim their new adherence to the market economy. As Lucas (2006) noted, "Confusingly, the ex-communists are now the most ardent capitalists and the ex-dissidents often sound authoritarian" (p. 5).

But the church faces great difficulty in defining its new role in the lives of Poles. It has embarked on large-scale church development in Poland and new churches are being built throughout the countryside. Many Poles resent this expansion of expensive churches when the needs of people are not being met as a result of changes in social services, for example, guaranteed employment or welfare, health care, and education.

While the church continues its role as a haven of resistance, it now serves as a moral barometer of the temptations of greed and selfishness that capitalism represents. This message has found some degree of acceptance, but many Poles, particularly the young, find that taking advantage of the opportunities that the market economy presents is alluring and key to their survival. Finally, many Poles feel the Church has lost touch with the Polish people. For example, in 1995 Lech Walesa, the president and once-popular leader of the Solidarity Movement who was strongly supported by the Catholic Church, was defeated in his re-election bid by a former Communist. His opponent's victory was accomplished with heavy support from Poles in the countryside, who voted against the church's wishes because they felt it just did not understand the issues they faced on a daily basis.

And yet the Polish people do not want to forget the past. In 1999, movies based on Polish history proved to be surprising box office successes, while American blockbusters faltered. Similarly, Poles love the opera, *The Haunted Mansion*, by Stanislaw Moniuszko, not only because of the unique Polish *muzurhas* danced in it but also because of its explicitly favorable treatment of rural Polish life. This opera, written in the Polish language, was considered so subversive that it was banned during the Partition period. As might be expected, Frederick Chopin, the 18th-century romantic and nationalist Polish pianist and composer, is revered, and the home in which he lived until immigrating to France is an oft-visited museum.

To understand Poland in entirety, it is important to comprehend the transition to modern economic life as represented by Warsaw, its modern buildings, and its modern hotels such as the Marriott. Just as important, however, it is critical to understand the traditional culture of Poland, which has withstood centuries of attack. There is no better way to understand this culture than to examine the traditional village church, using the features of the church's central plan in the village, the partitioning of Poland and its influence on Polish identity (particularly in terms of public and nonpublic behavior), and the survivor's adaptation during periods of abrupt social change and transition.

CHAPTER 8

Kimchi and Korea

I think kimchi practically defines Korean-ness.

—Park Chae-lin, curator of the Kimchi Museum
(quoted in Demick, 2006, p. A30)

As a child and young man in Seoul, Korea, I hated to eat kimchi and ran to McDonald's whenever possible. But then I studied for several years in the United States and sometimes had an uncontrollable urge to eat kimchi. Unfortunately the nearest restaurant serving kimchi was several hundred miles away, and many a weekend was spent traveling to and from it. I never realized the importance of kimchi and the feelings it engendered until I was deprived of it.

—Korean professor of economics
(personal communication, 2002)

Throughout its long history Korea has suffered terribly at the hands of many nations—China, Japan, Russia, England, and France—whose armies invaded and ravaged it over several centuries. After being occupied by British troops in 1860, the Koreans in vain tried to close off their borders to foreigners and, in the process, earned the epithet "the Hermit Kingdom." In 1910 Japan annexed Korea, which then suffered through 35 years of severe mistreatment until becoming independent in 1945. Although the Koreans were supposed to enjoy equal status with their Japanese counterparts, they were governed as a conquered people. For the first decade of Japanese rule Koreans were not allowed to publish newspapers or to develop any type of political organization.

During the 1920s Koreans were granted much greater latitude in voicing their opinions as a result of several large student demonstrations. The people gradually asserted themselves through various activities and organizations, including labor unions.

In 1937 the official Japanese policy of separate but equal treatment for Koreans came to an abrupt end, in large part because of the war between Japan and China between 1937 and 1945. Korea was mobilized for war and hundreds of thousands of Korean men were conscripted to help the Japanese army, both in Korea and China. The Koreans were required to communicate in Japanese both publicly and in private homes; they had to worship at Shinto temples; and Korean children were encouraged to adopt Japanese names. All the Korean-language newspapers ceased production. In effect, Japan was trying to assimilate Korea in every way imaginable, with the intent of destroying Korean culture and identity.

Fortunately, Japanese rule of Korea came to end in 1945 and Koreans reasserted their culture with full force. Even so, their freedom would not be assured until 1957, as Chinese and U.S. troops occupied the country during the Korean War (1953–1957) between North and South Korea.

This chapter concerns only South Korea, which became separate from North Korea because of the Korean War. At the time of separation, North Korea was the industrial powerhouse while South Korea was much more of an agricultural society. Today, however, years of Communist rule have virtually destroyed the economic base of North Korea while South Korea has flourished. It is possible and even probable that reunification will occur eventually, given the very poor economic condition of North Korea.

In comparison to most nations, South Korea is one of the most pure and unified cultures in the world. Admittedly, this purity is under attack because of globalization, but Korea is still recognized as the most Confucian nation in the world.

South Korea today is a small country, about the size of Indiana, with a population of about 48 million. It is mountainous and the country possesses few natural resources. As in many developing nations, South Korea's cities have become very crowded. About 22% of the people live in Seoul and, if the suburbs are included, this figure rises to 35%. Economic growth since 1970 has been incredible and South Korea is known as one of the five Tigers of Asia. While the 1997 Asian financial crisis created problems in the economy, especially in the financial sector, the economy recovered. However, the global economy began to deteriorate in 2008, and Korea, along with many other nations, is having some major economic difficulties.

Kimchi represents Korea in the same way that hamburger reflects the United States. Other than steamed rice, kimchi is the most popular food in Korea. Kimchi comes from a Chinese word translated as "immersing vegetables in a salt solution." There are literally hundreds of versions of kimchi.

Early forms of kimchi have been traced to the great wedding feasts held by King Shinmun, who ruled the country between 681 and 692 CE. Koreans probably created kimchi out of necessity. As mentioned above, Korea is a small country with few natural resources. The long winters made growing fresh vegetables very difficult. As a result, families would spend weeks in the fall harvesting their crops and pickling them in salt solutions (with other spices added for taste) to ensure they would ferment and be preserved for the winter, providing food for the family all year long. Kimchi includes ingredients such as Chinese cabbage, sea salt, sugar, crushed red chili, radish, chopped ginger, garlic, and green onions. Today Koreans simply buy ready-made kimchi in jars and cans. They still eat it with almost every meal, although fresh vegetables are now readily available at any time of the year.

There is a national museum in Seoul, the nation's capital, dedicated only to kimchi. Given such facts, it is appropriate that kimchi should serve as our cultural metaphor for Korea.

The 60th Birthday

A Korean's 60th birthday is important. Whereas in the United States, many people detest growing old and stop celebrating birthdays, Koreans tend to rejoice when reaching 60. Presumably all major life goals have been attained and it is now time to relax and enjoy life to the fullest. The Chinese zodiac cycle is 60 years long and Koreans believe that life runs in accordance with that cycle. Also, until recent generations, the estimated life span of those born in Korea was about 60 years. Thus the 60th birthday signifies that family and friends should gather to celebrate. Today the life expectancy of Korean men averages 74.5 years and the ritualistic birthday party is frequently repeated on a man's 70th birthday and beyond.

Often the birthday party is held at the *K'unjip*, or "big house," the eldest son's household, where he lives with his wife, children, and parents. Through the eldest son, the family genealogy is traced from one generation to the next. He is responsible for caring for parents and conducting rituals in their honor and his wife is expected to produce the male heir for the next generation.

Korea owes much of its values and culture to Confucianism, which was first introduced to Korea late in the Koryo dynasty, which ended in 1392. The succeeding Choson dynasty, which ruled until 1910, adopted Confucian philosophies as the state ideology.

Confucianism is built on five relationships defining a hierarchical system in which there is a very high degree of power distance. These relationships are as follows:

1. Father and son: governed by affection
2. Ruler and minister: governed by righteousness
3. Husband and wife: focused on attention to separate functions
4. Old and young: organized on proper order
5. Friends: faithfulness

These relationships are reflected in most activities in Korean society. It is significant that except for friends, the relationships are based on authority and subordination. Even the father-son relationship is governed primarily not by mutual affection but by the ability and willingness of the son to carry out his father's will. All of these relationships assume that the person in the superior position will act in a responsible manner, which will motivate the subordinate to react positively and in a similarly responsible manner. Confucius talks about the "ideal" man in this way, that is, striving to help others and the community but within a hierarchy. Thus Confucianism tends to create a particularistic culture in which people's behavior is motivated by their relationship with specific individuals. Koreans can be comfortable using different ethical values in different relationships and they may be very uncomfortable with universalistic values such as those prevalent in the United States, where general rules of behavior apply to all, regardless of the situational relationships at hand.

The Family Model

Perhaps the key model or unit of analysis of Confucianism is the family and the mutual obligations that are central to it. In 2001, Lee Jong Dae, then chairman of the insolvent Daewoo Motor Company, was forced to lay off 7,000 employees, but he took extraordinary measures to find employment for them, including holding a job fair, sending a personal letter to his counterparts at 26,000 companies begging them each to hire one of his fired employees, enlisting politicians to sell his cars, and abjectly bowing to a fired employee at the job fair and profusely apologizing to him for having to lay him off. Rarely if ever do American executives make such extraordinary efforts (see Choi, 2001).

If the son successfully carries out his obligations, which include hosting the 60th birthday party, he will be rewarded on his father's death, as he inherits the house as well as a greater portion of his parents' estate than his siblings. Naturally he becomes the family patriarch.

At dinnertime during the 60th birthday party, each of the five relationships is activated. The food is placed in the center of the table for all to share in a communal manner. When members of the family are seated on the floor cushions and begin to eat, the son's hard-working wife will fill the father's plate first and kimchi will definitely be on it.

Confucian relationships demand that the father should begin to eat first, followed by others in order of age. Ranking by age is critical and governs so much of Korean life that when Koreans meet a stranger, one of the first topics of conversation is each person's age so that each person can immediately assume the appropriate role in the relationship as the superior or the subservient. In comparison, asking a person's age in many Western cultures is considered intrusive.

With the exception of those who are friends, no one at this party (or anywhere in Korea) refers to others by name. Rather, titles are used to highlight hierarchical rankings. For example, the eldest son will be referred to alternatively as *son, big brother, husband, father, elder,* or *young one* throughout the day, depending on circumstances. The titles are so important that many at the party may not know the names of others. Supposedly some families separated during the Korean War could not reunite because the brothers and sisters did not actually know each others' names, making a search impossible.

Strangers by Day, Lovers by Night

Many aphorisms mirror the Confucian perspective on male-female relationships, including the one in this heading, which is focused specifically on husbands and wives. Given the subordination of the wife, it is logical to expect that the wives and their daughters prepare the birthday feast and clean up. Men are expected to be good workers and represent the family in public. The women are to provide children (in particular, at least one male), care for the family, and be subservient in public.

First and foremost, the relationship between a husband and wife is functional. Romantic love is an afterthought that is nice but not critical, for the purpose of marriage is to carry on the family lineage by producing a male heir. Even today, although less often, marriages are often arranged, frequently by matchmakers. As might be expected, the husband and wife may know each other only slightly, if at all, at the time of marriage.

Because of the importance assigned to the functions of husband and wife, Korean families still want to approve or veto any potential union. Families tend to believe that successful unions are created when both husband and wife are of similar social and economic status. Also, ancestral heritage is usually critical when evaluating potential partners, for members of the same family clan, or *tongjok*, cannot marry. A *tongjok* includes families that share a surname and have a similar heritage or source dating back for several generations.

Finding Partners

This emphasis on tradition causes many problems for younger people, as traditional Confucian practice is to separate males and females until marriage. In large cities such as Seoul, it is sometimes difficult to meet potential partners, and the more affluent young people sometimes go to "booking clubs," spending hundreds of dollars in one night to engage in a stylized form of matchmaking. Men and women sit at separate tables, and when a man sees a woman he wants to meet, he asks a waiter to bring her to his table. Frequently she acts as if she is resisting but reluctantly accompanies the waiter to the man's table; the waiter may hold her arm firmly, thus signifying her resistance. In the event that she does not like the man when they meet, she can politely leave (see Choi, 2002).

A newly married woman leaves her parents' household permanently to join her husband's family, in which she ranks at the bottom, below all the other siblings. If she marries the eldest son, she will be required to care for both her husband and his parents. Her standing improves if she delivers a male heir.

In public, women traditionally enter and exit elevators after men, help men with their coats, and follow dutifully behind them as they walk down the street. However, as is represented by the yin and yang of the Korean flag, wives do possess power, especially at home.

As this discussion implies, women face difficulties in the workforce. The gap between median earnings of men and women is 40%, in comparison to a gap of 21% in the United States, even though nearly 90% of females attend college and 52% of females work. Apparently the lack of opportunity has motivated many women to seek opportunities outside of Korea. For example, of the 138 women golfers in the elite category of the Ladies Professional Golf Association (LPGA), 34—almost a fourth—are Korean (Ramstad, 2007).

The Work Ethic

Korean men's status is a function of their occupation and the company employing them, and the strong Confucian work ethic requires that the men work long hours. They are often absent from home as a result. At home the wife has more power than the husband: She manages the family finances, runs all the activities in the home, and takes care of the children. Some men become helpless at home. Hence, Korean widowers often remarry almost immediately after a wife dies, just to keep the household operating effectively.

Korea is noted for its strong work ethic, as we might expect in a Confucian culture. Koreans spend more hours on the job than their counterparts in any other comparably developed nation. Also, it is quite common for men at all levels of business to attend nightly functions with their coworkers and business counterparts. These functions often include not only dinner but late evenings of drinking and karaoke-style festivities.

There is a downside to this excessive emphasis on work, although it is largely responsible for Korea's remarkable transformation from a rural to an industrial nation since 1970. Today agriculture accounts for only 8% of employment, industry 46%, and services 56%. However, the Southeast Asian Crisis of 1997 was accompanied by massive layoffs of workers and many of them lost faith in the relationship between hard work and success/status. Suicide is common and is the fourth leading cause of death, increasing from 11.8 per 100,000 people in 1995 to 26.1 per 100,000 in 2005; it is the leading cause of death for Korean men in their 20s. One sociological explanation is that Korea has created a modern society in which many of its members live by traditional rules based on family status and shame (Rodriguez, 2007). When there is too much tension between these traditional values and living in the modern world, some Koreans may feel that they have let down their families and experience a feeling of shame, both of which are supposedly precursors to suicide.

Similarly, women work very hard to fulfill their many duties. Because their functions revolve around the home, Korean women tend to be evaluated on their ability to run a tight ship. The Korean home is almost unfailingly well maintained. Some Korean wives who have emigrated to Western nations do not like returning to Korea too often because of the onerous family responsibilities that are automatically thrust on them when they arrive.

Korean children also possess the strong Confucian work ethic. They help out at home and are expected to work hard at school. Korean children attend school 6 days a week, and even the younger students will have 2 to 3 hours of homework. Like Japan, Korea has a literacy rate that is nearly 100%.

Kimchi's Public Role

The history of kimchi serves to highlight Korea's collectivist nature. As noted earlier, Korea has suffered many periods of foreign occupations, and during them, kimchi reflected the collectivist nature of Koreans, which was critical for survival. Families often had to hide food to protect it from foreign soldiers. Kimchi proved to be perfect for these resourceful people. Because it is a preservable food, Koreans would bury the kimchi underground to protect it from soldiers until the family was ready to eat it.

As in other Asian cultures, Koreans, when first introduced to one another, tend to ask "what company do you work for?" rather than inquiring about a person's title or occupation, which is of much more importance in the West. Also, people use their surnames first, followed by their individual names. Given such collectivism, it is easy to understand why Koreans have traditionally deemphasized individual stars in favor of team effort. However, an emphasis on individualism is emerging, as our discussion of Korean female golfers indicates. Hee Seop Choi (2005), first baseman for the Chicago Cubs, reflects this emphasis: "In spring training I'm in every newspaper (in Korea) every day. In the regular season, they watch it on TV. . . . My job is baseball. That's it. After that, I'll talk to my country. I'm happy to do that."

In Korea, what is good for the group is viewed as more important than what is good for the individual. This is a high-context society in which loyalty and the party line are valued more than the truth. Koreans are often uncomfortable with direct Western communication styles. Instead, from the highest levels of government to the household level, people are expected to tell the "truth," which is not facts per se but rather thoughts and feelings conforming to traditional ideals and norms. The messages delivered, then, can be found not in what is said but in the implicit message underlying the spoken words.

Men and women are expected to control their thoughts and emotions and express only that which serves society best, as we might expect in a Confucian culture. For example, when asked to lie under oath to save a friend, more than 90% of people in Western cultures, such as the United States and Canada, but only 37% of the Koreans, would refuse (Trompenaars & Hampden-Turner, 1998).

The Irish of Asia

Koreans can be compared to the Irish because they tend to be more emotional than their Asian neighbors. Kim Sun Yang, a food scientist, addressed this issue of emotionality: "Koreans are very much like kimchi—hot tempered. . . . On the outside, both kimchi and Koreans might seem rude, but once you get to know them, they become long-lasting friends" (quoted in Solomon, 2001, p. A1).

Korean kimchi tends to be more fiery than Japanese kimchi. Some estimates suggest that Japan produces 70% of the world's kimchi, which has motivated several Korean firms to develop and market creative brands of kimchi and sell them worldwide as the new salsa (Solomon, 2001). Any non-Korean company selling its kimchi in Korea leads to emotional reactions in defense of the native product. For example, a small Chinese-owned shop in Seoul began to sell kimchi at half of the cost of Korean brands and farmers demanded that the government pass a law mandating high tariffs. The government was reluctant to do so, as it might cause a trade dispute with China.

A good amount of this emotional expressiveness can be traced to Korean collectivism and long periods of domination by foreign powers, during which they spasmodically arose in fierce and open rebellion. But some of it is an outgrowth of Buddhism, which 25% of the population practices. The principle of the Middle Way that the Buddha advocated—keeping emotions under control, even to the extent of holding the body as quiet as possible under all conditions—is still dominant in South Korea. As discussed in "The Japanese Garden" (see Chapter 3), such practices in combination with traditional Confucian ideals tend to result in suppressed emotions that periodically flare up.

Given this perspective, it is easy to understand why Catholicism with its emphasis on rituals, collectivism, and shared emotional expressiveness has been much more successful in Korea than in other Asian nations. About 25% of the population, particularly those in the younger generation, practice it. Evangelical Protestantism is also popular, for similar reasons. Also, Christianity is viewed as less restrictive than Confucianism in terms of onerous family obligations, as exemplified by the activities at the 60th birthday party.

Harmony and "Face"

Collectivism and repressed emotional expressiveness are both related to *kibun,* the sense of well-being and harmony among people, which is central to the Korean way of life. The concept of *kibun* is closely related to "face." Taken literally, losing face means losing an eye, a nose, and ultimately the entire face and the person's life. Losing face is a very serious matter, and even minor criticisms in public are problematic. Unlike the Western concept of winning at all costs, face is the set of unspoken rules by means of which everyone's dignity and group harmony are preserved. Negotiations are frequently long and drawn out, simply to avoid losing face and to give face to others. Negotiations also involve a great deal of high-context behavior, including an abhorrence of saying no directly. Palliatives, such as "perhaps we can consider this proposal at another time" or "that would be difficult," may really be the equivalent of no. Like the context-specific Chinese, the Koreans tend to study an actual situation at great length before agreeing to a course of action, and that study includes getting to know prospective business partners well before doing business with them. Unlike Westerners, who tend to start negotiating by taking a position and then modify it as they proceed, the high-context Koreans are frequently comfortable with an ambiguous position at the start that gradually coalesces into a position or conclusion near the end of the negotiations. U.S. managers listening to an audiotape of Chinese managers making a decision reacted incredulously because of this pattern of behavior, as did Chinese managers who listened to U.S. managers making decisions (Twitchin, n.d.). Although no comparable analysis of Koreans and Westerners has taken place, we can reasonably expect similar findings, given the significant Chinese influence on Korea.

If Koreans like someone, they are prone to express this feeling more openly than their Asian neighbors. It is not unusual for a Korean businessman to take off his gold watch and give it to a Western visitor who has just said in passing that he likes the watch. Such actions are also rooted in Buddhism, which believes that it is the gift giver and not the person receiving it who is the main beneficiary.

Just as kimchi reflects collectivism, it also mirrors the Korean trait of emotional expressiveness. Positive emotions tend to occur not only during the 60th birthday party but also at any meal where kimchi is served. But when the high-context nature of Korean communication is violated, especially when kimchi is on the table, the affront is serious and may well lead to an unpleasant and jarring emotional confrontation.

Business and Culture

Business in Korea is a direct reflection of the cultural values highlighted by kimchi. South Koreans have organized their business activities in a manner similar to that of the Japanese, who resurrected the outlawed *zaibatsu* after World War II and reconfigured them as the modern *keiretsu*. A *keiretsu* is a large consortium of companies, perhaps 300, that have overlapping boards of directors, one large bank, an import-export arm, and subcontractors, for example, Matsushita. In South Korea these groupings, which are heavily family-based in terms of ownership, are called *chaebols*, for example, Samsung. However, the operations of the *chaebols* are more reflective of Chinese culture than Japanese culture because the Japanese stress employee participation to a greater extent. Like the Chinese, the Koreans tend to believe that power flows downward, and they act accordingly. This may be one reason why management and labor are more separate in Korea than in Japan (see Carroll & Gannon, 1997). Ownership in the *chaebols* is also less dispersed among non-family members than in the Japanese *keiretsu*. Because of the Asian crisis of 1997, Korea has become a more open economy, but the *chaebols* are still important.

At work, Koreans experience difficulty in taking a risky position or making an individual decision. Nearly all action is taken by committees, which provide a buffer to prevent any one person from drawing attention to himself for decisions that later prove to be problematic. Recently, for example, Martin Gannon made a presentation at a conference in Korea, before which he described a few procedural activities such as the placement of the podium that would facilitate interaction. Although several people supposedly responsible for such activity indicated that everything would be done, in fact, nothing was. In various forms this aversion to taking individual action is repeated in many contexts.

In the public arena Koreans provide others with very little personal space. While some of this behavior occurs because of crowded living conditions, it also is caused by adherence to cultural values. The respect for relationships and their governance of action apply only to known relationships. If a relationship has not been clearly defined, then Koreans act as if the relationship, or other person, does not exist.

When a relationship does not exist, Koreans will avoid action because they do not know the correct response. Instead, they will treat other people as if they do not exist, leading to no or limited personal space. For example, in such nations as Canada and the United States, lines are universally honored; in Korea, there is no respect for public lines. Instead, people push and jostle to get to the front of the line. In grocery stores strangers bump into each other with their carts, and they don't even notice this slight inconvenience. By comparison, a fistfight might occur over a similar encounter in the United States.

In the GLOBE study of 62 national societies (House et al., 2004), Korea ranked last in terms of practicing gender egalitarianism and 49th in terms of valuing it, findings that reinforce our previous discussions. Similarly, Korea ranked 18th on the practice of assertiveness, as we might expect from the Irish of Asia, although 31st on valuing it. These findings suggest that Korea is experiencing a conflict between values and practices, most probably because of the rapid transformation of the economy. Still, the value of institutional collectivism remains strong: Korea ranked 2nd on this dimension. Geert Hofstede's research (2001) generally confirms these results.

A Confucian Approach

In sum, kimchi's preservative nature and its historical role in safeguarding families in times of crisis and war make it an excellent proxy for the Confucian and collectivist nature of Korean society. Kimchi spotlights the Confucian ability to let time stand still, revere tradition, and create the hierarchical rules surrounding relationships. The use of kimchi to protect families reflects the importance of putting the good of the country and family ahead of the individual, protecting against risk, and understanding the value of the present moment.

However, this perspective on Korean culture may be too oversimplified. As globalization spreads, Koreans have tried mightily to preserve their heritage, but new international Korean superstars in sports, the increasing respect accorded women in the workplace, and the weakening of the financial system created by the *chaebols* in the face of the global economy and markets are slowly making Korea a new place.

Perhaps a supplementary cultural metaphor for Korea is the national flag, whose symbol is the *taeguh,* representing unity and harmony or what Korea essentially values. Still, the *taeguh* has two parts, the yin and the yang. In combination yin and yang connote harmony, but separately, they represent opposites. Possibly the opposites better define modern Korea, because the contrasts and tensions in the culture have become more pronounced. These contrasts include the subordination of women in family relationships versus their hidden power in the household; the need for self-control balanced against the explosive volatility of individual emotions; and the hierarchy of power created by the Five Relationships vis-à-vis the collectivism that officially preaches at least a degree of equality among members of the group. Last, but perhaps the most potent contrast, is the pull of history and tradition and the push of globalization. Still, kimchi and its symbolic meanings capture a large part of Korean culture and behavior and a visitor to this nation would be remiss to overlook its importance.

PART III

Scandinavian Egalitarian Cultures

In egalitarian cultures there is a high degree of individualism but a low degree of power distance (horizontal individualism). This type of culture emphasizes interval statistical scaling; that is, there is a common unit of measurement so that it is possible to say that Individual B is 100% more important than Individual A; that Individual C is 100% more important than Individual B; and therefore that Individual C is twice as important as Individual A on a particular dimension. But this type of culture does not recognize a true zero point such as zero money. Hence it is possible to compare individuals on one dimension at a time but not several dimensions, as there is no way of transforming values to one common unit of measurement and arriving at a final score for each individual.

Still, there are various degrees of egalitarianism. Both the Hofstede (2001) study and the GLOBE study (House et al., 2004) have identified Scandinavian nations as a separate group of egalitarian national cultures and reflective of the most extreme type of egalitarianism currently existing among nations. For example, in the United States and most other nations, a traffic citation for the same violation would uniformly apply to all, such as paying $200 for being over the speed limit by 15 miles. However, Scandinavian egalitarianism has been associated with the concept that those with more resources should pay more for a violation, even if the violation is the same for two individuals. In some instances the fine for driving 15 miles per hour faster than the speed limit has been more than $100,000. Why and how such concepts and practices have evolved in Scandinavia is the subject of this part of the book.

CHAPTER 9

The Swedish Stuga

Get a headache in Sweden and the pills you buy will be from a government-owned pharmacy. . . . With stakes in 55 groups employing 190,000 people, the state's huge corporate presence forms part of the famed Swedish model, a dovetailing of state and society whereby a large government and high taxes are mixed with liberal economic policies to provide Swedes with a high standard of living.

—David Ibison (2007, p. C4)

Few if any nations have manifested as firm a commitment to cultural values as Sweden. For 65 of the past 78 years, Sweden has been governed by the Social Democratic Party, which champions the values reflected in this quote. In the 1990s a center-right government briefly came to power and tried to remake Sweden along the lines of the highly competitive U.S. markets; in 2006 the Social Democrats once again lost to a coalition of four center-right parties that is moving cautiously to change the Swedish model. The government wants to make markets more competitive and to sell off its ownership in many firms. Simultaneously the government wants to retain its commitment to the welfare of all citizens, which has been the hallmark of the Swedish model.

Admittedly, there is much to admire in this model. For example, Sweden ranks fourth (tied with Luxemburg and Canada) on the Human Development Index, a measure combining three submeasures: gross domestic product, adult literacy, and life expectancy. Still, in 2006, although the official unemployment rate was only 4.3%, disguised unemployment—including those in state-run training programs, part-time jobs, and those on long-term sick leave—was at a highly uncompetitive 17%, especially given today's global economy. Thus economic forces and globalization are causing a reevaluation of these traditional values, which are effectively symbolized by the *stuga* or summer/weekend home.

Sweden is one of the "three fingers of Scandinavia," located between Norway and Finland. In size, it is just larger than the state of California. There are thousands of islands lining the coast and mountains form much of the northwest. Many rivers flow from the mountains through the forests and into the Baltic Sea. Sweden is dotted with lakes and more than half of the country is forested.

Although it rarely snows in the southern part of the country, the north has more than 100 snow days each year. The unique culture of the Swedes has been a result of the struggle to deal with and control harsh surroundings, and the outcome has been a stable society that, although it has experienced some difficulties in recent years, is economically and socially the envy of many others.

The population of Sweden is 9 million. About 88% of the people are ethnic Swedes. Immigrants represent about 12% of the population, one of the largest percentages in the world. Sweden's egalitarian and democratic culture has made it the country of choice of many immigrants. For example, about 10,000 Iraqis have been admitted to Sweden because of the invasion of Iraq by the United States and its allies; for the United States, given its large population, the comparable figure would be 5 million, although it admitted only 466 Iraqi refugees in 2007. Recently Sweden has tightened its immigration laws, in large part due to the need to fully service and integrate such immigrants.

In Sweden, about four out of every five couples are unwed. Under Swedish law couples have the same rights to property and inheritance whether they are married or not. If Swedes marry, it is generally on the birth of a child. The divorce rate exceeds 50%. Sweden is a highly secular society. Although 95% of Swedes belong to the Evangelical Lutheran Church, most people rarely if ever attend church.

The standard of living is high in Sweden, as indicated above. Slums do not exist and the crime rate is low. Free, compulsory education has served to raise the literacy rate to 99%. Sweden has a constitutional monarchy and the head of government is the prime minister. Members of parliament are elected for 3-year terms.

An appropriate metaphor for Sweden is the *stuga* or summer/weekend home. Before showing how this metaphor accurately reflects modern Swedish society and its cultural mind-set, we need to provide some background about its history and form of government.

Early History

Sweden's recorded history begins about 500 CE, with a continuing series of minor wars and feuds. The Svea tribe, the largest at the time, gave its name to Sweden or *Sverige*, which means the Realm of the Sveas. About 1000 CE, the famous Vikings formed a tribe of superb seamen known for their violence and success at ravaging Europe. But the Swedish Vikings soon turned their attention toward the east, setting up principalities in Russia and trading relationships with the Byzantine empire.

Domination of Sweden by German-speaking groups began in the Middle Ages. Missionaries from German-speaking regions, particularly those in the north, arrived in Sweden during the 9th century but were not successful in converting the Swedes to Christianity until the middle of the 11th century. German-speaking merchants conquered through commerce and gained almost total control over Swedish trade and politics in the 13th century. Although there was no formal nation of Germany until 1871, German-speaking regions ruled Sweden for the next few centuries using a modified form of European feudalism. Still, Swedish farmers held tightly to their ancient rights and privileges under Norse village ordinances and Viking democracy, where the chief was only the first among equals.

Sweden, Norway, and Denmark were united in 1397 as the Union of Kalmar, which was formed because their leaders felt that all of Scandinavia should be ruled by the same monarch, Queen Margareta of Denmark, to fight against German domination. The union was successful in defeating the Germans in battle and stripping some of their power, but German political and commercial influence continued for several centuries.

During the early 16th century, the Union of Kalmar was dissolved and Sweden became an independent nation. About the same time Sweden took the first steps toward parliamentary government. A national assembly (the *Riksdag*) was formed, consisting of four estates: nobles, clergy, burghers, and peasants. Gustav Vasa accepted the Swedish throne in 1523 and began the Vasa dynasty, which still rules the country. Gustav Vasa's achievements were remarkable and far-reaching. He supported the Swedish reformation to the Lutheran religion, succeeded in separating church and state for the first time, and strengthened the monarchy by making it hereditary. During the early years of the Vasa dynasty Sweden became a major military power and sought to expand in trade and territory. The following years were filled with battles. Sweden became a formidable military power and Sweden took over various parts of Europe.

Modern Evolution

The 17th century ushered in the Age of Enlightenment, a time of scientific discovery, growth of the arts, and freedom of thought and expression. In 1769 the Englishman Joseph Marshall, commenting on his travels through Sweden, reported that one could search "in vain for a painter, a poet, a statuary, or a musician" but admitted that "they were unrivalled" in the natural sciences (Jenkins, 1968, p. 63). Swedish leadership in the sciences extended beyond the Age of Enlightenment due to the Swedish talent for orderly classification and systematization. The world's first systematic registration of population statistics was begun in Sweden in 1749, and the census is still taken every year rather than every 10 years, as happens in the United States. In 1819 traveler J. T. James commented, "On the whole, with regard to science, there is no country in Europe which, in proportion to her numbers, has contributed so largely to its advancement as Sweden, and none where it is still so steadily and successfully pursued" (Jenkins, 1968, p. 78). Scientists today visit Sweden to gain access to her remarkable compilations of data. Among nations, Sweden ranks fourth both in terms of innovation and information and communication technology measures and third in terms of the number of patents per 100,000 people.

The Age of Freedom followed in the 18th century, during which the *Riksdag* introduced a constitution limiting the power of the monarchy. This Age of Freedom witnessed the start of the transformation of Sweden from an agriculturally based nation to a trading nation. In 1809 the Treaty of Fredrikshamn was signed with Russia, resulting in the loss of Finland as part of Sweden. Norway separated into a sovereign state in 1905. In 1912 the three Scandinavian countries of Sweden, Norway, and Finland declared their agreement on a policy of neutrality, even during world wars.

Movement toward equality came fairly late in Sweden, but Swedes were much more progressive than their European or American counterparts after the movement began. In 1842 the *Riksdag* introduced compulsory education and elementary schools. Universal suffrage, including women, was proclaimed in 1921. In 1971 the *Riksdag* became a single chamber and in 1974 a new constitution gave the ruling monarch purely ceremonial functions.

Sweden's late Industrial Revolution in the early 1900s contributed to its swift rise from poverty to prosperity. The country had already established solid educational and transportation systems, which supported industrialization. There was also money for expansion into new industries because of Sweden's valuable timber and iron, which were in great demand in Europe. Sweden had discovered early that a partnership of private and public interests was the best combination both for achieving economic success and benefiting as many people as possible. The evolution into a social democratic system had started.

Social Democracy

To understand modern Sweden, it is imperative to understand Swedish social democracy, which in Sweden is quite different from socialism or communism; it is a merging of the ideals of socialism and capitalism. This form of government is the hallmark of the Social Democratic Party. Other political parties accept many of the traditional ideas of social democracy within the framework of a more capitalistic and privatized economy.

Furness and Tilton (1979, p. 38) stress six fundamental values of Swedish social democracy: equality, freedom, democracy, solidarity, security, and efficiency. This form of democracy in Sweden is often called the Swedish model and sometimes "humane capitalism"; it relies heavily on the close collaboration of business, government, and labor. Social democracy has created a society free of the inequalities of capitalism and many of the inefficiencies of authoritarian central planning.

The Swedish word *lagom* is key to understanding the rationale behind social democracy. *Lagom* is untranslatable but essentially means both "middle-road" and "reasonable." The word (*lag* meaning group and *om* around) is said to derive from a circle of men sharing a single mug: It was essential that no one take more than his fair share in order to leave equal shares for all others. This word is representative of the foremost characteristic of Swedes: unemotional practicality. Swedes tend to believe that all problems can be solved rationally and satisfactorily through the proper application of reason. The Swedes have shown unusual thoroughness and ingenuity in translating their beliefs into social reality. The result of their efforts is a uniquely sensible way of life: a calm, well-ordered existence, part welfare, part technological advance, part economic innovation, and part common sense.

Still, Sweden has been reevaluating its unqualified commitment to social democracy. Today Sweden has begun to move toward a more competitive economy, a less generous welfare state, and lower rates of taxation for individuals and business firms. In 2006 the government took 51% of national income in tax, about double the percentage in the United States. The forces of globalization, as noted earlier, are causing many of these changes. Swedes seem to understand the necessity of these measures if the country is to remain prosperous and internationally competitive. Even the high rate of foreign aid, of which the Swedes are justifiably proud, has been adjusted downward. Nevertheless, the social welfare guaranteed to every Swede is far higher than that found in most countries, an outcome reflective of the model of social democracy or combining socialism and capitalism.

One advantage to this pursuit of *lagom* or rational thought and behavior is that arguments erupt less frequently than in most if not all other nations. When the rational Swedes begin the decision-making process, there is a strong tendency to find agreement. This allows the Swedes to reach a consensus quickly and act to solve the pending problem. In fact, Swedes are so zealous and efficient at attacking perceived problems in their society that they often will import not only ideas to solve problems but problems as well. Thus, John Kenneth Galbraith's (1984) *The Affluent Society* led the Swedish government to bigger and better spending programs to eliminate "public squalor," although it would be difficult to find anything in Sweden to fit Galbraith's image of American conditions.

Social democracy has been dubbed "the middle way" because it is a reasonable middle road between capitalism and socialism. Since social democracy is so consistent with the Swedish value of *lagom*, it has become much more than simply a model for economic planning. Social democracy is a functioning social system in Sweden as well as a political party. Thus, even when the Social Democratic Party is not in power, Sweden is still a social democracy. As Kesselman et al. (1987) explained:

Anyone who visits Sweden will quickly recognize that the influence of Social Democratic thinking extends beyond the 45 percent of the adult population who vote for the Social Democrats. To a remarkable extent, the Swedish Social Democrats have been able to define the problems on the political agenda and the terms in which these problems have been discussed by all major political parties. It is in this sense that social democracy might be described as the hegemonic force in postwar Swedish politics. (p. 529)

However, the downside of this model is that it is sometimes difficult to motivate employees, even to come to work. Labor costs, both direct and indirect, are high, and several large Swedish firms have moved some operations abroad to developing nations for this reason. Some firms have even moved their headquarters to another nation to avoid corporate taxes they regard as excessive.

The interesting paradox in Swedish social democracy is the seemingly contradictory goals of equality and efficiency. Modern Sweden, however, is a society that seems to have found a viable mean between equitable distribution of profits and economic performance, even given the problems described above. Milner (1989) believes that the "institutionalized social solidarity around them enables Swedes to feel secure and thus prepared to follow the market in the promising directions it opens up" (p. 17).

The ability to achieve equality and efficiency may be a result of certain circumstances in Sweden conducive to a successful social democracy. Sweden's small size and relative cultural homogeneity allow for social and cultural consensus. International competitiveness is valued universally but is also seen as the means to secure the cultural value of equality.

The Swedish Summer Home

The dream of most Swedes is to spend the summer in the family *stuga*, the summer home. The typical summer home is a small wooden house, painted the traditional reddish-brown color that is a by-product of copper mining; there are white trimmings around the door and windows; the facilities are modest; and the furniture is plain and simple. Only the necessities are found in the summer home. Often, there is an outhouse, but otherwise there is no disturbance to the surrounding area.

More than 600,000 *stugas* are scattered everywhere in Sweden, around the lakes and in the countryside. No fences or "no trespassing" signs are to be found. Usually Swedes cannot see another house from their *stuga*, but sometimes, clusters of homes make a small community. Most Swedes are only a generation or two away from the farm, so many can go back to the old family home. Some Swedes use the summer homes of friends or relatives and others may have access to a *stuga* owned by their company.

The ideal vacation is spent at the *stuga* in June or July, communing with nature. Outdoor activities range from river rafting to walking or just sitting underneath a tree and reflecting on life. Swedes usually like to spend time alone or in small family groups at the *stuga*. The idea is quite different from the American "the more the merrier" vacation or the German vacation, during which activities take place on a regular and predictable basis so that no time is wasted. Swedes want to use this time to get away from it all, be alone with their thoughts, rejuvenate themselves, and refresh their ties to nature.

One Swedish graduate student in the United States began to wax enthusiastic about her upcoming vacation in Sweden to her American friends. She planned to spend an entire month by herself at her family's *stuga*, taking along only a few small necessities, a radio, a small rowboat, and numerous books she wanted to read. The Americans were incredulous and one of them expressed the unspoken but unanimous group opinion: The month sounded more like a prison term than a vacation. While many Americans react this way, Swedes generally nod knowingly and approvingly when another Swede offers such an enthusiastic description of an upcoming vacation.

Sweden's best-known artist, Carl Larsson (1853–1919), focused primarily on the *stuga* and the informal country style it represents. His art reflects the admired lifestyle highlighted in Swedish magazines today and the simple lines of Swedish design that have made firms such as IKEA and Volvo world-famous. IKEA, the Swedish multinational, seems to be a perfect reflection of such values, and its simple designs and furniture that can be assembled easily by a homeowner have proved to be popular in many nations.

As this discussion intimates, Swedish culture and values are mirrored in the metaphor of the Swedish summer home, or *stuga*. The following characteristics of the *stuga* clearly reflect Swedish culture: love of untrammeled nature and tradition, individualism through self-development, and equality.

Love of Untrammeled Nature and Tradition

The national hymn in Sweden is "*Du gamla du fria*," which concentrates not on glory, honor, or warfare but on a land of high mountains, silence, and joyfulness. Gustav Sundberg, in his 1910 book on Swedish national culture, discussed this issue in the following manner:

> The most deeply ingrained trait in the Swedish temperament . . . is a strong love of nature . . . This feeling is equally warm among both high and low—albeit not equally conscious—and this strong attachment to nature, which in some cases may produce wild unruly emotions, is on the other hand the most profound explanation of the indestructible power and health of the Swedish nation. (p. 104)

The Swedes' love and respect for nature is a strongly held value that has led to many laws that protect wildlife, parks, and waterways. The country has been termed a "green lung"—a place where you can breathe and enjoy the untouched countryside. Swedes were environmentalists long before it became fashionable, probably because of their intense love of nature and the desire to preserve the way it was when families lived on the farm. Sweden was the first nation to pass an environmental protection law, in 1909.

This preoccupation with nature stems in large measure from a centuries-long battle against the hostile elements in a vast, damp, cold, and sparsely settled land. It is not surprising that engineers are held in high esteem in Sweden, as their efforts are directed toward combating and controlling the forces of nature. Similarly the Nordic invention of orienteering is a result of this desire to combat the forces of nature. The orienteer is set down in a remote, unfamiliar location and challenged to find his or her way out with the help of only a map and a compass.

One of the least densely populated countries in Europe with only 19.8 people per square kilometer, Sweden has the space to provide ample outdoor activities of many kinds. Walking is a favorite pastime of the Swedes, and a long walk on the lakeshore or in one of the many national parks is a preferred way to commune with nature. Fishing and picking flowers, berries, or mushrooms are also on the back-to-nature agenda. It is regarded positively when employees return to Stockholm on Monday morning with the stain of berries on their hands.

For a true back-to-nature experience, Swedes like to go up north with a backpack and walk in the mountains for days. Some hikers follow well-established trails and stay overnight in cabins provided for them. Other hikers prefer to stay overnight in tents that they carry with them, and these hikers use only a map and compass to guide them. Another back-to-nature experience that Swedes love is rafting. The journey starts at one end of a lake where there are many logs from Sweden's huge timber industry. For a fee, participants make their own rafts from the logs and find their way downstream, where the logs are returned for use in the timber mills. The trip takes several days and riders stop the raft at the side of the lake at night and make camp. This experience is the essence of the idea of getting back to nature and living off the land.

Until relatively recently, Sweden was an agriculturally based society. About 90% of families lived on farms until the Industrial Revolution in the early 1900s brought workers to the cities. Sweden has changed quickly into an industrialized, city-based country, with only 2% of the population employed on the land. Thus many Swedes remember life on the farm or have certainly heard stories about it. Ties to the farm are very strong and deeply personal. Although they enjoy everyday city life, most Swedes are still peasants at heart, who could easily return to the ways of their ancestors, for the past is not too distant. Back-to-the-farm and nature romanticism constitutes a major part of Swedish culture. Swedes long for an escape to the country where they can remind themselves of a simpler time.

Furthermore, Swedes generally value tradition and still have a village culture. The Swedes deal with this paradox by spending the summer back in the traditional village setting of the old family home. As Arne Ruth (1984) relates,

> The [Swedish] model presents the paradoxical spectacle of a nation possessing more of the outward trappings of modernity than any other society, yet at heart remaining the incarnation of Tönnies's mythical *Gemeinschaft*. With the cushions of tradition gone, the brutally modern forces of *Gesellschaft* throw the society's delicately balanced social structure out of gear. (p. 63)

Geert Hofstede's (2001) 53-nation study of cultures indicated that Sweden clusters with those countries emphasizing a small power distance between individuals and groups in society, a pattern that is typically found in a village setting. Swedes tend to be horizontal individualists who favor norms of equality over equity, in sharp contrast to the United States (see Tornblom, Jonsson, & Foa, 1985). Their individualism is manifested by their extreme self-reliance and avoidance of long-term relationships outside of the family. For example, Swedes tend to pay for a cigarette on the spot if they ask for one; living alone is common, even among the elderly; and Swedes who stay overnight at a friend's home bring their own sheets (see Daun, 1991). Thus, it is not surprising that about 90% of Swedes regularly report that they like to live "as they please."

The Swedes have traditionally had the characteristics of practicality and rationality. When practicality conflicts with other values, practicality tends to win. This can be seen repeatedly in the setting of national policy. Staunchly neutral since its last war in 1814, Sweden can be said to avoid war largely because its citizens view it as impractical and, in the long run, not beneficial to any of the participants.

Individualism Through Self-Development

The summer home is a place to go for solitude and quiet individualism. Swedes escape into the nature surrounding the summer home to spend some time reflecting on life and getting in touch with their inner selves. In the untouched countryside people can stroll for miles without coming across another human being. Great value is placed on this time for self-development. Managers complain about the unwillingness of workers to put in overtime so that they can pursue such activities. Currently most Swedes are entitled to 5 weeks of vacation a year, which they generally prefer to take in the summer. As noted above, absenteeism and excessive use of sick leave are problems for Swedish managers.

Swedes must work to afford to pay the high national sales tax and a dual income is essential for a family to maintain the quality of life desired. Although Sweden is a "doing" culture and Swedes are achievement-oriented, they tend to prefer jobs where they can develop as people, and they frequently look for jobs that are intrinsically interesting or that allow them to spend more time away from work. Decentralized decision making is valued and consistent with the ideals of equality and independence. Swedes are highly committed to quality of life and, thus, tend to value private time alone or with family more than or at least as much as work.

Much of the emphasis on self-development is motivated by individualism. There is a common misperception that Sweden is a collectivist society because it is a welfare state. Swedes pay high taxes to fund a system where education, housing, health care, and many other programs are available to all. It is true that Swedes value equality, but they have come a long way from the farm. Modern Swedes tend to place individual interests before collective interests. Gustav Sundberg, a Swedish statistician, observed in 1912 that "we Swedes love and are interested in nature, not people" (Jenkins, 1968, p. 154).

In Hofstede's (2001) 53-nation study of cultural values, Sweden clusters with those nations emphasizing individualism, but it is obviously more collectivistic than the United States, which ranks first on this measure. Swedish individualism is different than American individualism, however, because Swedes desire individualism to facilitate personal development (horizontal individualism) whereas American individualism is more competitive in nature (vertical individualism).

The emphasis on community and family has eroded with the increased industrialization of Sweden. As a visiting American observed (Heclo & Madsen, 1987),

> Officially Swedish ideology is very good. Their attitude toward society as a whole and the world is generous. But the funny thing is that people don't care much about their neighbors. Swedes are so insular, self-contained. It's easy to be lonesome here. (p. 4)

This American experienced the typical treatment of visitors to Sweden: They are cordially and generously welcomed, but then the enthusiasm quickly ends. At heart Swedes are basically loners who are not accustomed to thinking in terms of other individuals. After the original courtesies are over, it is usually difficult to pursue the relationship on a deeper level.

Ironically the success of the social welfare programs has fueled the move toward individualism. Many Swedes feel that paying high taxes absolves them of responsibility to other generations and to the needy. Likewise teenagers in Sweden are encouraged to be independent, and they frequently travel outside of the country with friends. As a result, Swedish teenagers tend to be much more mature than the typical American teenager. A weaker bond between the generations has resulted. As Hans Zetterberg (1984) explains, "The milk of human kindness therefore flows less frequently from one human being to another; instead, it is dispensed in homogenized form through regulations and institutions."

An amazingly large number of organizations in Sweden exist to promote the interests of their members in a variety of areas. Close to 200,000 membership organizations create a dense social network of 2,000 associations per 100,000 people. An apparent contradiction exists between the Swede's preference for working toward a goal as a group and the individual isolation that is a national characteristic. The organizational system is accepted only because it is seen as the most practical and efficient way to get things done. Interactions between members are generally kept on a formal basis, thus preserving the individual's isolation.

Because the views of a single person are seldom voiced without first being digested by a group, many Swedish organizations tend to promote conformity. It has even been argued that the end result is a lack of individual initiative and creativity, although such measures cited above such as the number of patents per 100,000 citizens suggest otherwise. However, it is a difficult jump in logic to state that organizations cause the typically reserved affect of the average Swede. Rather, it is more likely that the organizations have allowed the natural temperament of the Swedes to be retained while satisfying the practical need for a public voice for their concerns.

Swedish individualism is further supported by the insistence on individual rights, which seems to extend even to trivial levels. When it rains, pedestrians who are splashed by a passing car have recourse in the rules. They take the license number, go to the police station, and file a complaint, indicating that their clothes were damaged. If the court finds that the driver did not take appropriate precautions, he or she would be ordered to pay damages (in this case to have the clothes cleaned). Sweden has a form of legal welfare that allows those who cannot afford court costs to pursue a civil case.

Given this emphasis on individualism and individual rights, it is little wonder that the Swedes invented the concept of the ombudsman, an official whose job is to process and negotiate issues on behalf of others. There are several types of ombudsmen. These include the equality ombudsman, who is largely responsible for promoting equality between men and women and for opposing discrimination against immigrants; the justice ombudsman, whose role is to ensure that citizens are treated fairly by governmental authorities; and the consumer ombudsman, who is responsible for ensuring that companies treat consumers fairly.

Equality

The summer home brings into focus most of the important values in Sweden, including equality. The ancient tradition of *Allemansratt*, or "everyman's right," permits anyone, within reason, to camp anywhere for a night or to walk, ski, or paddle a canoe anywhere. Fishing is generally free, but licenses, permits, or papers are frequently required. Everyman's right also allows Swedes to pass over any grounds, fields, or woods regardless of who owns them. Swedes are careful not to abuse this right and generally do not disturb the areas close to an individual *stuga*. They use common sense, but it is not illegal to pick berries, mushrooms, or flowers from any source. There are no laws of trespass; "keep out" or "private" signs do not exist. Equal access to the greatly loved nature in Sweden is representative of the value of equality in other aspects of Swedish life.

The egalitarian passion that has been almost as important as a moral force is termed *jamlikhet*. Success in the creation of an egalitarian society hinged on rapid economic growth and social democracy, and the national wealth resulting from the Industrial Revolution was distributed to all citizens through the guidance of social democratic policies. The goal was to organize society much as the Swedes would construct a machine. In pursuit of this goal, Swedes are constantly looking for new approaches or ideas that will make the societal machinery run better. For example, Sweden's distinctive and cutting-edge retirement systems automatically adjust monthly payments both upward and downward so that retirees do not become wealthier or poorer than the working citizens.

An outstanding example of this egalitarian passion is the *dagsböter*, or "day fine," designed to equalize legal penalties in terms of income levels. A formula is used to calculate the fine based on income level, and this results in significant differences. For example, for the same infraction, low-income defendants may pay $1,650, whereas their high-income counterparts pay $12,150.

However, the Swedish machinery designed to achieve equality is quite complicated. Sweden could not be described as a simple and efficient machine. Rather, it is more like a Rube Goldberg invention: highly complicated and inefficient in the sense that many unnecessary steps and processes must be followed, but nevertheless the goals are eventually achieved.

The expression *krangel-Sverige* roughly means "red-tape Sweden," and it denotes a bewildering assortment of complications, regulations, procedures, and channels of authority. This does not imply that the system has broken down but merely that it is inconvenient for those who must deal with it. Swedes feel that the main function of the system is to operate; whether it gives pain, irritation, or pleasure to someone along the way is irrelevant to the main reason for its existence. Thus the Stockholm tramways regularly pass would-be passengers if the tramway is behind schedule because the function of a tramway is to meet the established timetable, and serving the public is only secondary. Similarly the medical care system is intended to deliver medical care. Because waiting in line does not seriously interfere with this goal, long waits to obtain medical care are viewed as not worthy of discussion.

In the past, the legendary Swedish *titelsjuka*, or title sickness, resulted in telephone listings that included the Swede's title. Today, about 50% of the listings have titles following the names, but this is done for the practical reason of distinguishing people. The fact that the practice has declined is evidence of the desire for a society of strict equality. Comparable opinion polls indicate that only 2% of Swedes desire high social status compared to 7% of Americans and 25% of Germans (Triandis, 2002). Most Swedes now label themselves middle class. The principle of universality means that all Swedes, regardless of social position, rely on the same network of services. Thus all have a stake in the quality and accessibility of services.

Historically Sweden has had high tax rates. In 2005 the government took 51% of national income in tax, about twice as much as in the United States. The taxes not only work as a leveling device but also pay for the extensive social welfare system. The government provides programs to make jobs available to all who wish to work, pensions for the elderly, sickness benefits for all workers, housing allowances, free education through college and university, almost-free medical care, and part-free dental care.

Family benefits are abundant in Sweden. The government provides child allowances for all children and gives advance payments for child support. Day care centers, called *dagis,* are provided by local authorities. These benefits are important in Sweden because about 90% of all Swedish women of working age are employed outside of the home.

Equal pay for equal work is another feature of egalitarianism in Sweden. Still, many occupations are clearly divided by gender. Swedish women face the same problems as those in the United States in trying to move up the corporate ladder. Sweden is also struggling with providing equality to immigrants. As indicated earlier, about 12% of the residents are immigrants, many of whom came for political and humanitarian reasons, and they have difficulty achieving the standard of living that native Swedes enjoy. The government has made tremendous attempts to integrate the immigrants and to offer equal opportunity, but there is a countermovement among some Swedes to restrict drastically the number of immigrants.

On Hofstede's (2001) masculinity-femininity scale, Sweden is extremely feminine when compared to other nations in that 53-nation study, and this finding was confirmed by the GLOBE study (House et al., 2004). Femininity is defined as the extent to which the dominant values in society emphasize relationships among people, concern for others, and the overall quality of life. The characteristic of femininity is descriptive of the Swedes, but in some respects the society is changing. Relationships among people are becoming more distant, and concern for others is eroding due to increased individualism.

Such forces as globalization and immigration have resulted in a change of the ideological climate but not in the moral content of Swedish social policies; the issue is really the extent of the state's prerogatives versus individual and family rights. As early as 1984, Arne Ruth pointed out that the ideals and policies espoused by the Social Democrats needed to be updated and reinforced by new models:

A rationalistic futurism took the place of religion; social and technological change was felt to be not only unavoidable, but morally imperative. Antitraditionalism became, paradoxically, the dominant tradition. . . . Industrial organization and technological innovation are among the oldest and most formidable currents in Swedish culture. . . . There will need to be a new source for the cultural myths necessary to release once again the social energy that has for so long characterized the Swedish model. (p. 93)

Thus, the opposition to the Swedish model by the new breed of Swedes appears to be the result of the loss of a sense of community and heritage that occurred due to rapid industrialization. Sweden is now preoccupied with such domestic problems as high taxes to support the welfare system, energy sources, a declining work ethic, the physical and social environment, globalization, and the purchase of Swedish firms by non-Swedish firms. Still, the unity of the Swedish mind-set is remarkable, as is the Swedes' devotion to untrammeled nature. We can expect that the Swedes will continue to emphasize the values and attitudes associated with their summer homes, particularly those of love of nature and tradition, individualism expressed through self-development, and equality.

CHAPTER 10

The Finnish Sauna

If sauna, tar, or alcohol doesn't help, you are sick to die.

—Finnish adage

Finns are reverential about the sauna. In the glow of the softly lit wood-lined space, they chat jovially or fall into a comfortable silence. The heat makes one welcome a dip in ice-cold water or a roll in the snow, as improbable as that sounds. Food and drink afterward never tasted so delicious. There is etiquette which has to do with practical and safety questions, but generous Finns will walk you through it. As is their method of rearing children, it's a window to the Finnish psyche.

—Anonymous Finn (quoted in Kaiser & Perkins, 2005)

Finland is located between Sweden and Russia on the Baltic Sea and has a population of 5.3 million. There are 15.4 people per square kilometer in comparison to 31.7 in the United States. With 2 million saunas in Finland, the sauna density is clearly greater than that of any other country in the world, even if most of the saunas are in the countryside where 70% of the land surrounding 60,000 lakes is forested. Finland also extends beyond mainland Finland into the Archipelago Sea with its 80,000 islands, which include the self-governing, Swedish-speaking Åland Islands province. Throughout Finland both Finnish and Swedish are official languages. However, most mainland Finnish-speaking Finns, who have taken obligatory Swedish courses in school, have only limited competence in that language.

The sauna began in the forests as a hole in the ground with hot stones and evolved into a small log cabin. The sauna metaphor is functionally and symbolically related to local and nature-focused values and customs. Over time Finns have moved from an agricultural environment next to the forests into urban environments. Urban living became significant only in the 1950s. As Finland has moved from an agrarian to an urban base, the countryside has remained not far away and more like a backyard down the road for many Finns. There are 415 municipalities: 67 urban, 74 semi-urban and 274 rural. Helsinki is the largest city with a population of 568, 531. Four other cities have a population close to 200,000 each, and 14 have more than 40,000 each. The Finnish government is currently integrating small towns into larger urban units to support an economic base for improving social services.

The manifest or seemingly obvious functional roles of the sauna as a place of birth and physical cleaning during life and after death belongs to history, but the sauna's latent or hidden or symbolic roles have become an integrated part of most urban living units. Finns in urban places have easy access to saunas, which are frequently attached to homes or businesses. However, the dream and actuality of returning every summer to the sauna in a summer cottage have remained a key part of Finnish reality.

Saunas are located almost everywhere and play a role in the local social context, for example, as a family weekly ritual and as a gathering place for friends to discuss the local gossip and politics. The sauna is also used as a place both to relax and do business. As an illustration, Finnish businesspeople may well have a beer as they cool off after a sauna and then call clients to finalize business deals. Or they may discuss important business issues that they have had difficulty resolving during the regular workday. There are saunas in other nations such as Sweden and Ireland. However, the distinction between the manifest or apparent functional roles and the latent, hidden, or symbolic roles of saunas in Finland is distinctive, if not unique, and beyond comparison.

In this chapter we employ the sauna as a cultural metaphor for Finnish values and practices. Using the sauna opens the door to an integrative discussion involving history, nature, the economy, social class structure, manager-subordinate relationships, communication norms, and so on. The three major themes associated with the sauna and explored in this chapter are: From survival in nature to political and economic success; apparent meanings and symbolic functions, that is, a secular "holy" place of equality; and communication and comfort with quietude.

As a Nordic and Scandinavian country, the Finnish welfare state overlaps with the Swedish metaphoric example focusing on the *stuga* of high taxes, universal health care, speeding fines related to income level, free education, and so on (see Chapter 9). Also, as equality-matching or egalitarian nations, Scandinavian nations have high scores in comparative rankings of nations on individualism and low scores on power distance (see Hofstede, 2001; House et al., 2004). Hence this chapter focuses on those aspects of Finnish culture that differ from those of other Scandinavian cultures.

From Survival in Nature to Political and Economic Success

We begin with Finland's historical roots that evolved into a Finnish Scandinavian society and then turn to different aspects of Finnish culture where the sauna is often important. The roots of identity that are still active in the Finnish mind and culture are clearly the sauna, nature/forests, and survival, the last of which means that Finland has created political and economic competencies for successful relations with both the neighboring countries that once ruled its territory and the globalizing world.

Whereas more than 90% of the world's nations are multiethnic, Finland is a relatively homogeneous country united by shared history, language, and religion. Finnish development has evolved via moving from "living at the mercy" of both nature and political domination by other nations toward "living in harmony" with nature and a changing political reality.

This nation has experienced historical periods during which both a harsh natural environment and political domination by other countries challenged it, especially domination by Russia and Sweden. It has turned these challenges into living in harmony with both nature and international political realities. Finland is a welfare society in which economic development has been a long-term process, bringing Finland to top international rankings as a very competitive, highly educated, and ethically outstanding society with a very low level of corruption. Its unique political history has evolved out of keeping the best from a centuries-long period of Swedish domination, including the maintenance of the Swedish legal system. In 1809, Finland became part of the Russian Empire as an autonomous region that operated in a relatively independent fashion until 1917, when it became an independent nation. During War II Finland defended itself against Soviet invaders, remained independent, and developed a postwar diplomatic approach to become the only Soviet neighbor during the Cold War that could remain as an integral part of Western capitalism and democracy.

History and Nature

To understand the roots of Finnish reality, it is important to begin with the relationship between "at the mercy of" and "in harmony with" nature. According to the mythic origins of the Finnish people, Väinämöinen, the poet eternal of the classic Kalevala songs, stood upon a barren island and asked Pellervoinen, the god of fields, to do the planting, sow the seeds, and sow them thickly in the treeless land. The trees grew into forests. In the forest lived ancient Finnish tribes led by men and women alike. Historically the forest and nature have been the foundation of life for Finns. Finns are well-aware of the saying "At the mercy of nature." They clearly recognize both the limits and opportunities when living in harmony with nature. Over time the traditional myths related to the forest have become less influential but the dream of most Finns is to have time in nature and to sauna close to a lake, at least every summer.

Out of the forest grew the sauna and a nation of sturdy crofters—landless farmers who were also "freemen"—in contrast to Russian serfdom. They kept their noses to the grindstone, struggling against an arctic frontier and a difficult geopolitical-cultural frontier in which Russia was dominant. As children of the forest, Finns gradually carved out a zone of comfort by processing raw materials from the forest to sell on foreign markets—first tar and timber to European naval powers and later paper to European and American newspapers. Today Finland is a leader both in the paper and shipbuilding industries, and Finns build every fourth big cruise ship in the world.

Prior to the mid-1800s, the forest was easily interpreted as a primitive area with limited hope for economic development. However, technological innovations capable of turning natural resources such as wood into exportable products played the key role in Finnish economic evolution starting in the mid-1800s. These innovations also created a high demand for the resources that Finnish forests can produce, which has led to expanding Finnish paper-related production technology abroad. To be in harmony with nature—to recognize limits and opportunities—opened the door to long-term technical and economic development beyond reliance on nature-produced resources.

If the forest represented the basis for and symbolism of innovation for a century and a half, the multinational cutting-edge high-technology Finnish firm Nokia currently symbolizes a message throughout Finland: Without innovation, there is no hope for economic success in a globalizing world. Living in harmony with nature now requires benefiting from both the development of Finnish human resources and of technology.

History and Foreign Control

Similarly, Finland has a long history of both living at the mercy of foreign domination and developing competence to create a successful degree of diplomatic harmony with other nations. The Swedes came from the west in the 12th century and gradually turned Finnish tribes into a subordinated part of Sweden for 700 years. After the Napoleonic wars in the early 1800s the European powers redrew the map of Europe and transferred Finland from Sweden into Russia. Unlike other parts of the Russian Empire, Finland benefited from an autonomous status as a grand duchy and retained the laws and Protestant religion from the Swedish era, which differed from Russian feudalism and Russian Catholic Orthodoxy.

During this period of autonomy in the Russian Empire, Finns developed their own culture and political institutions in the land where they had always lived. In essence, they created a Finnish—neither Swedish nor Russian—identity. The Finnish parliament gave every Finnish adult (male and female) the right to vote during the period of "Russification"—Russian attempts to reduce Finland's autonomy at the turn of the 20th century. This movement toward egalitarian democracy, which occurred before any other country in the world acted similarly, supported Finnish determination to keep its autonomous rights to govern local issues. "Before this reform 70% of all adult people were outside suffrage," Professor Borg, formerly of Finland's Tampere University and a member of parliament in the 1970s, told the BBC News website (Smith-Spark, 2006).

Movement toward equalitarian universal suffrage was a critical way to develop a sense of national unity to defend Finnish autonomous rights at that stage in Finnish history. Johan Sibelius, the most famous Finnish classical composer, reinforced the

Finnish identity by employing nature-focused motifs and patriotism to bring home these points. His symphonic poem, *Finlandia*, is the major symbol of Finnish determination to resist Russification.

Independence

Finland finally became an independent nation in 1917 after the collapse of the Russian Empire and the rise of the Soviet Union. The Finnish parliament declared its independence on December 6, 1917, and the Soviet government recognized Finland's independence on December 18. However, Finland was politically unstable and a civil war occurred between the "Reds" representing the ideology of communism and the "Whites" representing the ideology of conservatism in the period from January 27 to May 15, 1918. This civil war involving Finns versus Finns remains the most controversial and emotionally loaded event in the history of modern Finland.

After the civil war the existence of a Soviet neighbor and possible revival of leftist extremism led to the distribution of part of privately owned land to landless rural inhabitants. At that time the government believed that redistribution of farmland property was essential for preventing the growth of communism in Finland. One of the unique aspects of Finnish history is this decision of partial land redistribution by conservatives, in marked contrast to outright land transfer and confiscation in communist countries to gain leftist support. Today many if not all economists believe that widespread ownership of property is a key determinant if not *the* key determinant of successful economic development. Hernando de Soto (2000) titled his pathbreaking exploration of owning property, *The Mystery of Capital: Why Capitalism Triumphs in the West and Fails Everywhere Else*. Finland was decades ahead of many other nations because of the land distribution to the landless.

After becoming independent, Finland slowly developed an economic infrastructure that made it possible to sell paper to *The New York Times*, which enabled Finland to be the only European country that paid its debts to the United States during the 1930s. The Finnish image of financial prudence was so powerful that one American newspaper published a picture of a man responding to a bill collector with the following question: "Who do you think I am, Finland?" (M. Berry, 1990, p. 53).

Just prior to World War II Hitler "gave" the Soviet Union permission to "have" Finland. Only three European nations militarily involved in the war (the Soviet Union, Britain, and Finland, which was neutral) were not occupied during World War II. However, the Russian Army defeated the Finnish Army and confiscated some Finnish land, forcing 400,000 Finns or 11% of the population to move rather than be part of Russia. Still, Finland remained neutral and independent.

Against great odds Finland remained the only independent country on the postwar Soviet border by implementing a "harmony with political reality" strategy to cautiously move from "survival to success" (Jakobkson, 1987, p. 7). The nation moved toward the West by emphasizing cautious "sauna diplomacy," sometimes even using the sauna as a place for relaxation, goodwill, and some degree of equality between Soviet and Finnish diplomatic negotiators. For example, former president Kekkonen invited his Soviet counterpart to sauna, where the two national leaders could free themselves, to some extent, from outside concerns. They discussed and sweated out previous problems in the sauna so that the Soviet Union and Finland would be in harmony when Finland eventually entered the European Economic Community (EEC) in 1973. The EEC was the predecessor to the European Union (EU), which Finland joined in 1995. During the Finnish presidency of the EU in 2006 the concept of sauna diplomacy was still active: The Finnish minister of foreign affairs was encouraged to invite the Greek and Turkish sides of the Cyprus issue to negotiate in a Finnish sauna.

The Strength of People

Throughout its history Finland has been like a prickly juniper that bent when necessary but refused to break. Finland's strength has been in its people, particularly in their ability to accept limits when necessary and to maximize opportunity and cautiously move beyond those limits when possible. Shared survival against external political threats contributed to the creation of a social welfare society, successful economic development, a high degree of market competition, and a welfare state stressing social equality. In market pricing cultures, for example, the United States and Great Britain, social welfare cohesion and competition are often interpreted to be a counterproductive combination. In contrast, an integration of social welfare cohesion and governmental policies stressing competitiveness has become the basis for Finnish success against external political challenges and international competition. Such success is distinctive in a nation with such a small population.

All of this historical development required *sisu*—a combination of guts, determination, and optimistic realism with moments of pessimism to survive against great odds. Morrison and Conaway (2006, p. 162) indicate that this word is extremely difficult to translate and is sometimes translated as "never say die." *Sisu* has served Finland well but the current meaning of the term has changed, referring more to individual will and determination, for example, of an athlete who falls down but still wins the race. The current meaning of the term has also broadened to include brand names ranging from powerful trucks to one Finnish candy product labeled *sisu*.

EU membership is currently considered important but also as something that could interfere with Finnish national independence. Finns justifiably believe that within the limits imposed by climatic and geopolitical realities, they have historically done it all on their own, and they have performed spectacularly. Given the Finnish communication norm that modesty brings respect, Finns will rarely make this explicit when communicating with people from other cultures. Pride must remain active but under the surface. The Finnish saying, "running away from a wolf is running into a bear," resonates well with the deeply held Finnish value of finding ways to resolve seemingly impossible problems as soon as they appear and then to move forward.

The Social Welfare State

An egalitarian sauna approach to diplomacy was also integrated into the development of a social welfare state after World War II. Finland needed more children, healthy adults to work, attractive employment opportunities and wages to minimize the size of the small Communist Party in Finland, better education tied to future development, and a supportive governmental role in economic development. Today the differences between Finland and the other Scandinavian countries related to gender and social class equality are rather small. Still, there are other variations of equality-matching or egalitarian nations, including Canada and Germany, which is why they are treated in a separate part of the book (see Part IV). The GLOBE study (House et al., 2004) and the Hofstede study (2001) come to a similar conclusion about the uniqueness of the Scandinavian nations.

In 1871 writer and historian Zachris Topelius succinctly summarized the Finnish political objective, given its small size, harsh winters, and the desire to be independent of other nations: "To be neutral, to be self-sufficient, to have the freedom to look after one's own" (quoted in Mead, 1989, p. 15). Topelius's goals have been realized, as confirmed by international rankings of nations in terms of multiple indicators. Finland has consistently ranked at or near the top on each of them, for example, environmental sustainability, competitiveness, a low level of corruption, and the highest-scoring 15-year-old students in the areas of mathematics, science, and reading.

The sauna, found around every corner, allows the Finns to relax briefly and build up energy among friends and fellow citizens before moving forward quickly once again, to overcome almost overwhelming problems that they have attacked with patience and vigor. As a result, the economy did not develop in terms of short-term responses to "quarter to quarter" stock market pressures. In the United States, small groups make decisions in a quick huddling and act on the next best guess; in Finland, the decision-making process is longer and more deliberative (see Chapter 16). Finland's president Tarja Halonen (2007) summarized the Finnish innovative version of a "Nordic welfare society model": "In order to exploit the real benefits of research, we need to be patient and far-sighted and to invest in a sustained manner equally in education, science and technological development."

Sauna: A Secular "Holy" Place of Equality

Entering a sauna is often similar to entering a holy place full of the spirituality of nature. A Finnish proverb is that people should behave in the sauna as they do in church. The ideal associated with the sauna is a nonreligious cleansing of body and soul. Entering the sauna signifies leaving burdens and controversies behind, relaxing, and cleaning more than the surface of oneself. There is often a combination of privacy and togetherness that is both taken for granted by Finns and difficult for non-Finns to understand. For example, Finns often share comfort with quietude in the sauna, but at other times they also talk about private matters or humorously joke among themselves in a good-spirited fashion. In short, Finns are people comfortable with both quietness and with talking in different ways. Taking an argument into the sauna corresponds to taking one into a church service, which Finns regard negatively. However, discussing the matter as equals in a nonemotional manner, either after sauna or during it, is acceptable and frequently leads to successful resolutions. Similarly, in many cultures, including Finland, nudity and sex are not seen as integral. In a Finnish sauna, nudity has nothing to do with sex but with relaxation and the cleansing of the body.

The word *sauna* is the only Finnish word that is used throughout the world. It moved into the German language after publication of a sauna advertisement: "If you want to experience heaven and hell simultaneously, go to a Finnish sauna" (Laaksonen, 1999, p. 279). For a moment, one can be at the mercy of too much water thrown on very hot stones but can also be in harmony with the relaxation that follows. During a summer tour of Finland Robert G. Kaiser and Lucian Perkins of *The Washington Post* wrote about the sauna in an article titled "Finland Diary" and presented an unqualified positive viewpoint on the sauna: "This, our first sauna experience, was everything promised: We finished feeling refreshed and invigorated. The wrinkles and bumps caused by days of hard traveling quickly disappeared from our bodies" (Kaiser & Perkins, 2005). A description of the sauna does not end, however, with reference to throwing water on hot rocks to send steam throughout the sauna. The sauna is associated not only with roots in history and a secular "holy" place of equality but also with the expression of values that exist throughout Finnish society.

Sauna Practice

Many families grow up in the sauna. A weekly ritual tends to reinforce togetherness. All members of the family are together until the children are teenagers. Sauna is more than just a time together; it is also a time during which everyone learns to enjoy respecting the quietness of others, even if some would prefer to talk, and to have something nice to eat and drink after the sauna. Whenever the children return home as adults, it is common that the sauna is ready for them, especially in rural areas where it is heated with wood rather than electricity.

Sauna is a weekly ritual for many, and even a daily ritual for some Finns, and it is an essential ritual for Christmas and midsummer when the sun is visible between 20 and 24 hours each day. To have a midsummer celebration with family and close friends beside a lake in the forest without a sauna is incomprehensible. Even if most families live in urban areas with access to sauna, bringing sauna and nature together in the summer is the decided preference for summer vacations. Vacations out of Finland occur usually in the dark autumn and winter.

All people are equal when they enter the sauna and there are no visible symbols of social status. In business, there is a de-emphasis of social status and everyone sits where they want to sit, whether a janitor or a CEO. Moments of egalitarian relaxation contribute to better superior-subordinate and human relationships. There is also an implicit link between the concept of equality in sauna and the responsibility to carry one's own weight in work. As a consequence, there is an emphasis at work on the autonomy of subordinates. Once assigned a task, the subordinates become "self-bossed," possess individual responsibility for quality work, and contact the superior only when a problem arises. Such a participative managerial hierarchy with a "bottom up" dimension facilitates autonomous development and allows Finns to create a vision for the future shared by all when it is effectively coordinated from above.

As suggested earlier, Finland shares this emphasis on equality with other Scandinavian nations. For example, fines for legal infractions are related to income levels, as the assumption is that those receiving high incomes must suffer as much as those receiving low incomes. In one situation, a Nokia executive was fined $71,400 for driving 18 miles over the speed limit. A low-income person might receive a fine only of $50 for the same infraction (Stecklow, 2001).

There are different interpretations of what it means to be Finnish. An American interpretation is aptly illustrated in the following incident. During a training session for three groups—Americans, Finns, and Swedes in a Finnish subsidiary in the United States—an American came to the consultant during the first coffee break and mentioned that there were problems with Finns because they didn't talk much, if at all. This American didn't know what the Finns were thinking or hiding, or even if they ever thought independently or whether they were interested during group discussions. In contrast, he said, the Swedes were easy to deal with because they talked, even if not as much as Americans. During the afternoon coffee break the same American told the consultant that the Swedes were difficult because, after a decision had been made, they kept discussing whether it should be modified. In contrast, the American now felt that the Finns were easy to work with because they focused on the goal and, once a decision had been made, implemented it quickly.

The following day the Americans politely raised a question about Finnish inability to make decisions and the Finns politely responded that they felt the Americans also had a problem. The consultant then asked if there was perhaps a difference between "let's get the ball rolling" (which from a Finnish perspective might not be a good decision) and "let's plan carefully before the ball gets rolling" (which from an American perspective might prevent the ball from ever rolling). During the discussion the American response was the following: "For years we have been misunderstanding each other. Maybe long-term planning is what the Finns and Swedes have in common, but the Swedes just keep discussing after a decision has been made and the Finns seem to be silent during the planning process, yet there is no silence when they act on a decision" (Berry, 2008a, pp. 87–88). This interpretation is consistent with the discussion in the next section of this chapter, "Communication: Comfort With Quietude," namely the importance of listening, thinking, and acting rather than talking too much. Even if talking is expected, Finns would say, it should be done at appropriate times, which Finnish culture often defines differently than other national cultures.

Gender Equality

Gender equality is rather high in Scandinavian nations, even if it is an ideal concept when compared with the societal reality, at least in some instances. For instance, we encounter a paradox related to sauna and gender equality in business and political contexts: The sauna has traditionally symbolized equality but typically the sexes do not sauna together.

However, women have significantly increased their leadership roles in business and politics, including major leadership roles in the Finnish parliament. Finland has a female president, 42% of the parliament elected in 2007 is female, and the Finnish government has a tradition of gender balance for ministers, with more women than men in 2007. Recent studies have also demonstrated that female business leadership has been successful in many instances and that many of the leadership practices traditionally associated with women, including a greater emphasis on participative decision making and employee involvement, are appropriate for male managers seeking success.

Given the egalitarian emphasis in Finland, these results are not surprising. Even stories about ancient Finnish tribes living in the forest refer to leadership by men and women alike. As an autonomous part of the Russian empire, Finland was the second parliament in the world to give women the right to vote (New Zealand first) but the first to give women both the right to vote and to stand for election. Sixty years after World War II, a postage stamp dedicated to the defense of Finland during that conflict focused on the active role of women supporting the Finnish soldiers. Hence the traditional equality among men to make important decisions unilaterally without the active involvement of women during sauna negotiations has been reduced significantly.

Egalitarian Education

In contemporary Finnish society, the educational system is unique. It offers egalitarian educational opportunities regardless of the geographic location or the social background of the children. The Organization for Economic Cooperation and Development (OECD) conducts regular surveys as part of its Program for International Student Assessment (PISA). Combining the results from 2000 and 2003, Finland is at the top for learning skills among 15-year-olds for mathematics, science, and reading. In the 2006 PISA tests Finnish students were at the top for science. All education in Finland is free up to the doctorate level, school teachers are highly educated, competition to be accepted into the university department of education is very strong, and the salary for teachers is determined at the national level, with modifications according to the local costs of living.

This egalitarian focus in education is sometimes criticized for not challenging the most talented students enough. However, it provides a basis for advanced levels of education throughout society. In fact, educators from many nations travel to Finland to see firsthand "What Makes Finnish Kids So Smart" (Gamerman, 2008, p. W1). According to Gamerman, Finnish teachers still favor traditional methods such as chalkboards rather than PowerPoint® slides and other high-tech approaches, and the classes are somewhat chaotic and unstructured, but it is the intense focus on achieving high goals combined with strong teacher-student and student-student interactions within small classes that are hallmarks of the system.

Communication: Comfort With Quietude

Two Finnish academicians raised the intriguing question: "Silence: Myth or Reality?" and responded: "The terminology [of English] may . . . be highly misleading depending on the type of culture that it is applied to" (Sajavaara & Lehtonon, 1997, p. 279). Finland is not a silent culture; it is a culture full of people "comfortable with quietude."

An American television program, CBS's *60 Minutes*, produced a documentary, "Tango Finlandia," which commented negatively on the Finnish ways of living and communicating but from an American frame of reference, that of "discomfort with silence." Words used by Finns to describe themselves in "Tango Finlandia" include *shy, silent, private*, and *brooding*. These words overlap semantically with the words used by Finns when they explain Finnish ways of communicating to others. Voice-over terms used by the American commentator included *clinically shy, terminal melancholy, depressed, mourning, brooding*, and *isolated*, as well as the following judgment: "The national mission seemed to be to not be noticed, grimly in touch with no one but themselves . . . a difficult time making even the most casual social contact" (Berry, Carbaugh, & Nurmikari-Berry, 2004, p. 267). This American commentary epitomizes the incorrect ways many people from other cultures interpret Finnish comfort with quietness.

The Meaning of Silence

One of the challenges related to understanding other cultures and explaining one's own cultural ways to others is reliance on a shared international language, English, which carries multiple hidden cultural meanings for people from different cultures. As many Finns have emphasized in private conversations and writings on the subject, silence is not necessarily silence, even if the dictionary tells us so. There are various types of silences connoted by facial expressions, body motions, and even the length of each silence. Each type of silence expresses a particular meaning to those socialized in this fashion.

The English words used by Finns are often inaccurate when compared to what they actually mean in the Finnish language, especially when communicating Finnish comfort with quietude to people from cultures that are uncomfortable with silence. Finns see silence positively in terms of the following cultural meanings: communicating without words, thinking, reflecting, pondering before acting or using words in important situations, and being in one's own thoughts and/or respecting the privacy of others, even on occasions when surrounded by others. This Finnish interpretation of silence corresponds to the combination of privacy and togetherness in sauna.

Movement back and forth between both comfort with quietness and comfort with talking is a natural Finnish way of being. There is no Finnish comfort with quietness if there isn't also comfort with talking when one has something relevant to say. What others might interpret as a "disconnected social void" can be full of Finnish respect for others, reflection, and a willingness to stand behind one's word. Sauna is a symbol of cleanliness that can be associated with the "cleanness" of society based on the subtle clarity of individual moral and physical responsibilities, for example, concepts of politeness and honesty.

In Finland, people make promises without using the word *promise*. They feel that the idea of pledging to do something is implied by the message itself. Similarly, Finns rarely use the word *please* for that courtesy is implicit. The meanings that are associated with the English words are considered to exist in the statements or requests themselves. As a result, Finns might be puzzled or suspicious when words like *promise* or *please* are used. Conversely, native speakers of English might be confused or offended when English-speaking Finns fail to use such language.

Respect for privacy, communicating togetherness nonverbally, and being in harmony with a social or physical environment reflect deep Finnish comfort with quietness. For example, a U.S. American was driving a Finnish woman visitor through the Appalachian mountains to observe the beautiful fall leaves. She was silent and wrapped up in the enjoyment of the beautiful displays of leaves. Suddenly he stopped and demanded to know what was wrong. By contrast, a Finnish friend would have assumed that her quietness was a sign of deep feeling and reverence for nature. Listening to a recounting of this story, a Frenchman expressed the following opinion: "I guess we are a little like Americans. Perhaps we could be comfortable like her for just a moment but not that long" (Berry, Carbaugh, & Nurmikari-Berry, 2008, p. 48).

The Cell Phone Invasion

The Nokia era of mobile phones has added a new dimension to the Finnish communication style, even if the basics remain the same. Nokia's "connecting people" message can be interpreted by Finns in both positive and negative ways. One should not talk just to be talking, Finns believe, but a mobile phone call often means that the other person considers it important to talk. Therefore, they have a responsibility to make a full response. There is, however, a Finnish discomfort with the extent to which this modern connecting system also invades the traditional desire to enjoy one's privacy. A common solution in many cases is to send a text message if appropriate and to talk when necessary. In this regard people from other nations who keep their cell phones in their cars and strictly limit their use are most probably comparable to the Finnish preference.

Reference to being relaxed and in harmony with nature during a sauna in the forest still holds true. Still, as soon as people are out of the sauna, especially business people, the mobile phones can be polluting the harmony of the relaxing post-sauna environment and disconnecting the people who were together in the sauna. The phones allow business people to go anywhere at any time but also reinforce the concept that the modern world is not the same as the natural world with which they are most comfortable. On the other hand, Finns who go abroad without access to mobile phone connections can feel very disconnected—not because they feel the need to call often but because of the unease that others can't contact them about something important.

Factors other than comfort with quietude influence communication, especially in business. Charismatic leadership or expecting subordinates to follow leaders out of a blind faith in their exceptional power can be counterproductive in Finland. Finns value quality, honesty, equality, autonomy, independence, and responsibility. If charismatic leadership actively interferes with the expression of these values, Finns often react negatively.

Another factor influencing communication is that the Finnish norm is to allow other people to finish talking before saying something. Thus speakers do not overlap with one another, which demonstrates both a respect for others and a belief that quiet listening can often be a form of reflection. Hence silence, which is often understood in other cultures as agreement, is not necessarily the case in Finland. Finns might remain quiet if they disagree and they don't have a good alternative to put on the table. From childhood, they are taught to never interrupt.

Personal Style

Furthermore, Finns tend to stick to their plans, sometimes simply because of stubbornness but often because of the importance they attach to making a decision based on careful planning. To change a person's opinion in Finland, the sharing of quality information to consider when modifying decisions is more relevant than clever communication tactics.

Finns also prefer meetings based on premeeting preparation rather than brainstorming meetings. Those who talk the most are those who come well-prepared. Similarly, listening to different ideas and opinions before speaking is esteemed. Everyone listens to a person who is well-prepared but respectful of others' ideas.

Finnish values of quality can be seen in the sauna, especially when individuals are together but relaxing in their own ways, with the cleaning process going deep below the surface. This link can also be connected to the Finnish concepts of honesty and trust, which go beyond "don't lie/steal" to mean "stand behind one's words." For example, Finns tend to de-emphasize words such as *promise* or *please*, as discussed previously. The meanings are embedded in the communication norms, which indicate that such words are often missing when Finns are speaking English. If a Finn has to say *please* or *I promise* in Finnish, everyone assumes that there are hidden problems of which they have not been apprised. Unfortunately, the Finnish reliance on a "grammar" of honesty and nonverbal politeness can easily hide, at least initially, such deep values as independence, responsibility, and politeness.

Finnish comfort with quietude is one natural way to be in harmony with nature, one's self, and others. This comfort, which does not deny the importance of talking, frequently hides the complexity of nonverbal communication expressive of both togetherness and respect for others. Initially a short-term visitor can easily be confused. Similarly Finns might be initially confused when French people interrupt to show interest and Americans ask "how are you?" when walking past and not expecting an answer. The more we realize how we (and others) talk about ourselves when talking about others, the more people "uncomfortable with silence" and people "comfortable with quietness" can begin to move beyond being at the "mercy of intercultural communication discomfort" toward "being in harmony with multiple ways of communicating" (Berry, 2008b, p. 77).

Most Livable Nation

This, then, is Finland, a small nation with spectacular economic and political success that evolved out of a difficult natural and political environment. In 2007 Matthew Kahn, an environmental economist, ranked 141 nations in terms of a livability criterion, and Finland finished first followed by Iceland, Norway, Sweden, and Austria; the United States was 25th.

As this chapter has demonstrated, Finland is similar to other Scandinavian nations in terms of many values, but some of the differences are significant, and they evolved out of the history and geographical position of the nation. In many countries people refer to pessimism or optimism. The "Tango Finlandia" commentator from CBS's *60 Minutes* referred to "terminal melancholy" when describing Finns. As one Finnish student explained to exchange students,

> For us Finns melancholy means something like yearning, longing or even hope and desire. These are all positive words that live inside every Finn. Melancholy is a national characteristic rather than a symbol of depressed individuals . . . There may be something like pessimistic streaks in Finnish culture, but the word pessimism is too strong. Finns are more like realists rather than optimistic or pessimistic people. We have our feet firmly on the ground, are hard working, and look forward to the future. But we realize that life is full of complexity. We must be prepared for difficult times even if we have hope in our hearts. (Berry et al., 2008, p. 42)

Finland's unique history helps to explain why Finns tend to emphasize realism, granted that there are both very low and very high points. And the sauna, which combines aspects of heaven and hell, is a realistic institution rooted in history and nature. It remains central to the Finnish way of life and offers examples of Finnish values related to communication norms, the quality of action, social equality and autonomy, and social and organizational responsibility.

CHAPTER 11

The Danish Christmas Luncheon

Christmas in Denmark goes on and on. The Danes really enjoy seeing their relatives. Here in Denmark, there is no end to who you might meet for a Christmas lunch. . . . everybody was dancing around the tree, holding hands and singing their lungs out. Even 80-year-old Granny was hopping about.

—Tom Gilbert, English researcher
at Copenhagen University, 2007

Denmark, which has a population of 5.4 million, is an affluent and highly educated nation in which the gross domestic product (GDP) per person is $44,710, ranking it seventh among nations. According to *The Economist* intelligence unit, it is the best place to conduct business in the next five years followed closely by Finland and Singapore. In a recent survey of countries belonging to the Organization of Economic Co-operation and Development (OECD), Denmark was found to be the most wired among the members of the rich nations club with more than 30% of its inhabitants with access to broadband. In the United States, by comparison, the proportion was 22%, slightly above the OECD average of 20%.

Denmark is situated in northern Europe, jutting out north of Germany. Sweden and Norway are nearby. Denmark, Sweden, Norway, Finland, Iceland, and the Faeroe Islands in combination form Scandinavia. Although the Faeroe Islands and Greenland are part of the kingdom of Denmark, they are governed by home rule. The focus of this chapter is Denmark itself, which is composed of 406 islands. This peninsular country is linked by land to Germany and is flat, which has made the use of bicycles popular; former President Clinton was invited to tour by this means on one official visit.

The geographical location is accompanied by a variable pattern of weather, which can be cold, gray, and rainy throughout the year. Most probably because of its history, geography, and small population, this equality matching and egalitarian nation has developed a unique culture.

Our cultural metaphor for Denmark is the Christmas luncheon or *Julefrokost*. Although numerous nations have Christmas-related festivities and parties, Denmark's luncheon has become a unique institution that provides insight into the Danish culture.

All or almost all businesses and organizations in Denmark host a Christmas luncheon, which typically occurs on one of three Fridays in December set aside officially for this purpose. Some businesses schedule the luncheon for other times such as Saturday evening. Subways and railroads are open around the clock on these three Fridays to ensure that everyone gets home safely. The hosting organization generally provides the alcoholic and nonalcoholic beverages, while employees bring the food. The luncheon tends to begin about or after 1 p.m.; in many organizations it starts in the early evening and extends late into the night, frequently ending at 2 or 3 a.m. Traditional dishes include herring, rib roast, meatballs, rye bread, cheese, and rice pudding.

While the provision of public transit on these occasions suggests a nation that is conservative in terms of ensuring the safety of its citizens, otherwise the Danish Christmas party seems to differ little—at least superficially—from parties throughout the world at this time of the year. Denmark is a controlled culture in terms of emotional expressiveness and outlandish behavior when compared to many other cultures, although the Danes are more expressive than other Scandinavians, even to the extent of seeing themselves as the Italians of Scandinavia. During the Christmas party most inhibitions are overcome, including the inappropriate and possibly hostile expression of feelings toward the organization, its management, and other employees. Overt sexual overtures and even extramarital affairs related to the party are not uncommon. These activities may not be forgotten, but they are condoned, although a major repercussion may well be widespread gossip based on what transpired.

A working wife attends her luncheon by herself while her husband babysits the children one weekend, and then they switch roles on another weekend. In recent years spouses have tried to attend each other's luncheons, assuming they occur on different nights. In short, for one day a year, Danes are allowed to express thoughts and emotions that are seen as clearly inappropriate at all other times for all of the reasons provided in the rest of this chapter.

Interdependent Individualism

In some ways Denmark reflects many of the critical issues facing all small nations. Queen Margrethe I of Denmark united Denmark, Norway, and Sweden in 1397. Norway belonged to Denmark until 1814, as did southernmost Sweden until 1658. Schleswig and Holstein in north Germany were also part of Denmark until 1864. Periodic wars led to a gradual diminution of both the military power and the size of Denmark.

During World War II Germany invaded both Norway and Denmark. The German military demanded that all Jews in the nation wear armbands identifying them as Jewish. The Danish king, in a wonderful display of leadership, put on such an armband and the citizens quickly followed his example. The Danes then attempted to secretly transport all Jews out of the country to the safety of neutral Sweden by a series of dangerous activities, saving about 95% of the 7,000 Jewish residents.

Denmark is clearly an individualistic nation, similar to its northern European counterparts and is ranked 9th of 53 nations by Hofstede (2001) on this measure. However, its geography, small size, and historical experiences have helped to create a type of individualism that might best be characterized as interdependent. To some extent, this is borne out by the GLOBE study (House et al., 2004) in which researchers found that Denmark scored very low on in-group collectivism practices (indicating an individualistic orientation) and high on societal institutional collectivism practices (indicating a collectivistic orientation). Thus, we see some interesting differences on individualism at the individual and societal levels.

Denmark is a social democracy, like Sweden and other similarly inclined nations, and it provides all sorts of social services, facilitating a sense of equality among its citizens. Thus it is not surprising that Denmark is low on power distance and acceptance of risk; in the Hofstede (2001) study, it ranked 51 of the 53 nations on both counts and in the more recent GLOBE study (House et al., 2004), an almost identical set of findings on the two dimensions. It is, however, a caring nation, as its social democracy suggests, and the Hofstede study ranked it third highest on femininity. The GLOBE study placed it at No. 5 on gender egalitarianism, which is similar to Hofstede's masculinity-femininity dimension. Still, open and frank discussions and debates are encouraged, and there are many political parties representing various points of view, some of which espouse platforms outside of the social democratic framework. Such minority parties are generally weak.

Furthermore, as these extreme rankings suggest, it takes a delicate balancing to ensure that individual expressiveness and group norms are both respected. Thus, even if Danes are wealthy, they play down this aspect of their lives, thus deemphasizing the power distance between themselves and others—for example, by not having large and ostentatious residences—even at the expense of individualistic expressiveness. Part of this de-emphasis comes from the Lutheranism that pervades the culture, although few Danes attend church on a regular basis. The political system is rooted in social norms that are enforced by subtle peer pressure and the Lutheran heritage with its emphasis on community and distrust of ostentation, which reinforces modern social democratic policies (Kuttner, 2008). The Danes are tolerant about everything except intolerance, which is one of the reasons that they were very surprised at the extreme negative reactions from around the Muslim world in 2005 to the decision of the newspaper, *Jyllands-Posten,* to publish cartoons depicting the prophet Muhammad.

Interdependent individualism as practiced by the Danes was articulated by a few critical writers, including Nikolai Grundtvig (1783–1872), a Danish minister who studied in England. He observed that the educational community at England's Trinity College emphasized mutual respect, which was in marked contrast to the more authoritarian educational system then in place in Denmark. Fundamental to his way of thinking is that people must know themselves before they can know others, but this requires being able to say *we* and mean *I* at the same time. People must understand that they are unique and interesting without making the mistake of believing that they are the only individuals who are interesting.

However, other influential Danish writers have conceptualized interdependent individualism with a much more restrictive set of guidelines. Aksel Sandemose described the Law of Jante in such a way in a well-known novel, *En flygtning krysser sitt spor.* Briefly, the protagonist commits a murder in his hometown, but readers learn that the conformist values in this hometown are essentially responsible for the murder, as these values destroyed the protagonist's ability to act against them. The Law of Jante includes the following (cited in Thomas, 1990): Thou shalt not believe that thou art: something, as good as we, more wise than we, better than we, more knowledgeable than we, or greater than we. The other side of this concept of *janteloven* is that companies may be open to innovative ideas from their users; in business, this provides a fertile ground for citizen-driven product design. For example, the Danish company Lego is famous for tapping its customers to help develop its robotics kit (Fitzgerald, 2007).

Interdependent individualism is usually enacted in Denmark in a gentler manner than the Sandemose description suggests, but the essential point is that there is a tendency toward conformity, and the Christmas luncheon provides a major escape valve that allows average citizens to express forbidden thoughts and emotions. For many Danes, and for many citizens of northern European nations, this valve is not sufficient to ward off the melancholia that may well be caused by the dreary weather, which lasts for long periods each year, and the social pressures on Danes, as suggested by the Law of Jante.

Geographic Ambivalence

After the 1864 war in which it lost Schleswig and Holstein, Denmark was the second-smallest nation in the world. It could well have ended up as part of Germany after Germany became a nation in 1870, and Denmark certainly faced this threat again during World War II. It is well known as a small Scandinavian nation, but its culture can be considered a mixture of Scandinavian and German. Given all of the threats it faces, it is not surprising that Denmark joined NATO at the earliest opportunity and ahead of its Scandinavian neighbors; it also helped to found the United Nations after World War II. Its military is virtually nonexistent, composed in large part of a medical corps set up to deal with national emergencies. Denmark is a member, but a relatively reluctant one, of the European Union (EU) and was known as the "footnote" member because of all of the items to which it would not give full approval, thus requiring footnotes noting Danish disapproval.

Commentators sometimes portray Denmark as the Italy of Scandinavia, suggesting that it has a much more relaxed approach to life than its neighbors. It is no accident that the famous Tivoli Gardens are located in Copenhagen. This 20-acre amusement park has more than 20 restaurants and regular musical shows. One American commented on this relaxed atmosphere:

> There's a jovial warmth. Danes drinking their Tuborg or Carlsberg beer join for a song. Clothes and colors seem so varied and merry. Couples hold hands and eat their soft ice cream. Children run loose with pockets filled with candy. At those rare moments both the security and the spontaneity of life in Denmark cannot be exceeded. (Heinrich, 1990, p. 48)

In a recent study of international happiness involving 80,000 respondents, Denmark was the surprising first-place winner. According to Professor Kaare Christenson of the University of Southern Denmark, who was interviewed on the CBS newsmagazine *60 Minutes,* "What we basically figured out was that although the Danes were very happy with their life, when we looked at their expectations, they were pretty modest." Free health care and subsidized child and elder care may also play a role.

There is also a relaxed approach to sexual relations in the areas of nude bathing, free access to pornography, and communal marriage. Escort services are readily and legally available and help to supplement the Christmas luncheon in terms of emotional expressiveness.

Still, this relaxed atmosphere is counterbalanced by an emphasis on predictability, simplicity, and practicality. Danish lawns and flower gardens are carefully tended and parks are clean. Danish-modern furniture is world famous for its simplicity and practicality. Similarly there is an emphasis on science in the educational system. Denmark's most famous scientist was Niels Bohr (1885–1962), who was a Nobel laureate in physics; only seven other nations equal or surpass Denmark in terms of Nobel laureates in physiology or medicine.

One of the challenges that Denmark faces is an increasing shortage of labor. Young Danes are global citizens, who take advantage of the great education in Denmark and move to other EU countries where the taxes are comparatively much lower than the possible 63% upper limit in Denmark (Dougherty, 2007). This marginal tax rate results in income redistribution that makes Denmark almost the most equal society in the world but heralds its own set of problems. Low-skilled immigrants consume a disproportionate share of the public services and although the Danes get a lot back for their taxes, there is not much margin for complications like high unemployment, which may strain the budget in unexpected ways (Kuttner, 2008). In general, however,

Danish history, its geographical location, and its small size all contribute significantly to a feeling of ambivalence, particularly geographically. Denmark is a small but prosperous and highly educated nation that actively seeks to be Scandinavian, European, and part of the world community.

Denmark wants to retain its unique culture but recognizes the need to be flexible and adaptable in a rapidly globalizing world. One example is the concept of "flexicurity" in the workplace, which essentially provides substantial unemployment and welfare benefits but imposes very few restrictions on hiring and firing by employers. The five key elements of the flexicurity model are: full employment, strong unions recognized as social partners, fairly equal wages among different sectors, a comprehensive income floor, and a set of labor market programs supported by the government, which spends 4.5% of the GDP on customized retraining, transitional unemployment assistance, and wage subsidies (Kuttner, 2008). Denmark's need for control and group effort is clear, as is the need for individual expressiveness, which should help it to adjust successfully. The Christmas luncheon is a critical vehicle for maintaining the balance between these competing demands.

Coziness

The Christmas luncheon emphasizes many cultural values that Danes stress, and perhaps the most important is making a social situation comfortable, cozy, and intimate *(hyggeligt)*. This concept is closely related to the German concept of *gemütlichheit*, also translated into English in the same way. Coziness is a concrete demonstration of Denmark's geographic ambivalence, involving both Danish and Germanic features.

The luncheon is frequently elaborate, with everyone bringing dishes that are appealing and cooked at home, and the setting tends to include many lighted candles and sometimes a lighted fireplace. Everyone is encouraged to participate in the fun and enjoyment, not only for 1 or 2 hours, as is the custom in the United States, but for many hours. All willingly stay.

In Denmark most employees reside in the same city and work for the same organization all of their lives. This provides a sense of continuity and history that is lacking in many American organizations, and rituals and myths from the past are important. At the Christmas luncheon it is not unusual for such organizational rituals and myths to be emphasized, such as the company president leading a toast to good fortune in the coming year. Hilarious stories from the organization's past are often retold. All of these activities enhance the sense of coziness and commonality. This also reinforces Denmark's culture of workplace collaboration, which produces win-win outcomes for both corporations and their employees (Kuttner, 2008).

Throughout the year, and particularly during the Christmas season, Danes try to create a sense of coziness in their homes. Sometimes this sense of coziness is accompanied by an expression of patriotism befitting a small nation, for example, decorating the Christmas tree with the national flag. However, this is not done at the Christmas luncheon.

Originally the luncheon was a family affair held after Christmas. Sometimes an entire small village, of which there are many in Denmark, would hold a luncheon. Today, however, the Christmas luncheon has become the norm in business. Family-focused Christmas and after-Christmas luncheons at home, involving close friends and relatives, are similar to but different from the Christmas luncheon held at workplaces.

In summary, Denmark is similar to but different from its Scandinavian neighbors (see Chapters 9 and 10), and the features of the Christmas lunch highlight both similarities and differences. Its geography, small size, and history must be understood to fully comprehend the importance of the Christmas luncheon and of such key cultural values as interdependent individualism and coziness. While the luncheon has moved into the Christmas season and the workplace, it reflects values that are remarkably persistent and that are unlikely to change in a nation that is sensitive to its place in a globalized world economy.

PART IV

Other Egalitarian Cultures

In egalitarian cultures, countries have a high tax rate that allows them to offer generous support for many social programs. There is a belief in such cultures that government is responsible for the welfare of all of its citizens and can solve many problems that business firms by themselves cannot. In Part III we described Scandinavian egalitarian cultures separately because of the unique features found in them, such as the fact that legal financial penalties are related to one's assets, as in the case of Swedish day fines. In Part IV we describe other egalitarian cultures that have high tax rates supportive of many social programs. These national cultures include Germany, Ireland, Canada, and France.

CHAPTER 12

The German Symphony

Beat your wings and uproot trees. You are the wings. You are the tree. You are Germany.

—A recent German public service campaign
to raise national confidence

Germans, much like the Japanese and Chinese, represent a major challenge to American understanding, even though the largest group (about 30%) of Americans is of German descent. When Americans discuss Germans, they frequently describe them in terms of an excessive emphasis on rules and order. Sometimes, the rules stretch the bounds of credulity.

But rules in and of themselves provide only a very limited and stereotypical view of German culture. To gain substantial insight into the culture, it is important to understand the essentials of German history. Germany was not a nation until 1871, whereas many other European nations, such as Spain, England, Poland, Sweden, and France, had achieved this distinction several hundreds of years earlier. The Romans, whose empire was all-encompassing, feared and detested these "uncouth" peoples and left them alone after tribes led by Hermann the Great defeated them in 109 BCE. When other nations used the name *deutsch* or German, the connotation was pejorative (Schulze, 1998). Even the German mercenaries who fought in the American Revolution were objects of derision and considered ill-disciplined and lacking in courage. Finally, through the efforts of Frederick the Great and his father, a strong Prussia emerged between 1740 and 1780 and its military and governmental leadership led to the integration of many small kingdoms into the Germany of 1871. In fact, Germany operates as a federal confederation in which the 16 states are independent from one another on many issues, largely because of this historical background.

However, given its lack of an illustrious history, some Germans romanticized the 8th-century rule of Charlemagne, leader of the Holy Roman Empire, into what they called the First Reich. Such romanticism is found in most if not all nations as they seek to explain their past and the manner in which it relates to the present, but the grandiosity of the German perspective became distorted during Hitler's time. Although Charlemagne ruled almost all of Europe, his influence was brief and his three sons began the division of Europe that led indirectly or directly to the creation of several European nations. Germany, being part of Western Europe but bordering Eastern Europe, became host to a large number of ethnic groups who settled within its boundaries.

The Second Reich began in 1871. This period was marked by militarism but also by rapid economic growth, an emphasis on education and culture, and the development of the first modern welfare system. Several German universities were recognized as among the best in the world; there was ample government support of the arts, including the symphony; and even Otto von Bismarck, the Iron Chancellor, demonstrated his generosity by championing the first modern social security system for retirees, ensuring that they would receive a government pension. However, Bismarck selected the age of 65 for pension eligibility because almost all citizens died before that time.

Furthermore, although children study World War I in school, few realize that the Versailles Peace Treaty of 1918 imposed extremely difficult conditions on Germany, which resulted in great hardship and inflation in the 1920s. John Maynard Keynes, the great English economist, resigned from the Peace Commission because he foresaw the consequences of these conditions. A famous cartoon in the 1920s highlighted the economic breakdown by showing a man using a wheelbarrow of money to purchase a loaf of bread. The Depression of the 1930s only worsened the situation; numerous political parties vied for power, and street fighting involving these parties became common. Ironically Hitler, leader of the National Socialist Party, rose to power when Chancellor Hindenburg asked him to form a minority government, in large measure because most Germans at that time saw him as a source of stability and a counterforce to the street fighting and general uncertainty. This was the beginning of the presumptive 1,000-year Third Reich, whose atrocities and brief history have been abundantly documented elsewhere.

Postwar Evolution

Modern Germany began in 1945 when the Allies took control of Germany. The Anglo-American influence on Germany became evident in a flourishing democratic government and a small number of parties began to vie with one another in open and free elections. This influence was also evident in the support of labor unions built in terms of Anglo-American collective bargaining. Finally, the Marshall Plan of 1947, which provided the financial base for effective business and commerce, secured the nation's future. For the first time in modern history, a defeated nation was not subjected to undue hardship but offered the possibility of recovery. By the 1980s, Germany, although consistently portrayed as militaristic by the Western press, was in fact antimilitaristic. The standing joke in the 1980s in Germany was that the country had already won the Third World War: the economic war. There is a basis in fact for this assertion, as Germany far outstrips both the United States and Japan in terms of export sales per capita.

Today NATO, led by the United States, openly criticizes Germany for its lack of interest in military preparedness and low level of involvement in European military matters, such as the struggles in the Balkans. Germany is also facing considerable pressure from the United States to send more troops to Afghanistan as it has a 250,000-member military and is Europe's largest economy and one of America's staunchest NATO allies. However, as an outcome of its Nazi past, there is considerable pressure from within the country based on popular opposition to wars seen largely as U.S. problems and also the reluctance to send soldiers into battle (Whitlock, 2008). Also, Germany's system of social democracy may not be appropriate in a globalized world, as its extensive regulation of industry and commerce and high labor costs have motivated some German firms to move their operations into other nations. In particular, small and medium-size firms (the Middlestrand) have been adversely affected, and such firms have been the backbone of German industry since World War II.

In 2008, *The Wall Street Journal* argued that with the expected U.S. slowdown, Germany was the best hope for sustained growth in the Euro-zone economies and that domestic demand from German consumers needed to pick up to offer a critical prop to the other ailing economies in the region (Emsden, 2008). German behavior as viewed through the symphony suggests that, although change may be slow, it will occur systematically and logically. A few years ago, *The Economist* wondered if Germany would go the way of Japan and in late 2008, the world's fourth-largest economy slipped into recession along with the rest of the Euro-zone thanks to an overreliance on exports and limited consumption at home.

In Hofstede's (2001) study of 53 nations, the rankings of Germany were as follows, with 1 being the highest value: 43 on power distance, 15 on individualism, 9.5 on masculinity or aggressiveness and materialistic behavior, 29 on risk acceptance, and 11.5 (out of 35 nations studied) on long-term time orientation or the willingness to defer present gratification for future success. Similarly several surveys indicated that about 25% of Germans desire high social status, versus 2% of Swedes and 7% of Americans (Triandis, 2002). In the more recent GLOBE study (House et al., 2004), Germany scored in the middle on power distance, low on both in-group and institutional collectivism, in the middle on gender egalitarianism, low on risk acceptance, and high on future orientation. These findings are consistent with the cauldron of history out of which modern Germany emerged.

Explaining the Symphony

As it was developing economically and politically into a nation, Germany was attentive to its culture and arts, particularly the symphony. The essence of Germany can be experienced through the sounds and sights of the symphony, which was created in the German territories in the 16th century. Its origins can be traced to the early Italian opera (Copland, 1939/1957). What originally started as chamber music or operatic accompaniment, with a few musicians and a relatively uncomplicated score, matured into full sections of woodwinds, brass, strings, and percussion, the occasional piano, and a long list of creative sound effects. Certainly the most enduring achievements of composers from all areas have been accomplished in the symphony. Analogously, this endurance is reflected in the staying power of the German society and culture.

Haydn and Mozart developed many of the features of the symphony. Beethoven, however, single-handedly created a colossus that he alone seemed able to control (Copland, 1939/1957). The symphony lost all connection to its operatic origins. The form was expanded, the emotional scope broadened, and the orchestra stamped and thundered in a completely new and unheard of fashion. The composers following Beethoven added important innovations, including a cerebral emphasis and an atonal perspective. Today the symphony is established as firmly as ever, but it was and will remain unfinished as innovations enrich it.

As for the musical instruments, single instruments had been played for centuries. However, prior to the 16th century, musicians had not mastered the art of playing together (Schulze, 1938). Since that time nothing matches the capacity of the greatest of all musical instruments, the symphony orchestra, which includes about 100 instruments and uses about four fifths of the range of human hearing (Schulze, 1938). The symphony's power emanates through the sound of musical instruments collected from all over the world. Schulze (1938) captured the historical significance of the symphony's musical instruments in the following summary:

> Snake charmers from the Orient contributed the oboe. Ancient Greeks before Homer developed the primitive clarinet. Horns were probably first used in the religious ceremonies of the God-fearing Israelites. Conquering and militaristic Rome brought trumpets to a high point of favor, while fifty generations later this same land took pride in its superb violins. At a late date, tubas sprang from Europe, and still later the saxophone and sarrusophone. Africa is famed for its drums and Greece for its pipes of Pan. (p. 5)

The music and performers are brought together by the conductor. A skilled baton unites the disparate personalities and talents of the musicians so that they perform as one, at the literal level of the meaning of the phrase, "in concert." The various musical divisions of style and perspective, such as those between entire sections of strings and the brass and between the individual flutist and percussionist, are melded and molded by the conductor to produce a unified sound.

The orchestra, like society, is made up of individuals, each with his or her own likes and dislikes. However, for the greater good, which is the music, individual preference is subordinated to the demands of the conductor and the needs of the symphony. Not everyone can be a soloist, nor do all wish to be. In any event, the soloist's time of improvisation is brief and the conductor soon signals that it is time to return to the history of the piece. This discipline or subordinated individualism—the voluntary submission of individuals to the whole, the guidance of the conductor, and the shared meaning of the music—allows the symphony orchestra to flower and flourish.

Germans love symphonic music. Much as in the past, descendants of the aristocracy living in the old castles along the Rhine still arrange concerts of baroque and chamber orchestra music, played in secluded rooms illuminated only by intimate candlelight. These concerts are peaceful affairs that mentally transport listeners to an idealized Germany of old. Germans also frequent the symphony on a regular basis; the former West Germany, with its population of 66 million, boasts more than 80 symphony orchestras. Frequently visitors to small German towns and villages will be able to attend symphonic concerts given by local musicians in churches, typically on Sunday afternoons and weekend nights and particularly during and around religious holidays such as Easter and Christmas.

This societal and cultural love of music has produced many of the world's great composers, including Haydn, Mozart, Schubert, Schumann, Bach, Handel, Strauss, and, of course, Beethoven and Brahms. It has also produced several world-class conductors, including Herbert von Karajan and Rudolf Kempe.

Many Germans play musical instruments as a hobby, and of those, many belong to informal musical groups that carry forward the tradition. From the horns of the Alps to the brass polka bands and up to the apex of the magnificent operas of Wagner, music is an integral part of German life. German music is not only integral, it is serious; it is not generally an outlet for emotion and craziness as it is in the United States and other societies. Music is foreground, not background. It is meant as a collective experience intended to enrich life. Even the audience at German symphonies reflects this seriousness, as its members normally dress more formally than is typical in the United States. Members of the audience listen intently to the music and are silent until a major movement of a symphony is completed, at which time they tend to respond enthusiastically but with decorum.

In this chapter we emphasize the following features of the symphony as reflective of German culture and values: the diversity of the musical instruments, positional arrangements of the musicians, the conductor or leader, precision and synchronicity, a unified sound, and the unfinished nature of the genre.

Diversity of the Musical Instruments

Prince Metternich, a 19th-century Austrian statesman, once remarked that the characteristic feature of the Germans was not unity but multiplicity (Cottrell, 1986). This observation still has considerable validity today. Germany is divided into an astounding array of ethnic and religious groups. Historically and culturally each of these ethnic groups evolved separately, and each continues to have its own characteristics. Since the 1950s the population of Germany has become more diverse. Millions of foreigners have migrated to Germany seeking employment, citizenship, or asylum. The number of foreigners has grown from 500,000 after World War II to more than 6.7 million or about 8% of the population with another 7 million or so naturalized immigrants. Frankfurt is the country's most diverse city with 40% of its population either holding a foreign passport or coming from an immigrant background and this has made the city very tolerant, according to its head of the integration department (Siegele, 2006).

One fact is fundamental to understanding this varied array of ethnicity: Throughout its long history, Germany rarely has been united (Solsten, 1996). Charlemagne's 9th-century Holy Roman Empire was more symbolic than real within a generation. Medieval Germany was marked by division. The Peace of Westphalia of 1648 at the conclusion of the Thirty Years War left an exhausted Germany divided into hundreds of states. In fact, until the unification of West and East Germany on October 3, 1990, Germany had known only 74 years (from 1871 to 1945) as a united nation with one capital. As noted earlier, for most of the two millennia that central Europe has been inhabited by German-speaking peoples, the area called Germany was divided into hundreds of states. Like the musical instruments of the symphonic orchestra, the people of modern Germany can trace their ancestral lineage to a multitude of past and present countries and cultures.

The Force of Immigration

Historically immigration has been a primary force shaping demographic developments in Germany (McClave, 1996). After the first unification in 1871, a significant source of population growth was an influx of immigrants from Eastern Europe. Adding to an already ethnically diverse population, these immigrants came to Germany to work on farms and in mines and factories. This wave of immigrants was the first of several that would occur over succeeding decades, thereby increasing an already diverse population.

Foreigners began arriving in West Germany in large numbers in the 1960s after the construction of the Berlin Wall. Since then, partly due to unification, the foreign population in Germany has increased substantially. Several million ethnic Germans from Eastern European countries, especially the former Soviet Union, began migrating to Germany with the end of the Cold War in Europe. As of 2005, Turks made up the largest group (more than 2.6 million) of registered foreigners living in Germany, followed by immigrants from European Union (EU) member states such as Italy, Greece, Portugal, and Spain. Because of the higher birth rate of foreigners, about 1 of every 10 births in Germany is to a foreigner. Likewise, Germany's basic law offers liberal asylum rights to those suffering political persecution.

Compared to the United States and its population, Germany is not a melting pot society, and Germans are not mobile. Many have stayed in the same geographic region and even the same house for generations. Germans are not accustomed to meeting and interacting with strangers at home, partially because most people live in apartments or small homes. As a result, they treat foreigners with a certain amount of wariness (Hall & Hall, 1990). Still, Germany's foreign residents remain vital to the economy, parts of which would shut down if they were to depart. Also, Germany has the ninth-lowest birth rate in the world, far below replacement (2.1) at 1.3 children per woman. This is problematic in a social democracy offering generous retirement benefits that are dependent on the taxes levied on current employees.

Recently, Germany relaxed its citizenship laws. All children born in Germany automatically received citizenship at birth, provided that at least one parent was born in Germany or arrived before the age of 14 and has a residence permit.

Geographic Variation

Geographically Germany is also distinct. West Germany's absorption of East Germany enlarged its area by about 30%. The territory of the former East Germany accounts for almost one third of united Germany's territory. Now roughly the size of Montana, Germany is Europe's sixth-largest country although the absorption of more rural former East Germany caused the population to be more sparsely distributed.

Germany has had difficulty integrating East and West Germany since they were reunited in 1990, as the costs were greatly underestimated, and there has been a persistently high level of unemployment, about 10% or more, particularly in the East. This has led to tensions between the West and the East. Incorporating more diversity into the new nation automatically created complexity, much as the addition of musical instruments does in the symphonic orchestra. However, although former East Germany was under the communist regime of Eastern Europe for about 40 years, values that were integral to German society such as orderliness, straightforwardness, loyalty, and integrity remained strong, which is one of the reasons that the GLOBE researchers (House et al., 2004) identified a Germanic cluster that included both former West and former East Germany, Austria, Switzerland, and the Netherlands (Gupta & Hanges, 2004).

Various parts of the nation are also quite diverse. Sunny and heavily Catholic Bavaria, for instance, seems to be worlds away from the colder and heavily Protestant north in both temperament and culture. Frankfurt, the financial capital of the nation, is more like an American city than a German city. There are several other large cities, each with its own character. However, this diversity is muted in many ways, including the preference for *Dorfen*, or small villages, surrounding the cities in which large numbers of citizens live.

It is not to be denied that the large number of instruments in the orchestra creates complexity, in about the same way that the many forms of diversity (ethnic, geographic, religious) do in German society. The need to control this complexity is addressed by the other features of the symphony.

Positional Arrangements of the Musicians

When a symphonic orchestra is seated in front of the audience, the stage is crowded. To solve the problem of having large numbers of musicians on stage, and to make it easier for the musicians to play their instruments, the conductor arranges the seating so that there are four separate sections: strings, woodwinds, brass, and percussion. This compartmentalized arrangement of seats helps to maximize the musical instrumentation and the quality of the sound, as does the seating of musicians within sections in terms of skills and responsibilities, for example, first violinist, second violinist, and so on.

Such crowdedness is mirrored in German society. Germany is about 66% of the size of France, the largest nation in Western Europe, and only 4% of the size of the United States. More important, it is a crowded nation relative to the United States: 230.5 residents per square kilometer versus only 31.7 for the United States. Viewed from the perspective of both the symphony and crowded conditions, it is easy to empathize with the compartmentalization of German society, represented by their affection for privacy and respect for others. Apartment dwellers are wont to leave polite but anonymous notes asking their noisy neighbors to scale back the sound.

Just as the positional arrangement of the musicians in the orchestra separates them from one another, the home separates the individual from the outside world. Throughout the year—and most especially at Christmas—Germans greatly cherish the privacy and security of the home. In Christian Germany (about half of the Christian population is Catholic and half Protestant), Christmas remains a dutifully observed religious holiday, a time when *gemutlichkeit* strongly prevails over the people. This word translates simply "as 'comfort' or 'coziness,' but it has wider connotations—of the hearthside and deep content, of home cooking and family security" (Cottrell, 1986, p. 123).

Home as Haven

The home (*das Heim*) is at once a haven from the bustle and stress of the working world and a place where personal status may be more solid and unthreatened. It is a castle and a refuge from the outside world. Homes are protected from outsiders by a variety of barriers: fences, walls, hedges, solid doors, blinds, shutters, and screening to prevent visual and auditory intrusion (Hall & Hall, 1990). Front yards, although beautifully maintained, are rarely used. Outside activities such as sitting in the sun or visiting with family are restricted to the back yard, away from the street and the eyes of others. The home is the most important possession to Germans and life in the home with their family is treasured.

Germans are extremely prideful of their homes and German homemakers spend a lot more time than their U.S. counterparts keeping everything spotlessly clean. The second author was visiting close German friends in Germany when she was informed that they lived in mortal fear of neighbors dropping by, running their fingers over the tops of shelves and inadvertently finding dust. In turn, the second author's mother would make the servants clean and dust their home in India several times during the day when they had German visitors. In addition, it is remarkable to see some Germans vacationing on the seashore of the North Sea as they shovel sand on three sides to a height of 4 feet to define the space or sand home that they will occupy for only a few hours.

Like the home, the door "stands as a protective barrier between the individual and the outside world" (Hall & Hall, 1990, p. 40). Germans tend to keep doors closed. Doors are often thick and solid, and their hardware heavy and tight-fitting. The closed door preserves the integrity of the room, provides a boundary between people, and minimizes interruptions and accidental intrusions.

A related issue is the physical distance between people when talking to each other face-to-face. Germans keep more space between them than do individuals in nations such as France, Italy, and Thailand. Supposedly this distancing is a protective barrier and psychological symbol that operates in a manner similar to that of the home. In addition to physical distance, many Germans also possess a distinctive sense of aural distance. German law actually prohibits loud noises in public places during certain days and times of the week. In one town some residents filed suit against parents who allowed their children to play in the local playground during lunchtime and early evening. As noted above, Germany, particularly the western part, is a crowded country, and the emphasis on both physical and aural distances may be a reflection of this fact.

Respecting Formalities

The compartmentalization in German society also is actualized in the German penchant for formality. Within the German home family members abide by formal rules of behavior designed to give one another distance and privacy (Hall & Hall, 1990). These patterns of formality extend in adulthood to relations with friends and coworkers in the office. Germans tend to introduce themselves by their family name, as opposed to their first name. The use of formal and informal pronouns in the German language indicates the nature of this relationship. The familiar *du* (you) and first names are reserved for close friends and family. The difference between "close" friends and acquaintances is distinct. Those people at work or school whom Americans might label friends usually are referred to as colleagues by the more conservative Germans. This sharp distinction may be related to the socialization process throughout the German life course. Most Germans develop two or three close friends during their early years, relationships that often are maintained for life. It usually takes Germans much longer to categorize someone as a friend, but once this determination is made they will treat this individual in a very special manner.

Although the following illustration is extreme, it helps to bring these concepts into perspective. An Irish professor and a German professor worked side by side for 3 years at an international agency in Brussels. Although they were of equivalent social status, their relationship was impersonal. In Germany the professor was formally addressed as "Herr Professor Doctor" followed by his last name, while in Ireland the Irish professor was called "Professor" or by his first name. Over beer one night the German professor indicated that he and the Irish professor could now be more informal and friendly with one another and that the Irish professor should feel comfortable addressing him as only "Doctor" in the future.

On a more serious note, under an obscure Nazi-era law, anybody who has not earned a Ph.D. or medical degree in Germany is barred from using the title "Dr." Seven U.S. citizens were recently subject to criminal probes for "title abuse" because it was *verboten* to go by "Dr." although the legitimacy of the degree or the right to conduct research was not questioned. The correct form is to use the term Professor and add Ph.D. to one's name, followed by the university affiliation (Whitlock & Smiley, 2008).

A similar emphasis on compartmentalization is found in the degree of politeness that Germans extend to acquaintances and strangers. Germans are careful not to touch accidentally or to encourage signs of intimacy (Hall & Hall, 1990). Gestures and smiling are restrained and Germans do not generally smile at customers or coworkers. On the other hand, Germans do maintain eye contact in conversations to demonstrate they are paying attention. At times Germans may appear stiff, distant, and forbidding. However, this reserved nature should not mistakenly be interpreted as a lack of friendliness. They are usually gregarious after they get to know you.

Business Practices

These elements of compartmentalization, much like the positional arrangements of the orchestra, also extend to German business practices. Although Germans place a great deal of emphasis on promptness, delays in such areas as deliveries occur because of the compartmentalization of German life and business. Products may be delivered late without an apology. A salesperson may take an order and then leave on vacation without asking a coworker to follow up on the order (Hall & Hall, 1990). Likewise, problems in the shipping division of a German business may not be conveyed to the retailer and the customer.

Typical German business organizations tend to be more compartmentalized than American organizations, most probably because of the high value attached to respecting status and power within the hierarchy. The owner of an insurance company constructed a building in Frankfurt in the form of a hierarchy that is symbolic of this perspective. Each level of the building contained fewer people per square kilometer. The owner, surrounded by a few key staff people, occupied the top floor. The building itself looks like a line drawn at a 45° angle and is called "the ladder of success."

In recent years German business organizations have emphasized the use of cross-functional teams. Still, information flows less easily from one department to another in Germany than it does in the United States, as we might expect when privacy is so highly valued. As a result, quick decisions on business issues are difficult to reach. Inevitably, decisions often are made without the benefit of current information gleaned from fast-changing markets.

The importance of the German executive's office cannot be overemphasized. As in the United States and many other nations, the size of the office, having a personal secretary right outside of it, and related factors are key markers of status and hierarchy. But what is distinctive about the German office is that managers see it as inseparable from and as an extension of their personality (Hall & Hall, 1990). Hence the importance of closed doors, which visitors should generally avoid opening without permission. It is always a good idea to knock on a closed door and then wait to be admitted.

Furthermore, once inside the office, visitors should be careful not to violate the norms of physical space found in Germany by moving a chair too close to the manager's desk. Germans tend to stand farther apart from North Americans when talking so that moving a chair closer and thus rearranging the office furniture will be perceived as highly insulting. One German manager became incensed that visiting American salespersons would invariably move their chair closer to his desk, presumably to establish a physical sense of intimacy in trying to finalize a sale. This executive took the extreme measure of bolting the visitor's chair to the floor, thus frustrating any movement that a visitor might attempt.

As both the Hofstede (2001) and GLOBE (House et al., 2004) studies confirmed, Germans tend to be risk-averse and are less comfortable with strange situations and newcomers than people of many other cultures. They also generally believe that it takes time and meticulous planning to do a job properly. Just as the symphonic orchestra tends to be compartmentalized, so too the German penchant for analyzing projects in depth leads to a situation in which their levels of risk are accurately reflected. This risk aversion may account for the extreme lengths to which German executives will go when analyzing business opportunities before committing to them.

As in many nations, generalizations about such issues as German formality and physical and aural distance need to be tempered by the fact that these characteristics are less notable among younger members of the culture. For example, German university students are quite informal in their interactions among themselves and with others, arrive late for class, and so on. However, once in a full-time position in business, these students tend to revert to the more formal styles of behavior that have been a hallmark of German culture.

Conductors and Leaders

Like the general direction provided by the symphonic conductor in the creation of the single, powerful sound of the orchestra, German leaders have provided direction and guidance to an often fragmented state. One particular study of 22 German orchestras showed that conductors who exhibited a transformational leadership style were more likely to have orchestras of high artistic quality in the presence of a collaborative climate among the musicians (Boerner & Frieherr von Streit, 2005). Historically, however, with the notable exception of Adolf Hitler, Germans have not reacted favorably to charismatic leaders who are intent on leading them to a new world order without any questioning of their authority (Hall & Hall, 1990). Rather, Germans have preferred visionary leaders who are mature and strong enough to delegate responsibility and decision making to competent subordinates throughout the hierarchy.

Frederick the Great, who ruled Prussia from 1740 until 1780 and was largely responsible for creating the basis of the German nation, epitomizes the classic German ruler. He survived the rigors of his Prussian military education to become a philosopher-king, famous for his taste in art, music, and literature. Frederick wrote poetry and scholarly essays, played the flute, composed concertos and sonatas, and maintained a regular correspondence with Voltaire. To this day this ideal of leadership is emulated throughout German society. Corporate executives, for instance, tend to be more well-rounded than their American counterparts, and many of them are members of book clubs that make a serious study of science, literature, and philosophy; about 75% of the CEOs of German industrial companies hold a doctorate.

Clearly, the visionary leadership provided by Frederick resulted in reforms that transformed a previously fragmented state into a major European power. Similarly, corporate executives are expected to approximate this visionary but mature model of leadership and, because of that, to be able to provide the direction that organizational members desire. German CEOs will generally tell foreign marketing executives to see their counterpart responsible for making decisions in this area. German managers expect a great deal of their subordinates and test them by pushing them to perform, but they do not hover (Hall & Hall, 1990). Adler (2007) describes Germany as a "well-oiled machine," with a small power distance and strong uncertainty avoidance. German businesses reduce uncertainty by clearly defining roles. In contrast to high-power distance cultures—where employees want their managers to act as decisive, directive experts—Germans expect their managers to delegate responsibility and decision making to competent subordinates throughout the hierarchy.

Pluses and Minuses

There are, of course, some drawbacks to this style of leadership. Martin Gannon attended several briefing sessions in Germany focusing on the Daimler-Chrysler merger. Some of the differences between the German and American companies, even after the merger in 1997, were noteworthy. Daimler had only four organizational levels, which is consistent with its emphasis on low power distance and delegation of responsibility, whereas Chrysler had nine. However, it was much more difficult for Daimler executives to take action, as their subordinates demanded the right to participate in the decision and, at times, to veto it. Also, the Daimler hierarchy was much more rigid, even to the point that middle managers would not dare to jump organizational levels, which were to be respected by those who wanted to be viewed favorably. But the

American middle managers had no difficulty offering suggestions and ideas to upper-level managers; even Robert Eaton, the Chrysler CEO who was largely responsible for the dramatic success of this company in the 1980s and 1990s, could be easily approached without fear of consequence. In the end, the merger failed and Chrysler was acquired by a private equity firm, Cereberus Capital in 2007. Most analysts agree that the insurmountable cultural differences between Daimler Benz and Chrysler were partly responsible for the failure, although the lack of synergies as well as competition from the Asian automakers also played significant roles.

As indicated previously, the Allies strongly influenced Germany after World War II and this influence can be seen in the area of leadership. Germany developed a unique and democratic form of corporate governance termed codetermination. Unlike the single board of directors in American corporations, German firms with 2,000 employees or more by law must have two boards: a management board *(Vorstand)* and a supervisory board *(Aufsichtsrat),* the second of which must include some employees and labor union representatives. There is also a workers' council *(Betriebsrat),* which has the right to monitor and approve all corporate plans and actions. The *Vorstand* initiates an action, such as downsizing, but it must be approved by the *Aufsichtsrat* and *Betriebsrat,* and finally by the national union. Needless to say, making decisions is slow and complicated, but once decisions are made, they are implemented quickly because of the consensus.

Currently there is a good amount of discussion about this system, and many believe that a single board of directors that can operate quickly is more appropriate in a fast-changing and globalized world. Similarly labor unions are losing members, and only about 23% of the workforce is now unionized. In 2006, the German software company SAP took steps to establish a workers' council with nonunion employees to protect its start-up culture, that of a global high-tech company which mainly employed college graduates.

Precision and Synchronicity

Perhaps the most critical feature of the symphony is precision and synchronicity. Given the complexity of the music and performance, the conductor stresses these two related characteristics. Everything must be done perfectly and everyone must be willing to participate within the boundaries of the performance. Individual soloists are given ample opportunity to show their skills and abilities, but for the greater good of the orchestra, their solo time is short. Then they recede into the orchestra, once again merging their identities and personalities with other orchestra members. Unlike the opera, during which great singers clearly separate themselves from the chorus through their memorable performances of arias, symphonic performers must willingly subordinate their individual selves to the greater good. Similarly, Germans tend to be individualistic, but their subordinated individualism is different from the competitive individualism found in the United States or the egalitarian individualism of Sweden.

Business Rules

There are countless examples of precision and synchronicity in German business. Most Germans are quite conscious of time and how to allocate it efficiently. To many Germans there is no such thing as "free time." In one survey German respondents had difficulty with the category, "I had free time on my hands," when asked to indicate why they were seeking a part-time position. Germans see a sharp distinction between work and leisure, but in both cases, they prefer that the time be used rationally and efficiently.

In both social and business life, tardiness is frowned on. Everyone is expected to arrive at least on time for a meeting, and preferably 5 minutes early. Just as the soloist who comes in off-cue may lose a position with the orchestra, the businessman who comes late for a meeting may lose a client or miss important information. German meetings and negotiations are also long and tend to have well-marked stages. Unlike Americans, who like to get to the bottom line as soon as possible and follow the dictum, "keep it simple, stupid," Germans love historical background and charts.

One American executive who was raised in Germany recalls his presentation before an American board of directors when he was unconsciously using the German perspective. He originally had planned to give an historical overview of a project for about 25 minutes, but he stopped short after 5 minutes when he noticed the vacant and bored looks on the faces of board members. Similarly a German executive who has lived in the United States for many years likes to point out how Americans and Germans would approach the study of elephants, and we leave it to the reader to decide the national identity of the authors:

- The Elephant: A Short Introduction in 24 Volumes
- The Elephant: How to Make It Bigger and Faster in 20 Minutes

Precision and synchronicity also tend to influence communication. Germans communicate in a low-context manner, that is, they directly express their ideas both in written and oral form. Many American negotiators far prefer to deal with Germans than with members of high-context cultures who indirectly and vaguely present their initial ideas. But Germans tend to use a deductive way of thinking that relies on past history and theory, in contrast to the Anglo-American inductive style that tends to use cases and examples to back up an argument. The basic German approach to problem solving is to deduct from a theory or small number of principles, whereas the American approach focuses on the issue of operationalizing a solution as soon as possible.

American businesspeople are frequently surprised when their initial ideas are met with a seemingly hostile response: You are wrong. In reality, Germans must be convinced that they are incorrect, and they are open to such an approach, but the solution must be backed up by logic, data, and consistency. Objective facts form the basis of evidence, and feelings are not usually accepted in business negotiations. Similarly Americans must be not only well-prepared but also willing to discuss seemingly insignificant details. Precision and synchronicity must be honored, and one case study that is casually used to justify a course of action, or even a few case studies, are not ordinarily persuasive.

Similarly, words should mean exactly what they are intended to mean. An American might say "I'll call you for lunch" without really intending to do so, but the German preference for precision and synchronicity will require that this statement be honored. Failing to do so will mark the American as insincere, careless in thought and habit, and culturally insensitive. An extreme example of miscommunication occurred in the Daimler plant in Alabama. German workers assigned to it showed up at their supervisor's home after the first day of work because they had interpreted the statement "Let's have dinner" in a very literal manner.

Perhaps the most extreme example of the German tendency to take words literally involves the late avant garde American music composer, John Cage, famous for such revolutionary productions as *4'33"* in which the performers sit silently on stage for 4 minutes and 33 seconds. A group of German musicians was pondering what Cage meant by "as slow as possible" (ASLSP) in his organ composition, *Organ 2/ASLSP,* which was originally written for the piano. They decided to repair a long-discarded church organ in Halberstadt and, starting in February 2003, began to introduce the notes very, very slowly; intermission is scheduled for 2319 and the entire performance is scheduled to last 639 years. The shortest notes last six or seven months, while the longest last about 35 years (Wakin, 2006). The project is part serious endeavor, part intellectual exercise, and also a tourist draw in a town that is situated 2½ hours from Berlin.

Communication Styles

One distinctive feature of the German language is that the verb tends to be at the end of the sentence. Similarly the main point is often made at the end of a talk, meeting, or negotiations. Much like classical symphonies, meetings start slow, can last for hours, and ultimately build to a climax. At meetings and in negotiations, there is a slow but steady progress through easily recognizable stages during which timing, intonation, and control of voice, speech, and emotions are critical. Thus, in business as in music, sound, tone, modulation, and timing are key to a successful performance.

It is also helpful to demonstrate efficiency in communication, as this reflects precision in the manner in which the relationship will be established. For example, an American will tend to greet a German with, "Hello, Thomas, how are you?" It is probably better to say: "Hello, Thomas, what can I do for you today?"

Finally, the emphasis on precision and synchronicity also includes attire and dressing. Given their relatively conservative perspective, Germans dress more conservatively than Americans, and it is helpful for Americans to look professional and act accordingly. Sitting at ease but upright, wearing clothes that are fashionable but do not take attention away from the matter at hand, and politely listening and contributing insight but without constantly interrupting the precision and synchronicity of the discussion: These are techniques that can help to enrich the communication process and relationship.

The Education System

Germany's unity is also expressed through its educational system, which begins with the *grundschule* or elementary school, where one teacher is sometimes assigned to the same group of children as they progress from Grades 1 through 4 or, in some states, Grades 1 through 6; our discussion will assume Grades 1 through 4, although the total number of years of school is identical whether the *grundschule* extends through Grade 4 or Grade 6.

After graduating from the *Grundschule*, students are assigned to one of three different types of school. In some German states the teacher makes this assignment, while in other states it is based on the results of a standardized examination. Given the importance of this assignment, it is quite common for German elementary students to have tutors as early as first grade so as to improve their chances of success. The assignment to a type of school largely determines each student's career; no other society separates its children at such an early age. It is possible, but somewhat difficult, to change from one type of school to another after the fourth grade.

The first type of school is the gymnasium, or academic school, designed for those who want to pursue a university education. Students attend the gymnasium from Grades 5 through 13, but they can leave after Grade 10 for a non-college-oriented career. If they pursue this option, their degree is equivalent to that of a *Realschule* graduate as described below. Students in the gymnasium are required to pass a very difficult exam, the *Abitur*, before they can be considered for admission to the university, and admission is not automatic but dependent on the number of available slots. Even those who pass the *Abitur* and are admitted are not granted admission to the entire university but only to one of its faculties. Thus, German university students have fewer opportunities than their American counterparts. Also, as the system is very much rigged in favor of class, it is believed to be wasting a lot of the country's human capital.

After Grade 4, other students continue their education at the *Realschule* or "real world" school through Grade 10. Some graduates then serve as apprentices for 2½ to 3 years while attending school one day a week. German industry subsidizes these apprenticeships and hires most of these students after they have completed them. Other *Realschule* students attend *Fachoberschule* (middle school) for 2 more years after the 10th grade and obtain a middle school degree; of this additional 2 years, 6 months are spent as apprentices. These students can attend traditional universities if they pass the *Abitur* or attend the *Fachhochschule* (specialized universities) for at least one year. Normally, these graduates do not attend traditional universities but continue to pursue a university degree in these specialized universities, which emphasize four different concentrations: technical, social sciences, science, and business.

After the *Grundschule*, some students attend the *hauptschule*, and they can leave after Grade 9 without a degree. Most of them, however, leave with a degree and then serve as apprentices for 2½ to 3 years while attending school one day a week. These students also have the opportunity to attend the traditional universities or the specialized universities (*Berufsoberschule*) by going to school for 2 additional years and passing the *Abitur* examination.

In recent years this system, while logically consistent and unified, has become strained. German parents want their children to graduate from the gymnasium, much as American parents want their children to be in "gifted and talented" programs. Today more than 50% of students graduate from a gymnasium, but many of them cannot achieve the career success historically associated with such graduation because of their overwhelming numbers vis-à-vis available opportunities. Also, there are far too many graduates from traditional universities that specialize in the professions such as law and medicine and the universities are quite overcrowded.

Clearly, this educational system reflects the German penchant for order and unity. The sequence of schools mandates particular steps for students to attain the necessary education for whatever they prefer to do occupationally in support of the German effort. In a sense Germans subordinate some individuality to this system so that all of society may benefit. Like orchestra members, the order imposed by this educational system allows Germans to do their individual parts to make the German effort successful.

At present there is a great debate over the relative merits of the German educational system when compared to the American approach. In 2001, Germany experienced what has come to be known as the "PISA shock." PISA refers to the Program for International Student Assessment, a survey done by the Organization for Economic Cooperation and Development; it compares educational achievement in different countries. Germany scored 21st in reading and 20th in math and science among the 31 countries assessed (Siegele, 2006). This was largely attributed to the performance of its "migration-background" students. Reforms are under way and it is believed that in the long run, Germany may have to do away with its three-tier system because of demographical changes and the increased demands from German business. In contrast, the American system seems much more responsive to changing needs, and it is much easier to change career directions in the United States than in Germany.

A Pattern of Order

The subjugation of the individual for the good of the whole also is evident in policies to procure German military personnel. Even in light of geopolitical changes in post-Cold War Europe, Germany continues to maintain a military force structure based on a policy of conscripted service. Although conscription has long been rejected by a majority of the younger generation, the *Bundeswehr* (Army) is accepted by the majority of Germans as an instrument for defense and maintaining peace (Fleckenstein, 1999). This is largely due to the right of conscientious objection and civilian alternative employment. National military service and civilian alternative service are both regarded as valid options open to conscripts. In fact, one of the many arguments for the continuation of conscription is that German society would not be able to sustain the loss of the social contributions provided by these objectors. Clearly this ability of German youths to engage in a form of national service for the public good is analogous to the subjugation of the separate instruments to the unified sound created by the symphonic orchestra.

Germans even manifest their unified sound or unity in their love for flowers and gardens. The window sills of most German houses are lined with flowers. German gardens often contain a complete variety of vegetable plants and fruit trees, which are displayed and placed in a systematic way. Germany as a whole is dotted with thousands of small villages and sprawling urban centers, separated by fields or grasslands plowed and maintained in a systematic and orderly manner (Cottrell, 1986). The same can be said of the vineyards that dominate much of the land in southern Germany. This ordered beauty and sense of unity are most appreciated throughout Germany.

Furthermore, the extremely efficient rail and road networks tend to emphasize this sense of unity. There is a clear parallel between the efficiency and unity of these networks and the predictability, regularity, and unified sound of the symphony. Chronometer-like timing, precision, and conformity underlie both music and transportation activities. On seeing these German transport systems—more than 4,500 miles of autobahn (high-speed highways) and the *Deutsche Bundesbahn* or nationalized railroad, with its luxurious Trans-European expresses and fast, reliable intercity services—it is difficult to believe that, for centuries, German transportation systems lagged far behind those of other West European countries (Cottrell, 1986). In addition, foreign travelers to Germany will find that trains rarely leave their originating station late or fail to arrive at their destination on time.

Festivals and Celebrations

Just as the symphonic orchestra unites the individual musicians, the old tradition of folk festivals unites the German population. There are literally thousands of festivals in Germany each year. They range from village feast days to "elaborate civic displays of organized chaos" (Cottrell, 1986, p. 118). Typical of festivals held all over Germany is beer drinking, abundant food, brass bands, dancing, and colorful parades. Although at first it may seem that carefree festivals run counter to the dour and serious German stereotype, such communal activities actually are among the most popular and oldest of traditions, and they reinforce the feelings of group solidarity.

Most German festivals have a religious origin. *Karneval* is a Catholic celebration, a period of frivolity and deliberate self-indulgence before the period of Lenten fasting. In the 15th century, when ascetics criticized Germany's wild and unrestrained pre-Lenten celebrations, a churchman reasoned in defense that "a wine barrel that is not tapped will surely burst" (Cottrell, 1986, p. 118). To this day his comment remains as good an explanation as any of the extraordinary way millions of Germans can, once a year, suddenly cast aside inhibitions and everyday formalities and throw themselves wholeheartedly into the madcap fun of *Karneval* (also called *Fasching* and other regional names).

Although not directly traced to religious origins, there is nothing else in Germany to compare with the sheer overindulgence of *Oktoberfest*. This festival and many wine festivals celebrate the earthly yield of crops. Throughout the wine-growing districts, there are concerts, fireworks displays, dances, and nonstop pouring of wine. Germans cling fondly to old traditions designed to give the season a truly magical air and thus they are reminded of their unity as a people. Today, *Oktoberfest* celebrations are almost as ubiquitous as St. Patrick's Day festivities across U.S. towns and serve to bring together U.S. Americans of German origin and their non-German counterparts.

Similarly, festive events that bring Germans together regularly are sporting activities. In contrast to the United States, sports are not sponsored or organized by the schools. Instead, completely independent of schools, youngsters must join sports clubs for such competition. Soccer and tennis attract millions of participants and spectators. There are hiking trails and maps for trails all over the country. One of the more enjoyable activities in Germany is to hike up to a mountaintop restaurant. Even traditional U.S. sports, such as football and basketball, are beginning to attract more participants.

Two common elements run through all of these activities. First, they are enjoyed at all levels of society; second, they tend to be organized into, or at least enjoyed, in groups. These aspects of sport may provide relief from workplace stresses while providing the Germans the sense of group identity that they tend to enjoy. Unlike their U.S. counterparts, typical German workers have ample leisure time to enjoy these many activities, as the average number of hours devoted to work is much lower in Germany and holidays are much longer and more frequent.

The Unfinished Symphony

Many of the great German composers, such as Beethoven and Brahms, bequeathed to posterity unfinished symphonies, and scholars love to debate what the composers intended and how they planned to realize their intentions. Similarly Germany is a work in progress.

Because of the unification of East and West in 1990, one of the chief aims of the constitution, national self-determination for all Germans, has been achieved. However, Germans of the now-enlarged Federal Republic feel as if they inherited a huge estate that is proving to be more of a liability than an asset, at least in the short run. Unification may be described as an expected marriage of two partners from totally different backgrounds who started with a short honeymoon followed by a period of recriminations (Kettenacker, 1997). West Germans are said to expect gratitude; East Germans, an attitude of charity and caring. Both, however, are unreasonable propositions.

In recent years, the east has gradually been catching up to the west and the surprising divide now both in the east and the west is between the north and south with a wealthy and growing south and a struggling north, which mimics pre-war patterns of economic development (Benoit, 2007). Thus, Germans have embarked on a search for a "new" national identity that will shelter them from their worries and anxieties and overcome their loss of direction. Soul-searching, always a favorite German pastime, can no doubt be recognized as more justified than ever. In 2005, the German government launched a campaign to raise national self-confidence with the motto, "*Du bist Deutschland*" (You are Germany) featuring famous Germans like race car driver Michael Schumacher and Olympic skater Katarina Witt. Angela Merkel's charge as Germany's first woman chancellor is to put Germany back up on top by reforming the education system, creating service jobs, integrating immigrants, reducing unemployment, and making it a top tourist destination in 10 years.

Germany's comparatively short history as a nation-state since 1871 is the master key to understanding its political culture, from which a new identity must be distilled (Kettenacker, 1997). Clearly World War II and its aftermath changed Germans in some very material ways, as we have seen. Almost all of these changes have been positive, including German leadership in the EU, its economic prosperity, and its successful implementation of new organizational approaches in government and business. But like many European nations that are now part of the EU, Germany has dual allegiance, both to the nation and to the EU. At least some surveys have suggested that proportionately fewer Germans identify primarily with the EU than do citizens of other nations. There is nothing wrong about such identification, provided that Germany continues to maintain its expected leadership role, both in the EU and NATO. Otherwise, the peace and prosperity of the world could be jeopardized.

Looking through the lens of the characteristics of the symphony, we can speculate on the nature of German culture and changes within it over the next several years. While the population will continue to diversify, other factors will also be critical. There are clear generational clashes in Germany, as in many European nations confronting an aging population crisis. Currently, the proportion of Germans over 60 years old is 21%, but this proportion will probably rise to about 34% by 2040. About 18% of Germany's immigrants are over 55. In an interesting twist to this problem, Germany is struggling to provide for its immigrant Turkish pensioners, who unexpectedly decided to stay on in the country instead of returning to their homeland (Kresge, 2008).

Also, there are clear ideological conflicts that are reflected in the platforms and even the names of the Christian Democratic Party, the Democratic Socialist Party, and the Green Party. Chancellor Merkel campaigned in 2005 vowing to curtail the power of unions and roll back the welfare state but had to form a coalition with the left-of-center Social Democrats that has produced plenty of gridlock. Furthermore, the Left (party of ex-Communists) is making a stunning comeback by playing upon Germans' anxieties about globalization and welfare cuts. German business leaders have been tarnished by Germany's own version of the Enron and Tyco scandals and this has had an impact on the decades-old concept of co-determination, with workers trusting their leaders less and less (Esterl, 2008). Whereas a few years ago business leaders enjoyed twice as much trust as politicians, now they are in the bottom rung of the scale. The issue of Germany's leadership role both in the EU and NATO will probably be a continuing source of tension.

Still, it seems safe to predict that Germany will continue to emphasize the characteristics of the symphony, such as positional arrangements, compartmentalization, subordinated individualism, precision and synchronicity, and unity or a unified sound. Although parts of Germany are changing, and other parts still cling to the past, we can expect that there will always be a symphony. Surely, the country that gave the world Beethoven's "Ode to Joy" will soar once again and become a model for Europe and the world.

CHAPTER 13

Irish Conversations

We Irish have had a long, unyielding love affair with language.

—Edna O'Brien (quoted in Arana, 2002, p. 10)

We were walking across the university campus in the United States after lunch which, as usual, was marked by a lively and wide-ranging discussion among colleagues and friends. I casually mentioned that another longtime faculty member, well-known for eating lunch at his desk, had decided that he was going to relax a bit and begin to join us in the near future. Steve Carroll, an Irish American with decidedly Irish leanings, spontaneously expressed the Irish love of language by saying: "What a pity. He's already missed 25 years of great conversation."

—Personal conversation

Cultural tendencies travel across nations, and so it is little wonder that the two sentiments above—one expressed by a prominent Irish writer and the other by an Irish American professor—are consistent with each other. To understand this consistency, however, one needs some historical background.

The three pillars of Irish culture are its language, rural heritage, and the Catholic Church. For hundreds of years Ireland was essentially an English colony and this domination significantly influenced how these three pillars interacted with one another. Even as late as the 1950s most of the Irish lived on farms or in small communities, and Dublin was a very provincial city in comparison to other European capitals. But today Ireland is a Celtic tiger whose economy has experienced explosive growth.

In 1950 the gross domestic product (GDP) per capita was only $2,500; today it exceeds $45,000. This rapid and successful transformation has occurred for several reasons, including Ireland's movement away from a high-tax socialist economy beginning in 1990, its willingness to attract large foreign multinational corporations through tax breaks and the lowering of tariffs, and its entry into the European Union (EU) (Peet, 2004). (The EU subsidizes emerging national economies with the expectation that economic success will lead to higher tax revenues, some of which will then be allocated to other emerging economies in the system.) An Irish visitor to the United States explained to an American who had lived in Ireland 40 years ago but had not returned, "You won't even recognize the Ireland of your memories. We are European now." The decline of the rural heritage is especially pronounced in and around Dublin, which generates 40% of Ireland's GDP and includes about a third of the total population. Crime, drug use, and other urban ailments are now prevalent.

Similarly, although 44% of the population attends Sunday Mass regularly and almost 90% still claim they are Catholic, the Catholic Church no longer has the power and prestige that it flaunted and exercised for generations. There has been a marked decrease in the number of individuals studying to be priests and nuns. Some causes that are cited for this situation include secularism, prosperity, and priestly abuse of children (Daniszewski, 2005a).

Only the Irish language, Gaelic or *Gaeilge*, is increasing in importance after becoming nearly extinct. The Irish refer to this language as Irish to distinguish it from the Scottish form of Celtic or Gaelic. In the *Gaeltacht* or Gaelic-speaking area of Ireland, which includes 60,000 people, all English versions of geographic names and place names were replaced by Gaelic versions in 2005 and the English language no longer has official standing on signposts, legal documents, or government maps. All students in Ireland must study the Gaelic language for 13 years and receive more than 1,500 hours of instruction in total. There are 1.5 million who say that they speak Gaelic and 337,000 use it daily. About 5% of all Irish children attend schools where only the Gaelic language is used. There is a Gaelic-language TV station with 100,000 viewers daily and people listen to pop music on a 24-hour Gaelic-language radio station. Unlike most nations, Ireland has two official languages, English and Gaelic. Thus, the hope of sustaining Irish culture may be vested in its language (Daniszewski, 2005b).

It is a truism that the use of language is essential for the development of culture and that most if not all cultural groups take great pride in their native languages. Thus, it is not surprising that voice is one of the four essential elements of the metaphor for Italy: the opera (see "The Italian Opera" in this book). In the case of Ireland, the brutal English rule extending over several centuries essentially made the Irish an aural people, whose love of language and conversation was essential for the preservation of their heritage. More specifically, the intersection between the original Irish language, Gaelic, and English has made the Irish famous for their eloquence, scintillating conversations, and unparalleled success in fields where the use of the English language is critical, such as writing, law, politics, and teaching.

Early History

The Celts or Gaels conquered Ireland and its native inhabitants, the Firbolg, several centuries before Christ. They split Ireland into various kingdoms whose petty kings usually ruled over a local group of clans. During this period the belief in magical powers was the guiding force of the civilization, and even today many of the landmarks and traditions in Ireland can be traced back to this time of magical creatures.

Ireland escaped the rule of the Roman Empire due to its distance from Rome. St. Patrick, who had previously been captured and enslaved, returned to Ireland in 432 CE and had great success in Christianizing it. When the Roman Empire fell in 756 CE, Europe plunged into its Dark Ages, but Ireland's Golden Age of learning and scholarship began and lasted until the 11th century. As Thomas Cahill (1995) showed in his best-selling history, the Irish essentially saved European civilization through this emphasis on scholarship and its preservation. In 795 CE, the Vikings from northern Scandinavia started to invade Ireland, founded settlements along the coasts, and established Dublin as their capital. Eventually, the Irish united under the high king, Brian Boru, and defeated the Vikings at Clondarf in 1014 CE. The Vikings were allowed to remain in their seaport towns, intermarried with the Irish, and in time were absorbed by Irish culture.

Starting about 1160 CE, the English, sometimes at the invitation of the petty kings or rulers, became increasingly involved in Ireland's affairs. However, the English who settled in Ireland were quickly absorbed by the culture through joint business ventures, intermarriage, and a perception that England was too distant from Ireland to be controllable. In 1366 the English passed the infamous statutes of Kilkenny as a defensive reaction against this absorption, and they outlawed intermarriage and concubinage between the English and Irish and the use of Gaelic by both the English and "the Irish living among the English" (Beckett, 1986). The English were even forbidden the use of Irish names and dress, and laws excluding the Irish from positions of authority were passed. While many of these statutes were strictly enforced in later centuries, they tended to be disregarded in Ireland when first passed.

In 1534 Henry VIII began to exert strong English control over Ireland. His successors initiated a plantation policy, seizing Irish lands and encouraging Protestants to settle in the country in an attempt to make the nation Protestant. There were several Irish revolts, but they were beaten down. The 10-year revolt starting in 1641 was particularly important, as 600,000 people died during it, and eventually Oliver Cromwell, the Puritan leader, put it down mercilessly and brutally. Cromwell kept a diary describing the joy that he experienced at inflicting so much pain on the Irish for the greater glory of God. While some of the Irish still resent the English, they hate Cromwell. Ironically, today the English and Irish are working together amiably in the EU, and the Irish GDP per capita is higher than that of England.

English Oppression

The Irish eventually sided with James II, a Roman Catholic who became King of England but was forced from the throne in 1685, at least in part because he had abolished many anti-Catholic laws. Although he tried to regain his throne, he and his Irish supporters were defeated in the Battle of the Boyne in 1690. For the next two centuries conditions among Irish Catholics, which had been deplorable, got worse. Additional land was seized by the English, and by 1704 the Irish held only about one seventh of the land in Ireland. A series of penal laws were passed so that Catholics could not purchase, inherit, or even rent land; they were also excluded from the Irish Parliament and the army and they were restricted in their right to practice Catholicism. Also, Catholics were barred not only from the one university, Trinity, but even from most of the schools in which English was the language of instruction.

To counteract the assault on their culture, the Irish employed illegal schoolmasters who taught in Gaelic in "hedge schools" or schools hidden from sight behind hedges; sometimes the classes convened in the homes of the students. Because schoolmasters did not have access to many school materials or even classrooms, almost all of this education was oral in nature, involving memorization, oral presentations, and debates.

Between 1782 and 1798 the all-Protestant Irish Parliament ruled the country, and it restored to Catholics their right to hold land and lifted the restrictions on the practice of Catholicism. However, the Irish still did not have political rights and the wretched lot of most of the Irish did not change. Moreover, the Irish Parliament was dissolved in 1801, and from that time until 1922, when Ireland attained independence, the nation was ruled from Parliament in London, which included a small number of Irish MPs. In 1828, largely through the efforts of one Irish MP, Daniel O'Connell (The Great Liberator), Catholic emancipation occurred, and the penal laws governing the Irish were relaxed.

There were periodic rebellions: in 1798, 1803, 1848, and 1865. Although unsuccessful, they helped to lay the foundations for final independence in 1922, largely sparked by the Rebellion of 1916. Although all of these rebellions were unsuccessful, they reflect the Celtic trait of accepting nobility in death for a righteous but doomed cause.

In 1831 the English established a national school system in Ireland in which only the English language could be used. Irish children had to wear a wooden baton around their necks. If they were heard talking in Gaelic, the teacher would make a mark on the baton, and their parents usually beat them at home because the parents' opportunities to work and thus even to live were completely in the hands of the English. Many of the Irish began to discard Gaelic, at least in public. Patrick Pearse, an Irish patriot who was later executed because of his leadership of the 1916 uprising, characterized the educational system as a "murder machine" designed to stamp out Irish language and culture. Furthermore, Britain had set up an agricultural system that made the Irish dependent on one crop, potatoes. When that crop failed, the Great Potato Famine of 1845–1849 occurred. The English refused to provide any major assistance to the Irish because of their belief in the workings of a free market economy and thus allowed about 1 million people to starve to death out of a population of 9 million.

This event has often been compared to the Holocaust during World War II. Even before the famine, conditions among the Irish had become so deplorable that the satirist, Jonathan Swift, argued eerily in "A Modest Proposal" that Irish babies should be killed and eaten. Some of his modest proposals, such as using the skin of the babies to make gloves and boots, closely resemble what actually happened in the Holocaust.

Ireland as we know it today began to emerge in 1916, when a small group of Irish patriots commandeered the General Post Office Building in Dublin on Easter Sunday. The English executed the 15 leaders of the rebellion, and this sparked a war against the English that led to the modern division of Ireland into two parts, the Protestant north with a minority Catholic population and the Catholic south. The focus of this chapter is the Catholic south, which occupies five sixths of the land.

Identifying Links

It is said that everybody knows everybody else's business in a country village. Ireland and its culture still reflect this village-like perspective and mentality. Whenever the Irish meet, one of the first things they do is to determine one another's place of origin. The conversation usually helps to identify common relatives and friends. Given the wide circle of friends and acquaintances that the Irish tend to make in their lives, it is usually not difficult to find a link.

Ireland's size is little more than one hundredth of that of the continental United States. It lies to the west of Great Britain and is economically tied to that country. Ireland has four major cities: Dublin, Cork, Limerick, and Galway. In the early 1970s more than 60% of the workforce was employed in agriculture, but today only 7% can be found in that line of work; 28% of the workforce is in industry and 65% in services. Ireland is a high-tech leader in the EU and its workforce is young, although it is beginning to become more mature due to prosperity and higher life expectancies. More than 20% of the people are under 15.

The importance of conversation to the Irish makes it a fitting metaphor for the nation. However, to understand the metaphor fully, we need to explore the intersection of Gaelic and English, after which we can focus on an essential Irish conversation, praying to God and the saints. The free-flowing nature of Irish conversation is also one of its essential characteristics, as are the places where conversations are held.

Intersection of Gaelic and English

The Irish are a people who tend to enjoy simple pleasures, but the complexity of their thought patterns and culture can be baffling to outsiders. They generally have an intense love of conversation and storytelling, and they have often been accused of talking just to hear the sound of their own voices. The Irish use the English language in ways that are not found in any other culture. They don't just give a verbal answer; rather, they construct a vivid mental picture that is pleasing to the mind as well as the ear. When the transition from Gaelic to English occurred, the Irish created vivid images in Gaelic and expressed them in English; the vivid imagery of many Irish writers originated in the imaginative storytelling that was historically a critical part of social conversation.

Gaelic was, to the Irish, a graphic, living language that was appropriate for expressing the wildest of ideas in a distinctive and pleasing manner. Frank McCourt's (1996) popular *Angela's Ashes*, with his description of growing up in extreme poverty during the Depression in Limerick, is a perfect example of this use of language. Almost everyone who has read this book comes away with one major impression: Readers feel as if they are actually with McCourt as he goes through his dire and seemingly hopeless experiences.

As indicated previously, Ireland remained relatively bucolic until the 1980s. Eamon DeValera, the prime minister for many years, was an advocate of rural life, and he and his colleagues resisted efforts to industrialize and modernize Ireland. Given this situation, it is easy to see how the talent for conversation as an art form was maintained. And, while the Irish cuisine has improved markedly in recent years, the Irish conversation among many of the Irish is at least as important as, if not more important than, the food being served. Visitors sometimes notice that the Irish seem to forget about their food until it is almost too cold.

If, however, an Irishman admonishes a countryman for eating too much or too quickly, the witty reply is frequently to the effect that one never knows when the next famine will occur. Such "slagging matches"—in which a comment is made at the expense of another person, then a witty retort is made, followed by a series of such interactions—are much loved by the conversational Irish. Moreover, this emphasis on the primacy of conversation is in contrast to the practices found in some cultures, such as the Italian and French, where both the food and conversation are prized.

Important Writers

If the size of the population is taken into account, it seems that, since 1870 or so, Ireland has produced many more prominent essayists, novelists, and poets than any other country. This prominence reflects the intersection of the Gaelic and English languages and the aural bias of the Irish. They have also produced great musicians, who combine music and words in a unique way. Conversely, the Irish have not produced a major visual artist equal to those of other European countries and their achievements in science are modest.

Countless examples could be used to illustrate this intersection, but the opening words of James Joyce's (1914–1915/1964) *Portrait of the Artist as a Young Man*, in which he first introduced the technique of stream of consciousness, aptly serve the purpose:

Once upon a time and a very good time it was there was a moocow coming down along the road and this moocow that was coming down along the road met a nicens little boy named baby tuckoo. . . .

His father told him that story; his father looked at him through a glass; he had a hairy face.

He was baby tuckoo. The moocow came down the road where Betty Byrne lived: she sold lemon platt.

> O, the wild rose blossoms
> On the little green place.

He sang that song. That was his song.

> O, the green wothe botheth. (p. 1)

Several points about this brief but pertinent passage deserve mention. It expresses a rural bias, befitting Ireland, and it reflects the vivid Gaelic language in which Joyce was proficient. Also, it immediately captures the imagination but leaves readers wondering what is going to happen: They must read further if they want to capture the meaning, and it seems that the meaning will become clear only in the most circuitous way. Furthermore, although the essence of the passage is mundane, it is expressed in a captivating manner. Readers are pleasantly surprised by the passage and eagerly await additional pleasant surprises. In many ways this passage is an ideal example of the manner in which Gaelic and English intersect. Although some of the modern Irish and the Irish who have immigrated to other nations such as the United States and Australia may not be aware of these historical antecedents, their patterns of speech and thought tend to reflect this intersection.

Perhaps the most imaginatively wild of the modern Irish writers who incorporate the intersection of the Gaelic and English languages is Brian Nolan, who wrote under the names Myles na gCopaleen and Flann O'Brien. Nolan wrote some of his novels and stories in Gaelic and others in English. Even the titles of his books are indicative of this imaginative focus, for example, *The Poor Mouth: A Bad Story About the Hard Life* (O'Brien, 1940/1974). "Putting on the poor mouth" means making a pretense of being poor or in bad circumstances to gain advantage from creditors or prospective creditors, and the book is a satire on the rural life found in western Ireland. Nolan's masterpiece, *At Swim-Two-Birds* (O'Brien, 1961), sets the scene for a confrontation between Mad Sweeny and Jem Casey in the following way:

Synopsis, being a summary of what has gone before, FOR THE BENEFIT OF NEW READERS: Dermit Trellis, an eccentric author, conceives the project of writing a salutary book on the consequences which follow wrong-doing and creates for the purpose

The Pooka Fergus MacPhellimey, a species of human Irish devil endowed with magical power. He then creates

John Furriskey, a depraved character, whose task is to attack women and behave at all times in an indecent manner. By magic he is instructed by Trellis to go one night to Donnybrook where he will by arrangement meet and betray. . . . (p. 1)

The remaining characters are sequentially introduced in the same imaginative way.

Memorable Sounds

In the area of music, the Chieftains, who have been performing together for more than 35 years, represent the distinctive approach of the Irish to music. Their songs, played on traditional instruments, are interspersed with classic Irish dances and long dialogues that sometimes involve the audience. Similarly Thomas Moore, who lived in the 19th century, is sometimes cited as the composer who captured the essence of the intersection of the Gaelic and English languages in such poetic songs as "Believe Me If All Those Endearing Young Charms," which he wrote for a close friend and beautiful woman whose face was badly scarred in a fire:

> Believe me, if all those endearing young charms,
> Which I gaze on so fondly today,
> Were to fade by tomorrow and fleet from my sight,
> Like a fairy gift fading away,
>
> Thou wouldst still be adored, as this moment thou are,
> Let thy loveliness fade as it may.
> And upon the dear ruins, each bough of my heart,
> Would entwine itself verdantly still. (Moore, 1859)

A reminder that the Irish are different from the English and Americans is their brogue. When the conversion from speaking Gaelic to speaking English was occurring, this brogue was an embarrassment for many Irish. The English looked down upon these "inferior" people who were unable to speak "proper" English (Waters, 1984). Today the brogue is prized by the Irish and appreciated throughout the world, but it is clearly in decline as the Irish move away from their rural heritage.

Given that Ireland is a nation in which unhurried conversation is prized, it is logical that there is a balance between the orientations of "being" and "doing." Although the Irish have enthusiastically accepted change in recent years, they still express astonishment at the "doing" entrepreneurial activities of their 44 million Irish American counterparts, who have made St. Patrick's Day, a holy day in Ireland, into a fun-loving time for partying that embraces all people (Milbank, 1993). The Irish generally take life much more slowly than Americans, who tend to watch the clock constantly and rush from one activity to another. No matter how rushed the Irish may be, they normally have time to stop and talk.

However, urban living has significantly influenced this placid pattern of behavior. In 1969 Martin Gannon visited a slow-moving but entrancing Dublin in which people would walk you to the building when you asked for directions. During a recent visit to a now-crowded and busy Dublin, he experienced difficulty obtaining similar information from pedestrians, some of whom clearly signaled their unwillingness to talk through lack of eye contact and hurried steps.

The Irish also tend to place more importance on strong friendships and extended family ties than Americans do. When they first arrived in the United States, many early

Irish immigrants settled near other friends or relatives who had preceded them to the United States and developed a reputation for being very clannish. But slowly the Irish love of conversation and curiosity about all things led to their interaction with others and their Americanization.

Prayer as Conversation

Prayer or a conversation with God is one of the most important parts of an Irish life. As noted above, more than 90% of the population is Roman Catholic and regular attendance at Sunday Mass is 44%, which is one of the highest figures for regular weekly attendance among nations. Many Catholic households contain crucifixes and religious pictures, which serve as outward reminders of the people's religious beliefs and duties.

Furthermore, prayer is accompanied by acts of good works that stem directly from the strong ethical and moral system of the Irish. They are recognized as having made the highest per capita donation to relief efforts in countries such as Ethiopia, and they are quick to donate their time, energy, and even lives to help those living in execrable conditions. The extent of the crisis in Somalia in the early 1990s, for instance, was first reported to the United Nations by Mary Robinson, then president of Ireland, and an Irish nurse was killed after arriving in Somalia to help out.

The Irish state relies on the works of the Catholic Church to support most of its social service programs. For example, most of the hospitals are run by the Catholic Church rather than by the state. These hospitals are partially funded with state money and are staffed with nuns, when possible. The state has very little control over how the money is spent, especially because it lacks the buildings and the power to replace the church-run system that was in place when the state was formed.

Religious Influence

The national school (state) system is also under the control of the various religious orders in Ireland. It is the primary source of education for primary school children. The state funds the system, but the schools are run by local boards, which are almost always controlled by the clergy. There is a separate national school for each major religion. Local Catholic national schools are managed by local parish priests, while Protestant vicars have their own separate schools. Many of the instructors in these schools are nuns or brothers who work for little pay; thus, costs are much lower than in a state-operated school. In contrast, in the United States, church and state are clearly separate and it is no longer lawful to pray in public schools. In exchange for these lower costs, the Irish state has relinquished control. The schools are really the church's last line of defense because it has the ability to instill Catholic morality and beliefs in almost every young Irish child in the country.

Many Irish begin and end their day with prayer. This is their opportunity to tell God their troubles and their joys. One of the more common prayers is that Ireland may one day be reunited. This act of talking to God helps to form a personal relationship between the Irish and their God. It is difficult to ignore the dictates of God because He is such a personal and integral part of the daily Irish life. God is also present in daily life in the living personification of the numerous priests, and religious brothers and sisters found in Ireland.

Entering the religious life is seen as a special calling for the Irish. In the past, when families were very large, it was common for every family to give at least one son or daughter to the religious life. An Irish mother's greatest joy was to know that her son or daughter was in God's service, and this was prized more highly than a bevy of grandchildren. Vocations to religious life have decreased in recent years, and many of the churches rely on priests from such nations as Mexico and Nigeria.

In the Republic of Ireland, unlike Northern Ireland, there are few problems between the Catholics and other religious groups. In fact, the Catholics enjoy having the Protestants in their communities and treat them with great respect. In one rural area where the Protestant congregation had dwindled, the Catholic parish helped the Protestants with fundraising to make repairs to their church. This act of charity illustrates the great capacity for giving that the Irish possess. Generally many of the Irish are not greatly attached to material possessions and are quite willing to share what they have with the world, although materialism is increasing as prosperity grows.

A Free-Flowing Conversation—Irish Hospitality

Conversations with the Irish are known to take many strange turns, and participants may find themselves discussing a subject and not knowing how it arose. Also, not only what is said but also the manner in which it is expressed is important. The Irish tend to be monochronic, completing one activity before going on to another one, yet they cannot resist divergences and tangents in their conversations or their lives. They often feel that they are inspired by an idea that must be shared with the rest of the world, regardless of what the other person may be saying. The Irish tend to respect this pattern of behavior and are quite willing to change the subject, which can account for the breadth of their conversations as well as for their length.

Like their conversations, the Irish tend to be curious about all things foreign or unfamiliar, and they are quick to extend a hand in greeting and to start a conversation, usually a long one. It is not unusual for the Irish to begin a conversation with a perfect stranger, but for most of the Irish there are no strangers—only people with whom they haven't had the pleasure of conversing. The Irish do not usually hug in public, but this in no way reduces the warmth of their greeting. They often view Americans as too demonstrative and are uncomfortable with public displays of affection. They tend to be a hospitable, trusting, and friendly people. Nothing illustrates this outlook more than their national greeting, *Céad mile fáilte* or "One hundred thousand welcomes," which is usually accompanied by a handshake.

In addition, the Irish are famous for their hospitality toward friends as well as strangers. As Delany (1974) points out, "In the olden days, anyone who had partaken of food in an Irishman's home was considered to be secure against harm or hurt from any member of the family, and no one was ever turned away" (p. 103). This spirit of hospitality still exists in Ireland. In the country the Irish tend to keep their doors unlocked and open. People who pass by or ask for directions may be invited inside to have something to eat or drink. It is not unusual for the Irish to meet someone in the afternoon and invite that person to their home for supper that evening, and this happens in the cities as well as the country. Meals are accompanied by great conversation by both young and old.

Many of the Irish don't believe in secrets and, even if they did, it would be hard to imagine them being able to keep one. They seem quite willing to tell the world their business and expect their visitors to do the same. However, the Irish and other nationalities feel that Americans are much too willing to divulge information. As the Irish saying goes, "You can get a lift with an American and by the time you separate, you know much more about the person than you care to know."

However, the Irish are often unwilling to carry on superficial conversations. They enjoy a conversation that deals with something of substance and are well-known for breaking the often-quoted American social rule that one should not discuss politics or religion in public. The Irish enjoy nothing more than to discuss these subjects, hoping to spark a deep philosophical conversation.

Places of Conversations—Irish Friends and Families

There is no place where the Irish would find it difficult to carry on a conversation. They are generally quite willing to talk about any subject at any time, but there are several places that have a special meaning for the Irish. Conversation in the home is important for an Irish family. Conversation is also one of the major social activities of an Irish public house or pub.

Family Dinners

The typical Irish family tends to be closely knit, and its members describe their activities to one another in great detail. Mealtime in the Irish household should not be missed by a family member, for the conversation if not the food. Supper is the time of day when family members gather together to update one another on their daily activities. The parents usually ask the children about their day in school and share the events of their own day. One of the most famous Irish American families, the Kennedys, stayed true to this tradition in their new homeland. Dinner table conversations at Joe and Rose Kennedy's table became dinner seminars on politics and policies, and the children were quizzed on what they had learned that day. This practice has continued down through generations of Kennedys.

Education and learning have always been held in high regard by the Irish. Teachers are treated with great respect in the community and their relatively high salaries reflect their worth to the community. Ireland has a literacy rate of 99% due to compulsory national education.

A frequent topic of conversation at family dinners is news of extended family, friends, or neighbors. The Irish have an intense interest in the activities of their extended family and friends, but this interest isn't purely for the sake of gossip. They generally are quick to congratulate on good news and even quicker to rally around in times of trouble or need. When someone is sick, it is not unusual for all the friends and family of the person to spend almost all of their time at the hospital. Friends help the family with necessary tasks, and they entertain one another with stories and remembrances. Many of the Irish have a difficult time understanding the American pattern, in which the nuclear family handles emergencies and problems by itself.

Weddings and Wakes

This practice also holds true whenever there is a death in the community. Everyone gathers together to hold an Irish wake, which combines the viewing of the body with a party that may last for 2 or 3 days. There is plenty of food, drinking, laughing, conversation, music, games, and storytelling. Presumably the practice of a wake originated because people had difficulty traveling in Ireland due to poor roads and nonmechanized means of transportation, and the wake afforded an opportunity not only to pay respect to the deceased but also to renew old friendships and reminisce. Although the problems of travel have been solved, the Irish still cling to this ancient way of saying good-bye to the deceased and uplifting the spirits of those he or she has left.

An event that is as important as the wake is a wedding; it is a time of celebration for the entire family and neighborhood. Customarily a big church wedding is followed by a sit-down dinner and an evening of dancing and merriment. Registry office weddings are very rare in Ireland, as might be expected in this conservative and Catholic-dominated nation.

Irish parents tend to be strict with their children. They set down definite rules that must be followed. Irish children are given much less freedom than American children, and they usually spend all day with their parents on Sunday and may accompany them to a dance or to the pub in the evening. Parents are usually well acquainted with the families of their children's friends and believe in group activities. The tight social community in which the Irish live makes it difficult for children to do anything without their parents' knowledge. There is always a third cousin or kindly neighbor who is willing to keep tabs on the behavior of children and report back to the parents, some of whom have even managed to stretch their watchful eyes across the Atlantic to keep tabs on children living in the United States. This close control can sometimes be difficult for young people, but it creates a strong support network that is useful in times of trouble.

Pub Life

A frequent gathering place for men, women, and children is the local pub, given that the drinking age is not enforced throughout most of Ireland. There are two parts in most pubs: The plain, working-man's part and the decorated part, where the cost of a pint of beer is slightly higher. In the not-too-distant past, it was seen as unbecoming for a woman to enter a pub; there are still some pubs in which women are comfortable only in the decorated part, and women typically order half-pints of beer while the men order pints.

Pubs tend to be informal, often without waiters or waitresses and with plenty of bar and table space. Young and old mingle in the pub, often conversing with one another and trading opinions. The Irish are raised with a great respect for their elders and are quite comfortable carrying on a conversation with a person of any age or background. They tend to be a democratic people by nature. Although they may not agree with a person's opinion, they will usually respect him or her for having formed one.

Irish pubs are probably the sites of the liveliest conversations held in Ireland. Ironically, the number of pubs per capita has declined dramatically in recent years, in large part because of the movement against excessive alcoholic consumption, new and strict laws, and the powerful association of pub owners, who lobby against new licenses (Zachary, 1999). The Irish tend to be a sociable people who generally don't believe in drinking alone. This pattern of behavior has often resulted in their reputation for being alcoholics. In fact, the Irish consume less alcohol per capita than the average for the EU, but there are at least two distinctive groups: those who do not imbibe at all and those who do so, sometimes to excess.

Admittedly many Irish drink more than they should, but the problem often appears worse than it is because almost all of their drinking takes place in public. Furthermore, coworkers and their superiors frequently socialize in pubs, and they tend to evaluate one another not only in terms of on-the-job performance but also in terms of their ability to converse skillfully in such a setting. The favorite drink of the Irish is a strong black beer, Guinness. It is far more popular than the well-known Irish whiskey.

Even more important than a good drink in a pub is good conversation. The Irish are famous for their storytelling, and it is not unusual to find an entire pub silent while one man holds sway. It is also not unusual for someone to recite a Shakespearean play from memory in its entirety or to quote at length from the writings of such Irish writers as James Joyce and Sean O'Casey. In Ireland and in other nations to which the Irish have emigrated, there is a tradition of reading all of James Joyce's famous and path-breaking book, *Ulysses*, on Saint Patrick's Day in pubs.

Besides stories, many a heated argument can erupt in a pub. The Irish seem to have a natural love of confrontation in all things, and the conversation doesn't even have to be about something that affects their lives. They are fond of exchanging opinions on many abstract issues and world events. During these sessions at the pub, the Irish sharpen their conversational skills. However, while these conversations can become heated, they rarely become violent.

Irish friends, neighbors, and families visit one another on a regular basis. As indicated above, rarely if ever is one turned away from the door. Family and friends know that they are welcome and that they will be given something to eat and drink. It is not unusual for visitors to arrive late in the evening and stay until almost morning. Such visits are usually not made for any special purpose other than conversation, which is the mainstay of the Irish life no matter where the conversation occurs.

Ending a Conversation

A conversation with an Irishman can be such a long and exciting adventure that a person thinks it will never end. It will be hard to bring the conversation to a close because the Irish always seem to have the last word. Ireland is a country that welcomes its visitors and makes them feel so comfortable and accepted that it is hard to break free and return home after an afternoon or evening of conversation.

Geert Hofstede's (2001) research profiling the value orientation of 53 nations includes Ireland and his analysis confirms many of our observations. Ireland is a masculine-oriented society in which sex roles are clearly differentiated; it ranks 7th of the 53 nations. However, the more recent GLOBE study (House et al., 2004) contains a more nuanced profile. Ireland ranks 3rd of 62 national societies on a comparable *value* measure, gender egalitarianism, with higher rankings indicating greater male domination. But in the *practice* of gender egalitarianism, Ireland ranks only 39th, thus suggesting that changes in lifestyles such as families in which wives work full-time are leading to new behaviors. Further, Ireland is not an acquisition-oriented society, as Hofstede's classification might suggest but a society in which there is a balance between "being" and "doing."

Also, Hofstede's (2001) study of cultural values indicated that Ireland clusters with those countries emphasizing individualism, ranking 12th of 53 nations, as we might expect of a people who are willing and eager to explore and talk about serious and conflict-laden topics. Individualism is expressed through conversation and views on issues that affect society; it is also expressed in other talents such as writing, art, and music. As suggested previously, music offers its own means of conversation, and Ireland reportedly has one of the highest numbers of musicians per capita of all countries. Still, the Irish tend to be collectivistic in their emphasis on the family, religion, a generous welfare system, and the acceptance of strong labor unions.

In short, the Irish tend to be an optimistic people who are ready to accept the challenges that life presents, although there is a melancholic strain in many of the Irish that is frequently attributed to the long years of English rule and the rainy weather. The Irish usually confront things head-on and are ready to take on the world if necessary. They can be quite creative in their solutions but also quite stubborn when asked to compromise, and they tend to be truly happy in the middle of a heated but stimulating conversation. Given their history and predilections, it is not surprising that the Irish prefer personal situations and professional fields of work where their aurally focused approach to reality can be given free rein, even after they have spent several generations living in countries such as the United States and Australia.

CHAPTER 14

The Canadian Backpack and Flag

Many Canadian nationalists harbour the bizarre fear that should we ever reject royalty, we would instantly mutate into Americans, as though the Canadian sense of self is so frail and delicate a bud, that the only thing stopping it from being swallowed whole by the U.S. is an English lady in a funny hat.

—Will Ferguson, Canadian writer and novelist

The U.S. is our trading partner, our neighbour, our ally and our friend . . . and sometimes we'd like to give them such a smack!

—Rick Mercer, Canadian comic

Canada is the world's second-largest nation in size, the first being the new Russian Republic. Its population is about 32 million; about 90% live within 125 miles of Canada's border with the United States. Canada is a nation of open spaces. The climate in most Canadian cities is comparable to neighboring regions of the United States, but the empty spaces of northern Canada are very cold. Throughout the nation winters are cold and summers brief.

An apt metaphor for Canada is the Canadian maple leaf flag, which countless Canadians unobtrusively sew onto their modest backpacks and suitcases. Amid the bewildering variety of suitcases at the baggage claim of any major international airport, one occasionally sees a Canadian backpack, sturdy but a bit frayed around the edges. Its most notable feature is a sewn-on patch displaying the Canadian maple leaf flag; such a bag symbolically and emphatically proclaims: I am definitely not a U.S.

American. This is a typically indirect way of Canadian communication, especially when contrasted to that of U.S. Americans, and only rarely will one encounter a straightforward statement of hostility like the one found in the opening quote.

But this backpack also conjures up other messages. The bag's modesty reflects the Canadian bias for egalitarianism, and its flag suggests a personality that is indirect and suggestive in communication. The backpack also suggests that Canadians are willing to accept others as equals, even if different. In all likelihood the average Canadian probably owns more copies of the national flag than the average citizen of any other nation. A 2007 poll by the Dominion Institute conducted before Canada Day showed that most Canadians feel attached to the maple leaf (87%), the beaver (74%), hockey (73%), and the "mountie" (72%) as unifying Canadian symbols. The percentages were smaller for the French province of Quebec, where only 66% indicated that they felt attached to the maple leaf. Canadians older than 55 were also more likely to display the flag than those aged 18 through 34. The Canadian flag appears on lapel pins, cereal boxes, bumper stickers, bathing suits, and of course, army surplus backpacks. This may seem curious to those who know Canadians to be polite and somewhat reserved in nature, but the need to sew their flag onto their backpacks and elsewhere is an undeniable feature of Canadian culture. In short, the bag and its flag combine to form an apt metaphor for our Canadian neighbors.

In this chapter we will examine three aspects of this metaphor: egalitarianism and outlook, the Canadian mosaic, and clear separation from their U.S. neighbors. Before doing so, however, we briefly trace some historical facts without which Canada is difficult to understand.

Historical Background

The history of Canada cannot be separated from the history of other nations, particularly Britain and France. The British formed one of the first chartered trading companies in the world, the Hudson Bay Company, in 1670 as a way to organize and profit from trade as well as to promote exploration and settlement of Canada and other lands. There were numerous battles between French and English colonists over the territory that is now Canada until France officially ceded its claims in the 1763 Treaty of Paris. During the American Revolution thousands of British loyalists fled to Canada, drawing the colony closer to the British crown.

In 1867 the Dominion of Canada was established as a self-governing member of the British Commonwealth. Until 1965 the official Canadian flag was the British Union Jack, at which time the maple leaf flag replaced it. The modern nation of Canada was forged by the experiences of World Wars I and II; there were more than 60,000 Canadian soldiers killed in World War II out of a population of 8 million.

Because of World War II, Canada and the United States began to cooperate closely. Their soldiers fought side by side for 3½ years, and afterward, these two nations were among the small number that had not been completely shattered by the war. Most important, Canada and the United States worked together to rebuild Europe and supported one another through trade. Canada and the United States share the world's largest undefended border and do more business together than any other two nations. The United States accounts for about 85% of Canada's exports and 59% of the country's imports. In fact, the trade in the North American Free Trade Agreement (NAFTA) area (which includes the United States, Canada, and Mexico), accounted for 8% of all international trade in 2004, a decade after its formation. Similar findings elsewhere suggest that international trade is more regional rather than global (see Rugman, 2005).

Egalitarianism and Outlook

The Canadian soldier's backpack was part of his standard gear in warfare, able to hold the bare essentials for life under extreme conditions: blanket, tent, rain gear, cup, cooking pot, can opener, rifle cleaning kit, and little more. The modern version includes much of the same equipment. Although today more than 80% of Canadians are city dwellers, the easy access to splendid scenery and challenging trails has made hiking, trekking, and canoeing national pastimes. Many Canadians, especially among the older generations, have worked in the interior as loggers, road builders, miners, and related occupations in Canada's resource-based economy. Even today, city-dwelling Canadians in the younger generation will spend a summer or two planting trees in the nation's parks or working at some other job in the wilderness requiring life to be lived out of a backpack.

Such experiences have contributed to a sense of egalitarianism and a conservative outlook, both for Anglo and non-Anglo Canadians. Similarly, other than the presence of the prominently displayed maple leaf flag, the most striking feature of the Canadian backpack is its lack of striking features. Discerning U.S. backpackers might go to REI or a similar store to choose from a number of models featuring the latest lightweight waterproof fabrics, snazzy color schemes, and a $500 price tag. Most Canadian backpackers will simply go to the closet and pull out the trusty and well-used model. They silently rejoice in its utility and the fact that they do not have to spend $500 like their U.S. counterparts, which appeals to their generally frugal nature. Canadian backpackers will, of course, make a purchase if necessary, but then they will select the plainest possible color. Canadians avoid flashy displays of status or wealth, instinctively avoiding material possessions that would seem like an attempt to be better than others.

The conservative English Canadians also avoid open expressions of emotions and feelings but the French Canadians may be less reserved, gesture more expansively, and may stand closer during a conversation. Although they are extremely polite and friendly, Canadians tend to be more like the English in their preference for internalizing problems and concerns as opposed to the U.S. tendency to "let it all hang out." It is no accident that the national animal of Canada is the beaver, given the nation's vast natural resources. Beavers survive primarily by working together in colonies and building impregnable lodges. The colonies, or lodges, consist of one or more family groups, and beavers in one colony generally have little contact with those in another colony except when danger approaches. Then the beavers loudly warn neighboring colonies and try to escape through one of the two escape tunnels they have constructed.

Canadians tend to be more diplomatic than U.S. Americans, often preferring to accept situations that are not to their liking rather than to call attention to them with complaints and demands for better treatment. Canadians who are served the wrong dish at a restaurant might go ahead and eat it, whereas U.S. Americans would demand the correct dish. Similarly, a Canadian book was published several years ago with the modest title, *Your Guide to Coping With Back Pain*. The publisher decided that the book would never sell in the United States using this title, so in the United States, the title was changed to *Conquering Your Back Pain*.

Canadians emphasize good manners and conflict avoidance but not at the expense of truth. This approach may not be fully desirable in a business meeting where open expression of needs and concerns is vital to the identification and resolution of problems. Similarly Canadians may find it somewhat difficult to initiate an unpleasant conversation with their superiors or business partners. Although polite and indirect, however, Canadians are more direct in communication than their Asian and Latin American counterparts, without the complexity of meanings that are characteristic of these cultures. In this sense Canadian communication patterns overlap with and are similar to but different from those of the more direct U.S. Americans.

Another manifestation of the Canadian sense of egalitarianism and fairness is the country's support for universal health care and the social safety net. Compared with the United States, services are much more evenly distributed in Canada, which is more in line with the social democracies of Western Europe. By contrast, about 50 million U.S. Americans have no health insurance and it is common for a hospital emergency room to refuse admission to seriously ill patients. Michael Moore, in his celebrated documentary 2007 *Sicko*, focused on these differences, suggesting that life expectancy was higher in Canada, Britain, and France than in the United States because of the health care system.

He did not point out, however, that a Supreme Court Justice in Canada had opined in 2005 that many Canadians suffered from chronic pain while they waited for treatment and several died before they received it. The other downside is that Canada's total tax burden is much higher than that in the United States, and this progressive tax system encourages highly paid individuals to cut back on their workloads or to move out of the country. Still, most Canadians accept this tradeoff, which reflects their egalitarianism and conservative outlook. To ensure that people receive adequate health care, they prefer using modest backpacks rather than spending money on flashy items.

The Canadian Mosaic

Canadians are taught from an early age that no person is inherently any better than others, and this sense of equality pervades much of the Canadian perspective on social status and Canada's place in the world. Canadians have a strong appreciation of foreign cultures and have actively encouraged immigration from around the world for the past several decades. In fact, Canada is often dubbed "a global village in one country." Although Canada has much open space to fill, immigration also stems from the Canadian people's sense of equality and fair play. Canadians are sensitive to the suffering that occurs in overcrowded, poverty-stricken lands. Canada receives most of its legal immigrants from three main sources: people with skills who immigrate for economic reasons, people with relatives already in Canada, and finally, refugees from around the world that Canada traditionally accepted as a result of its fairly liberal policy on political asylum. These immigrants help make up the famed Canadian mosaic.

The diversity of cultural backgrounds in Canada is breathtaking. Its immigrants come from more than 200 separate countries and territories of birth. At least 16% of the Vancouver area's 1.8 million people are Chinese, and airport signs and oral instructions are in both Chinese and English; another 14% of Vancouver residents come from other parts of Asia. Similarly more than 10% of Toronto's population is either Chinese or Filipino. Canada has large numbers of immigrants from China, India, Southeast Asia, Eastern Europe, the Caribbean, and so forth. Toronto and Vancouver have two of the biggest Chinatowns in North America. Immigration represents about 60% of the population growth in Toronto and Vancouver and is expected to continue. Canadian businesses are focused on reaching these ethnic markets, especially the Chinese and South Asian markets. It is estimated by Stats Canada that by 2017, roughly 1 in 5 Canadians will be a member of a visible minority. Thus it is not surprising that Toronto has not only a very large Chinatown but also large neighborhoods that are overwhelmingly Greek, Indian, Italian, Portuguese, Caribbean, and Ukrainian.

In some ways Canada is changing the new arrivals, but they are also changing Canada. The distinctive Canadian personality lives on through the values passed down through the schools, businesses, and other institutions helping to shape the newcomers into Canadian citizens. However, Canada emphasizes the creation of a mosaic of cultures rather than the U.S. melting pot (Andrews, 1999). This approach is officially recognized in the Canadian Multiculturalism Act of 1988 and Canada is proud of the fact that it became the first country in the world to adopt an official Multiculturalism Policy in 1971.

Some Anglo Canadians resent the large influx of immigrants, but their extremely polite and friendly nature for the most part allows them to neutralize this feeling. In 2006 Toronto reported 162 hate-related crimes, a relatively low number in a city with residents from 169 different nations speaking more than 100 languages. This tolerance is especially prevalent among the younger generation of Anglo Canadians, who have grown up with immigrants from around the world in their classrooms. Outside the major cities the population is more uniformly Anglo or French, and there is greater suspicion of newcomers. Most important, the Canadian economy has advanced significantly because of its large number of immigrants, and immigration has made

Canada one of the richest countries in the world. Other nations, such as Japan and Italy, are suffering economically because of declines in their populations and an increase in the average age, but Canada readily accepts people who can contribute as workers, skilled professionals, managers, and consumers. Multiculturalism in Canada is a doctrine and part of the national Charter of Rights and Freedoms, which is very different from the situation in Britain or France or the Netherlands. Of course, integration is a much greater challenge for "visible" minorities, and according to government statistics, "satisfaction with life" decreases from first to second generation for these minorities.

Canadians as Non-U.S. Americans

Although Canadians and U.S. Americans cooperate significantly with one another in many areas, Canadians dislike being mistaken for their neighbors to the south. Many of them are not happy that U.S. fast food chains, books, and films are much more popular than their Canadian counterparts, and the Canadian government periodically considers legislation that would restrict the sale of U.S. books, magazines, music, and films. In fact, Canadian TV and radio stations are required by law to carry a specified amount of "Canadian only" material, which is defined as being at least partly written, produced, presented, or otherwise contributed to by people from Canada. The government subsidizes significantly the activities of Canadian artists and companies, but the North American Free Trade Agreement (NAFTA) and the World Trade Organization preclude exclusion of non-Canadian goods and services.

A 28-year-old Canadian actor became a cultural hero in the late 1990s when he spontaneously spoke the following words, subsequently made into a Molson Beer ad, at a coffeehouse open-mike night:

> I am Canadian. I'm not a lumberjack or a fur trader. I don't live in an igloo, eat blubber, or own a dogsled. I don't know Jimmy, Suzie, or Sally from Canada, although I'm certain they're very nice. I have a prime minister, not a president. I speak English and French, not American. And I pronounce it "about," not "a-boot." I can proudly sew my country's flag on my backpack. I believe in peacekeeping, not policing; diversity, not assimilation. And that the beaver is a proud and noble animal. A tuque is a hat, a chesterfield is a couch. And it's pronounced zed. OK? Not zee. Zed. Canada is the second-largest land mass, the first nation of hockey and the best part of North America. My name is Joe, and I am Canadian. ("Joe's Rant," 2000)

Throughout the nation Canadians lined up at open mikes to shout the rant or their version of it into microphones set up on the ice between periods during hockey games. The Joe Canada discussion was a daily feature in the nation's editorial pages and most

Canadians agreed that a little national pride was a good thing. Ironically, however, Joe's day of fame was deflated when he moved to Los Angeles to seek work as an actor due to the poor job prospects in Canada. Some critics felt that Joe should have carved out a separate Canadian identity instead of focusing on the "not American" aspects. However, the ad spawned many parodies and imitations in Canada and around the world. In 2007 the Muslim American Society created a video entitled "I am a Muslim," which took many lines from the Molson beer ad and applied it to Muslim stereotypes.

Similarly a popular Canadian TV weekly show, *This Hour Has 22 Minutes*, satirizes U.S. ignorance by interviewing unsuspecting U.S. Americans with such questions as the following:

> Do you think Canada should join North America? It's a big story up north. Care to comment?

> Should Canada outlaw the slaughter of polar bears in Toronto?

However, the 31-year-old creator of this show and its interviewer, Rick Mercer, comments tellingly on the show's appeal:

> Canadians spend a huge portion of their social life trying to define what it means to be Canadian. Americans never spend any time trying to define what it means to be American. Canadians have an identity crisis. We look like Americans. We sound like Americans. We know everything about Americans. They know nothing about us. . . . We find that funny. (Brown, 2001, p. C7)

There could be several explanations for this non-U.S. American (but not anti-U.S.) bias. Canada was formed principally by those loyalists who left the United States during the American Revolutionary War. The more activist U.S. outlook contrasted sharply, and still contrasts sharply, with the more conservative Canadian outlook. Whereas the U.S. constitution protects life, liberty, and property, Canada's constitution emphasizes "peace, order, and good government." U.S. comics frequently use Canada as the butt of their jokes as in Robin Williams's famous rendering of the song, "Blame Canada," from the movie, *South Park: Bigger, Longer & Uncut,* or Comedy Central's Jon Stewart's comment, "I have been to Canada and I have always got the impression that I could take the country in two days."

Also, it is difficult to be the much smaller and less prosperous neighbor of the United States. U.S. teams have won the Stanley Cup, hockey's highest honor, consistently since 1993. Given that hockey is Canada's national sport, this is particularly difficult for Canadians to accept. However, most of the players on the U.S. teams are Canadians, which assuages the damage. Unfortunately, the greatest hockey player of all time, Wayne Gretzky, was traded to the Los Angeles Kings in 1988 because the Edmonton Oilers could no longer afford his salary; in fact, many Canadians considered it a national calamity. Whatever the explanation or explanations, Canadians are definitely not U.S. Americans. This may well be the essence of Canadian culture, defining itself in terms of what it is not.

However, the similarities between Canadians and U.S. Americans are much greater than the supposed differences. Both cultures tend to be individualistic, and both prefer a direct, low-context form of communication in which accomplishing goals is more important than process or getting to know one another before doing business. The manners of English-speaking Canadians are very similar to those of English-speaking U.S. Americans, but French Canadians may exhibit less reserve and wealthy Hong Kong Chinese with Canadian citizenship have their own business etiquette reflecting their identity.

Some Canadian firms have a well-deserved reputation for being aggressive and successful in international business. In recent years, in the context of the volatility in the world's oil prices as a result of growing demand and instability in some key oil-producing countries, Canada's vast reserves of oil sands in the province of Alberta hold considerable potential for North America. Free trade between the United States and Canada has benefited both, particularly Canada, in terms of balance of net exports. A large number of Canadians have become successful in the United States (e.g., entertainers William Shatner, Celine Dion, and Martin Short).

Still, Canadians revel in pointing out the differences, even to the extent of sewing the maple leaf flag in the middle of their backpacks. Presumably, Canadians see themselves as the middle point of a distribution in which the brash U.S. Americans are on one end and the withdrawn English on the other, although the reality is much more complex.

And the complex reality of Canada is epitomized in the difficult relations between the Canadian English and Canadian French. Most of the generalizations in this chapter apply particularly to the English, but there are two French-speaking provinces, Quebec and New Brunswick. The locus of the hostility, which has been increasing since the 1960s, is Quebec, which historically has been much poorer than the English provinces. In 1995 Quebec held a provincial referendum about separation from Canada and the final vote was extremely close: 50.5% to 49.5%. Victory would have cast doubt on the concept of the nation of Canada and it was only through a heroic flag-waving effort that the Quebecois were convinced to stay. Subsequent opinion polls indicated a fall in support for independence and the Parti Quebecois, which was pro-independence, was defeated in the 2003 provincial election.

Ironically, 1995 was the year of a well-financed campaign celebrating the 30th anniversary of the Canadian maple leaf flag. The issue of separation is still lurking in the background, but the continuing influx to Quebec from non-French cultures and an improved economy have blunted or at least postponed the threat of separation. Furthermore, in 2006, parliament agreed that the Quebecois should be considered a "nation" within Canada in a gesture that was largely symbolic.

Anglo Canadians have been economically dominant since their arrival in Canada, and the new mosaic (including the French component) will probably help to enlarge the economic pie and prosperity but increase cultural diversity. As this happens, new and perhaps conflicting cultural symbols will emerge, such as the Chinese family altar (see Chapter 26). However, according to the Dominion Institute poll conducted in 2007, 7 out of 10 Canadians believe that their common history, heroes, and national symbols are responsible for their success as a nation. Therefore, at least for the foreseeable future, we believe, the maple leaf flag sewed on the modest backpack will serve as an effective cultural metaphor for this vast land and complicated culture.

CHAPTER 15

French Wine

Wine making is really quite a simple business. Only the first 200 years are difficult.

—Baroness Philippine de Rothschild
(quoted in Rachman, 1999, p. 91)

To describe French culture, it is useful and perhaps essential to consider French wine, the best of which is considered preeminent throughout the world. French wine agriculture has produced many of the leading wine practices for centuries and has created many of the most well-known grape varieties, including cabernet sauvignon, chardonnay, pinot noir, syrah, and sauvignon blanc. Wine has played a vital role in determining the economy, traditions, and attitudes in France. It has helped shape the country's disposition, weaving a common thread through all the varying walks of French life. Just as there are more than 5,000 varieties of French wine, so, too, the French have a variety of idiosyncrasies and personalities, many of which are puzzling to outsiders. Beneath these differences, however, is a culture that unifies the people and their institutions. Accordingly, wine appears to be an appropriate metaphor for describing and analyzing the French culture.

In many ways the French wine industry represents the classic tradeoff between globalization and cultural identity. France has been the leader in wine production among nations for centuries and is still No. 1. But serious harm has been caused by governmental regulations of French wine making; the fragmentation of the industry into many small growers, who cannot take advantage of large economies of scale found in other wine-growing nations; the confusingly large number of wine categories; and poor and inadequately funded marketing campaigns. The wine industry is growing in many nations, such as Australia and Chile. The Bordeaux region of

France by itself has 20,000 different producers; Gallo, the world's largest wine company, one that is based in California, spends more than twice as much on advertising as all the Bordeaux wine companies together. Still, the preeminence of French wine, particularly of the finest vintages, is universally acknowledged, and the perception is that French wine is the best, at least in the category of the best but most expensive wines. Furthermore, French companies have established joint ventures in such nations as China and India that now produce superior wine at a reasonable price, and it is a mark of distinction to carry the French name on the wine bottle.

To focus this discussion within the metaphor of French wine, five principal elements of wine will serve as a guide: pureness, classification, composition, suitability, and maturation.

Pureness

To more fully appreciate the metaphor, we must first understand what wine is, what its origins are, and how wine is made. A wine's characteristics are the summation of its past as evidenced by the environment from which it develops. Viniculture is the precise and patient process that further shapes the destiny of the grape's transformation into wine. Fine wine is considered to be the distillation of 2,000 years of civilization. It is understandably viewed as an object of great pride, a survivor of nature's caprices and uncertainties, a product of patience and modesty, and, most important, a symbol of friendship, hospitality, and *joie de vivre*.

Vital to the quality of the wine are the soil, climate, vine type, and the viniculturist who tends the wine-making process. The complex interplay of these factors determines whether pride or disappointment reigns once the wine flows from its bottled womb. Soil and climate nourish growth in mysterious ways that defy chemical analysis. Contrary to what might be expected, the vine thrives on the type of soil on which little else will grow. Climate and vine type must complement each other so as to produce healthy grapes, and the viniculturist must be careful to make sure that the timing of harvest and maturation is meticulous and accurate.

Vines are also interbred, making it difficult to classify the wine produced because the offspring may resemble their parents or display entirely new characteristics. Furthermore, what is recognized as the world's greatest wine—wine that costs more than $100 a bottle—is grown in only a few select vineyards in France. The art and science of viniculture also influence the wine's personality. Experience, diligent patience, and effort are required to propagate the wine through its various stages of cultivation. The *vendange*, or harvest period, brings a sudden preoccupation with time. A delay of even 12 hours between the harvest and preparation for fermenting can spoil both taste and aroma. After transfer from vats to bottles, the wine continues its aging process. Age, then, influences personality development, with great wines requiring more than 50 years to mature to perfection.

A Perfect Land

The French, much like U.S. Americans, have a romantic view of their country as being special and unique. Like a flawless bottle of vintage wine, France displays perfection in the land and its people. The French have mentally massaged the image of their borders into a hexagon, perfectly situated midway between the equator and the North Pole, balanced in soil and climate. Symmetry, balance, and harmony—it all coalesced in one great land because the French supposedly willed it. This perception of symmetry and unity in the physical dimensions of this geographically diverse land is more wishful thinking than reality.

The north and the south disagree as to the origin of the French people. In different parts of the country people have different identities: The northeast identifies with the German and Swiss, the northwest with the English, the southwest with the Catalans and Basques, and the southeast with the Italians. Perhaps, then, the eternal struggle for unity has deep and stubborn roots in the past that dictate France's destiny as an aggregate of individuals struggling to forge themselves into a single entity.

Thus the French tend to give the impression that France is the center of the universe around which the rest of the world rotates. One can quickly learn to resent the French belief in their cultural superiority along with their lack of immediate friendliness. But they readily defend their position in part by noting that international business and diplomacy were conducted in the French language until World War I and that French art, literature, and thought remain pervasive in modern education and society. The French also band together by having shared a long and illustrious history of crusades, wars, and devastation.

The extent to which the French defend this supposed superiority is extraordinary. A commission must decide which foreign words can be a part of the official language; there is a requirement to have proceedings of conferences published in French, even when English would suffice; and the seventh summit devoted to the French language was held in 1997 in Vietnam, which France formerly ruled but in which only about 1% of the population speaks French. In the 11th summit held in Bucharest in 2005, former Prime Minister Jacques Chirac warned of a unilingual world in which the dominance of English would limit cultural creativity and variety.

Historical Roots

The earliest traceable ancestry of the French begins with the Celts, a Germanic people who later became known as the Gauls. They planted an extensive empire, which was subdued by the invading Romans under Caesar in 52 BCE. For the next 500 years the Gauls were repeatedly subjected to invasions by other Germanic tribes and, last, by the Franks, who emigrated westward across the Rhine. Out of the Middle Ages and its feudal system emerged the Renaissance (rebirth), a time of increasing wealth and power for France but also of continuing turmoil.

A golden age burst forth in the person of Louis XIV, the Sun King, who proclaimed, "*L'état, c'est moi*," or "I am the state." He orchestrated his rule from the extraordinarily lavish Palais de Versailles, and during his reign, France expanded and built an impressive navy. Its language was spoken and its culture was emulated all over Europe.

But the Sun King allowed a widening gap to develop between the wealthy and the poor, and in an increasingly financially stressed economy, that gap exploded in the French Revolution of 1789. The quest for *liberté, égalité, fraternité* first resulted in the brief rule of Napoleon III. He allowed a more organized France to take shape by enforcing the Code Napoleon, which gathered, revised, and codified a vast, disorderly accumulation of laws, both old and new. Unfortunately the oppressive restrictions that Napoleon III placed on laborers, peasants, and women are still felt today.

Moving closer to the modern era, France was also heavily affected by World Wars I and II. Although a victor in World War I, France lost nearly 2 million men. World War II still lives in the memories of many French people as a time of hopelessness and disgrace under German occupation.

Troubled Years

The history of French wine has also endured notable difficulties, one of which occurred from 1865 to 1895. During these years a disease called *phylloxera vastatrix* destroyed virtually every vineyard in France (Vedel, 1986). This was a tragedy of major proportions and adversely affected thousands of wine growers. But, even in such dire circumstances, the French did not lose their resolve. After toiling for months, they discovered by trial and error that a viable solution was to graft French vine strains onto American stocks that were resistant to the disease. Forced into a corner, the French came together to save the industry. In subsequent years the crops regained their health and the wine industry flourished once again.

Even with this brief overview, it becomes apparent that the lives of the French people have been planted, uprooted, and replanted throughout history. The conquering Romans left the French with a sense of pageantry and grandeur as well as an affinity for control and bureaucratic organization. The Romans introduced the concept of centralization and a complex bureaucracy, which have taken root in French hearts and minds. France's present concept of grandeur was first thrust on the French people in the time of Louis XIV. From this epic era emerged the idea that the French were guardians of cherished universal values and that their country was a beacon to the world. The French saw themselves as favored, as possessors of ideas and values coveted and treasured by the rest of mankind. As the French poet Charles Peguy wrote, "God loves the French the best."

It is true that the purest, finest wines must be grown in special soil. In this sense the French consider France to be a pure and proud country. Accordingly, those not born and raised in France need to guard against forming quick and negative first impressions without understanding past trials and tribulations as they relate to the formation of the French culture.

Classification

As noted earlier, France produces 5,000 varieties of wine precisely classified so that impostors cannot pass for superior wines. This insistence about nomenclature bestows on wine a pedigree that is displayed on each label. Through fermentation and the aging process in the bottle, wine develops its final personality, blend, and balance.

Although wine is classified in excruciating detail, those who are not connoisseurs might dare to divide it into four major classes: the *Appellation d'Origine Contrôlée* wines, which are the best and most famous; respectable regional varieties, known as *Vins Délimités de Qualité Supérieur*, which are quite good for everyday use; the *Vins de Pays*, which are younger, fresher wines suitable for immediate consumption; and the *Vins de Table*, which are truly wine but lacking in taste and pedigree.

Similarly French society is also clearly stratified and divided into four principal and generally nonoverlapping classes: The *haute bourgeoisie*, which includes the few remaining aristocrats along with top business and government professionals; the *petite bourgeoisie*, owners of small companies or top managers; *classes moyennes*, or the middle classes—teachers, shopkeepers, and artisans; and *classes populaires*, or workers.

People may know their place in society but that does not imply that they feel inferior to others. The French are comfortable accepting and living within the confines of this classification system rather than resisting it. Workers in the *classe populaire* are just as accepted for their contribution to society as elitist officials in the haute bourgeois class. This norm extends to French attitudes toward outsiders. Visitors to France sometimes have difficulty relating to the French, who can be just as ethnocentric as anyone else, for the French tend to behave independently of others who mistakenly expect tacit cooperation from them in an undertaking. The reason for this outlook is quite simple. In the French mind, France and a person's own particular social class tend to come first. The French interpret the phrase *tout le monde* (literally, all the world) to mean all who are French or, even more specifically, all who are in their particular social class. Outsiders are countenanced but not openly welcomed.

Creating Order

Rules, regulations, and procedures give certainty, definition, and order to French life along with guaranteeing the preservation of a particular quality and tradition in the lifestyle. Still, *savoir-vivre* facilitates life, meaning there is a certain way to do something no matter what the situation is or how trivial it may appear. The rules provide an avenue for security where a threat is presupposed and certainty where fears and doubts exist. Consequently nothing is left to chance. Even as Napoleon codified civil law, another Frenchman codified gastronomy in 12 volumes. Nothing has been omitted or forgotten; even slavery was codified, and it resulted in more humane treatment of slaves in France and its colonies than in England or Spain and their colonies.

On the other hand, *savoir-vivre* can lead to preoccupation with form over substance, transforming every aspect of life into a ceremony. Preoccupation with form is evident in the French sense of style and fashion and a flair for elegance. Some first-class French hotels have decrepit furniture, yet the rooms have such utter charm that all is forgiven and overlooked. Image, and the stress on the sensual, is more important than the facts in business advertisements as well. French advertisements tend to be attention getting. They concentrate on creating a mood or a response instead of informing (Hall & Hall, 1990). French advertising campaigns must first produce pieces of art before trying to be effective; this is why many of the French believe that the products in U.S. advertising are not very good—the messages are too straightforward for French tastes. Similarly it is no accident that the French subway system has reproductions of classic paintings on display at major stations.

That the French love classifying things is apparent not only with regard to wine but also with regard to the classification of titles and in their fastidiousness with regard to and insistence on politeness and attention to social forms. Like labels pasted on wine bottles, labels that are applied to people will stick. There is little room in French society for impostors and little fluidity in crossing class barriers. The French pay attention to status and titles and expect others to do likewise. Correct form must be followed. For example, when introductions are required, the person who makes an introduction must be of the same status as the person being introduced. In a business meeting it is essential that the person with the highest rank occupy the middle seat. The importance paid to social standing is so great that even salary takes second place. Likewise, when honor is an issue, keeping one's word has more value than profit (Hall & Hall, 1990). Given this situation, it is little wonder that many people who seek social mobility have migrated to countries such as Canada and the United States in recent years and that the rates of innovation and entrepreneurship in France are low when compared to nations such as the United States.

Cartesian Legacy

Like the viniculturist, the French generally have a need to control and refine life and to order the universe. A significant portion of this desire can be attributed to the thinking of Descartes, whose desire to make man the master of nature led him to ponder a rational meaning of the universe. Because he elected to discover this meaning without leaving his study, due to his belief that "I think, therefore I am," his findings were not always accurate. But that did not matter to him as long as he was convincing. He has been described as the "intellectual father of the French preoccupation with form" (De Gramont, 1969, p. 318). One example of Cartesian thought is the general who devises a perfect battle plan with incomplete knowledge of the enemy's strength and capacity and suffers defeat—but with style and elegance.

The legacy of the Cartesian method can also be witnessed in life's most commonplace occurrences. Preoccupation with shaping, organizing, and magically transforming raw material into a work of art is evident in food shop windows, where displays of colorful and elaborately prepared casseroles and desserts could vie with paintings in the Louvre in terms of master-crafted creations. In sum, Descartes, who is frequently described as the founder of modern philosophy, left nothing unexplained, and neither do the French today.

French businesspeople are no less concerned with form in business life. Presentations are given from the heart so that the French display eloquence much more than U.S. Americans do. Their obsession with form shows in their belief that how one speaks makes as much an impression as what one says. The French love to discuss abstract and complex ideas spontaneously, in detail, and at length so that agendas, time factors, and conclusions appear to have less importance.

In addition, French business tends to be highly centralized because of the long-lasting influence of the Romans. In an autocratic and bureaucratic way, the person atop the hierarchy is most important and wields power in many if not most decisions. This norm encourages the French to bypass the many intervening layers and to appeal directly to the pinnacle of power. The French manager in authority demonstrates it physically; the manager's desk is placed in the center of the office. Those with the least influence are relegated to the far corners of the room. It is little wonder that change in French organizations tends to come from the top downward, not from the bottom upward (Crozier, 1964). Behavior in France today mirrors Crozier's findings, at least anecdotally.

Managerial Style

Such a centralized social structure lends itself to the acceptance of autocratic behavior. Managers, therefore, display almost total control over their subordinates. In fact, French managers are often accused of not delegating authority, instead sharing vital information only within their own elitist network. As a result, it can be very difficult for lower level managers to move up the corporate ladder.

This tight inner circle of upper management can be likened to the tight inner circle of the highest quality wines that win awards year after year. Only those wines from the best regions, with the best color, brilliance, and taste, can gain the status of *Appellation d'Origine Contrôlée*—the highest distinction available. In the same way, gaining a top management position frequently requires education at the finest of schools and upbringing in the most affluent regions.

This high level of centralization can also be regarded in a positive light, as it serves to maintain unity. The French remain a diverse people who tend to be proud of and loyal to their respective regions of origin. Throughout history the French have strived for unity, but they tend to detest uniformity, for they are a people who generally love to differ. For example, when the United States went to war in Iraq, several European nations were lukewarm in support of this action. Among these nations France was most prominent, even to the point that there was an organized boycott of French wine in the United States that led to a significant decrease in sales. The answer to this dilemma of integrating diversity of opinion and unity of action may be in the modern autocracy of centralization. People who exist in a hierarchy of niches can find their own niche, security, and sense of belonging among members of the same social class.

This knowledge of France's hierarchical business structure can be instructive to American business managers. Just as the best bottle of wine is saved for presentation only to the most distinguished guests, so too should the most polished proposals be saved for presentation to a top manager in a French corporation. Efforts to persuade lower and middle-level managers may prove frustrating, as few final decisions can be made without approval from the top.

Another advisory note in business relations is to be aware of how the French handle the uncertainty of unknown colleagues at a typical business lunch. Uncertainty brings unease and implies a lack of control, for it entails a threat that must be thwarted. The lunch, then, gives the French time and opportunity to reflect, study, and learn who outsiders are and how they can be expected to behave in various situations. Pleasure and business are intertwined, for there is no better way in the Frenchman's mind to open up communication and understanding than with a leisurely repast of food and wine (Hall & Hall, 1990).

Interestingly, however, the average amount of time devoted to lunch during the busy day has decreased from 90 to 40 minutes, suggesting that the globalization of business has influenced even the diehard French, Similarly, the number of French brasserie cafes has decreased from more than 200,000 in 1960, when the population was 46 million, to about 50,000 today, when the population is 61 million. Even the average number of gallons of wine consumed per French person per year has declined from 26.5 in 1960 to 14.5 gallons in 2000, according to Onivins, the national wine office. In France it is now common to drink wine at special occasions or when dining out or on the weekends, not at every lunch and dinner.

Classifying Behavior

Perhaps the best attempt to categorize and classify different types of French behavior was completed by the anthropologist Edward T. Hall (1966). According to Hall, France is at the middle of his context dimension, with Japan and the Arab countries representing high-context behavior while Germany and the United States represent low-context behavior. That is, the French are high-context in the sense that they frequently do not need explicit and/or written communication to understand one another. However, the French tend to emphasize low-context behavior in the form of excessive bureaucratic rules and regulations. This seemingly contradictory behavior reflects the Roman emphasis on centralization and bureaucracy and, at the same time, the innate desire of many French people to know one another deeply before transacting business.

Hall's framework may help to explain why the French tend to evoke strong emotional reactions in many short-term and long-term visitors to their country.

Frequently visitors complain about the rudeness that greets them in France, a complaint not unlike those voiced by some visitors to the United States. This rudeness or directness in speech among the French seems to represent both low-context and emotional behavior, which is interpreted, correctly or incorrectly, as rudeness. Conversely, other visitors paint an opposite picture of the French, describing them as concerned about their feelings and welfare and as going to extreme lengths to demonstrate hospitality and friendship. This may represent more high-context and emotional behavior.

In both instances it is best to reserve judgment, as initial external manifestations of behavior can be misleading. For example, the French tend to smile at someone only after they know a person for some time, a cultural practice that is centuries old. This can be a major problem in any situation but particularly in a service-related industry. Today it is common to teach "surly" waiters and clerks in France to smile and speak in a friendly tone, even to new customers (Swardson, 1996).

Composition

Wine authority Alexis Lichine states that "wine is an extraordinary, intricate, and inconstant complex of different ingredients," and so it can be said of the French. Who they are has been determined by their ancestry, the region of the country where they were raised, and the social and educational systems that influence the kind of people they become. Their society is changing, and so are they. Perhaps their adaptability is an outgrowth of their complexity and inconstancy. The French tend to do many things at once, and they do everything with alacrity— especially in an urban setting such as Paris, where the pace is rapid. In this sense the French are polychronic. This pace encourages quick decision making and contributes to their impetuousness. For example, in a highly centralized structure, many businesspeople try to skirt cumbersome intervening layers of hierarchy to accomplish their objectives, even when there is some risk to the progress of their own careers.

To the consternation of U.S. Americans trying to do business in France, the French tend to tolerate disruptions for the sake of human interactions because, to them, it is all part of an interrelated process. This toleration makes planning difficult, even for the French. After all, given life's uncertainties, one never knows what obstacles may prevent promises from being kept.

Still, Edward T. Hall (Hall & Hall, 1990) points out that the French tend to be monochronic—doing one activity at a time—once they have defined a goal they wish to attain. This simultaneous emphasis on polychronism and monochronism is, once again, a reflection of the fact that France is at the midpoint of the high-context/ low-context continuum. Given this intermediate position, it is easy to understand why foreigners have difficulty comprehending French behavior.

Work and Play

The French work hard and prefer to be their own bosses, although few have this opportunity, given the relatively rigid social structure and governmental regulations. At the very least, most French people would like to be recognized leaders, even if what they lead is the smallest unit in the organization. In fact, too much entrepreneurial spirit is considered disruptive to normal business activity in many organizations (Taylor, 1990).

By law the French must devote 5 weeks, including all of August, to vacation, with about 40% of them migrating to vacation spots such as the Côte d' Azur. Holidays, like food and wine, are taken seriously, and the French tend to prepare carefully and meticulously for them.

Weekends tend to be devoted to family matters. Of the 85% who ascribe to Catholicism, only about 15% attend Mass, usually on Saturday evenings or Sunday mornings. Saturday afternoons are often reserved for shopping. In regard to religion, some argue there is an anti-Church sentiment; others disagree, asserting instead that the French have no particular persuasion on religious matters, that they are more areligious than antireligious. This is probably the true state of affairs, given the justified pride that the French manifest toward their magnificent cathedrals such as Chartres, which are found throughout the country.

Geert Hofstede's (2001) analysis of 53 nations in terms of cultural dimensions tends to confirm the profile of the French and the apparent contradictions that emerge when the metaphor of wine is employed. Not surprisingly, the French tend to accept a high degree of power distance between individuals and groups in society and to dislike uncertainty, preferring to be in familiar situations and to work with long-term colleagues. But the French also tend to be individualistic and even iconoclastic, and they cluster with other nations who value a high degree of individualism.

Still, Hofstede (2001) shows that France is a feminine society in which aggressiveness, assertiveness, and the desire for material possessions are of much less importance than the quality and pace of life. The French have deep-seated needs for security and getting along with insiders, colleagues, and family members. In short, the French accept centralization and bureaucracy, but only insofar as it allows them to be individualistic and buffers them from life's uncertainty so that a high quality of life can be maintained. The GLOBE study (House et al., 2004) is in remarkable agreement with the Hofstede study in terms of all of these findings.

Conversational Style

An integral part of maintaining this high quality of life in France is conversation. Like French business, however, conversation is not without classifications, rules, and hierarchical structure, which is quite different from the free-flowing Irish conversational style. Similarly, just as the process of making wine is complex, intricate, and meaningful, so, too, is the art of French conversation. There is an innate restlessness in the French to explore every conceivable issue or topic through lengthy and lively conversation. Whether it is politics, weather, history, or the latest film, contrasts and controversy challenge the French intellect and heighten morale.

Many consider the French to be argumentative—no one more so than the French themselves. They can be quick to criticize, but this is often only to stimulate discussion. Few Frenchmen or Frenchwomen are satisfied by mere superficial discourse. If a conversation is worth beginning, like the production of a tasty Burgundy, it is worth cultivating into a meaningful discussion.

For example, late one evening in the small town of Dijon, a U.S. American stepped into a cab to save himself a 25-minute walk. No sooner had he sat down than the cab driver quickly blurted out, "*Ah, vous êtes Américain?!*" Not expecting such quick questioning, the American cautiously replied, "*Oui, je suis Américain . . . pourquoi?*" The cab driver responded excitedly that he had never been to the United States and wanted to learn more about it. Conversation continued briskly and was quite philosophical at times. The cab driver was mostly interested in comparing the respective styles and mannerisms of the U.S. Americans and the French. Meanwhile, time—and the money meter—was ticking away. The passenger knew pretty well the different ways to get home, and the cab driver's route was not one of them. Finally, after 30 minutes or so, the cab arrived at its destination. The U.S. American was a bit angry that the ride had taken so long and was worried that he didn't have enough money. Just as he reached for his wallet, the driver said abruptly, "*S'arrête! Je ne veux pas d'argent. Merci pour la conversation*," and, after a pause, "*Bonne nuit.*" What did this mean? "Stop yourself! I don't want any money. Thank you for the conversation. Good night." Other Americans have had similar experiences.

The finest wines result from following carefully a detailed, meticulous set of rules (Johnson, 1985). If one of these rules is forgotten or ignored, the quality of the wine will be greatly diminished. The same is true of speaking and conversation in France. Even small mispronunciations have the unnerving effect of fingernails scraping on a blackboard. Conversation is a highly developed art and follows very specific rules.

Not surprisingly, the French language is governed by a seemingly endless (and annoying, in many eyes) set of rules. For example, all French nouns are either masculine or feminine, and articles and adjectives must agree. Verbs have so many different endings and participles that they are almost impossible to keep track of. Proper sentence structure often places the verb at the end of the sentence, as opposed to its beginning or middle, a practice that is sometimes difficult for English speakers to understand.

In conversational circles there are two very different forms of addressing a person in French. The second-person singular, *tu/toi*, is reserved for only the closest of friends and family members of the same age or younger. *Vous*, which is the second-person plural as well as singular, is used on a more formal level. Care needs to be taken in using *tu* and *vous*, as the wrong usage can spoil a conversation or jeopardize a relationship at an early stage. Until you know a person well, the *vous* form should be used. As you become better acquainted, an occasional usage of *tu* is not considered offensive. But only much later, when friendships have thoroughly developed, should you use the *tu/toi* form regularly—but not with elders or superiors. Among younger people these formal rules are being rapidly replaced by a more informal conversational style, although they revert to the more formalized style once they join the full-time workforce.

Making Friends

A helpful guide to building meaningful friendships in France is to recognize the parallels of this process to cultivating wine: Don't rush the process and allow quality to improve over time. That is, approach the relationship in a high-context manner. For Americans venturing into France for the first time, it is important to acknowledge that the American conversational style is quite different from the French conversational style. Whereas Americans tend to have many small conversations with a number of people, the French prefer fewer conversations on much deeper levels.

Another inherent conversational rule is that to smile at someone you don't know and say hello is frequently considered provocative, not friendly. On the other hand, to pass a friend on the street or bump into family acquaintances without offering conversation would be considered rude (Taylor, 1990). These differing interaction styles for different levels of friends in France are not so common in the United States, where a friendly hello and small talk are the norm.

Like a glass of vintage Bordeaux, the family tends to be important to the French because it is a source of acceptance, nourishing them in the midst of life's vicissitudes. The French often look to their families for emotional and economic support. Similarly, as when the horrible disease struck the wine industry in the late 1800s, wine growers look to each other in times of trouble. Likewise, ordinary French people vigorously supported the resistance of wine growers who eventually forced the large American firm, Robert Mondavi, to sell the land on which it planned to harvest wine grapes in southern France in 2001. Relationships among family members are close. Family bonds are strengthened by eating weekend meals and taking extended holidays together. These are times for catching up on family matters, planning, and simply enjoying each other's company.

Paradoxically, however, although French people can be very romantic about love, the concept of marriage and children tends to be approached in a businesslike, practical manner. Children are considered the parents' obligation, and children's behaviors directly reflect proper (or improper) upbringing. French parents are not as concerned about playing with children as they are about civilizing them (Carroll, 1987). This parental guidance continues through adolescence and often through university. In fact, it is not uncommon for parents to help their children out considerably with housing or other expenses, even after marriage. One foreigner who had lived in France most of his life described this orientation in the following way: "The French support their children until they are stepping on their beards."

The French are very private with regard to their homes, so an invitation to dinner implies a high level of intimacy. The home is mostly reserved for family or very close friends. Restaurants are used for acquaintances and first-time get-togethers. One should never ask to visit a French home if not invited, and if you want to stop by for some reason, telephone first.

Women's Role

Furthermore, similar to a fine vintage selection patiently awaiting its proper time for opening and enjoying, the women's movement in France has been in no great rush for full equality in societal roles, even though many women are now prominent in business, politics, and the professions. French women see themselves, and are regarded as the equals of men—equal but different. Presented with opportunities to play the same roles as men, women have shied away from doing so. Accordingly, many milestones in women's rights came about much later in France than in other countries. Only in 1980 did the Académie Française admit its first woman member. Until 1964 a wife still had to obtain her husband's permission to open a bank account, run a shop, or get a passport; and only in 1975 and 1979 did further laws remove inequalities in matters of divorce, property, and the right to employment. The negative U.S. American reaction to then-President Bill Clinton's relationship with Monica Lewinsky was puzzling to the French, and one prominent French politician formed a group to support Clinton. The French easily countenance the fact that politicians such as former President François Mitterand had a second, unofficial family.

So, given the Frenchwoman's tendency to prefer femininity over feminism, it is not surprising that the growth of the women's liberation movement has been relatively slow in France. This movement will most likely be just as successful, or even more so in France than in America, but like the superior wines of France, its growth will be patient and organized.

Suitability

The most meaningful occasions in the French person's life often center on food and drink. Certain wines "marry" certain foods; furthermore, the type of wine served dictates the shape of the glass. The wine must be drunk properly with one's hand on the stem so as not to warm the liquid. A fine wine must be gently swirled, checked with a discerning eye for clarity, sniffed to detect bouquet, and tasted critically before it can be accepted for guests. Meals can last for hours, consisting of several courses served in proper sequence and with appropriate wines to match each course.

This challenging and often controversial business of putting wine together with the proper food requires much mating and matching, contrasting and complementing (Johnson, 1985). In the same way the French people constantly struggle to find a political system that is the best match for the desires of the nation. During the Third Republic alone, from 1872 until 1940, France had 102 governments, while the United States had only 14 over the same period. In the 12 years of the Fourth Republic, from 1946 until 1958, there were 22 governments.

People and Politics

On three separate occasions since 1981, France was governed by what the French themselves term "cohabitation": a president, responsible primarily for defense and foreign policy, from one political party and a prime minister from another party. In all three instances the two were directly at odds with each other on many issues, particularly in the debate about nationalizing private industry versus privatizing a number of state-run organizations.

Part of the reason for such political instability is that the president, with the consent of the prime minister, has the power to dissolve the national assembly at any time except during a crisis. This would be the equivalent of a U.S. president telling the 535 members of the Senate and House of Representatives that their jobs have been terminated. The dissolving can also operate another way: That is, the president or prime minister can resign voluntarily, causing a new round of national assembly elections to take place. Such upheaval and constant change make matching the government (wine) with the people (food) extremely difficult.

Internationally, France has been careful to guard its individual sovereignty. For example, in 1966, President Charles de Gaulle informed the North Atlantic Treaty Organization (NATO) that France was going to withdraw its land and air forces from NATO "to regain her whole territory and the full exercise of sovereignty." In 1982, when U.S. President Ronald Reagan was planning an air strike on Libya, French President Mitterand refused to allow U.S. aircraft to fly over French air space. Similarly France was vigorously opposed to the Iraqi war that the United States led in 2003, ostensibly because it viewed the action as unauthorized by the United Nations. As indicated previously, this created anti-French feelings in the United States and anti-American feelings in France, even to the extent that french fries were renamed in some U.S. restaurants. However, Nicholas Sarkozy, who was elected prime minister in 2007, has taken several steps to smooth over the relations between the two nations.

Paradoxically governmental instability and the penchant for controversy are welcomed by the French people. They tend to be proud of their wide range of active political parties—six in all. Abundant political choice is consistent with the concept of French devotion to individual freedom. Furthermore, the political process is helped along by a free press and a love of political discussion. It is little wonder that France averages about 10,000 public protests of all types a year (Fleming & Lavin, 1997).

The related issues of immigration and racism have been in the political spotlight. Although the percentage of immigrants in France has remained relatively constant at about 7% during the past 25 years, immigrants are disproportionately found in the economically deprived and unemployed categories. There is tension between Catholics, who make up the majority of the population, and Muslims and Jews, some of whom do not want to assimilate into French culture. In 2002 Jean-Marie Le Pen, an avowedly racist politician, actually won the first round of the presidential election, although he ultimately failed in his quest. This provoked emotional articles through-out the world with such titles as "France's Shame" and "After the Cataclysm." However, numerous nations ranging from India to the United States to Israel face similar issues.

The French seem determined to maintain their governmental system even though this may often result in controversy. Many observers point out that no matter where you set foot on French soil, you will find yourself engaged in political discussions. Thus, although the wine and food of French politics don't always suit each other, their very unsuitability stimulates conflict, controversy and a spirited involvement of the citizens in the political arena.

The Maturation Process

Viniculturists strive for disciplined growth, tirelessly pruning and training the vines to conform to their will (Carroll, 1987). This disciplined growth is also reflected in the French educational system, which is strictly controlled by the state and where children begin their education at the age of 6, with many enrolling in preschool by age 2. Unlike other countries such as Japan, where promotion from one grade to another is generally automatic, French students do not advance until they attain certain skills. The result is that children of widely varying ages are in the same class. French education is known to be rigorous, with 30 days of the school year tagged for examinations. But unlike the *vendange* or harvest and its intense preoccupation with accomplishing its mission in 12 hours, young students must focus their developing minds on 35 intense hours of instruction each week. Mercifully, the French have the shortest school year in the world.

For children who continue in the French school system, controlled educational growth signifies growth that is directed toward a certain diploma. The teachers who tend their crops of young students make the decisions that guide their paths—and ultimately their careers. There is a tendency to stress mathematics because profi-ciency in that area is seen as a key to success and guarantees treatment as a member of the elite. Today the most able children are being trained as scientists, technologists, and businesspeople to operate in an international culture. Some observers are critical that this preoccupation with mathematics reaches into inappropriate realms of study such as music classes, which also require mathematics proficiency.

Destiny Determined

The destiny of less able or privileged children is to glean careers from the leftovers. Parents of these children lack the education and influence to help them mold a bright future for their offspring. In a real sense, however, social status more than anything else determines children's educational opportunities or fate. The elite, who understand the importance of education, begin to prepare their children early in life to attend the *grandes écoles* (elite universities similar to American Ivy League universities) and *universités* or *facultés* (specialty colleges and universities, e.g., in journalism), and this practice tends to reinforce the sharp social class differences.

Educational choices are limited in other ways. It is difficult for those who do not belong to the upper social groups to have any freedom of choice in their vocations. Rather, the system early on categorizes them and shapes and defines their destinies. One young Frenchman, son of a successful flower shop owner who had built his business selling flowers on the street, did not want to carry on the family business. Because of the intense competition in his country, he was convinced that he had no chance of going to the *lycées* (elite secondary schools) or *universités*, *grandes écoles*, and *facultés*, which most privileged students attend. As a result, he pursued educational opportunities in the United States, which were better than his choices in France. Immigrants to France, such as those from Algeria, have reached the same conclusion, believing that their educational and career opportunities are very constrained. Living in the poorer districts, they have been involved in several violent riots recently to protest against the system.

In recent years this elite approach to education has been widely criticized, as it tends to lead to a narrow perspective on problems, both in business and government. For example, the 5,000 graduates of the elite École Nationale d'Administration (ENA) tend to dominate government bureaucracies and then, later in life, frequently become chief executives in firms they once regulated.

The French tend to take many things seriously, even life's joys. Among the French, one of life's greatest pleasures is friends; therefore, friendship is taken seriously, as might be expected of a society emphasizing high-context behavior. Friendship must be carefully cultivated and tended over the years. The growth of friendship is a slow and deliberate thing. The French are critical of the quick and seemingly casual manner in which U.S. Americans make and discard friendships. Such behavior is uncouth—not unlike chugging down a glass of wine when a sip is appropriate. A friendship, then, is not to be taken lightly, but like a carefully selected wine, it is to be savored and enjoyed to the fullest. A good wine and a good friend—they are the joie de vivre.

For example, a U.S. businessman had been in France just over 3 weeks, working on a 1-year project, when he came into contact with a French colleague critical to the success of the project. The two businessmen naturally had to spend a great deal of time together. The U.S. American appreciated this opportunity to brush up on his French-speaking skills, as his colleague was a traditional Frenchman who supposedly did not speak English. The U.S. American did his best to communicate effectively but often felt frustrated that he couldn't express himself clearly and feared that he might convey the wrong message unintentionally. Then suddenly, after 4 or 5 weeks of such effort, the two were preparing for a business meeting when the Frenchman said, "You can speak in English if you would feel more comfortable." The U.S. American was astonished. The Frenchman then explained that the French are cautious in dealing with foreign business colleagues and are careful not to overexpose themselves before a more serious, respectful relationship is established. In recent years many politicians and businessmen have become comfortable speaking English to their monolingual American colleagues.

Health and Fitness

Another area where maturation, as well as adaptation, appears in France is in the country's gradual response to the modern health and fitness craze sweeping from the United States into European countries. Traditionally the French have shown a disdain for such trends, choosing continued adherence to national pastimes such as fine wine and cuisine. But times are changing. Nowadays, joggers are everywhere: in the Parc de la Tête in Lyon, or on the coastal roads of Brittany, or at the foot of the Eiffel Tower. Smoking, which has traditionally been popular in France, has become decidedly unchic and tobacco consumption has dropped sharply. The government has taken advantage of this social change by passing a law that makes smoking more difficult in public places.

By most accounts the health craze began about 1990. In 1979 the Gymnase Club opened its first gym. Today there are Gymnase Clubs in many cities, all filled with people eager to trim down or get into and stay in shape. As if this were not enough, today's French are actually concerned about what they eat. They are subscribing to nutritionist services in record numbers, and the book *Eat Yourself Slim* was a best seller for more than a year. Thus, it is not surprising that overall wine consumption has declined, even while—in true French fashion—consumption of the higher quality wines has risen.

The Changing Portrait

As our discussion implies, the French have a predilection to think in terms of greatness. This is understandable, given the glories of their past. Greatness also applies to the reputation the French have developed and maintained in the wine industry. French wines are the best, a fact that the whole world acknowledges, and for the French, that acknowledgment tends to extend to their other accomplishments and to themselves as well. The assumption of greatness tends to be inbred in the French, down to the most humble villager.

Undoubtedly the French imagine greatness to be part of their future. In an era of technological upheavals, they are excelling in high-tech areas. De Gaulle exhorted his countrymen to shape their own destinies independent of the United States; to find their own place in the sun; and to become a nuclear power in their own right. This they have done with lightning speed, making rapid advances in numerous areas of importance. They enjoy an impressive network of highways; more than 75% of the nation's energy is provided by nuclear power.

Economic Constraints

Sometimes, however, the French penchant for going it alone hurts the economy. For example, the French developed Minitel, which just a few years ago was the world's largest database, connecting every conceivable service in Paris and some other cities, as well, by computer. However, the French were slow to use the World Wide Web as an alternative. Today use of the Web is widely accepted and employed.

With the building of the Concorde, the French married high-tech grace and excellence; the supersonic jet is, indeed, a work of art on wings. Ironically the Concorde was a joint project with the British, but people commonly attribute the accomplishment to France, not Great Britain. Could it be because prestige and grandeur seem to be typically French concepts? However, due to poor market demand, the Concorde was decommissioned in 2003.

The famous *Train à Grande Vitesse* (TGV) is the bullet-shaped orange train that streaks past vineyards and villages at more than 170 mph. Riding the TGV at night with blackened windows is to experience complete absence of motion. To incorporate both transport and motionlessness in this train is a typical sort of French creation: mastering nature, striving for perfection, and blending unlikely and diverse elements.

In spite of its various wrinkles, the societal system in France functions quite well, giving its citizens a safety net that cuddles them like a blanket at birth and softens the unpleasant jolts life can bring in old age. It provides 98% of its citizens with a level of medical care and benefits unknown to U.S. Americans and many other nationalities. Income is guaranteed to those over 65 even if they have never worked, and the wealthy collect, too, for it is their right under the law.

The French would revolt if this system were taken away from them, but it is costly. While benefits are being cautiously pared to reduce a large accumulated federal deficit, pension benefits are almost equal to an average wage, and mothers and babies alike get free care to boost the sagging birth rate and elevate standards of living.

An Evolving Culture

Even as wine has a maturation or aging process that alters its personality, so do the French. Also, just as the flavor of wine changes with different blends of ingredients, France's cultural composition is evolving. The French have experienced many changes in their history, and more will follow. One out of 12 people in France is a foreigner; 1 out of 20 is a Muslim. France is one of the few European countries that actually encourage assimilation by granting citizenship to all who are born on French soil. More recently, many have emigrated from France's former colonies, with more than

1.5 million coming from northern Africa alone. Far more than 100,000 political refugees have been allowed into the country. There is resentment, prejudice, and discrimination against these newcomers, as noted previously. Still, it is a fact that France is one of the few nations that has welcomed so many outsiders and refugees.

However, while the face of France is slowly changing in numerous and complex ways, short-term and long-term visitors to this country will most likely continue to react to the French in strong emotional terms. The French most likely will continue to spend a great amount of time nurturing relationships, being wary of outsiders, being sensitive to social class differences, accommodating themselves to the centralized bureaucracy, and being individualistic and iconoclastic. While such activities may appear on the surface to be contradictory, they reflect an approach to life that welcomes, and even thrives on, many different and contrasting ideals. The secret of being able to accommodate such difference is moderation, and wine offers good training in the exercise of moderation. Healthy in itself, it must be taken in proportion, as the excess of it can cause serious problems.

In more than 2,000 years of wine cultivation, French wine makers have accumulated a wealth of experience from which they have established successful techniques and procedures. Throughout the history of viniculture and cultural development in France, the quest for quality has mobilized their collective energy. While nature ensures that each vintage will be different, it is the human element at each step along the way that determines the ultimate outcome. The best wines are objects of great pride and, at the same time, offer lessons in patience and modesty. In this respect the French obsession with rules, procedures, classifications, and form certainly help to develop a product—and a culture—that is world renowned. Thus, whatever paths the people of France choose to follow, the metaphor of composing a fine wine will continue to give insights into their fundamental motivations and system of values.

PART V

Market Pricing Cultures

Market pricing cultures assume that it is possible to compare individuals using ratio statistical scaling, that is, there is a common unit of measurement and also a true zero point. In this way, it is possible to compare individuals on several dimensions simultaneously and even to transform all of the scale values into one final score for each individual. The true zero point is zero money, that is, everything is judged and evaluated in terms of this point. A stark reminder of this cultural orientation occurred in the opening days of the 2003 invasion of Iraq. For a day or two the Allied Coalition (the United States, Britain, and Australia) suffered few if any casualties; the strategy was dubbed "Shock and Awe"; and the bombing of Iraq was intense. On CNN and other TV stations, war coverage was immediately followed by an optimistic report about the U.S. and world stock markets, which increased significantly because of the assumption that the war was going to be short. By the third day, however, Allied casualties and Iraqi resistance began to become significant, and the war coverage was immediately followed by a pessimistic report on the stock markets, which had lost significant value as investors began to sense that the war would be longer than anticipated. Regardless of one's stance on the war and the suitability of following war coverage immediately with coverage of the stock market, it is clear that market pricing mechanisms were accorded great prominence.

CHAPTER 16

American Football

Indeed, the growing complexity of business makes many corporate managers shy away from baseball as a metaphor. Baseball, more than most other major sports, is structured in ways that promote the emergence of improbable heroes. . . . Many business leaders see their game as more like football, with its image of interdependent players with multiple skills cooperating to move the ball down a long field 10 yards at a time.

—Jonathan Kaufman (1999, p. A8)

The game of football, though, simply explodes. From the moment the ball is passed backward through the legs of the center to the quarterback, 22 players are in constant motion trying to execute complex instructions, each intimately involved in the way the play will be resolved. No writing can really capture such intricacy; it must be seen.

—"American Football" (2006, p. 95)

The game of life.

—Headline of article by Patrick Goldstein (2004)

When you don't understand the rules, it just looks like big men running around wrestling.

—Alan Paul (2007, p. W6)

There seems to be unanimity among business writers and pundits that football (U.S. version; soccer is called *football* in most of the world) is the appropriate cultural metaphor for understanding U.S. Americans and U.S. business behavior.

Geoffrey Colvin (2002), a *Fortune* columnist, analyzed the large number of U.S. corporate scandals occurring since 2000 by comparing the complex rules of football to the complex accounting rules and laws that U.S. executives and citizens would like to evade. By comparison, devotees of European soccer are faced with a simpler game and only half of the number of rules, most of which are stated as principles that the referees use to make calls, in a manner similar to the more general accounting principles governing European business. Appropriately Colvin titled his article "Sick of Scandal? Blame Football!" In 2003 Colvin followed up with an article titled, "For Touchdown, See Rule 3, Sec. 38," in which he reiterated his beliefs:

> We Americans like the idea of a national pastime, so let's hold a retirement party for baseball as it relinquishes the title. . . . We all know America's real pastime is NFL Football. . . . The reason is that NFL football is the most legalistic, rules-obsessed, argumentative, dispute-filled sport on the planet, and that means we Americans simply cannot get enough of it. (p. 34)

The intent of this chapter is to demonstrate that, if you don't understand U.S. football, you will have difficulty understanding U.S. culture and the manner in which business is practiced by U.S. Americans. Social critic Camille Paglia (1997) presents a similar argument by encouraging women to study football rather than attend feminist meetings. Shiflett (2000) echoes these sentiments by suggesting that presidential candidates should simply insist that football viewing be made mandatory for U.S. youth, as it "embodies and promotes the very values many parents seek for their offspring" (p. A10).

Apparently both U.S. males and females are in agreement that football is the most popular sport in the United States: At least 68% of the nation's sports fans, and 57% of female sports fans, love football (Conty, 1999). As *The Economist* ("In a League," 2006, p. 63) states: It remains the most popular of the four big American sports on almost every measure, from opinion polls to television ratings. Super Bowl Sunday has replaced Christmas as the national holiday on which families and friends gather together for parties to watch the winners in the two football conferences, the National Football Conference and the American Football Conference, play a single game for championship of the National Football League (NFL). The trial of O. J. Simpson on charges that he murdered his wife and her friend was billed as the Trial of the Century in large part because it involved a football idol.

However, not everyone is enamored. Columnist George Will declared that football manages to combine two of the worst aspects of national life: violence and committee meetings. Journalist Alistair Cooke described it as a cross between medieval warfare and chess (see "Punctured Football," 1993).

In this chapter, there is a description of U.S. football in terms of the following topics: the tailgate party; pre-game and half-time entertainment; strategy and war; selection, the training camp, and complex plays/the playbook; individual specialized achievement within the group; aggression, high risks, and unpredictable outcomes; huddling; and the church of football and celebrating perfection.

The Tailgate Party

A professional U.S. football game typically starts with a tailgate party, a uniquely U.S. phenomenon. Some fans drive hundreds of miles to attend the game, while others drive only for a few minutes, but everyone arrives at the parking lot fully prepared for an outdoor party. They quickly unpack their barbecue grills, food, beer, and soda. Friends may cluster together, but they are sure to share their food and thoughts about the home team and its opponent with other fans in the vicinity. After the party almost everyone is ready to attend the game. But not everyone. Many times in freezing weather, some fans stay in the parking lot in heated vans to watch the game on TV, presumably because they can see the game better, although they could have watched the game on TV at home.

High school football in some states such as Texas is, as the saying goes, the game of life. People in the small towns and cities talk constantly and sometimes obsessively about each game for days and weeks before it occurs. In 2005 one father in Canton, Texas, severely wounded the football coach, Gary Joe Kinne, ostensibly because the coach had cut his son from the team. Several years ago one Texas mother planned the assassination of her daughter's chief rival for a coveted spot on the cheerleading team. Such incidents, however, are rare, and the town's citizens look forward to cheering their sons onward and participating in the tailgate party, for which they prepare elaborately.

At the college level the tailgate party takes on additional social functions. For instance, at a University of Michigan tailgate party, fans who have never attended the university but identify closely with it wear blue to reinforce the team's motto, Go Blue. Entire families, from grandparents to 5-year-old children, will arrive at the tailgate party with everyone dressed totally in blue, and some motor vans are covered in blue as well. Frequently a part of the parking lot is set aside for special alumni groups, such as the Class of 1980, and not only are friendships renewed but financial contributions to the university are encouraged.

After the game all fans hurry to get out of the parking lot as quickly as possible. They drive safely but are not above cutting off another driver they met for the first time and with whom they had been friendly during the tailgate party. All conviviality is forgotten, especially if the team has lost. Fans return to the normal fast pace of U.S. American life in which getting ahead, even at the expense of cutting off another driver, is a strong cultural value.

Pre-game and Halftime Entertainment

Before the game and during the halftime, there is entertainment on the field. Let's look in on one particular case. There must be at least 300 performers on the field before the game, each with a flag and a lavish smile, all synchronized to the slightest move. The sun is glittering on the dancers' shiny outfits, and the wind makes the flags dance to the beat of the drums and the sounds of the horns. The sky is clear except for a Goodyear blimp circling high above the outdoor stadium where the New England Patriots, one of the most successful teams in professional football history, play their home football games. Although it is freezing and snow begins to fall, the fans are so responsive that they react spontaneously to anything and everything that occurs. They cheer the cheerleaders! They cheer the announcer! They cheer other fans! They even cheer the beer man! These fans are seriously committed to having a good time at the football game this Sunday afternoon.

The crowd seems to be homogeneous; fans are wearing the same colors, supporting the same team, and hoping for the same outcome. Yet every fan has added a personal touch to this weekly extravaganza. One man has painted his face with the team's colors. A younger guy in front of him has a huge banner with a witty message that indicates his frustration with the coach. Everybody is eating, yet no one is particularly rapt in this food-feast. They are just continuously munching and crunching while focused on doing something else.

The cheerleaders gather on one side of the field and intensify their dance routines. The music is suddenly blasted, shaking one's body. A bomb, yes a bomb, explodes on the field, igniting a thunderous roar from the crowd. Smoke fills the air, and the middle part of the field becomes virtually concealed. As the smoke clears, a car seems to appear on a podium right on the 50-yard line. Yes, it is a car. Now that the view is unclouded, one can see a Honda standing gallantly on a wide podium surrounded by dancers and models. The Honda jingle is played on the stadium's loudspeakers; the crowd is singing along; and then the announcer proudly proclaims Honda as one of the major sponsors of this week's game. Just as the announcer ordains, the reliable Honda Civic is driven away followed by the dancers who are still persistently frolicking and prancing in a marketing celebration.

Then the mood becomes serious. The military color guard marches with the U.S. flag to the center of the field and plays the national anthem while one of its members sings the words, accompanied by the voices of many fans, all of whom stand respectfully. During periods of war, this period of silent awe takes on renewed energy. If this mood is violated, fans are not happy. For example, at the various Super Bowls different renditions of the national anthem have been offered by popular singers selected for each occasion, and periodically these singers are roundly and publicly criticized for disrespecting the anthem and what it is supposed to convey, namely awe, respect, and honor.

At halftime, the Patriots' marching band comes onto the field to perform a spirited and complex series of moves and themed musical numbers under the leadership of the band's director. Accompanying the band are others, including a person carrying the team's flag, cheerleaders, and twirlers. When the second half starts, the home crowd is ready to cheer on its beloved team.

The halftime entertainment, which is similar to the short breaks found at work—the coffee break, the celebration and presentation of awards for outstanding and/or long employee service, and so on—is over. It is now time to become serious once again and return to the action on the field.

Strategy and War

After the sponsor of this game is announced (and never before), the stage is set for the actual game of football. A sport that captures many of the central values of American society, football has steadily become an integral component of the community. Football is not only a sport in the United States but also an assortment of common beliefs and ideals; indeed, football is a set of collective rituals and values shared by one dynamic society. The speed, the constant movement, the high degree of specialization, the consistent aggressiveness, and the intense competition in football, particularly professional football, all typify the American culture.

Strategy is fundamental in football, war, and American business. Zbigniew Brzezinski (2000), a native of Poland who became a U.S. citizen and eventually former President Carter's national security adviser, was initially puzzled by football until he was able to define the major roles of the various actors in terms of war and strategy. First, there are the owners of the teams, who host special guests in elaborate skyboxes during the game. They can be nasty dictators, aloof monarchs, and variations thereof. Next are the coaches or commanders in chief, who are the CEOs or generals responsible for the team's strategy and overseeing all activities. The head coach is assisted by 9 or 10 assistant coaches, each of whom specializes in a particular area, such as defense and passing. Other coaches sit high in the stadium to provide detached and immediate observation and information to the head coach through their electronic headsets. The head coach calls most if not all of the plays that the quarterback and the offensive team execute. Then there are the quarterbacks, who are the field commanders making last-minute tactical decisions. Finally, the home front or spectators play a key role in cheering on and energizing the home team while demoralizing the enemy.

Football is more like chess than the Chinese game known as Go (see Chapter 25). The objective is to wear down and destroy the enemy or opposing team totally. Just as the game of chess ends with the loser's pieces removed totally from the board, so too in football the winner is exalted while the loser is quickly forgotten. In Go, the objective is to conquer as much space on the board as possible while rendering the opponent powerless. However, saving face or the dignity of the loser is also important, as the winner may someday need his or her help, and it is considered best not to create an eternal enemy. For this reason the powerless opponent still has pieces on some of the board. In contrast, saving the dignity of the loser is of minor concern in football.

In business, it frequently takes several years for a firm to create a new business model against which other firms have difficulty competing. Examples include the Wal-Mart cost advantage of 3% to 5% provided by its efficient supply chain or General Electric's movement from commodity/low-profit products to high-margin differentiated products starting in the 1970s. Similarly football has had several such business models, including the no-huddle offense pioneered by the Buffalo Bills and the one-back offense (freeing up the other back to become a receiver) pioneered by the Washington Redskins.

Bill Walsh, the coach of the San Francisco Forty Niners for several years, created one of the most successful strategies by having the quarterback use many more short passes, which had a higher probability of being caught and a lower probability of being intercepted. He and other coaches also realized that they needed to protect their quarterbacks more strongly, as their injury rate was alarming and finding talented recruits for this position is difficult. Soon the left tackle protecting the quarterback's blindside rose in stature and compensation, and in some instances his salary is higher than that of the quarterback, usually the highest-paid player (see Lewis, 2006).

This approach is consistent with the U.S. emphasis on linking pay directly to performance, especially in using a strategy that will result in complete and total victory.

Selection, the Training Camp, and Complex Plays/the Playbook

It is often said that selecting the right people for the job and training them about company expectations is the best guarantee of firm success. Companies devote enormous resources to selection and training. For example, U.S. companies spend more on training and education than do all of the U.S. business schools combined. This emphasis on selection and training is mirrored in the practices of professional football teams. Their coaching scouts tour the entire country and the world evaluating college and semi-professional football players, and even high school players. A draft is arranged and managed by the NFL. Sequentially each team in turn, starting with the one with the poorest record in the previous year, selects candidates. All of these candidates have been thoroughly evaluated in terms of each team's needs and strategies and are selected accordingly.

Once selected, they arrive at each team's summer camp to learn the numerous complex plays. There are three specialized teams, each with 11 players: The offense, the defense, and the special teams, some of whose members are extremely specialized. For instance, one player may only kick field goals. The field is 100 yards long, and after the opening kickoff, the offense is given four plays in which to win a first down by advancing 10 yards, gaining them the right to another four plays. Each field goal is worth three points and is not attempted unless the offense faces a fourth down and a low probability of scoring a touchdown, which is worth six points; a successful kick after a touchdown is worth one point, although a run or a pass into the end zone is worth two points.

Every player is a specialist and responsible for specific actions on a particular play. These plays are complex and numerous, sometimes numbering 200 or more. Each coach has a playbook describing each play in detail, and during summer camp, players are expected to learn these plays thoroughly. At times some players have complained that the plays are so complex and numerous that they are confusing and hard to remember. Still, for the team to be successful, each player must know not only the intricate rules governing the game but also the complex plays that must be executed successfully.

Individual Specialized Achievement Within the Team Structure

Some researchers such as Hofstede (2001) have ranked the United States ahead of all other nations in terms of individualism, that is, identifying one's needs and interests and making decisions accordingly. The opposite end of this same scale is collectivism, that is, identifying primarily with the group and its needs and making decisions accordingly. Such a perception is only partially accurate. Americans are also group-oriented and being part of a group or network and identifying with it is essential for success in almost all instances. Within the group structure specialization is exalted and everyone is expected to add value to the final product or service because of it.

Similarly it is no accident that the undergraduate business and MBA degrees originated in the United States. American business schools moved away from the prevailing model stressing such areas such as liberal arts and religion to emphasize specialization in the functional specialized areas of business such as finance, accounting, and marketing.

The emphasis on the group does not mean that everyone receives the same rewards and compensation. Depending on special qualifications, some players such as the quarterback and the left tackle protecting him are rewarded much more handsomely than others. Similarly, rewards in U.S. business are much more unequal than in all or most other nations, with some CEOs making 500 to 600 times more than the average salary in the firm. This trend has accelerated since 1960, at which time the comparable ratio was about 30 to 1. Americans believe in equality of opportunity, but not equality of outcomes. Life, liberty, and the pursuit of happiness is the ideal. In comparable surveys involving five nations (the United States, Germany, France, Britain, and Italy), U.S. respondents were far less likely to agree that it is more important for government to guarantee that no one is in need and far more likely to indicate that government's role is to provide freedom to pursue goals (Parker, 2003). Still, if the company or team is successful, everyone benefits financially.

There has always been a belief in the Horatio Alger myth in the United States, that is, one can lift oneself out of poverty through individual efforts and become a CEO or U.S. president. Football strengthens this belief, as many of the stars are assumed to have unbelievably superior skills and abilities that lead to success; they will often play the game even when suffering from major injuries. Ironically, a special report in

The Economist ("Meritocracy," 2005) reviews the mounting evidence that since 1970, it is more difficult to move from a lower social class to a higher class in the United States than in Europe. In 2008 the Organization of Economic Cooperation and Development issued a study of economic inequality in 20 developed nations, which confirms that the United States now leads all of these nations in the level of inequality among its citizens.

Football is a team sport, yet the individual is glorified and celebrated. The extent of individualism in football seems to be unsurpassed in any other team sport. All the major trophies in football are named after individuals who have contributed to the sport. There is the Heisman trophy, the Vince Lombardi Super Bowl trophy, and many other awards that extol the individual. As indicated previously, every player has a particular role to play. The play's success depends on how well all the players perform, yet there is frequently one player who exerts extra and unusual effort. This distinguished player is seen as making the play happen and receives most of the accolades for doing so. However, typically he does not receive all of the accolades, as clearly team coordination and effort are required for success, and so other players and sometimes the entire team are lavishly praised. There is, in effect, individualism within the team structure, and the average professional player receives more than $1 million per year in compensation. Hence all players benefit, but to unequal degrees.

The New England Patriots exhibit the fusion of individuals within a team structure more than other teams. Its members have refused to be photographed individually but will pose for team photos. Similarly, Tony Nicely is the CEO of the highly successful GEICO Insurance Co., part of Berkshire-Hathaway, and he has acted in a similar manner. The esteem in which Nicely is held is captured by the legendary Warren Buffett's public comment that he is a shareholder's dream.

Extreme Specialization

Professional football teams are actually multimillion-dollar corporations subdivided into departments and divisions, each with a large, highly specialized staff. Each member of this football "organization" has one very specialized task. Each squad has its own coach or coaches, and there are also the medics, the trainers, the psychiatrists, the statisticians, the technicians, the outfit designers, the marketing consultants, and the social workers, all with specific duties and assignments. There is even a person assigned to carry the coach's headphone wire throughout the game so that he will not trip on it! Professional football epitomizes perfect specialization for many Americans. The plays in football are complex and precisely executed athletic routines; the players' movements and maneuvers are designed to achieve a high level of physical perfection; and the NFL, in general, embodies all that is impeccable and optimum in the American mind: profits, fame, and glory.

Even the equipment is highly specialized; no other sport in the world provides such specialized equipment. Similarly, it seems as if every American carries around a gadget of some sort to do some kind of task. Look at the football player! He wears a helmet, shoulder pads, neck pads, shin guards, ankle pads, and thigh guards to protect himself; he wraps antistatic nonadhesive tape around his wrists and fingers for support; and he wears state of the art astro-rubber shoes for artificial turf or evenly spiked fiber-saturated shoes for grass fields. Then each type of player has his own distinctive equipment: The receivers wear grip-aligned synthetic gloves to catch the football; the linebackers wear tinted-glass face masks (shatterproof, of course) to protect their eyes from the glare; and the cornerbacks wear ultra-light reinforced plastic back pads to maximize their speed. Even the coaches have their own specialized gadgets: They use sensitive cellular devices to communicate with the statistician and the assistant coaches watching the game from seats high in the stadium.

The "families" within a single football team are the different groups of players forming a "squad," each of which is one family that includes a group of players with similar attitudes and traits. Each player relates primarily to his squad. As indicated previously, there are three squads on every football team, each with its own distinguishable characteristics and values: The defensive squad is usually the most aggressive and violent; the offensive squad includes the higher profile, higher paid players; and the special teams squad is characterized by big plays and high intensity. It is amazing how players try to fit into each of the particular squads' cultures, even to the extent of using outlandish nicknames such as "the Hogs" to describe a squad.

Even though football's rules and regulations are constantly changing from one season to the other, the basic values and ideals of the sport have changed very slowly over the years. In U.S. society innovation and modification are encouraged and sought, but usually not when it comes to values and ideals. U.S. values, like those of football, have developed rather slowly, and few radical shifts in ideals have taken place over the past two centuries, although coaches such as the late Bill Walsh who develop innovative strategies are esteemed, as are entrepreneurs and CEOs such as Microsoft founder Bill Gates and Warren Buffett. Equality of opportunity, independence, initiative, and self-reliance are some of those values that have remained as basic American ideals throughout history. All of these values are expressive of a high degree of specialized individualism, but within the structure of the team.

Competition as a Goal

As this discussion implies, competition seems to be more than a means to an end in the United States and has apparently become a major goal in and of itself. Just as over half the rules and regulations in professional football deal with protecting and enhancing competition in the league, so too American antitrust laws and regulations were essentially created to safeguard competition and equality of opportunity for individuals and groups, and these are deeply rooted in U.S. ideals and values that trace back to the European immigrants who came to the United States. These immigrants represented diverse groups who were basically at war with each other. Although these hostilities and negative feelings were intense when the immigrants first came to the

United States, they were not usually displayed violently because the immigrants were tired of wars and bloodshed. Rather, these feelings manifested themselves in competition, specialization, and the division of labor.

Each geographical area in the United States specialized in a particular category of production: the Northeast in manufacturing, the Midwest in agriculture, and the West in raising cattle. Even within parts of the country that specialized in agriculture, there was further specialization. The more fertile regions of the United States (for example, Idaho) specialized in farming crops that are different from those of the less fertile (but more populated) south. The distinct, highly specialized immigrant communities were still competing with each other economically, but their use of the law of comparative advantage propelled them to focus their efforts on one particular domain of production in which they excelled.

Thus the communities were involved in a new kind of war, one that once and for all was designed to settle the score between Protestants and Catholics, Poles and Jews, English and Irish, and Germans and Danes. Competition prevailed, and the battles of the Middle Ages were refought in the United States but with new specialized weapons and for new competitive endowments. Each community frequently believed that it carried the burden of proving its own superiority through bigger dams, larger statues, more crops, and greater wealth. The legacy of competition continues to flourish in U.S. society today, but with more legitimacy.

The European immigrants, although distinct and dissimilar, shared deep suspicions of authority. They assumed that authority impedes competition and foils specialization, both for individuals and groups. Their main reason for fleeing Europe in the first place was corrupt and oppressive authority and/or government. Systems of checks and balances were developed in the United States primarily to protect the people from rulers who might control the economy, dictate religion, or dominate political power.

Similarly, the NFL uses a system of checks and balances during a sporting match. There are two sets of referees, one on the field equipped with whistles and flags and the other in the review booth equipped with a videocassette recorder and a color TV. If one team does not like a call that a referee on the field makes, it can appeal to the review judges, who watch the play on video and make a final judgment.

Technology and Tools

Technological development is a catalyst to competitive specialization. Technology is the ingredient that provides competitive specialization with the efficiency required in an intensely capitalist society such as the United States. Similarly technological development plays a key role in the NFL. The weight machines that professional football players use, the cameras and satellites that follow them, and the specialized equipment that they wear are integral to the U.S. fascination with tools and machines.

The reason behind this fascination with tools is quite simple: America has historically been short on labor and long on raw materials. To use the abundance of raw materials

U.S. Americans had to substitute machinery and equipment for unskilled labor. Influenced by the success of their highly mechanized industry in the late 1800s, U.S. Americans were more and more shrewdly induced to use the power of machines and technology. To U.S. Americans technology was empirically proven to stimulate growth and success, and their dependence on machines grew deeper and deeper with every increase in the number of U.S. patented inventions. A recent manifestation of this focus is the continuing success of Silicon Valley and the emergence of high technology industries.

U.S. Americans are typically not influenced greatly by extended kinship or family groups; the nuclear family is the locus of activity and identification. In many ways U.S. families are like the three squads in football mentioned earlier. A football player relates primarily to his squad and only secondarily to the team. Similarly U.S. Americans relate to society through nuclear families and not kinship groups. As a general rule, U.S. children are taught to relate essentially to their nuclear family at a young age; they learn through the example of their parents that the nuclear family is the integral part of their lives. U.S. Americans normally encourage independence, self-reliance, and initiative in their children. Children are raised to believe that a rich, healthy, happy, and fulfilling life can be attained by almost anybody as long as they are willing to follow certain steps and procedures. The U.S. egalitarian spirit is nourished in children's spirits and accentuated in their minds by parents eager to make their children successful. As a result, when children have matured to the adult stage, they tend to believe that success (wealth, health, and happiness) is an individual's responsibility and duty. People who are poor, unhealthy, or unhappy are frequently viewed by U.S. society as people who failed to avail themselves of the opportunities offered.

As indicated earlier, U.S. Americans believe in equality, but only equality of opportunity; personal successes and failures are attributed directly to the individual (see Stewart & Bennett, 1991). Many if not most U.S. Americans tend to see poverty and misery as self-induced, at least to a large extent. Stewart and Bennett (1991) point out that this makes life difficult for the average U.S. American, who must constantly achieve to meet such high expectations. U.S. Americans are not honored for past achievements but for what they are currently accomplishing; in this sense their personalities tend to be constantly in a state of flux and evolution.

Openness to Change

One of the major results of this orientation is that U.S. managers are generally very open to change and are constantly introducing new programs with which they can identify and for which they can claim much of the credit; it is not sufficient for them to build on the programs of their predecessors (Kanter, 1979). They tend to jettison programs just as quickly, however, and the United States is famous for both the fads that it introduces and the short-term orientation of its managers.

In short, competitive specialization, both for each individual and the groups in which he or she participates, appears to be the most evident feature of the United States. Generally speaking, the notion of "specializing to compete" is the principal ideological ideal that U.S. Americans endorse, practice, safeguard, and promote worldwide. Competitive specialization is the tool with which U.S. Americans tend to tackle life's main challenges. This tool is often serviced and maintained with high levels of emotional intensity and aggressiveness. From this perspective it is not surprising that most U.S. Americans, when asked what they do, immediately describe their occupation or profession, unlike the Japanese, who tend to respond with the name of the company in which they work (see Chapter 3).

Aggression, High Risks, and Unpredictable Outcomes

The similarities between real life in the United States and a professional football game are astounding, especially within the warlike atmosphere that pervades this sport. Football closely parallels the actions that take place in the extroverted, and sometimes even belligerent U.S. society. According to the Myers-Briggs Type Indicator®, which is the most widely used personality scale in the United States, 75% of American males and females are extroverted and aggressive in personal relations (Keirsey, 1998). While such extroversion is in large measure a positive feature of U.S. life (Barnlund, 1989), the United States is a leading country in terms of indicators that profile the negative aspects of extroversion and aggressiveness. For example, there are an estimated 240 million guns owned by U.S. citizens in a population that is just over 300 million. As the noted political scientist James Q. Wilson (2007) points out, this fact in and of itself does not confirm violent aggression. However, he also points out that the non-gun homicide rate is three times higher in the United States than in England.

U.S. Americans recognize instinctively the link between life in America and what happens on the football field. Violence and aggressiveness are part of football's appeal to U.S. society, and they both relate football to actual life. Aggressiveness, which is often interpreted as energy and intense motivation, is encouraged in the United States. Analogously, aggressiveness is a celebrated characteristic in football. The teams compete with one another, players on the same team compete for starting positions, and even the fans compete for better tickets.

In the early part of the 20th century, there was a movement to outlaw football as a college sport because of the large numbers of permanent injuries and deaths. The late U.S. President Woodrow Wilson, then president of Princeton University, led a group that changed the rules to make the game safer, thus saving college football from extinction. Similarly, while some U.S. states have moved to outlaw the popular brutal mixed martial arts contests that take place in a cage with few rules, defenders of these regulated struggles point out that 63 deaths have occurred in football since 1993 and only two deaths from the caged contests, with these occurring in unregulated contests in Korea and Russia (Schrotenboer, 2006).

Speed dating, which originated in the United States, is a concrete example of such tendencies. Young and not so young unmarried adults do not have the time for traditional dating, so they attend sessions at which each of them talks consecutively to 30 or 40 members of the opposite sex, each for 3 minutes. An official sounds a bell after 3 minutes to signal that it is time to move to talk to another person. At the end of the evening each person is free to accept or reject an offer for a longer meeting. *The Economist* fittingly titled an article on this phenomenon "McDating" (2002). Similarly Paul Fahri (2002), a columnist with *The Washington Post*, argued that females tended to watch *The Bachelor* on TV because it was a female version of football, with the brutal selection process going from 25 females to one supposedly fortunate female.

Survival of the Fittest

This type of individual competition and aggressiveness seems to be particularly suited to the United States. Richard Hofstadter (1955) graphically describes why social Darwinism or the survival of the fittest individual was a social philosophy that appealed to U.S. Americans at the turn of the 20th century:

> With its rapid expansion, its exploitative methods, its desperate competition, and its peremptory rejection of failure, post-bellum America was like a vast human caricature of the Darwinian struggle for existence and survival of the fittest. Successful business entrepreneurs apparently accepted almost by instinct the Darwinian terminology which seemed to portray the conditions of their existence. (p. 44)

Given the violence and aggression, it is difficult to see why football has such pervasive popularity in the United States. There are, however, compensating factors, including the expression of deeply held values. Also, the complexity of football intimately involves football aficionados in every play and they easily see analogs in their daily and business lives. Fans and experts compare statistics on each player and team along numerous dimensions, and they have developed a highly specialized vocabulary for doing so. A week before a major game, sports radio programs featuring call-ins from fans with decided opinions and vast knowledge are popular, and there is a follow-up program after the game using the same format. Such knowledge creates a sense of in-group solidarity, and the clueless spectator is ignored.

Just as important is the highly unpredictable nature of the game. One team may be in total control of the game in the first half, only to wither and die through attrition and the superior offensive and defensive/aggressive play of the opponent in the second half. Football is one of the most unpredictable sports games in the world. For example, before the final two minutes of the game, there is a two-minute warning, after which it is not unusual for a seemingly defeated and discouraged team to score two touchdowns, creating a tie. In the overtime, this team may win. Similarly, one team may be losing 35 to 0 at half-time, only to come back and win. In the 2008 Super Bowl, the Patriots were seeking a historic finish to an undefeated season of 19–0, but they lost in the final minute of the game to the underdog New York Giants using a modified no huddle offense to save time (see huddling below).

Expect the Unexpected

Furthermore, although each player is expected to perform a particular action in each play, the unexpected happens so frequently that the individual player must be creative, that is, changing the strategy of the play and implementing a tactic that is inventive and hopefully successful. For instance, a play designed for a run may be altered to include a pass when the designated runner is surrounded. All of the players, both on the offense and defense, must then adapt to the new situation.

There are, as we have seen throughout this chapter, many reasons to study football and watch it played. Among these reasons complexity, risk taking, and unpredictable behavior rank high on the list.

Huddling

There is another characteristic that differentiates football from any other sport in the world: the huddle, a feature in which offensive teams come together as a group before each play to call a certain plan into action. There is no other sport in the world where one sees the huddle after every play. In the huddle there are different players from diverse backgrounds and with various levels of education. All have agreed that the only way to accomplish a certain task is to put differences aside and cooperate objectively to achieve a certain goal. After the game every player returns to his own world, living his own life in his own unique way. That is the essence of the melting pot, a diversified group of people who forget their differences temporarily to achieve a common goal. As early as 1832 Alexis de Tocqueville drew attention to this facet of the U.S. perspective (as quoted in Miller & Hustedde, 1987):

> These Americans are the most peculiar people in the world. You'll not believe it when I tell you how they behave. In a local community in their country a citizen may conceive of some need which is not being met. What does he do? He goes across the street and discusses it with his neighbor. Then what happens? A committee comes into existence and then the committee begins functioning on behalf of that need. And you won't believe this but it is true. All of this is done without reference to any bureaucrat. All of this is done by the private citizens on their own initiative.

Can a football team afford the luxury of eliminating the huddle? In most cases, no. However, the 1991 Super Bowl, one of the best Super Bowls of all, featured the Buffalo Bills, who did not huddle after each play, and the New York Giants, who did. It was a very close game that was decided only in the last minutes. The no-huddle Bills were able to use this approach only because of the expertise of their outstanding quarterback, Jim Kelly. Once again an individual was able to shine within the group context. Still, the final result was that the huddling Giants won. The final verdict on this issue has not been made, as other NFL teams have since employed the no-huddle offense, but only sparingly thus far. Similarly most if not all groups and organizations in the United States employ a version of huddling to handle their problems and achieve their objectives. Wal-Mart, for instance, pioneered the daily early-morning meeting at which all stand so as to get down to business quickly, shorten the meeting time, and then go out and execute agreements made. This company also pioneered the Saturday morning motivational breakfast meetings featuring awards to individual employees and managers and company cheers such as "In Sam we trust," in honor of the late Sam Walton, its founder.

Business Groups

The U.S. concept of huddling to coordinate activities is quite different from the Japanese sense of community within the organization. The Japanese normally socialize with coworkers after work, and if a major problem occurs, they will sometimes go off-site for an evening of drinking, dinner, informal camaraderie, and finally the discussion of the problem at hand and how to address it. Periodically they will repeat such sessions until the long-term problem is solved. U.S. Americans, on the other hand, tend to huddle together in a business meeting specifically to address and solve the problem at hand, after which they scatter to complete their other work-oriented activities. If additional meetings are necessary, they are normally conducted in the same fashion.

As Daniel Boorstin (1965) has so persuasively shown, the lone cowboy or lone frontiersman is a poor metaphor for the United States. Rather, as adventuresome U.S. Americans moved westward to pursue a better life, they came together frequently to form temporary associations or teams to solve specific problems. However, given the rapid mobility of U.S. society—a characteristic still dominant today—relationships among members of these groups tended to be cooperative but only superficially friendly; there was no time to develop deep friendships. The United States is the classic "doing" society whose members are primarily interested in building and accomplishing goals; the focus in the United States is on accomplishing present goals to ensure a safe future; little attention is given to past activities and history. Europeans and Asians frequently complain that it is very difficult to establish deep, personal relationships with U.S. Americans. As a general rule U.S. Americans intensely commit themselves to a group effort, but only for a specified and frequently short period of time. Unlike many Europeans, who live, work, and die within 30 miles of their birthplaces, U.S.

Americans frequently huddle in temporary groups as they change jobs, careers, geographical areas, and even spouses throughout their lives.

In football the huddle divides the game into smaller sets of tasks that, when accomplished, successfully lead to the fulfillment of victory. The game itself is divided into independent jobs that are separated by short periods of position reassessment and are fueled by continuous tactics. The huddle phenomenon in U.S. culture—meeting together to subdivide a large task into smaller related jobs that are accomplished one at a time—is illustrated in the way that U.S. Americans tackle problems. Any intricacy is broken down to smaller issues addressed one at a time. U.S. Americans tend to believe that any problem can be solved, as long as the solution process has a specific number of steps to follow and questions to answer. Likewise in football, no matter how complicated a situation may be, teams are convinced that they can overcome the complexities through a standardized planning process.

Manufacturing System

The U.S. American system of manufacturing (ASM), a system that developed as a result of Frederick Taylor's work on time and motion study at the end of the 19th century, is the huddle of American economic history. It reflects how the U.S. mind is tuned. Just as the huddle in football allows a standardized planning process for a specific situation, the ASM allows a standardized manufacturing process for different products.

The ASM emphasizes the simplicity of design, the standardization of parts, and large-scale output. U.S. Americans introduced the concept and use of mass production as a direct consequence of the intense use of machinery. Whereas it took a group of German workers one week to masterfully produce 10 high-quality shotguns in the 1800s, the same number of U.S. workers, with the help of standardized parts and ready-to-assemble components, produced in one day many more shotguns at a cheaper price. U.S. products throughout history have generally been known not for their elegance but for their utility, practicality, and cheapness. The U.S. culture is one that respects machines and considers them critical to civilization. In fact, one of the major U.S. contributions to the development of society has been its emphasis on ingenious tools and giant machines, including the creation of the personal computer, the telephone, and the Xerox machine.

U.S. society directly relates standardization to the ranking of individuals. There is a great dependency on ranking via a standardized process (often based on statistical analysis) in the United States. Usually there is no time to judge people subjectively. Stewart and Bennett (1991) point out that many "being" or high-context societies rank or rate employee performance in an absolute sense so as to save face; for example, Jones is a superior or good performer. However, U.S. Americans in their "doing" mode typically disagree with this approach and compare individuals to one another when evaluating performance. Standards are relative and not absolute, and an employee may well be replaced when someone else ranks higher than he or she does.

General Electric, for example, employs such a forced rating system: 20% receive above-average rewards, 70% average rewards, and 10% few if any rewards, with some of the bottom group being advised to leave if earlier warnings did not lead to positive changes in performance. General Electric is also the only multinational firm in the world that has a public tournament for the position of CEO, which must be vacated at age 65. Four or five long-time GE executives are selected and announced two years ahead of the current CEO's departure and their behaviors are closely scrutinized. Once the new CEO is announced, the other candidates typically depart. This Darwinian system, however, has encouraged other large firms to hire the rejected CEOs because GE thought so highly of them that they were asked to participate in the tournament.

Academic Competitiveness

Even in an academic setting, students in the United States are often evaluated and compared relative to each other. Students' grades are plotted as a curve of relative performance (usually a bell curve). This curve determines the students' final reported grades.

As might be expected, U.S. Americans normally want to know what the bottom line is so that they can objectively make a decision. This perspective is particularly disconcerting in U.S. schools and colleges where many students want to know only what will be on the final examinations. The egalitarian character of the United States, together with the fast pace of life, necessitates one form of standardized ranking or another when assessing a certain situation. Such rankings are accepted by individuals who still huddle and work together within the group structure.

Although many U.S. Americans consider mathematics boring and tedious, all forms of standardized rankings involving figures and numbers are used. When ranking the quarterbacks, for instance, the NFL standardizes the ranking process by defining different numerical categories that cover all aspects of the position. Each quarterback has certain numbers and figures ranging from "percentage of pass completions" to "interceptions over touchdowns" to "number of yards passed per game." The quarterback's livelihood depends on those numbers. To negotiate a raise a quarterback has to "improve" his numbers, and when a quarterback is benched it is often due to "unsatisfactory" numbers. Similarly, about 13 million Americans belong to virtual football leagues in which participants select players from different professional teams; at the end of the season the participant whose virtual team amasses the most points, defined by each virtual league differently, wins the jackpot. Such leagues even have a draft at the beginning of the season.

Numbers have an immense impact on the decision-making process in American society, whether it is a financial market analysis, a college recruitment program, or a political decision. U.S. politicians are very sensitive to polls (which are nothing but bottom-line numbers), even to the extent that there is always a statistician on any major political staff. Similarly the dependence on aptitude tests is crucial in the U.S.

education system. Academic institutions normally require that students take one kind of standardized test or another. Although the admission decision is based on a larger number of criteria, the aptitude test score is extremely influential. Just as a college football player's "numbers" are the arguments on which he is judged for professional recruitment, so too the student's aptitude test scores frequently become the decisive factors for college acceptance or rejection.

Saving Time

This whole notion of standardized ranking evolves from the U.S. perception of time. Time is not thought of as a continuous and abundant commodity. There is frequently no time for conducting numerous specialized or personalized tests when judging recruits. There is only one standard test, to be taken once every so often, because naturally time is limited. Analogously, the sport of football is based on the notion that time is limited, and teams are continuously trying to beat the clock, particularly near the end of the game when the 2-minute warning is given.

There is almost always a time limit in the United States, and the huddle phenomenon in football reflects this time shortage. There is a specific time limit for the huddle before each play. Football players often rush into or out of the huddle in a quick attempt to conserve time. Similarly U.S. Americans constantly have a stressful feeling that time is running out, and therefore they must talk quickly, walk quickly, eat quickly, and even "rest" quickly. When eating, U.S. Americans sometimes attempt to consume the largest quantities of food in the shortest time possible. Where else in the world can a mere human being eat a "double whopper with cheese," accompanied by massive amounts of potatoes and a cup of soda so big one cannot lift it, in less than 10 minutes?

When talking, U.S. Americans have mastered the art of developing acronyms for time-consuming words like economics (econ for short). The concept of extended formal names—for example, Herr Professor Doktor—is not only anathema to egalitarian U.S. Americans but also runs against the grain of their time-saving efforts. Any combination of words that forms a name or a title of some sort is promptly diminished to a fewer number of letters so as to save time when saying them. The National Football Conference and the American Football Conference are never referred to as such, but rather they are efficiently called the NFC and the AFC. The Grand Old Party is the GOP, madam is ma'am, President Kennedy is JFK, the federal bank is the "Fed," amplifier becomes amp, and best of all, "howdy" is the time-conscious way to say "How do you do?"

Due to the U.S. notion affirming scarcity of time, when improvident news events occur, especially scandals such as the O. J. Simpson trial or former President's Clinton sexual mischief, they are vigorously discussed and energetically debated in U.S. society, but not for long. U.S. Americans do not have much time to dedicate to one news event. As Edward and Mildred Hall (1990) have shown, U.S. Americans are monochronic, doing one activity at a time rather than several activities. After intensely analyzing one major event for a short period, U.S. Americans become distracted and begin to focus on a new event.

Time is also limited in the United States because there are so many things to do in a lifetime. The society develops technologically at horrendous speed, and it is difficult to keep up. One has to be continuously on the move. This is the United States; there is little time for contemplating or meditating. Ideally one succeeds at an early age, and success in the United States is often impersonal and lonely. And the trend has accelerated: In 1985 U.S. Americans shocked pollsters when they indicated that they each had on average only three personal friends; in 2006 the number of friends had decreased to two (McPherson, Smith-Lovin, & Brashears, 2006). Similarly, de Mooij (2005) points out that Japanese advertisements employ single-person ads presumably because there is a desire for individualism while in the United States the opposite situation for ads prevails, that is, groups at a party or a bar are emphasized because of too much individualism and loneliness.

Football provides that sense of belonging and the brotherhood/sisterhood atmosphere that success lacks, and the huddle is the ideal time-efficient approach for handling problems either in football or at work. The popular belief that the top is often lonely, however, does not inhibit most U.S. Americans from pursuing higher levels of achievement because life, they believe, is a test of self-reliance and independence.

The Church of Football and Celebrating Perfection

Robert Lipsyte, a well-known journalist, titled his 2007 article, "The Church of Football." This arresting title captures the esteem and awe in which football is held by most U.S. Americans. The legendary Vince Lombardi, who took over as coach of the Green Bay Packers after the team had posted a 1 and 12 win-loss record in 1960, is an example of a leader who linked hard work, religious beliefs, and success closely. When he met the team for the first time, he thanked the Packers for allowing him to be their coach. Then he followed up by saying (Schaap, 2008):

> Gentlemen, we are going to relentlessly chase perfection, knowing full well we will not catch it, because nothing is perfect. But we are going to relentlessly chase it, because in the process we will catch excellence. (p. 8)

The adage "Winning isn't everything, it's the only thing" is also attributed to Lombardi. This relentless focus on perfection and winning led to the team capturing five NFL championships from 1961 through 1967 and the first two Super Bowls in 1967 and 1968. Today Lombardi is revered by long-time football fans and coaches of all stripes.

Similarly the new immigrants had a utopian vision for the United States. The ideals that the nation symbolized were and still are considered sacred and perfect.

Unlike other major societies, the sources of U.S. values are written materials that are believed to be inviolate and perfect, namely the Constitution and the Declaration of Independence. U.S. history is an ongoing battle designed to preserve these perfect values and utopian ideals incarnated in the Constitution and the Declaration of Independence.

With that in mind, one can understand why the portrait of the United States in the minds of the immigrants was actually a utopian image. Charles Sanford (1961) encapsulates the essence of this image in the following way:

> The Edenic image, as I have defined it, is neither a static agrarian image of cultivated nature nor an opposing image of the wilderness, but an imaginative complex which, while including both images, places them in a dynamic relationship with other values. Like true myth or story, it functions on many levels simultaneously, dramatizing a people's collective experience within a framework of polar opposites. The Edenic myth, it seems to me, has been the most powerful and comprehensive organizing force in American culture. (p. vi)

For centuries the United States was "utopianized" by people fleeing persecution, subjection, tyranny, and oppression. America the concept was actually an attempt to create utopia. America the value encompassed all that is perfect: strength, wealth, philanthropy, family, children, and glory. This is utopia. Analogously football personifies that unblemished portrait of utopia. Professional football is a symbol of that perfect U.S. utopia.

Founding Documents

It is very important to note that the Declaration of Independence and the Constitution, the basic written sources of U.S. values, were created by individual human beings, not gods, prophets, sacred apostles, or holy emperors. From this fact one can understand why the United States ceremonially glorifies the individual. The common belief in the United States is that individuals are capable of doing anything they want to accomplish. Individual achievements, whether earning a degree or scoring the highest number of field goals for one game, are considered precious human deeds and are entitled to commemoration in one type of ceremony or another. Ceremonial celebrations of more significant accomplishments of U.S. individuals, such as winning the Democratic Party presidential nomination or the Super Bowl, become automatically a reflection of the U.S. pursuit of perfection and utopia.

When retired NFL players are selected to the Football Hall of Fame, the most prestigious honor in professional football, the induction ceremonies celebrate both the individuals selected and the country that bestowed such perfection. There is a high sense of nationalism in professional football that is ceremonially reflected before the start of each football game, as we have seen. The national anthem is played and sometimes sung by a celebrity, and the fans proudly sing along; the flag procession precedes every football game, and representatives from the Army, Navy, and the Air Force carry American flags and assemble on the field. Apparently there is nothing worth celebrating more or nothing more perfect than the United States. In the United States, sporting events in general and professional football games in particular are in essence ceremonies that celebrate how perfect teams can be, how spectacular the nation is, and how well the system works.

It is little wonder that the Super Bowl, watched by about 100 million U.S. Americans and by millions of other fans throughout the world, has become the major family holiday in the United States. At the first Super Bowl in 1967, there were vacant seats. Because of such factors as adroit marketing and promotion of this event, it is very difficult if not impossible for the average person to purchase a high-priced ticket, even if they can afford it. The shortest TV advertisements cost more than $1 million per showing during the Super Bowl, and the half-time show is spectacular. In the city where the contest is held, parties begin a week before the game and invitations to them are scarce. In fact, there are so many glitterati, public relations specialists, and businesspeople who attend the parties but skip the game itself that *The Wall Street Journal* recently headlined its article on the event, "What Game?" (2008). In essence, the Super Bowl epitomizes the relationship between basic cultural values and material success. Russell H. Conwell—the first president of Temple University, a Methodist minister, and famous orator—provided an apt expression of this relationship in his Acres of Diamond sermon, which he delivered more than 10,000 times throughout the United States in the late 19th century:

> I say you ought to be rich; you have no right to be poor. . . . I must say that you ought to spend some time getting rich. You and I know that there are some things more valuable than money; of course, we do. Ah, yes. . . . Well does the man know who has suffered that there are some things sweeter and holier and more sacred than gold. Nevertheless, the man of common sense also knows that there is not any one of those things that is not greatly enhanced by the use of money. Money is power; money has powers; and for a man to say, "I do not want money," is to say, "I do not wish to do any good to my fellowmen." It is absurd thus to talk. It is absurd to disconnect them. This is a wonderfully great life, and you ought to spend your time getting money, because of the power there is in money.
>
> Greatness consists not in holding some office; greatness really consists in doing some great deed with little means, in the accomplishment of vast purposes from the private ranks of life; this is true greatness. (quoted in Burr, 1917, pp. 414–415)

The celebration of nationalism is evident not only in football but in most if not all areas of socializing in the United States. On television, which U.S. Americans watch while socializing, many commercials are tied directly to nationalism, and innumerable marketable products or services are related to patriotism. In general, U.S.

Americans perceive their nation to be young, successful, and prestigious. By associating products with the United States, marketers aim to relate the youthfulness, beauty, and sex appeal of the United States to their own products. For example, here are the words to a Miller beer commercial:

> Miller's made the American way,
> born and brewed in the U.S.A.,
> just as proud as the people . . .
> who are drinking it today,
> Miller's made the American way.

The jingle is accompanied by pictures of smiling U.S. American faces and happy U.S. American children, all of whom are joined in a dramatic celebration of the perfect beer brewed by the perfect country. In the background and throughout the commercial, the U.S. flag is flying gracefully, symbolizing flawlessness. There is basically nothing stated about this beer other than that it is pure American. Well, that's enough for many U.S. consumers. The utopian U.S. character is partial to success, popularity, and prestige.

Nationalist Beliefs

Sometimes nationalism generates ethnocentric behavior in the United States. In football, for instance, the Super Bowl is referred to as the world championship in spite of the fact that only U.S. teams compete in it. To many U.S. Americans, the United States is the world or at least the best part of it. The globe revolves around the United States, and the poorer members of the international community are protected by U.S. financial and military force. John Parker (2003) captures this sense of exceptionalism in his survey of the United States, "A Nation Apart." The United States stands out with 45.7% of the world's military spending, 27.5% of the world's gross domestic product, and only 4.6% of the world's population ("Briefing American Power," 2007). Simultaneously, the United States stands out among developed nations because of its higher emphasis on regular weekly church attendance and traditional values. Nations of the world must, therefore, follow the U.S. course and act the U.S. way or else suffer the consequences of failure ("Living With," 2003).

However, such confidence has been questioned in both the United States and elsewhere in recent years. As globalization has proceeded, other nations such as Brazil, China, and India are rising in influence. Even the U.S. preeminence in technology is under duress: 14 of the 25 most competitive information technology firms are now located in Asia, and almost 50% of the patents filed in the United States come from foreign-owned firms and foreign-born inventors (Hitt, Ireland, & Hoskisson, 2007).

Still, the economic success and military might of the United States often induce egotistical reactions to international events among its citizens. These narcissistic feelings are frequently evoked by the media's persistent portrayal of and fascination with the world's disasters and mishaps. To the U.S. media "no news is good news," and therefore the outside world is often reported as volatile, violent, and miserable. The positive features of foreign countries are rarely highlighted by the U.S. media. Because average U.S. Americans are continuously bombarded with news of famines, wars, violence, and political turmoil from the outside world, they are frequently not aware of the pleasant and fascinating features of other nations. This ethnocentrism is so extreme that many U.S. Americans believe that their country is the safest and most prosperous country in the world, even when the facts do not support such a conclusion.

Religious Affiliation

The U.S. utopia is not complete without the practice of religion. More than half the U.S. population attends church regularly. Only a few other nations such as Ireland and Poland surpass the United States on this measure. There are more than 400,000 churches in the United States appealing to all different types of religious beliefs. Although the country is primarily Christian, there are numerous other denominations. Also, billions of dollars are spent on church-related activities. For example, it is often said with some justification that the Vatican would be in grave financial difficulties if U.S. Catholics decreased significantly their level of financial support. Gallup polling indicates that 68% of the population reported they were members of a church in 2003, although this was a decrease from 75% in 1960 (Parker, 2003).

In many ways the church of football incorporates the values of all of U.S. churches, even when there are conflicting dogmas and creeds that impede the possibility of mergers. During games and in interviews, coaches and players invoke God and religion as the source of their success. Coaches have followed the late Vince Lombardi's lead by espousing a football gospel of hard work, team integration and effort, religious themes, perfection, and worldly success. However, if a coach or player violates the ideals, members of the football church renounce him, as in the case of the fabled player, O. J. Simpson.

Just as team managers try to relate to and identify with the society through religious ceremonies, so too U.S. Americans participate in religious activities to belong socially to a group. For the early immigrants who left their families and possessions, religion became the basic element that reestablished their social life in the United States. To these new immigrants, moving to a different society entailed a weakening ethnic identification and rootlessness. Religion thus became a means of identification and belonging for U.S. Americans, and churches and religious organizations frequently formed political pressure groups to achieve goals that helped their members and were compatible with their beliefs.

By way of summarizing this discussion of U.S. culture, it is possible to relate this football metaphor to Hofstede's (2001) five dimensions of culture described in the first part of this book. As indicated previously, of the 53 countries in Hofstede's original study, the United States ranked first on the importance that its respondents attached to the belief in individualism. As we have seen in this chapter, however, it is aggressive individualism within the rules and structure of the team or group. The United States also clustered with those countries that accepted and even relished a high acceptance of risk in everyday life and that manifested a high degree of masculinity or an aggressive and materialistic orientation to life. Also, U.S. Americans demonstrated a preference for informality, low power distance between individuals and groups, and weak hierarchical authority. In light of our discussion in this chapter, these findings are not surprising.

The United States is still viewed by many throughout the world as far better than the countries in which they live, if not as a utopia. It still attracts large numbers of immigrants who are willing to put up with the glaring problems of U.S. cities because of the opportunities that are available to those who work hard and strive for success. When they arrive, they tend to experience a distinctive form of culture shock that occurs because of the interplay among the factors described in this chapter. Much of this culture shock is encapsulated in common U.S. sayings and aphorisms, and we end by highlighting some of them:

- Life is just a bowl of cherries.
- Let's get together and work on this problem.
- It's not violent. It's energetic, dude!
- It's lonely at the top.
- Get out of my face!
- Gotta run man, maybe we'll talk later.
- America . . . Love It Or Leave It!
- Where there's a will, there's a way.
- One step at a time.
- It's never too late.
- You're from Nigeria? Is that a city in Germany?
- Hi! How are you? (As he or she walks by not really wanting or expecting an answer)

CHAPTER 17

The Traditional British House

An Englishman's home is his castle.

—old English proverb

We shape our dwellings, and afterwards our dwellings shape us.

—Winston Churchill, speech on rebuilding
the House, October 28, 1944

In the early 1990s, there was an extended discussion about a possible cultural metaphor for Britain in an MBA class on cross-cultural management and behavior. Try as they might, two students—one U.S. American who had lived in Britain for 10 years and one Brit married to a U.S. American—had been unable to identify one possible metaphor, even after a semester's work on the topic. Finally, the U.S. American started to compare her experiences in the United States and Britain. She likened U.S. Americans to independent atoms that periodically come together to accomplish a specific goal, and the British to individuals connected through some sort of "invisible glue." According to the U.S. American, the British appear to understand the manner in which their destinies are inextricably linked; there is little if any need to be explicit about their basic values and why they behave as they do. The British student immediately saw how this invisible glue was closely associated with the traditional British house, and serious work on this cultural metaphor then began.

Britain's long and illustrious history lends credence to this cultural metaphor, and the heroic manner in which the British people rallied in World War II, when the nation was bombed and heavily damaged, tends to reinforce it. Britain is the only major industrial nation that does not have a permanent written constitution; it does not even have an official flag. Supposedly everyone knows what to do, and they do it. But there have been some critical changes. Britain, once the most centralized of European nations, is going through a devolution of power, as evidenced by the fact that a parliament in Scotland and an assembly in Wales were recently created. Also, membership in the House of Lords is no longer a matter of heredity. And former Prime Minister Margaret Thatcher, who held office in the 1980s, led a movement toward privatization that is still continuing.

David Cannadine (1998) depicts three models of Britain, the first of which supports the invisible glue concept: It is a nation whose people are linked closely together by shared values and expectations. However, the second model focuses on a nation divided along three class lines: the upper class, the perpetually striving middle class, and the lower class. Cannadine's third model is the bleakest: Britain is really two nations, one very privileged and one socially disadvantaged. In this chapter we explore the relevance of these models to modern Britain.

The Traditional House

First we need some understanding of a traditional British house, which is built to stand the test of time. It reflects the local materials which tend to be stone, brick, flint, or combinations. Houses hundreds of years old and a few buildings thousands of years old are still standing and functioning in their original capacity. These houses were not built overnight, nor were they meant to last just a short time. A few properties proudly display signs indicating that they were mentioned in the Domesday book, the great land survey commissioned by William the Conquerer in 1085 after he invaded Britain in 1066.

Much of the essence of Britain and her people can be sensed from the fortitude and long-lasting style of their buildings. The design of a British house is traditional and blends with the others around it; the foundations are firm and strong; the floor plans are often similar to one another. Like the British people, the traditional house holds few surprises. A British person could find his or her way around a three-bedroom "semi" (semi-detached house or duplex) anywhere in the country. They are often based on the same tried and tested design. The British way of life is reflected in their traditional houses. They change very slowly except for some gradual chipping away or eventual weathering over the years. There is usually one "right" way to do things and most if not all citizens know what it is—no one has to be told.

Putting up a modern California-style wood house with an open and unusual floor plan surely does not meet this need. Perhaps due to the legendary British weather, a timber-frame house of any sort would tend not to meet with approval and could make financing and insurance difficult to get. In fact, even when the population was sparse and forests abounded, houses were made of stone, not wood. A house, like a way of life, should have strong foundations, be familiar, unchanging, and built in tried and tested ways, the British believe.

Close Connections

Former Prime Minister Thatcher was fond of referring to U.S. Americans as her first cousins, and U.S. Americans reacted positively to this and similar statements. In fact, this strong sense of identification is justified, as surveys focusing on many countries show that the British and U.S. Americans tend to cluster together in terms of basic values and attitudes. As Winston Churchill once shared with a group of Harvard University students, the great Otto von Bismarck, who brought several countries together in a united Germany, is said to have observed toward the close of his life that the most potent factor in human society at the end of the 19th century was the fact that the British and U.S. peoples spoke the same language.

These values result in concrete actions. For example, U.S. American law, government, and social mores are rooted in their British antecedents. Britain and the United States have been staunch allies for more than 100 years, the "special relationship" cemented during the Second World War through the extraordinary collaboration between Prime Minister Winston Churchill and President Franklin D. Roosevelt. In recent years, the war in Iraq has turned the British public's attitudes against the United States; the Pew Global Attitudes Survey indicates that favorable opinions slipped from 75% in 2002 to 56% in 2006. However, there are significant cultural differences that surprise and confuse U.S. Americans when they consider their "cousins," the British.

It will be helpful to examine the British culture in terms of the metaphor of the British house. We will examine the "laying of the foundation of the house" evident in their history, as well as the political and economic climate of today. Next, we will look at the "building of the brick house," which naturally includes the various elements of growing up British. Last, we will describe "living in the traditional brick house," including some of the cultural patterns in business and social situations.

History, Politics, Economics—Laying the Foundations

First, let us identify Britain geographically. A number of terms are used when discussing this area, such as the British Isles, Great Britain, Britain, and the United Kingdom. We will be discussing Britain, which is synonymous with Great Britain, consisting of England, Scotland, and Wales. These make up the larger of the two islands that lie off the northwestern coast of Europe. The smaller island is made up of the Irish Republic (a separate country) and Northern Ireland, which is still under British rule. Britain reluctantly increased its control over Northern Ireland in the early 1970s, imposing direct rule in 1972 because violence between various Protestant and Roman Catholic elements had been increasing for more than 30 years. Recently some of the animosity has declined, and Northern Ireland now has a government in which

Catholics and Protestants share power. Although the term United Kingdom includes Northern Ireland (since 1922), Great Britain does not.

Britain's early history is a succession of invasions and immigrations from tribes such as the Celts, Romans, Angles, Saxons, and Jutes, many of whom were warriors, barbarians, and pirates. The date when Britain was first inhabited is unknown, but parts of a woman's skull were found at Swanscombe in Kent and dated at about 300,000 years ago.

The Years of Empire

Britain's history from the middle 1500s to the 1950s is a story of empire building and domination that many nations have tried to emulate without success. For many years their empire was so vast that "the sun never set on the British empire." According to historian Niall Ferguson (2002), the British empire, begun as a counterbalancing force to other European empires like Spain, was nevertheless a great modernizing force in history, which led to the spread of liberal capitalism and parliamentary democracy around the world.

The British have left their mark on many of the foreign lands over which their flag has flown in a variety of ways. One of the most obvious is the fact that the accepted international business language or *lingua franca* across the world today is English. More than 350 million people speak it as their first language and another 450 million claim it as a second language. Many former British colonies set up their parliamentary and legal systems to emulate the British system, including freedom of the press, and still look to the British for advice and consent in their own affairs. In fact, the Magna Carta, which was signed in 1215 in Runnymede, England (considered to be the birthplace of modern democracy), has influenced one of the most important documents in world history, the U.S. Constitution.

The rate of immigration to the home country of Britain from former colonies and commonwealth countries continues to rise and experts predict that Britain's population will increase to more than 70 million in the next 20 to 25 years. Many of these immigrants have settled permanently in Britain, and more than one quarter of the population of central London is of African, Asian, or Caribbean origin. Britain has lagged behind the United States in passing equal opportunity laws and opening up job opportunities for ethnic minorities. However, many Brits do not mind living close to others of different social classes and frequently marry them, a result supportive of the invisible glue perspective: Often council houses (social/low cost housing) are to be found in the affluent parts of town.

Most 19th-century Britons were better educated, better behaved, and much richer than most other Europeans. Across Europe people set out to dress like the British, talk like them, and imitate their good graces. Even the fashion of wearing black, which survives in our contemporary evening wear, came from the British style in the 1830s. "It's their confidence," Aldous Huxley (1951) said of them in *Antic Hay*, "their ease, it's the way they take their place in the world for granted, it's their prestige, which the other people would like to deny but can't" (p. 38), that made them admired models. The adoption of British ways was so universal that it was unquestioned. People automatically chose the best, and the best was British (Barzini, 1983, p. 36).

For hundreds of years, there was a tacit admission of British supremacy in almost every regard. Furthermore, there was universal admiration and envy for their wealth, power, sagacity, and brutal ruthlessness, whenever necessary. There was a certainty that Britain knew best; that its overwhelming power and wealth would take care of most of the world's military, political, and economic problems; that private problems could be controlled by the British moral code and rules of good manners; that Britain could prolong the status quo indefinitely; and that there was nothing to worry about (Barzini, 1983, p. 37). This is the heritage of the British people—the foundation on which the rest of the traditional house is built.

As noted in Chapter 1, culture can vary along several dimensions, including time, or whether a society is past, present, or future oriented. It is easy to see why the British would emphasize the past far more than U.S. Americans (Trompenaars & Hampden-Turner, 1998, p. 130). The traditional pageantry and ceremony of Britain's political structure is built right into the walls and shell of the house—without it, the house would certainly not stand as tall and proud.

Government and Politics

Britain's government is considered one of the world's most efficient, and it is based on Acts of Parliament. There is a written constitution (although it is sometimes said that Britain has an unwritten or uncodified constitution), which is quite different from the permanent U.S. Constitution in that it consists of the Acts of Parliament, court judgments, and European treaties existing at any one time. Within Parliament, the House of Lords basically acts as a review body and can delay but not veto acts passed by the House of Commons. Thus the real power rests with the House of Commons. Precedent or common law is used only when there is no legislation concerning a major issue or no clear legislative intent. Common law, however, is widely used. It is frequently assumed by the British that everyone "knows" the rules, and so the application of common law is positively received by most citizens. Contrary to the fundamental U.S. American assumption that supreme authority must never be vested in a single institution, the British Parliament exercises supreme power over both the figurehead monarchy and the courts. Like a load-bearing wall, it accepts full responsibility for governing in Britain. More recently, treaties with the European Union (EU) make EU law, in some cases, superior to British law.

In a different way the monarchy is a vital part of the political scenario; panoply and pageantry distinguish every ceremonial phase of royal processions. This enduring display of pomp simultaneously manifests the people's unity and strengthens it.

The official opening of Parliament involves a ceremony over which the queen presides, and each week she meets with the prime minister to discuss current national issues. The past, present, and future are united in the crown and the monarch. It is one of the things that makes Britain unique, and even many of its former colonies welcome the royal family with open arms and a unique respect. Without the presence of the royal family, some of the mortar that holds the traditional British house together might surely decay. Periodically the monarchy is criticized, particularly in terms of being a feudal institution not appropriate in modern times, but the public strongly supports it. For example, 1 million or more citizens turned out to watch the funeral procession for the queen mother in 2002, and many more observed the funeral on television.

There are two main political parties in Britain: Labor and Conservative (Tory). There is also a third party, the Liberals (started in 1859) and now known as the Liberal Democrats (or Lib Dems) who work toward social justice and have recently espoused the idea of making Britain carbon neutral by 2050. Britain has a representative form of government. In a national election people vote for their constituency member of Parliament (MP), and that person represents them. It is quite possible and has happened that the majority of votes go to a party that has a minority of seats in the House and so it does not win the election because winning is determined by the number of party-affiliated MPs elected. Elections can be called by the government at any time within 5 years of the last election. This policy gives the party in power a political edge because it can within limits arrange that an election takes place when public support for the government is greatest. To offset this, a vote of "no confidence" in the prime minister and his or her party can force a reelection at any time.

Voting is often along party lines, and personalities can be less important. Voting is also often allied to social class, and so it is fairly common to vote the way your parents did and the way your neighbors do. In addition, the political parties are generally farther apart ideologically than U.S. American parties, and changes in government can result in noticeable changes throughout the country. For example, the election of a Conservative government led by Mrs. Thatcher resulted in the privatization of nationalized industries, the reduction of income taxes, and the rewriting of trade union legislation. Prime Minister Tony Blair's government, however, adopted the "Third Way" and billed themselves as New Labor, adopting some free market and Thatcherite policies after losing four general elections between 1979 and 1997.

In addition to the elected officials, there exists a whole matrix of official and social relations within which power is exercised called "the establishment." This elite web includes such diverse interests as Oxford and Cambridge universities, the British Broadcasting Corporation (BBC), and wealthy, high-society individuals known as the "great and the good." It is a difficult network to break apart because of the different forms it may take under different circumstances. Some say that this network has lost some of its clout, whereas others feel that it just keeps changing shape. U.S. Americans are also familiar with the civil service consisting of long-suffering officials fighting against the whims of ministers in changing governments, portrayed so effectively in the BBC series, *Yes Minister*.

Economic Divisions

There is some support for the belief that social structure at the top levels of society has changed. *The Economist* has compared the educational backgrounds of those who held the top 100 jobs in Britain in three time periods: 1972, 1992, and 2002. In 1972 and 1992 the percentages who had attended the elite public schools (67 and 66) and either Oxford or Cambridge (52 and 54) were very close, but in 2002 the corresponding figures were 46% and 35% (see "The Ascent of British Man," 2002; "How the Elite," 2002). A YouGov poll conducted for *The Economist* in 2006 revealed that the key identifying markers of class were occupation, address, accent, and income in that order. Only 28% expected to end up in a different class than the one into which they were born. In general Brits in 2006 placed themselves in the same categories as they did when asked in 1949 but the flow of immigrants and the fast-growing ranks of celebrities like Posh (Spice) and Becks (David Beckham) have served to blur the class distinctions (see "But Did They Buy," 2006).

Economically, Britain can be divided into two parts. The south (the area within a few hours drive of London) is considered by those who live there to be superior in sophistication, wealth, and social status (much to the chagrin of the others). This is somewhat accurate due to the predominance of service, high-tech, and other growth industries in the south. The rest of Britain, in the north, is associated with heavy industry, engineering, mining, and unemployment although a lot has changed with revitalization of cities like Manchester and Liverpool, the latter of which was anointed the European city of Culture 2008. High unemployment has resulted in some parts of the north, where traditional industries such as steel, coal mining, and textiles have contracted in size. The differences between north and south were celebrated in the acclaimed social novel *North and South* by Elizabeth Gaskell, published in 1855 and adapted for television in 1975 and 2004.

Long after their empire was dissolved by granting independence to former colonies, the British were still considered the moral leaders of Europe. After World War II, they felt they had earned the right to be the third superpower in the world since the country had paid so dearly in terms of human life, physical destruction of property, and courage. However, in 1955, they stood aloof from or outright objected to attempts at unifying Europe. While British opposition has softened in recent years, there is still resistance to the idea of a unified Europe. Britain is a member of the EU, but it has not accepted the use of the Euro, the common currency.

Evidently the British could have collaborated on many occasions to formulate the recommendations for a unified Europe in their own interest, but many British leaders found it unthinkable to join with other nations and risk losing their full identity. Britain has tended to be a jealous guardian of its freedom of action, proud of its solitude. The British prefer to be victims of their own mistakes rather than to trust the judgment of other people (Barzini, 1983, p. 61). In fact, the UKIP, the fourth-largest party in the UK, is dedicated to "withdrawing Britain from the European Union" because it believes that it is bad for Britain's economy and prosperity.

The British tend to admire the balance of power in the United States, and they especially point to the U.S. Congress in which there are two strong entities, the Senate and the House. It is no longer possible to be a hereditary member of the House of Lords, but this body is still of minor importance.

We have taken a look at the British character as exemplified in their traditional, rigid, but long-lasting houses and suggested that the strong sense of history is the foundation on which society rests today. Some of the traditions and shared beliefs represent the mortar that binds people in a national identity. With devolution or perhaps because of the sentiment among the various constituent peoples of the British Isles, national fervor now often runs along Scottish, Welsh and Irish lines. The English tend to be more reserved in their expression of nationality, as this often invokes unpleasant images among the others. The Welsh Dragon and the St Andrews Cross of Scotland fly with pride. Now we will look at the social mechanisms—socialization processes and the educational system— that sustain these feelings.

Growing Up British—Building the House

Due to the prestigious foundations of British history, foreigners still expect the British to show something of their ancient firmness, resourcefulness, diplomacy, leadership, and certainly their good graces. How do the British continue to instill some of these admired virtues into their society? How do they put the bricks together atop the foundation to grow up British?

It has been said that all Britons have a few ideas firmly embedded in their heads that exactly and universally give them the answers about how to be perfectly British. This has allowed them, throughout history, to know exactly what their countrymen would expect them to do and how to go about doing it. For the most part there is only one right way to do just about anything—from greeting guests to waging war.

Children Behave

Above all else, most British families bring their children up to behave. Inquiries about children will often be about whether they are well behaved rather than whether they are happy. It is assumed that they need to be controlled and not spoiled. A great importance is placed on learning proper manners. George Mikes, the Hungarian-born British humorist, is supposed to have observed, "On the Continent people have good food; in England people have good table manners."

The British are taught at an early age to control spontaneity, and because children can be embarrassingly spontaneous, the emphasis is on keeping them quiet so they don't bother adults. For example, it is rare to see children in a restaurant in the evenings. Children who do not know their place are thought to be not well brought up. Behavior equated with showing off or boasting meets with disapproval. After all, children are meant to be seen and not heard. This emphasis on manners and the importance of being aware of one's responsibilities increases with age.

A result of this restraint is in the well-known British reserve in adults. In fact, following a survey in 2006, the British Heart Foundation suggested that the famous British reserve motivated 42% of the people polled to adopt a "wait and see" attitude before calling for help with chest pains. This resulted in a massive billboard campaign showing a man with a belt tightening around his chest and the caption, "A chest pain is your body saying call 999." Britons have many things in common, and a respect and strong desire for privacy are primary among them. This sentiment is so strong that the British often appear to others as distant and aloof. It is as if they are building walls between themselves and others to create their own space within the brick house.

This need for privacy is a defining factor and one of the dimensions along which cultures differ. The phrase, "we like to keep ourselves to ourselves," is particularly British (Glyn, 1970, p. 176). Glyn notes that even in the case of child care, babysitters need to be hired because relatives will generally not help or interfere, although rising costs and modern life have a way of changing social mores. He also observes that even on trains, the British will frequently not sit near someone when a more secluded seat is available. In a crowded fast-food restaurant, if people are forced to share a table, they will often act as though the other party is not present. Striking up a conversation with a seatmate would generally be considered an invasion of personal space, although this is more characteristic of people living in the south of the country.

There are practical reasons for the British insistence on private, personal space. Britain is a crowded country, and space must be used efficiently. The average population density in Britain is about 245 people per square kilometer, which is considerably higher than most of its Western counterparts (e.g., the United States has 32 people/square kilometer) and also most of the EU countries (e.g., France has 111 people per square kilometer).

Unlike other Europeans, the British prefer houses and are not great flat (apartment) dwellers, but their houses are small by U.S. standards and closer together than in the United States. Just as there are partitions of rooms within the house, the British also grow hedges around their gardens to separate their space from neighbors and to keep out the eyes of the passersby.

A good neighbor is one who is friendly at a distance and who does not intrude. Because physical distance is not possible, the only available protection of personal space is psychological distance. One should always phone ahead before visiting a British home—dropping in unannounced would generally be unwelcome. Also, frequent social telephone calls tend to be regarded as an intrusion.

Emotional outbursts (except at times of genuine crisis) are seen as "soppy" or as evidence of an unstable personality. Unique to the British, however, is the idea that many a problem can be solved over a good cup of tea. The English love their tea. From childhood onward the British are admonished to keep a "stiff upper lip," referring to pursing the lips to prevent an outburst. Even good friends may never be as intimate as in many other cultures. However, this may apply differently to the different classes, with genteel restraint being seen as definitely more upper-class than lower-class British. John Cleese (of Monty Python fame) made this point entertainingly in the film *A Fish Called Wanda* (Shamberg & Crichton, 1988) when he said to his U.S. girlfriend, played by Jamie Lee Curtis:

> Wanda, you make me feel so free. Do you have any idea what it's like being English? Being so stifled by this dread of doing the wrong thing, of saying to someone "Are you married?" and hearing "My wife left me this morning. . . ." or saying "Do you have children?" and being told they all burned to death on Wednesday.

There is a lot of weight from their lofty heritage that the British must endure. Although the foundation is strong and deep, the walls and especially the exterior of the British house and personality must seem to be effortlessly indestructible and imperturbable.

Privacy and Pubs

Inside the house, private space is also cherished. The British do not generally like open plan living areas. Instead, rooms are separated and divided by walls and doors that close. Although houses are relatively small, family members are thus ensured private space. Still, Edward T. Hall (1966) points out that the smallness of the houses effectively precludes children from having their own rooms, as is the norm in the United States. He argues that as adults, the British do not have as great a need for separate offices as do members of some other cultures. Even members of Parliament do not always have their own separate offices.

All of this talk of privacy and reserve should not be taken to mean that British people are unfriendly. They are by and large friendly people and make the most gracious hosts. If you ask people for help, even on the street, most will give it cheerfully. U.S. Americans, on the other hand, are often seen by the British as coming on too strong and too gushy, as being insincere, and as getting too personal too soon. Although the British enjoy the privacy of their own small homes, they also enjoy the camaraderie of the local pub. Nearly every large and small town throughout Britain has at least one pub, many of them hundreds of years old. This is where many British do most of their socializing with coworkers and friends. It is a favorite pastime to share a few beers together after work, even with their bosses, although wine is slowly but surely replacing beer with the advent of the gastropubs. In fact, pubs can almost be thought of as a substitute for the living room in a traditional house because Britons tend not to entertain in their own homes as frequently as U.S. Americans.

Pubs generally serve various kinds of beer—ale, stout, lager (having replaced ale as the largest-selling British beer), or bitter—by the pint or half-pint. Beers are served at room temperature and never cold, which would disguise the flavor. Most other drinks are served without ice and, if you ask for ice, you will be given a cube or two.

Orderliness, patience, and unexcitability are hallmarks of British behavior, although in recent years these traditional characteristics are being strongly challenged by hordes of beer-guzzling football (soccer) fans who regularly invade other countries with fervent displays of nationalism. The British prefer to see each task through until completion, regardless of the priority one might assign to it. For example, if you approach a hotel desk where a clerk is alone attending to some paperwork, he or she will finish the current task unhurriedly before attending to you. There is nothing between the ordinary and a true state of emergency for most British.

The Class Structure

There is a wide range of living standards in Britain, from some of the worst slums in Europe to grand estates. Roles and statuses are generally well defined. Still pervading this stratified society are distinctions between the working class, the middle class, and the elite upper class. In numerous surveys over the past decades up to the present day the British themselves reinforce these divisions by assigning themselves more than 90% of the time to one of these classes without prompting, and generally more than 50% of the British assign themselves to the working class. By contrast, most U.S. Americans, including many who would be considered working class in Britain, describe themselves as middle class.

One is born into a class in Britain, and it is difficult to move from one class to another. Once again, tradition rules, whether it be in one's house plan or social class. Family background, especially the accent and use of the language, will tend to determine the type of education and hence the qualifications earned for a career. In the past it was not unusual to see job advertisements that explicitly required a public school (the equivalent of U.S. private schools) education. Indeed the playing fields of Eton contributed to most of the officer corps of the British Army in World War I. As noted above, social mobility is decreasing for the population overall, even if the social structure of the top 100 positions in Britain has changed rather dramatically in recent years.

Although real disposable income overall has increased in recent years, so too has income inequality. This same pattern holds for other nations, especially those involved significantly in the global marketplace. In 1979 the top 20% of British households had 43% of all earned income whereas the poorest 20% had only 2.4%. In 1996 the corresponding figures increased to 50% and 2.6% respectively. In 2006/2007, according to the National Statistics Office, the original income of the top fifth of households was 15 times greater than that of the bottom fifth before taxes (£72,900 per household compared with £4,900) and four-to-one after accounting for taxes and benefits (£52,400 and £14, 400 respectively). Such facts reinforce Cannadine's (1998) third model, in which Britain is divided into two separate nations: the haves and the have-nots. Even Cannadine's second model focusing on a three-class nation seems to have a good amount of support. If anything, the distance between the top 20% and everyone else seems to have increased.

The British are experts at classifying each other by tiny details of speech, manners, and dress. George Bernard Shaw wrote (in *Pygmalion*) that "the moment one Englishman speaks, he makes another Englishman despise him." *My Fair Lady*, one of the most popular musical comedies ever and an adaptation of *Pygmalion*, humorously but accurately emphasizes the importance of language and accents. Although Shaw's writing was done in the 1800s, the British can still identify regional accents and therefore where the speaker is from and his or her class. The discrimination that this once produced may be fading because regional accents can now be heard in professions and walks of life that were once reserved for the upper-class only. For instance, while there was a time when only "proper" upper-class English would be heard on the airwaves, particularly the BBC, regional accents are becoming more acceptable and are heard in all areas of broadcast. This may be indicative of a broadening and deepening of the middle class in society, raising the level of some of the working class.

Language Speaks

Even so, the use of the language (and other behavior) can still be telling of class. Alan Ross (1969) published a popular version of his article, "Linguistic Class Indicators in Present Day English" under the title "U and Non-U," referring to upper-class and non-upper-class language. Many of the distinctions have faded with time, but some are as true today as ever. For example, "I worked very hard" is U, and "I worked ever so hard" is non-U. "Half past ten" is U, and "half ten" is non-U. In terms of style, rugs or a plain carpet are U, and wall-to-wall patterned carpet is non-U; putting milk into tea cups first is non-U, but offering it after the tea has been poured is U. Another non-U expression that tends to puzzle foreigners is saying, "he wasn't half angry," which means he was extremely angry; or "he isn't half handsome," which means he was exceedingly handsome (Braganti & Devine, 1992). Even the use of terms for meals of the day may differ depending on one's social class; non-Us sometimes employ the terms *dinner* to indicate lunch and *tea* to indicate dinner.

In this regard, T. R. Reid (2000) has pointed out that the British admire U.S. pragmatism and hire U.S. firms such as Bechtel for difficult projects such as rebuilding the Royal Opera House at Covent Garden. But Reid also points out that U.S. Americans possess a distinctive advantage: The British cannot classify them by their speech, which tends to cut across class distinctions. As he states, "In Britain, the way you talk speaks volumes about you—your economic status, your school, your ancestry, your career prospects" (p. A26).

An interesting recent development has been a negative reaction to the proper English that has so distinctively separated the social classes. It is estimated that there has been a decline from 5% to 3% of speakers who use the clipped manner of speaking that is frequently associated with Oxford and Cambridge universities. However, there is a significant increase in the demand for "better" speech that masks or softens regional speech patterns, particularly in business settings and presentations (see "Accents: We Want to Talk Proper," 2002, p. 55).

Of course, the British seem to believe that U.S. Americans do not speak English at all and find their language a sloppy adaptation of it. According to Professor David Crystal, one of the world's foremost experts in English, the future will usher in a situation where people will have to learn two varieties of the language, one spoken in their home country (e.g., India and China) and another that is a standard international version of English. Apart from the strict use of grammar and pronunciation, the British and U.S. Americans can confuse each other with their use of different words. For example, in Britain, you'll hear *flat* rather than *apartment*, *chemist* rather than *drug store*, *lift* rather than *elevator*, *solicitor* rather than *attorney*, *ring up* rather than *call* (which means to visit in person in Britain) or *phone*, and *boot* rather than *trunk* of a car.

Some segments of the population have developed entirely different speech patterns that may not be easily understood by other British (or other English speakers). The best known is the working-class Cockney population, officially those born within hearing distance of the Bow Bells that ring from the Church St. Mary le Bow in the east end of London. These people have developed cockney "rhyming slang," some of which has passed into general usage. If a man talks about the "trouble and strife," he is referring to his wife. To say "I didn't say a dicky bird" means "I didn't say a word." Understanding cockney rhyming slang is yet more difficult when the conversion is one step further removed. For example, "loaf of bread" stands for head, but the whole phrase is not used, just "loaf." So if someone tells you to use your loaf, you are being told to think clearly or to use your head. Another popular expression is "Blimey!" which translated means "God, blind me if I tell a lie," and is used throughout Britain as a common exclamation.

The British tend to keep work and leisure completely separate, much as the living areas and bedrooms of the traditional two-story house are separated by downstairs and upstairs. Bungalows or single-story dwellings are often favored by senior citizens who prefer not to climb stairs.

Educational Policy

British educational policies have changed considerably over the past 25 years, but the old system and its influence are still evident. Tenure has been abolished at the universities, which are ranked on a series of criteria used for determining future funding. As noted above, the number of university graduates has increased, but it is still extremely difficult to get into the top-rated universities, and many graduates are dissatisfied with their jobs.

Today many children attend U.S.-style "comprehensive" high schools or secondary schools for children of all abilities. But the nature of the secondary schools is still very important. Whereas U.S. Americans tend to identify a person's social standing at least in part by the prestige of the college attended and rarely think about the secondary school, the British give approximately equal weight to the college and secondary school. Public secondary schools (which in the United States are called private schools) existed and still exist for children of the financially able or those able to win scholarships. These were often founded by trade organizations; for example, the Merchant Taylors' boys school in London and both girls' and boys' school founded by The Worshipful Company of Haberdashers. Others, including Harrow and Eton, are well known, and they provide more than their share of entrants to the prestigious Oxford and Cambridge universities. There is an ongoing clamor for the provision of more seats in these universities for state school graduates. Data show that in 2007, Oxford and Cambridge, with 53.7% and 57.9%, respectively, had among the lowest participation from state school students among universities ("More State Pupils," 2007). The reputation of the school attended accompanies one through later life, and the more prestigious it is, the more doors it opens. These more prestigious schools often have ties or scarves with certain colors or patterns associated with them. Wearing them is considered an earned honor, and it would be a major faux pas to wear a tie or scarf to which one is not entitled.

Although there has long been a strong connection between church and state, outside Northern Ireland religion plays little part in the everyday lives of the British. King Henry VIII declared himself the equivalent of the pope when he created the Church of England, but about 38% of the British population claim to have no religion. The rest of the Christian population, if pressed, would probably say that they are C of E (Church of England, or Anglican). Roman Catholicism is set to become the dominant religion of the country for the first time since the Reformation thanks largely to immigration (Gledhill, 2007). Among ethnic minorities, Islam is poised to take over as the most practiced religion, with more than 2.7% of the population identifying themselves as Muslim.

Just as the mortar holds the solid British house together, so too do the socialization processes, class distinctions, and educational system described above represent the floor plan or design of the traditional British house. It is within this framework that the British people live, work, and play. But once any house has been built, people live in it in their own way. We will now look at some of the dimensions of British everyday life, such as work habits, etiquette, humor, social customs, and leisure activities.

Being British—Living in the House

Most travelers expect to find the British stuffy and starched. Indeed, that is one side of the coin: the gentleman wearing pinstripes and a bowler hat, carrying a furled umbrella, greeting another with "I say, old chap" and the like; and the oh-so-proper lady shopping at Harrods and having afternoon tea with dainty little cups and rich pastries. Flip the coin, however, and you will find goths, grunges, emos (short for "emotional"), who fit the category based on musical, clothing, and lifestyle preferences, as well as other less noticeable eccentricities. Indeed the modern young generation has certainly become more adventurous in their sartorial tastes.

Furthermore, you may offend the Scottish (or Scots) or the Welsh if you do not respect their differences, even though Britain encompasses England, Scotland, and Wales. Many of the variances appear superficial, such as food preferences and specialties, but others are more substantial. For instance, Scotland has its own legal and educational system, which is patterned after the German system (see Chapter 12), and Wales finds its identity in its own language with the official documents and sites being bilingual.

Despite this wide spectrum, the British are considered a relatively homogeneous group of people. We will focus here on the similarities between them, not their differences. For instance, they all share the same constantly wet weather, which often serves as a convenient conversation starter. The benefit of this is that it produces some of the lushest, greenest land in the world. Perhaps to enjoy this aspect of life, the Scots invented the game of golf in the 14th century and still have some of the most desirable courses in the world. At one point the Scottish king outlawed golf because the citizens were devoting too much time to it and not enough time to archery and preparation for battle.

Maybe this constant wet weather is what the British mean by "making the best of a bad lot." If things go wrong (or the weather happens to be lousy), the right thing to do is to make the best of it. This will be admired. The British tend to think highly of anyone who suffers setbacks and perseveres; whether success comes is less important.

"It is not whether you win or lose that is important, but how you play the game." The British really mean this and will even applaud the opponent for making a good play in a cricket match. In fact a lot of conversational phrases have been spawned from the game of cricket. For example, in Britain—"it's just not cricket" is used to indicate that something is handled unfairly. Draws (ties) are quite acceptable in British cricket matches. This is something that is totally disheartening and rare in the United States where the rules are often such that tie-breaking is essential. Also, the heart of the British tends to go out to the underdog, the player who is in a lost and hopeless position such as that of the man playing for a draw in cricket, who must stay at his post until the close of play.

Sports and Leisure

Again, however, these generalizations must be tempered. In recent years the British have not won many international events; it is more than 70 years since an Englishman, Fred Perry, won the men's singles at Wimbledon. And the loutish behavior of some English soccer fans caused *The Economist* to identify them as "England's Shame" (1998); the phenomenon has been labeled as "The English Disease" in recent years.

According to Professor Eric Dunning, an expert on football hooliganism, the behavior is a relic of the empire and an example of aggressive nationalistic posturing. Still, most of the British follow the pattern set by their ancestors when it comes to fair play.

Whereas cricket tends to be a middle- and upper-class sport, popular in public schools, soccer (it is called football in Britain) is a working-class sport. And, just as elite public schools, clubs, and universities have their own scarves or ties associated with them, which are earned and worn with pride, so too scarves are available in soccer team colors. These scarves may not be earned, but they can be seen trailing from car windows on Saturdays as enthusiastic fans travel to watch "away" matches.

Regarding sports as a leisure activity, the British tend to be keener spectators than active participants. They invented many of the sports the world plays today: football (soccer), golf, tennis, badminton, and rugby. Indeed, the all-American sport of baseball is based on the old British city game of "rounders." The British tend to be loyal fans and as sports mad as U.S. Americans. They talk about it, read about it, gamble on it, and turn out in the worst weather to watch their local teams.

Those who have the financial resources spend much of their leisure time "in the country." It is considered the utmost luxury to have a house in the country, and most Britons yearn for this. In the smallest towns and villages throughout Britain, houses may still be addressed by their names instead of street numbers. These names are usually descriptive and may include the name of the original owner. Examples might be Hawthorne's End, Cadbury's Cottage, or Dunroamin (where the well-traveled person has finally settled). This is a quaint carry-over from the days when people referred to addresses this way and streets were unpaved and unnamed.

An important leisure priority for the British is a respect for nature in the form of the countryside and gardens. The British generally like to walk, sometimes around the block, but in a garden or country setting is better. Another way of getting close to nature is to create your own garden. However modest the patch of ground in front or back of the brick house, there is another traditional element, the English garden. It is not called a *yard*, which signifies a dusty concrete patch like a school yard, but a *garden*, which implies flowers and grass. The British are renowned for beautiful gardens, which may be elaborate or small but usually consist of a patch of lawn surrounded symmetrically by colorful flowers and shrubs. Of course, these gardens do not suffer from lack of water and tend to be wonderfully lush.

Celebration and Ceremony

Ceremonies and holidays offer an interesting example of how the past-oriented British hold on to age-old attitudes and traditions. A large proportion of Britain's oldest surviving customs come from towns and cities and these customs are bound with a long and complex history (Kightly, 1986). The celebrations are widespread; there are more than 7,000 fairs annually, and many a village has a yearly festival celebrating the day of the patron saint to whom its parish church is dedicated. Founder's days are common as well, with institutions such as schools, hospitals, and alms houses honoring their benefactors annually. Such celebrations reinforce the same conservative traditions that the British house embodies.

One favorite tradition is Guy Fawkes' Day, which is celebrated each November 5 and is dedicated to the memory of the fate of a traitor who tried to blow up the houses of Parliament. Bonfires are set, a straw doll to symbolize Guy Fawkes is set atop, and spectators cheer as it burns to ashes. At a Guy Fawkes bonfire people chant the rather morbid poem: "Penny for the Guy, poke him in the eye, stick him in the fireplace and watch him die."

British humor seems to run the spectrum of possibilities. Despite their well-recognized reserve, the British have a basic aversion to seriousness and prefer to lighten most events with humor. On the one hand, there is the dry, satirical humor that one would associate with Noel Coward. At the other end of the spectrum, the British love the broadest slapstick, replete with rude jokes and outrageous behavior, such as in *The Benny Hill Show* and *Monty Python*. Even in dignified settings such as the Parliament, high-spirited bantering finds its way into many otherwise dull negotiations. In one celebrated instance, an MP had to apologize to the Parliament after its members voted that he must retract the statement, "Half the members of Parliament are asses." He apologized by saying that "half the members of Parliament are not asses" and then sat down. This skillful display of biting humor disarmed his adversaries.

Perhaps this love of humor comes directly from their culture, where a direct display of feelings is suppressed. Humor distracts from embarrassing or tense situations that might otherwise be difficult for the British to handle. Just as the windows in a traditional home bring in light and fresh air, humor changes the atmosphere and relieves tensions.

One of the more popular British sitcoms is *Keeping Up Appearances*, both in Britain and in the United States where it is carried on public TV. This sitcom focuses on only one topic, with variants thereof: social class differences. Some commentators, both British and U.S. American, have heaped unqualified praise on this show, and the reader is encouraged to see it not only for its great humor but also its incisive depiction of social class differences.

Fashionably Late

In social and business situations alike, one should be right on time or up to several minutes late, but never early. The British institutionalized the idea of being "fashionably late," usually interpreted as being 10 to 20 minutes late. If the British arrive early for an appointment, they will often wait in their cars or outside until the agreed upon time. However, with the advent of mobile phones (and the ability to make apologetic calls or send a text message), almost two thirds of Britons admit to being regularly late for appointments. According to David Holmes, a social psychologist at Manchester Metropolitan University, the modern generation considers lateness to be more of a norm than a social breach to be avoided.

The old notion of the British being unable to relate to you unless you have been properly introduced by a third party is out of date. Increasing informality in this area is becoming more common, but the British reserve may still shine through, giving the impression of coolness or indifference. Nevertheless, first names are used almost immediately in business and social situations among people of the same level. People of a much lower status in business or socially will usually address their superiors by an appropriate title and surname. In some cases men may be heard calling each other by their last names only, but this is simply a carryover from public school days and need not be imitated.

The phrase, "How do you do?" is a common greeting in Britain. It is not an inquiry but evokes the same response, "How do you do?" and nothing more. Also, the British think using the phrase, "Have a nice day," on leaving someone is strange and see it as a form of command—although with the invasion of many American fast-food chains, the phrase is being heard more often. The British tend to be more circumspect in their conversation than others. They avoid being direct for fear of offending someone. In fact, they often phrase definite statements as questions. For instance, they may say, "It gets dark in the evening, doesn't it?" No one is expected to answer the question. These examples confirm the off-quoted adage that the British and the U.S. American share a common culture but are separated by language.

Politeness and modesty are the hallmarks of the British in social conversation. Do not be surprised if, after you pontificate at length about the history of Poland to a new acquaintance, you find out much later that he or she is an expert scholar in this area. It would be impolite and immodest for the acquaintance to point that out to you at the time. Most likely you will have to learn these facts from another mutual acquaintance.

In all forms of communication, whether oral or written, subtlety, imprecision, and vagueness are typical. Exact facts and figures are avoided. Anyone can find obvious examples of this in the local newspapers, where trends are expressed in "more" or "less," but not in terms of how *much* more or less. U.S. Americans, conversely, prefer precise figures and rankings (see Chapter 16).

Work Habits

The British view of work tends to be pragmatic, but less so than the U.S. view. Britons prefer "muddling through," which usually results in finding the most expedient rather than the most innovative solutions. The French, who prefer to emphasize pure theory, sometimes experience difficulty understanding this "muddling through" perspective, which their more empirically inclined British counterparts manifest. Similarly the Germans, who prefer a deductive form of thinking, contrast sharply with the inductive British.

Paradoxically, the average Briton will obey any rules that are spelled out or stated in exacting detail. Their strong sense of order and tradition dictates that they do what is right, and written instructions and legal signs are an indication of what is right. Unlike many other countries where rules and laws may or may not be obeyed depending on the circumstances, it is a common sight in Britain to see lines of people and cars waiting patiently in a way that would make members of other more hurried and anxious cultures cringe. It is understandable, then, that the British would shy away from having too many rules and written laws because they would tend to obey them strictly.

Most Britons prefer to work for other people rather than for themselves. Only about 10% are self-employed. Overall, in the business environment there is more formality between different levels in the hierarchy, which is more rigid than in America, and formality in dealings between employee and superior is greater. Adler (with Gundersen, 2007) gave an example of a U.S. executive who went to London to manage the company's English office. This executive noted with annoyance that visitors had to go through several people—the receptionist, the secretary, and the office manager—before seeing him. The English explained that this was usual procedure, and without it the executive's status would be compromised.

Similarly, Laurent (1983) studied diversity in concepts of management among executives in 12 Western countries. Twice as many British as U.S. managers felt that hierarchical structures exist so that everyone knows who has authority over whom, and twice as many British thought that there was an authority crisis. The British were also more in favor of well-defined job descriptions, roles, and functions. This is consistent with living in the traditional British house, with its floor plan divided into many small rooms, each with a designated purpose.

Trade union membership has declined in recent years, but about 28.4% of the British workforce is unionized. The relationship between unions and management tends to be adversarial, but in recent years the unions have deemphasized the use of the strike weapon.

The most important abilities of managers are seen as conducting meetings efficiently and having good relations with subordinates. It is a convention that instructions should be disguised as polite requests. Combined with British reserve, this makes for a distant relationship in which both sides are constantly on their guard. Fairness is the most important arbiter of management style.

Meetings, in general, are a significant part of the workday in Britain. Business decisions are typically made jointly and usually discussed, ratified, or implemented at a meeting. These meetings are generally informal, beginning and ending with social conversation, and individuals are expected to make a contribution, even if it is only in the form of questions. Ideas and opinions are normally encouraged, but their value to the group depends heavily on the status or seniority of the person stating them.

Egalitarian and Individualist

In addition, Hofstede's (2001) study points out that the British accept only small power distances between individuals and believe that all people should have equal rights. The GLOBE study (House et al., 2004) confirmed that England had undergone a significant cultural change since the 1970s, moving from a traditionalist society to one that is being challenged by a liberal culture where individuals are expected to succeed on the basis of merit. The GLOBE researchers suggest that business leaders of the future will not succeed by using an authoritarian approach but by understanding how to provide energy, vision, and inspiration.

Although the British are seen as more individualistic than collectivist in Hofstede's study, individualism in the British sense tends to find its form in eccentricity and non-conformism rather than initiative and competition.

Also, the British are often uncomfortable and unwilling to take a stand unless they know that the group consensus will support them. A concern to avoid disharmony among the group members will smooth over all but the most fundamental disagreement. For this reason, even as consumers, the British will frequently accept indifferent treatment from businesses with little complaint. Such acceptance may reflect social class differences. If a customer does complain, the response will quite likely bring a patient explanation of why the customer is wrong and how the business's actions are justified.

The British also normally prefer to work in the security of a group within an established order with which they can identify. Motivation comes when they see the work as useful to themselves and view others also striving to achieve a common goal. The basis of social control in Britain as in most Western nations is persuasion and appeal to the individual's sense of guilt at transgressing social norms and laws. Indeed, this works particularly well in Britain because of the strong sense of tradition, the right way to do things and "not letting the side down."

Most British will identify hard work, education, ambition, ability, and knowing the right people as the methods used in getting ahead. However, these factors often must be accompanied by regular company moves to achieve successful results. This is due to the fact that in many organizations, one must still wait for a higher up to be promoted, move on, or pass away to make space for an employee promotion.

Women make up nearly 50% of the workforce in Britain, which is much higher than in other European countries even though Britain provides low maternity benefits and little child care support. A study in 2003 concluded that among EU countries, women in Britain have the third-lowest maternity benefits after Greece and Luxembourg. Economic necessities have driven women into the workforce. Because they are paid less than men and many are willing to work part-time, companies are happy to have them. Do not expect to find a significant number of them in the higher levels of management or in technical careers. For instance, only about 20% of the information technology workforce is female. However, one of Britain's strongest leaders in modern history, Mrs. Thatcher, was a woman who was known throughout the world as the "Iron Lady." Women who find themselves in positions of authority can and do demand respect. In fact, far more women than men in Britain are training to become doctors and women are found to be better at generating financial returns ("Women in the Workforce," 2006).

The Royals

Speaking of respect, no one draws more than the queen. There is a magical and mystical quality about the royal family, and especially the queen. As might be expected in a tradition-bound society, the most glamorous and somber events involve royal processions. The queen's coronation was complete with decorative pomp and pageantry, the ritualistic anointment with oil and crowning.

More recently, however, the scandal pages and gossip newspapers have been filled with the less noble actions of various members of the royal family. This is nothing new: Throughout the history of the royals (or any family so unrelentingly examined), there have been some demonstrations of less than perfect behavior. However, in a democratic society, it is also no surprise to find that there is a constant questioning of the necessity and future of the monarchy—should the throne continue to exist, should the royals pay taxes, and so on. In answer to these charges, one must remember that in the queen are embedded notions of history, tradition, civility, and national pride, and one third of the British population still dream of meeting her (Michon, 1992). However, the royal family did begin paying yearly taxes in 1992, which implies a narrowing of the gap between the royal family and the ordinary citizen.

The royal family and the tradition of the monarchy are held very dear in the hearts and minds of the British people, and the queen and her nobility are constant reminders of the brilliant past of the British empire, the worldwide respect and awe afforded her throughout the world, and the hope and dignity of the future. The untimely death of Princess Diana in 1997 reinforced the popularity of the monarchy. It also caused many British to display their emotions much more than they normally do, and commentators are still trying to discover why the British were willing to let down their "stiff upper lip." By and large, this emotional display was probably more of the exception than the rule.

In conclusion, our discussion of the British suggests that they tend to be steadfast and traditional in orientation, and the traditional British house is an apt metaphor for understanding the country and its people. A report, specifically commissioned by Prime Minister Gordon Brown in 2008 to assess the strategic challenges facing Britain in the coming years, concludes that the country is uniquely positioned to succeed if it harnesses the skills and talents of its people and provides them with opportunities from childhood to the adult years to face the challenges of globalization. However, change does not come rapidly, and it must not do violence to the traditional, favored ways of the past. Due to this, there is a strong coherence among the people to a set of cultural values that are uniquely British, that have served them well in the past, and that—with modification—should help them adapt successfully to a rapidly changing world.

PART VI

Cleft National Cultures

A cleft nation is one in which the major ethnic groups are so clearly separate from one another in terms of values that it is difficult to form a national culture. These groups tend to be insulated from one another, particularly if they do not share the same or similar religions and value systems. In some cases, a cleft nation may have difficulty identifying a cultural metaphor expressive of the major values of all of the ethnic groups living there. However, it was possible to identify cultural metaphors for all of the nations treated in this part of the book.

We have categorized Italy as a cleft national culture not because of ethnic groupings but, rather, because the north and south are radically different from one another in terms of value orientations such as collectivism and individualism.

CHAPTER 18

The Malaysian Balik Kampung

We solve problems through unity and friendship. Malaysia is one nation with one people. Our strength lies in our unity and is drawn from our determination to involve Malaysians from all religious and ethnic backgrounds in the progress and prosperity of our nation.

—Prime Minister Abdullah Badawi,
campaigning on March 2, 2008

A few years ago in a faculty seminar at the Universiti Kebangsaan in Malaysia, Martin Gannon was explaining the concept of the cultural metaphor, that is, an activity, phenomenon, or institution with which all or most individuals of a cultural group closely identify and that reflects underlying values; in the case of a nation or national culture, these individuals would be its citizens. As explained in Chapter 1, ideally the activity, phenomenon, or institution should be unique or distinctive, and preferably difficult for outsiders to understand how it expresses underlying values. Some faculty members immediately observed what many foreign visitors to Malaysia also experience, namely that the three major cultural groups—the native Malays (about 60% of about 25 million people), Indians (8%), and Chinese (25%)—seem to be so different from one another that it is nearly impossible to identify such a national cultural metaphor.

Indeed, Samuel Huntington (1996) has argued that a national culture such as that in Malaysia is cleft: Its various ethnic groups are so separate that it is difficult to unite them into a common culture. Gannon pointed out that if a nation's citizens cannot identify one cultural metaphor about which there is general agreement, the possibility that the nation will survive in its current form is highly uncertain. From this perspective former Prime Minister Mohammed Mathadir's emphasis on Asian values, although sometimes anti-Western in tone, represents a search for cultural metaphors.

After a spirited discussion, Rozhan Othman suggested that the *balik kampung*, the periodic return to the village, would probably be a representative cultural metaphor. As he explained, the patterns of behavior associated with it tend to be defined and predictable across all three cultural groups. For instance, when the nation was waging war against internal Communists a few decades ago, police would wait outside a village and allow a prominent Communist to have the festival dinner during *balik kampung*, after which they immediately arrested him.

Malaysia gained independence from the British in 1957. It has evolved from a rubber and tin economy to one of the world's emerging industrialized nations in a little more than 50 years. It is similar to many developing nations undergoing rapid industrialization, particularly those in Southeast Asia where ethnic groups are so sharply delineated. Almost half of the nation's population lives in urban areas, but many of these city residents are newly arrived or have lived there for a relatively small number of years. Hence their attachment to traditional and rural values is understandable. In fact, the capital city of Kuala Lumpur with its world famous and iconic Petronas Twin Towers has been described as a city whose sections loosely represent groups from specific rural areas and even villages. In 1995, to ease pressures in the overcrowded capital city, the government decided to build Putrajaya (named after the country's first prime minister), as a garden city of the future about 25 km from Kuala Lumpur. The city, which will play the role of the federal government administrative center, is expected to be completed by 2015 with more than 70% being devoted to greenery and water. Today, Malaysia is one of the world's most advanced Muslim nations and America's 10th-largest trading partner.

Malaysia's history greatly influenced the occupational structure and separation of its various ethnic groups. The British, as part of their divide-and-rule policy, brought in the Chinese to work as laborers in the tin mines, and the latter were not allowed to acquire land. The most promising alternative was to engage in trading, and today the Malaysian Chinese are part of the worldwide group of expatriate Chinese who reside in many nations. Most of the Chinese are both Confucian and Buddhist. In contrast, the Indians were recruited to work either as laborers on the rubber plantations or as laborers and clerks in the railway service. Indians working on the plantation live in villages in estate housing apart from the Malays; those in the railway service tend to live in towns. Hinduism is their main religion. The Malays as a general rule live in villages and work either as rice farmers or as fishermen. Almost all of them are Muslims.

Thus these three ethnic groups were separated not only by their languages, religion, and traditions but also by their occupations and geographical concentrations. These separations were magnified by the British practice of opening up English schools mainly in the towns; rural schools, which tended to be Malay and Indian, used the native languages, Malay and Tamil. Before 1957 only those who had acquired an English education had the opportunity to advance to tertiary education.

This requirement restricted the possibility of achieving social mobility through education for Malays and Indians.

As might be expected, these sharp divisions facilitated the race riots of 1969 against the Chinese, who over time had become wealthier than the Malays and Indians. In response, the government devised a system of equal opportunity called *bumiputras* or "sons of the soil" to help the Malays; the Indians were not included. This policy seems to be working for the Malays, at least partially, although cronyism is a major problem in the awarding of government contracts. Companies seeking a listing on the local stock exchange are required to have a *bumiputra* shareholding of at least 30%. In 1970 *bumiputra* companies accounted for about 2.4% of corporate equities; by 1990, that share reached 19.3%, where it has remained in the early years of the 21st century largely because of the Asian crisis of 1997. The crisis, which put a greater share of the equity in foreign hands, occurred in part because of rent-seeking behavior, the practice of buying up the reserved quota of shares in initial public offerings and selling them at a profit, by the Malays themselves ("Race in Malaysia," 2005).

In contrast, 3% of the population in nearby Indonesia is Chinese, but they controlled about 70% of the Gross National Product (GNP) until their companies were nationalized following the deadly and widespread ethnic riots directed against the Chinese in 1998; similar riots had occurred in 1965. It is hoped that such riots can be averted in Malaysia through government policies and an understanding of common values or cultural metaphors with which all or most citizens closely identify.

However, many Indians—who on average have higher incomes than the Malays—have experienced great difficulty; the rubber plantations no longer provide employment for most of them, and they have been forced to live in shantytowns near major cities where they can obtain work. In recent years, the country has seen the emergence of a radical protest group called the Hindu Rights Action Force or Hindraf. In late 2007 and early 2008, Hindus staged boycotts and protests to complain about the government's treatment. They also resent the "creeping Islamization" in the country as evidenced by the demolition of "unauthorized" Hindu temples built during the colonial era.

Political parties tend to be based on these three ethnic identifications, and the attitude seems to be "let sleeping dogs lie" because the economy by and large has been prosperous in the past 25 years. There are, however, some worrying signs; for example, the Chinese withdraw into ethnic enclaves and send their children to Chinese schools, thus depriving them of interactions with children from the Indian and Malay cultures. About 90% of Chinese children attend Mandarin-language state-financed schools and 70% of young Indians are taught in Tamil. Malaysians, therefore, become used to segregation from an early age. In fact, 50 years after independence, it is sometimes lamented that the different races eat, learn, work, and socialize separately, giving the lie to the idea of one united Malaysia.

Three characteristics of the *balik kampung* are highlighted in this chapter, namely returning to nearby roots, authority ranking, and reinforcing basic values and behaviors. We treat each of them in turn.

Returning to Nearby Roots

Immigration has many effects, among them physically separating individuals from their root culture by thousands of miles. This can be a heart-wrenching situation, as the plaintive songs of immigrants to the United States and other nations during the 19th century attest. For almost all such immigrants there was no hope of ever returning to the root culture, and so they were forced to adopt new values and behaviors to meet the demands of the new situations. In contrast, Malaysia represents a nation undergoing rapid industrialization but one in which the immigrants to the cities are from the nearby rural areas. This pattern is widespread in many developing nations throughout the world. Thus describing a cultural metaphor for Malaysia should prove helpful for such nations as they modernize.

Returning periodically and regularly to their rural roots nearby is our first characteristic of *balik kampung*, for ironically the diversity of Malaysia's ethnic groups provides a metaphor apt enough to describe the Malaysian culture. In many societies differences in religion and tradition are sources of divisiveness. But in Malaysia the religious and traditional celebrations provide rituals common across all ethnic groups. During such festivals, *balik kampung* is very prominent, although it obviously occurs at other times. During these times, major cities, which are normally teeming with people, are transformed into veritable ghost towns.

For example, from 1996 through 1998 the major Muslim festival, Hari Raya ("the day of celebration") and the Chinese New Year fell on consecutive days, and cities were deserted in favor of the villages for several days. During the Hindu Deepavali festival or Festival of Lights a similar pattern emerges. *Balik kampung* is more than just traveling away from the hectic city to enjoy a few days in a serene village. It represents a critical integrating element of the Malaysian value system, that is, the attachment that Malaysians have to their origins and families. The rituals performed during the religious festivals help to shed light on the importance of *balik kampung* and the values associated with it. A significant feature of *balik kampung* is that it provides an occasion for the entire extended family to get together, and family members will typically gather under one roof for the celebration. In most Malaysian families this means three generations: grandparents, parents, and their children. For some families the living room is transformed into a huge bedroom.

Religious rites play a critical role in the celebrations of the three ethnic groups. The festivals provide a time for reflection, atonement for past sins, and personal reformation. Among the Malays the Hari Raya represents a victory over self and temptations after fasting during the whole month of Ramadan. As Muslims, the Malays pay the *zakat*, a form of alms, to purify their wealth and property; they also seek forgiveness from each other on this day. Among the Indians the Deepavali Festival or the Festival of Lights commemorates the victory of Lord Rama, who is reunited with his wife Sita following a 14-year exile and after having killed the demon king Ravana. This is the most popular interpretation and is recounted in the epic, the *Ramayana*. Another interpretation for the celebration is the success of Lord Krishna, one of the three main Indian gods, in killing the demon Narakasura. The common theme is the triumph of good over evil. Indians begin this festive day by taking a bath in oil as a symbol of cleansing away their sins.

A ritual common to all three ethnic groups during these festivals is the audience with the elders. Among the Malays, the children take turns kneeling before the parents to pay their respect and seek forgiveness. The Chinese take turns bowing before the parents and wishing them a happy New Year. In some traditional Chinese families, each child presents the parents with a small cup of tea and wishes them good health. Among the Indians the children kneel before their parents and seek their blessing. In all three traditions the eldest child begins the ritual and the other children then join. Age is important in determining status and position in the family.

In all three traditions children are given gifts of money. The Malays and Indians have adopted the Chinese practice of putting the money in small paper envelopes called *ang pow*. Gifts are also exchanged among relatives, neighbors, and friends. Among the Malays and Indians the gifts are usually in the form of food and special delicacies made by the family during the festival. Among the Chinese a favorite gift is mandarin oranges. Even deceased parents and grandparents are explicitly remembered. Many Malays will stop to clean their parents' or grandparents' graves and offer prayers after going to the mosque. Among the Chinese and Indians, family members pray for the departed ones at the family altar in the home.

After performing the prayers, having breakfast together, and greeting the elderly, the three ethnic groups begin visiting others in quite a similar manner. On the first day of the celebration they mainly visit their relatives and neighbors. Only then do they visit friends, and the Malays generally do not do such visiting until the second day.

A common practice across all ethnic groups is having an open house during the festivals at which Malaysians from all ethnic groups and walks of life are welcome. Anyone can come for a visit without making a prior appointment and be assured that he or she will be welcomed. In many instances the festival dinner becomes a multi-ethnic celebration involving friends, colleagues, and business acquaintances from all three ethnic groups. The Hari Raya celebration lasts 1 month, the Chinese New Year 15 days, and the Deepavali 1 week. Although some Malaysians have adopted the practice of having visitors on specific days during these periods, visitors are welcomed at any time. Even the king, prime minister, and ministers have open houses on specific days.

Although we have been emphasizing the festivals, many Malaysians do *balik kampung* monthly and even weekly because it is relatively easy to return to their nearby roots. Parents use these periods to remind their children of their humble origin and teach them the courtesies of traditional Malaysian culture.

Authority Ranking

Using Alan Fiske's (1991a) typology, we can describe the three ethnic groups as cultures emphasizing authority ranking; Triandis and Gelfand (1998) employ the term *vertical collectivism* in a similar fashion. According to Fiske, members of authority ranking cultures are able to rank individuals in terms of power and prestige, but there is no common unit of measurement, that is, the data points are ordinal rather than interval. For example, 2 is greater than 1 and 3 is greater than 2, but 3 is not 2 times greater than 1. This tends to give rise to powerful leaders whose orders are accepted without question, as the power of leaders becomes exaggerated in the minds of followers, who have difficulty calculating the exact amount of power that a leader possesses. However, such leaders are required to take care of group members in return for such unquestioned loyalty. Behaviorally, authority ranking cultures manifest themselves in elaborate bowing in which the lower-ranked person bows lower than the superior. However, in recent years the handshake, which signifies equality, has replaced or supplemented the traditional bow. This is true of most Asian cultures. There is a much stronger psychological relationship found in authority ranking organizations than in Western organizations where only rational-legal authority bonds individuals to one another (Weber, 1947).

Thus Malaysians are very conscious and sensitive to status and position in an organization's hierarchy. This is consistent with both Hofstede's (1980a, 1991) finding that Malaysia is a high power-distance nation and the recent GLOBE study (House et al., 2004) where Malaysia scores relatively high on power distance. In business dealings it is customary to send someone of equal rank to deal with the counterpart from another organization. Failure to do so will normally be interpreted as not showing proper respect to authority and will tend to vitiate future business dealings.

Similarly the importance of age as an authority ranking device is reflected in the honorary title *Dato,* which accompanies an award bestowed by the king; it literally means grandfather. In Malaysia this indicates that the person is wise and honorable. Even when the person is relatively young or middle-aged, Malaysians are expected to call him by this honorary title. It is difficult to imagine a middle-aged Western person who would be comfortable with such a title.

Also, other phrases tend to emphasize the importance of age. Among the Malays the elders are considered as *lebih awal makan garam,* which translates as "having tasted salt earlier." The Chinese say *wo' che' yen tuo' kuo ni che' fan,* which translates as "I have eaten more salt than you have eaten rice." Furthermore, it is expected that the younger members of a family will not start eating until the elders have begun. There are even rules for order of movement when walking, although these are not as strictly enforced today. In all three communities politeness also requires that a person should never sit with feet pointed directly toward someone, especially someone older.

Even within the extended family, all three ethnic groups employ specific titles for family members according to ranking and position. These titles are much more complicated than those used in Western nations. For instance, the Malays refer to their uncles and aunts not by name but by rank: The eldest uncle would be termed *Pak Long,* the second uncle *Pak Ngah,* and so on. These and similar terms tend to reinforce the concept of authority ranking and high power distance. Addressing Malaysians properly while doing business is also a complex matter for Westerners who are unfamiliar with the naming practices in Malaysia's various ethnic groups. Most business people should be addressed by their proper title and name and it is better to err on the side of formality and follow the lead of your Malaysian host.

Reinforcing Common Values

As countless immigrants have discovered, it is difficult to maintain the values of the root culture if thousands of miles separate them from home. It is important to reinforce these values, at least on a periodic basis. As our discussion has already indicated, such reinforcement is possible with the *balik kampung*, particularly in the area of authority ranking symbols. However, the *balik kampung* goes far beyond authority ranking, as it is a dynamic forum in which common values and behaviors are reinforced. Some of these are family-focused collectivism, maintaining harmony and preserving face, reaffirming religious traditions, and adjusting to other cultural groups.

Family-Focused Collectivism

The individual's relationship with the family is considered critical in all three groups, and this translates into a close relationship between the family and community. Among the Malays living harmoniously and cooperating with other families in the village are important, for in the easily remembered past the only way that families could work the rice fields was by sharing and pooling their labor and resources. Such working together is known as *gotong royong* among the Malays, and it is also practiced during weddings and funerals. Neighbors will work together to prepare the food and setting for the wedding celebration. Likewise, during funerals neighbors will help one another prepare the grave and make funeral arrangements. This is one reason why there are so few professional funeral services in Malaysia.

As migrant communities, the Chinese and Indian families tend to emphasize the importance of their relationships with the clan and trade association. In the recent past it was only through such self-help organizations that these two groups were able to pool their capital to help each other start their businesses. Thus events such as weddings and funerals tend to reinforce basic values common across all three groups and constitute a major reason for periodically completing a *balik kampung*.

The importance attached to collectivistic values is particularly highlighted by the *pok chow* system found among Chinese laborers. *Pok chow* literally means gang contracting, and it represents a group of workers who get together by mutual consent. A *pok chow* group is essentially a work group whose members decide on their own rules, division of work, the remuneration given to each worker, and even the election of the leader (Sendut, Madsen, & Thong, 1989, p. 61). In recent years Western firms have employed a similar system, self-managed work teams, which collectivistic systems have emphasized for centuries, and it is little wonder that both

Hofstede's (1980a, 1991) study of 53 nations and the GLOBE study (House et al., 2004) of 62 cultures indicated that Malaysia is highly collectivistic.

Most Chinese and Indians in Malaysia are third- or fourth-generation Malaysians and cannot be described as classic first-generation immigrants. Likewise, many Malays today live in urban areas and no longer rely on agricultural activities. Still, studies indicate that a great emphasis on collectivism remains in all three groups (Md. Zabid, Anantharaman, & Raveendran, 1997). Thus it is not surprising that Malaysians tend to prefer a participative approach to decision making, both at work and outside of it. In negotiations, compromise and collaboration are preferred to confrontation and a "winner take all" approach, which seems to be common in the United States. This emphasis on moderation is called *chung* among the Chinese (Wang, 1962, p. 12), and it reflects the Koranic concept of *ummatan wasata* among the Malays. As indicated previously, authority ranking reinforces collectivism, and all of the age-related rituals during *balik kampung* tend to keep it alive.

One of the most interesting results of collectivism in Malaysia is that Malaysians have a tendency to hold an entire group, and sometimes a nation, responsible for the misconduct of a person or a few people from that nation. As such, criticism by the media or nongovernmental groups from a foreign country about human rights or environmental issues is attributed to the country as a whole. Westerners tend to find such behavior as bordering on paranoia. To Malaysians, the experience of being exploited and humiliated during colonial rule has made them wary about the loss of face from the criticisms meted out by foreigners.

Harmony and Face

While *balik kampung* reinforces family-focused collectivism, it also serves as a means for maintaining harmony and preserving face. In a highly collectivistic culture great importance is attached to maintaining harmony and building relationships; self-centered individualism, which would detract from these activities, is normally viewed in a negative manner. One of the important ingredients in building harmony is to avoid doing anything that can disgrace and humiliate others.

Preserving respect and dignity is fundamental to understanding Malaysians. Among the Malays this is termed *maruah*, while the Chinese refer to it as "face," and the Indians use the closely related term *thanmanam*. Face is the unwritten set of rules that everyone is expected to follow to preserve individual dignity and group harmony. Whereas the West and East both seek to protect and preserve face for all, people from Asian nations such as Malaysia also attempt to *give* face so that everyone is a winner, at least to some degree. One never knows when additional interactions with others will be necessary, which is another reason for giving face.

The *balik kampung* provides an important venue within a social context in which the act of preserving face can be accomplished. If a personal offense or slight has occurred, an individual can attempt to correct it on a face-to-face basis during *balik kampung*, as happens during family meetings involving younger adults and their elders. At the very least, the individual can offer prayers and alms as a way of alleviating a sense of shame. Members of collectivistic cultures tend to respond more to a sense of shame at having hurt others than to a sense of guilt, which seems to be more important in Western nations.

Malaysians are sensitive about maintaining face and will try to avoid disgracing others, particularly in public. In extreme situations some Indians consider suicide preferable to accepting a loss of face, which motivates others to protect them from experiencing it. Similarly it is common for a Chinese businessman to withdraw from negotiations if he feels that the outcome will make him lose face. Westerners need to be aware that sometimes they may receive a quick "yes" when it comes to negotiations and decision making while their Malaysian counterparts find an appropriate way to deliver a polite and final "no" at the end of the negotiation without engendering a loss of face for anyone.

Given that all three ethnic groups value age, Malaysians frequently defer to the more senior or elderly member of the organization, who will generally be the first to speak at a meeting. As suggested above, however, the most senior person is expected to be responsible and should respect his juniors. At least partially for this reason heated debates tend to be uncommon, and it is expected that the most senior member will consult with subordinates, although he may withhold critical information if he feels it is in the best interests of the group. If you are a member of a delegation, it is important to line up in the order of importance so that the most important or senior person is introduced first. Speaking in calm, gentle tones is highly valued and it is never appropriate to point at anyone with a forefinger or to pound a fist into the palm of a hand to make a point.

To maintain harmony and preserve face, Malaysians have developed elaborate courtesies and rituals. During the *balik kampung* and before and after it, the Malays seek forgiveness from one another, yet they are not expected to mention, and usually avoid mentioning, any specific mistake or error. In this fashion harmony is maintained while debating over specific issues is avoided. Malays leaving their village or hometown for a long time will usually seek such forgiveness from parents, relatives, and friends. Sometimes this is also done when someone is returning home during the *balik kampung*.

Avoiding Confrontation

All three ethnic groups emphasize the need to be polite even while disagreeing. An abrasive and confrontational style is problematic. Speaking with a raised voice, shouting, and swearing are inexcusable, as the person subjected to this kind of behavior will suffer a loss of face and the instigator will experience a loss of respect that is difficult if not impossible to recover.

Thus Malaysians engage in a great deal of high-context communication, and this includes avoiding the word *no* when disagreeing with another person (Hall & Hall, 1990). Westerners commonly believe this is dishonest and annoying, but Malaysians are simply trying to avoid damage to a relationship. As stated earlier, there is a deeply ingrained culture of politeness which demands that a Malaysian not disagree openly and therefore, the word *no* is rarely heard. Differences tend to be stated in a tactful and polite manner. Heated debate tends to be uncommon. Often, a polite but insincere *yes* is simply a way to avoid giving offense and preserving face. Likewise Malaysians sometimes use subtle and indirect words to convey their meaning, for example, using proverbs and idiomatic expressions. It is little wonder that Westerners frequently employ third parties to clarify the meanings involved in communication and negotiations. "No" is often conveyed through these third parties.

Malaysians also use third parties, but they do so when they expect that the outcome of a particular negotiation can be potentially embarrassing. Among the more traditional Malays it is common for arranged marriages to occur, and the man's parents will send the third party to the woman's parents to inquire about her availability. If for some reason the woman does not want to consider the man as a potential groom, both families have avoided an embarrassing situation. Only after the woman's parents have responded positively will the parents from both sides meet so that his parents can propose that a wedding take place.

An anecdote may help to highlight the elaborate and complex ways in which Malaysians communicate. A hostess was entertaining her guest at a dinner at home. The guest was hungry and was hoping to have another helping. However, she noticed that there was no more food at the table. Rather than asking for another helping directly, she complimented the hostess on the beauty of the plate that had been used to serve food. The hostess noticed that the plate was empty but did not want to say that there was no more food. Rather, she showed the pot used to cook the food to the guest. She then explained to the guest that she had cooked the food in the equally beautiful pot, thus showing the guest that the pot was empty. This enabled the guest to understand that the food was finished without the hostess being forced to say so. Of course, this anecdote is exaggerated but it does indicate the lengths to which Malaysians will go to maintain harmony and preserve face.

Similar communication patterns can be found in many situations. For example, during *balik kampung* both the visitor and host know that serving food is part of the festival. Yet the visitor will always make it a point to tell the host not to trouble himself or herself with preparing something. Likewise the host will always apologize for not making adequate food preparations, even though the meal is elaborate and sumptuous. Malaysians also engage in the practice of using rhetorical greetings similar to the British "How do you do?" Malaysians will ask, "Have you eaten?" or "Have you taken food?" even if they don't expect to offer food, and the polite response is "Yes," whether or not you are hungry. Similarly, you are not expected to reveal the details of your itinerary when asked, "Where are you going?" A simple "nowhere important" will suffice.

Religious Values

We have already suggested that the *balik kampung* is particularly important during religious festivals, and religious practices tend to reinforce common values. Malays will offer special prayers on the morning of the Hari Raya, as will the Chinese on the eve of the Chinese New Year and the Indians on the morning of Deepavali. All generations of a family will participate in these religious rites.

As Muslims, Malays pray five times each day, and this practice is continued during *balik kampung*. Malays engage in prayer services when someone moves into a new house or when a firm occupies a new building. Likewise, a Malay organization usually celebrates its success by having its employees offer a thanksgiving prayer to Allah. Many Chinese shops and offices will have an altar for the owners to offer prayers and, as indicated in Chapter 26, Chinese homes frequently have family altars. At some work sites, some Chinese workers will build their own altar.

Among the Indians the same attachment to religion is evident. In some factories a Hindu altar featuring several prominent Indian gods is erected so that Indian workers can pray at work. The Indians are similar to the Chinese in that they tend to offer prayers in the morning and evening, and they are similar to the Muslim Malays in that Friday is a special prayer day. Many Indians either fast or observe a vegetarian diet on Fridays, even during *balik kampung*.

We have already described gift-giving practices during the festivals. Westerners are frequently surprised by the degree of gift giving that occurs in Asia, and many Western businesspeople have belatedly contacted their home offices to airmail a large number of gifts that they can give to their counterparts at various ceremonies and dinners held while they are staying in Asia on a short-term or long-term basis. All three ethnic groups engage in a great deal of gift giving, as it is viewed as an ideal way to maintain harmony and save and give face. As described earlier, gift giving is a common practice among Malaysians during *balik kampung* at least partially for these two major reasons.

There is, however, an element of reciprocity in this relationship. The person given the gift cannot refuse it, as this would involve a loss of face. At the same time, he or she is now obligated to the person giving the gift. Among the Malays this is called *hutang budi*, which roughly translates into a debt of deed. The person receiving the present need not reciprocate immediately but is expected to remember the *hutang budi* when the giver asks for assistance at a later date.

This custom of reciprocating is important in building long-term relationships. Non-Malaysians may find this practice exploitative, especially when practiced by some Malaysian businesspeople seeking favors. However, refusing a gift is ill-advised. Rather, the recipient should reciprocate immediately by giving a gift of similar value, which signifies that *hutang budi* has been paid.

Of course, religious sensitivity and custom should be observed when giving gifts. Malays do not appreciate alcoholic beverages as gifts or food containing pork. Meat should come from animals slaughtered in accordance with Islamic rites. Similarly the Chinese give gifts in even numbers, for example, a pair of pens rather than a single pen. However, the Indians prefer their gifts in odd numbers. Among the gifts that should not be given to the Chinese is a clock, which is considered a bad omen, as the Cantonese word for clock also means "to go to a funeral" (Craig, 1979, p. 37). There are other "dos and taboos" that can be easily learned by consulting books specializing in this area that are available in almost all bookstores or can be ordered online (see, for example, Morrison & Conaway, 2006)

Coming Together

Thus far we have been emphasizing practices and behaviors that reinforce common values and behaviors across the three groups but that occur only within each group. But *balik kampung* also displays a certain paradox about the Malaysian culture, for each of the three groups reinforces not only its identity but also the larger identity of being Malaysian. Malaysians of all groups and religions will join their fellow Malaysians to celebrate the respective festivals. There is evidence that certain customs formerly practiced by one ethnic group during the festivals have been adopted by the others. In a similar fashion a Malaysian comfortably switches languages in one sentence. For example, the first two words may be Tamil, the next three English, the next three Malay, and the last one Chinese.

In employment, more organizations that formerly hired from only one ethnic group have expanded to include two or three of the groups. This is partly because of the governmental *bumiputra* policies but partly because a nation undergoing rapid industrialization needs to employ its citizens as productively as possible. Prime Minister Badawi told his pro-Malay party in 2004 to put down their crutches because they may "eventually end up in wheelchairs." The cost of the *bumiputra* policy is becoming too high because of the backlash and brain drain among the other ethnic minorities and lack of sufficient innovation, research and development, and entrepreneurship. In 2008, a half century of political domination by the powerful National Front coalition was dealt a massive blow when voters ensured that it lost its two thirds parliamentary majority for the first time in 40 years (Trofimov, 2008). This represents a strong wakeup call for the ruling party and its policies of the last several decades.

Despite these deep rumblings, the general approach taken by Malaysia in nation building also reflects an emphasis on the mutual adjustments of the three ethnic cultures so that harmony is preserved. Before his retirement in 2003, long-serving Prime Minister Mahatir Mohamad declared that Malaysia should attempt to forge a united "Malaysian nation" by 2020. To this end, the government initiated a new national service program to bring thousands of young Malays, Chinese, and Indians together to do community service, intermingle, and discuss the future of their country.

Another example of mutual adjustment is that while the Malay language has been adopted as the national language, Tamil and English are still widely used. Mandarin and Tamil vernacular schools continue to operate, and many are funded by the government. The state-owned TV station has regular news broadcast in Malay, Mandarin, Tamil, and English. Although Islam is the state religion, freedom to embrace other religions is guaranteed by the constitution. Anyone traveling through Malaysia will come across mosques, temples, and churches spread throughout the entire nation. Unlike some nations in which only the festivals of the majority are public holidays, the following are classified as public holidays: the Muslim Hari Raya celebrations, Chinese New Year, Buddhist Wesak, Hindu Deepavali, and Christmas.

This, then, is Malaysia, whose underlying values are expressed in the *balik kampung* of the three major ethnic groups. In many ways Malaysia and, to a lesser extent, other Southeast Asian nations such as Thailand and Singapore represent ideal laboratories for demonstrating that ethnic groups seemingly radically different from each other can be forged into a common identity over time. The *balik kampung* with its emphasis on returning to nearby roots, authority ranking, and reinforcing common values is critical in this endeavor. Thus far the *balik kampung* has helped to integrate rather than separate the three groups and aided Malaysia in its pursuit of what Berry (1990) has identified as ideal: allowing individuals to maintain their own cultural identities and simultaneously to adopt new values supportive of the cultural identity of a larger group such as a nation. The ideal of a united Malaysian nation is still a work in progress but values and practices are in place to address the grievances of the minorities and propel this multiethnic democracy toward that goal, hopefully sooner than 2020.

CHAPTER 19

The Nigerian Marketplace

A stroll through the alleyways of any West African market makes clear that most goods bought and sold by the "mama benzes" and "marché mamas" aren't exactly part of the technologized global marketplace. Rather, they're the products of West African, and particularly Nigerian, industry.

—Amy Chua (2003, p. 108)

Nigeria represents both the hope and the despair of Africa, a continent ruled by European colonial powers for much of the 19th and 20th centuries. Nigeria itself was ruled by Britain until 1960 and has been an independent nation for 48 years. Although the country is rich in natural resources, its ruling classes have squandered them irresponsibly. Corruption is widespread; Transparency International, in its annual survey of corruption among nations, consistently ranks Nigeria as among the most corrupt nations. Nigeria's 127.1 million people constitute a little more than 25% of the total African population and it is Africa's most populous country. Gross domestic product (GDP) per person is about $570 per year, which is just about 12% of South Africa's GDP. Nigeria is among the lowest in national rankings on the World Bank's Human Development Index with a ranking of 45.3 in 2007. Furthermore, Nigeria's tropical weather subjects its citizens to a higher incidence of disease than residents of the more temperate parts of Africa. Along with many of its African neighbors including South Africa, Nigerian has a low life expectancy of about 44 years as compared to the 75 and more years in the world's developed countries. Most government revenue comes from oil, making the economy particularly vulnerable to fluctuations in demand for oil. Furthermore, the trade in stolen oil has fueled violence and corruption in the Niger Delta and its key oil city, Port Harcourt, where gang membership, according to an article in *The Economist*, is considered "a democratic right: the only way to get a slice of the oil state's cash" ("An Eerie Lull," 2007). However, the World Bank forecasts substantial trade surpluses in the next few years based on high oil prices and the performance of the oil sector.

For almost all of its 48 years, the nation was ruled by the military. When he became president in 1998, General Abdulsalam Abubakar forthrightly described the nation's plight in the following way: "Currently, we are the world's 13th poorest nation. Given our resource endowments, this sorry state is a serious indictment." In 2000 he handed over power to a democratically elected president, Olusegun Obasanjo. Ironically, it was Obasanjo who, as military ruler in 1979, supported the first democratically elected government in Nigeria's history. Unfortunately that government was short-lived, and military rule returned after 1983. The April 2007 election of Umaru Yar'Adua had many logistical problems but marked the first time that political power was transferred from one civilian government to another and bodes well for postindependence Nigeria's future.

As a modern political entity, Nigeria came into existence in 1914 when the British amalgamated three of their West African colonial territories: the Colony of Lagos, the Southern Nigeria Protectorate, and the Northern Nigeria Protectorate. Each of the three territories, in turn, had been constructed by the British out of a diverse collection of indigenous kingdoms, city-states, and loosely organized ethnic groups through treaties or outright conquest. Thus Nigeria as a geopolitical entity was, in effect, a creation of British imperialism. Nigeria remained a colony of Britain until it gained independence in 1960.

Archaeological evidence attests to millennia of continuous habitation of parts of what is now Nigeria. One of the earliest identifiable cultures is that of the Nok people who inhabited the northeastern part of the country between 500 BCE and 200 CE. The Nok were skilled artisans and ironworkers whose abstractly stylized terra cotta sculptures are admired for their artistic expression and high technical standards (Burns, 1963, p. 25). High-quality bronze art at Ife and Benin City in the southwestern parts of the country also predated the arrival of Europeans in the region. Over the centuries successive waves of migration along trade routes into Nigeria from the north and northeastern part of the African continent swelled the population, which eventually became organized along tribal lines into kingdoms, emirates, city-states, and loosely organized ethnic groups. This was the setting the early European adventurers, notably Portuguese navigators, found when they first made contact with the people of the coastal regions in the 14th century CE.

The marketplace is an appropriate metaphor for understanding the culture of the people of modern Nigeria. In this context the marketplace refers to the physical areas of a city, town, or rural village where indigenous commercial activities are concentrated. This metaphor has a special significance because the bulk of the history of the region up to modern times can be summed up in one word: trade. Many of the ancestors of modern Nigerians first migrated to the region along trade routes; the slave trade was primarily responsible for the arrival of Europeans in large numbers; and after the abolition of the slave trade by the British in 1808, the need to enforce the abolition and to replace the trade with legitimate commerce was a primary motivation for the British to maintain a presence.

For centuries before the arrival of the Europeans, the marketplace was the community center for a town and the surrounding settlements. It was the center of commerce, the local town hall, and the main social center all rolled into one. This characterization is still true in the rural areas where about two thirds of the Nigerian population still lives. Market networks link many towns and villages together. The networks consist of major markets that meet on specific days and attract traders for miles around. In the past such networks were the conduits through which new commodities, outside influences, and immigrants reached local communities.

The characteristics of the Nigerian marketplace that stand out include its diversity, dynamism, and the balance between tradition and change. Each of these characteristics strongly reflects modern Nigerian society.

Diversity

The typical Nigerian marketplace, especially in the bigger towns, is a sprawling, usually bustling place where virtually every commodity is available. As a general rule the market is divided into rows of covered stalls, with a narrow aisle separating each row. Each row of stalls is assigned to a specific commodity, ranging from perishable foods to household electrical appliances. The marketplace is where the vast majority of the people shop for their daily needs. This is the case even in the urban centers where there are modern department stores. The roadside stalls called *Buka* also represent the best opportunity for a good meal; the best local cuisine is to be found here. Commodities in plentiful supply include fresh meat and fish, vegetables, household goods, electrical appliances, jewelry, building materials, clothing, and traditional medicinal herbs.

Sellers of similar commodities often unite to form market associations to look after common interests. Nearly everything is available either wholesale or retail. The major marketplaces are usually centrally located, with a network of roads linking most parts of the town to the market. In addition to commodities, the marketplace offers services, which include privately operated taxi and bus services to any part of town, heavy hauling, appliance repair, custom tailoring, and secretarial services. Intercity transportation hubs are also usually located in close proximity to the main markets. Diversity refers to the wide range of goods and services available and to the size, relative importance, and modern nature of each market compared to others in the area.

The diversity of the Nigerian culture is one of its hallmarks. In a country just about the size of Texas and New Mexico combined, there are about 300 ethnic and subethnic groups, with as many distinct languages and dialects. Often a dialect is clearly understandable only to the inhabitants of a town and its immediate environs. Just as the marketplace is clearly divided into sections by commodity and into subgroups by market associations, so too is the Nigerian society clearly delineated by ethnic and language differences. But just as the separate commodity sections form an integral whole to serve a common clientele, so too does the Nigerian society try to forge a national identity out of the widely diverse ethnic groups artificially joined together into a nation by the British.

The heterogeneity of Nigerian society is physical, social, religious, and linguistic. Each ethnic group is concentrated in clearly defined geographical areas that its people have occupied for centuries. The same is true of the subethnic groups. Two large rivers, the Niger (from which the name of the country was derived) and the Benue, form a Y that divides the country into three parts. Each of the three parts is dominated by one of the three main ethnic groups: the Yoruba, the Hausa-Fulani, and the Ibo (or Igbo). The Yoruba are the only major ethnic group with a traditional aristocracy and constitute the most articulate of the Nigerian populace. Although the power and influence of the aristocratic class has diminished over the years, they are still a major social force in Yoruba society. Yoruba-speaking ethnic groups also dwell beyond the Nigerian border in the Benin Republic.

The Ibos are primarily located to the east side of the Niger. Although they are more densely concentrated in their primary geographical area, they are more fragmented politically than the Yoruba or the Hausa-Fulani. The Ibos are notable for their business acumen and strong work ethic. The Hausa-Fulani occupy most of the area to the north of the Niger and the Benue, an area more than twice as large as the southern regions combined. In most cases differences in physical features among ethnic groups are not significant enough to reliably distinguish a member of one ethnic group from another.

In the past, permanent facial markings were used to distinguish members of one ethnic group from the others. Their primary purpose was to identify friend or foe in warfare rather than to serve as bodily decoration. The practice of facial markings has been largely abandoned in modern times. Today names and attire are more reliable indicators of ethnic membership than physical attributes.

The Social Structure

With so many different ethnic groups and languages, it would seem that defining the contours and structures of this highly heterogeneous society is impractical. Admittedly it is difficult, but it is not impossible. There are several reasons for this. First, despite the great number of ethnic groups, four of them—Yoruba (21%), Hausa and Fulani (29%), and Ibo (18%)—comprise nearly 70% of the Nigerian population. As the Hausa and the Fulani have intermixed over the centuries, they are now commonly regarded as a homogenous group.

Second, the correspondence between language and ethnic group varies. Several groups may speak essentially the same language and have similar cultural characteristics but prefer to identify themselves as separate entities in terms of differences that span centuries. For instance, the Yoruba have at least 20 subethnic groups, each of which vigorously protects its separate identity.

However, modernization and the federal government's adoption of policies that promote a sense of national identity have gradually resulted in the emergence of some common national characteristics. Examples of such government intervention include the establishment of English as the official language of government and commerce throughout the country, the practice of posting some federal employees to states other than their own, and the requirement of an immediate 1-year national youth service of all graduates of postsecondary institutions. In recent years the influence of the mass media, especially television, has also helped to break down cultural barriers among ethnic groups.

Nigeria has a long way to go in terms of interethnic harmony, however. Interethnic mistrust is deep and is a primary reason for the repeated failure to establish a viable Western-style democracy in Nigeria since independence, giving the military an excuse to establish authoritarian regimes for long periods of time. Although most Nigerians are patriotic and genuinely want their country to work, deep-rooted ethnic allegiance often takes precedence over national allegiance. As a typical example, the Egba are among the 20-odd subethnic groups of the Yoruba, but when the chips are down, an Egba considers himself an Egba first, a Yoruba second, and a Nigerian last. In the distant past, members of the subethnic groups fought bitter fratricidal wars among themselves. Territorial expansionism, control of trade routes, and the desire to cash in on the lucrative slave trade were among the reasons for these wars.

International Comparisons

Geert Hofstede (2001), in his landmark analysis of the cultural differences across 53 countries, did not include a single black African nation. This was not because the multinational corporation in which the analysis was completed (IBM) had no subsidiary in any of them, for there was one in Nigeria during the period of the study. But IBM, along with several other multinational corporations, later chose to leave the country rather than comply with a new federal decree requiring 40% Nigerian equity in certain classes of foreign business investments in Nigeria. This illustrates an important aspect of the Nigerian psyche: a strong sense of national honor and self-worth. While Nigerians in general admire and actively attempt to emulate Western economic and social development, they are quick to take offense at real or perceived condescension on the part of Western expatriates.

The recent GLOBE study (House et al., 2004) included Nigeria and found it to be high on power distance, high on both institutional and in-group collectivism, medium on gender egalitarianism (classified in the same band as China, Germany, and the

United States), and relatively high uncertainty avoidance or risk aversion. Traditionally, there is a large power distance between social classes and between superiors and subordinates in most Nigerian ethnic groups except the Ibos. If the four major ethnic groups are ranked in order of power distance, the Hausa-Fulani would most probably score the highest, followed closely by the Yoruba, and trailed by the Ibos. The Hausa-Fulani are overwhelmingly Sunni Muslims who have adopted much of the Islamic world's rigid social order and way of life. The Yoruba's high power distance is tempered considerably by opportunity for upward mobility based on individual achievement. The Ibo were noted as an egalitarian society for centuries and have largely remained so.

Collectivism and high uncertainty avoidance are more evenly applicable to the bulk of Nigerian society regardless of tribal affiliation. Virtually all the disparate ethnic groups believe in the extended family system. Hometown associations are formed with the expressed purpose of improving the infrastructure and facilities in the community. For example, during the early years of national development it was not unusual for a community to sponsor the overseas studies of promising local students in the hope that they would return to serve the community. Nigerians also score relatively high on uncertainty avoidance or risk aversion, preferring to deal with people and situations with which they are familiar.

However, some research suggests that the business-oriented Ibos are more individualistic than the other major ethnic groups. That is, they explain their success or failure in terms of their own abilities whereas other Nigerian ethnic groups emphasize luck or the situation (see Bond & Smith, 1998). Given this research, we would expect that the Ibos would have a longer term time orientation and would be more willing to sacrifice short-term enjoyment to attain long-term goals. The GLOBE study classifies Nigeria as scoring relatively high on a future orientation toward societal practices among the countries included in the project. The authors were surprised to find that many developed countries scored lower on future orientation than developing countries. They argue that emerging and lower income nations may "see a stronger need for taking a long-term perspective and sacrificing for the future because they must cope with scarce and limited resources" (Ashkanasy, Gupta, Mayfield, & Trevor-Roberts, 2004, p. 305)

In general, Nigerian society is heavily patriarchal and both its dominant religions proclaim the superiority of males to females. One of the outcomes of this is the status of Nigeria as the country with the fastest rate of HIV/AIDS in West Africa with women reportedly making up 60% of the infected (Isiramen, 2003). Because of the taboos associated with open discussions of sexuality among women and girls, many women suffer silently from sexually transmitted diseases. However, there are some notable differences among the three major ethnic groups. The Hausa-Fulani, true to their Muslim religion, tend to consider women as quite subordinate to men in virtually every respect. This is not the case among the Ibos and the Yoruba, where women have traditionally faced relatively little opposition to their entrepreneurial spirit.

Social Dynamism

Nigerians have often been characterized as resourceful, pragmatic, entrepreneurial, and energetic (Aronson, 1978, p. 137). In this context social dynamism refers to the energy, vigor, and adaptability of the Nigerian society. These qualities are readily observable at the Nigerian marketplace, which bustles with activity, especially on Saturdays when most families do the weekly grocery shopping. Traders hawk their wares in sing-songs particular to their commodity. The bargaining is energetic, with the burden on the customer to negotiate a fair price. While many items have fixed prices, most do not. This means that the smart shopper must come to the market fully armed with knowledge of the latest fair prices for the commodities of interest. The shopper must also be savvy about the various shady characters who are attracted to the marketplace. The typical seller, in turn, has perfected the art of keeping a poker face in order to extract as much profit from the sale as possible without antagonizing the customer.

Many shoppers have favorite stalls that they repeatedly patronize. Such repeat buyers are often taken care of quickly, with the minimum of negotiating necessary to arrive at a mutually acceptable price. In recent years, sellers seem to have concluded that the time and effort spent on haggling are not worth the extra profits and have started to place a fixed price on as many items as possible. Prices are usually set by market or trade associations, and changes in them are frequently communicated by word of mouth several times a day.

Doing things quietly is not the Nigerian way, whether it is an argument, a political discussion, a celebration, or a sad occasion. Just as the market bustles with a cacophony of sights and sounds, so too do Nigerians complete most activities with gusto. The dynamism of Nigerian society is reflected in the business and political landscape, the attitude toward education, the manner in which the people celebrate holidays and family events, and their favorite leisure pursuits.

At independence, Nigeria's economy was engaged almost exclusively in the production of food for home consumption and commodities (e.g., cocoa, groundnut, and palm oil) for export. The vast deposits of oil and natural gas had not yet been discovered. Since then, successive governments, including the military regimes, have made the development of the industrial sector of the economy a priority.

Oil-Based Economy

Just as the local marketplace bustles from dawn to dusk with activity, so too Nigeria's more formal economic scene bustles with complicated deals and contracts. Nigeria has considerable crude oil and natural gas resources and has been a member of the Oil Producing and Exporting Countries (OPEC) since the 1960s. This natural wealth has provided the resources to modernize the economy, especially during the boom years of the 1970s, when vast amounts of new crude oil deposits were being discovered in the coastal areas at the same time that prices of crude oil were steadily rising in the world market. In the 1970s the federal government took steps to ensure greater Nigerian participation in such key economic activities as banking, insurance, manufacturing, and oil production, which were then dominated by foreign companies or their local subsidiaries. According to some accounts, although the performance of the non-oil sector is growing, the country is still largely dependent on oil and gas earnings, and the origin of the non-oil growth is difficult to identify. As long as oil prices continue to rise, it will be difficult to estimate the long-run sustainability of Nigeria's upswing.

Unfortunately, only a small fraction (less than 10%) of the population controls much of this wealth. The wide gap between the rich and the poor is a major concern for Nigeria's economic planners. For many years the federal government dominated and closely controlled the economy, but in recent years it has recognized the detrimental effects of meddling too much in the economy and has moved toward the establishment of a freer market.

The Nigerian political arena is not for amateurs or the faint-hearted, just as the marketplace is not for amateurs who are not versed in the fine art of haggling. Politics in Nigeria is a rough and tumble game and only the fittest survive. Politicians, including ex-presidents, are sometimes put in jail for questionable reasons.

However, unlike the situation in many African countries, no Nigerian political leader has ever tried to impose a single-party state. The electorate is too sophisticated for that. Modern Nigeria is a federation of 23 states and a new federal capital, Abuja, which is located near the geographical center of the country. The political framework is similar, by design, to the federal system of the United States, but the institutions and requirements reflect the needs and experience of the country. The federal government consists of an executive branch, a legislative branch, and a judicial branch. Each of the 23 states is similarly organized.

Officially Nigeria is a Western-style democracy, but the threat of authoritarian rule is never far away, and there have been several military coups. Nigerians are accustomed to waking up to a brand new government that has suddenly taken shape overnight. They have learned not to let the political instability interfere too much in their daily lives and business activities. One of the challenges for President Umaru Yar'Adua will be to bring greater stability to the turbulent Niger Delta region where kidnappings of foreign oil workers, gang warfare, and militant activity are quite common.

The Education System

Many of the traders and craftspeople in the marketplace are graduates of apprenticeships in their specific areas. In the same manner, Nigerian society believes that a sound educational system will provide the energy that will propel the country into the industrialized age. Parents who have the means hire private tutors, send their children to prestigious schools that may be far away from home, and give up their own comforts so that their children can do better than they have. Like the political system, the education system is patterned after that of the United States. One reason for this is that many of the policymakers in government were educated in the United States. However, Nigeria's federal government performs abysmally in the area of education.

Education is free and compulsory up to the sixth grade but most choose to go further, if they have the means. English is the official language of instruction. It is introduced at the primary school level and used exclusively at the secondary and postsecondary levels. The government, recognizing the value of a high level of literacy to rapid development, has tried to take advantage of the oil wealth to implement a free education policy at all levels. It has not been able to achieve this objective partially due to the sheer size and cost of the educational system, which is government controlled at all levels. Admission to postsecondary institutions is limited and extremely competitive. Although these institutions continue to expand, they are currently not capable of admitting all the students who qualify for admission.

Furthermore, Nigerians are not shy about making a public scene when the need arises, just as the haggling in the marketplace is sometimes done with rancor whenever either party steps over the line of decorum. Arguments are typically carried on in a loud and animated manner, but they rarely lead to physical combat. Bystanders eagerly offer their own loud opinions about the ongoing dispute. The urban centers, especially in the south, are overcrowded, and there is a constant influx of students and job seekers from the rural areas. Competition is intense for nearly everything. Jobs, housing, transportation, and other daily necessities of life are in scarce supply in the cities even though they receive a disproportionate share of federal development budgets.

So believing that it is a jungle out there, the smart urbanite steps out each morning prepared to survive another hectic day. Western expatriates usually but not always are untouched by all this hustle and bustle of daily life because they tend to live in the quieter, wealthier neighborhoods that had their origins in colonial times when the government and the larger foreign companies built residential reservations for their expatriate staff.

Leisure Activities

Not everyone goes to the market with the intention of shopping. Some go to take in the scene, browse, or meet friends. Nigerians take their leisure seriously and try to find the time to indulge in their favorite leisure activities. They are keenly interested in sports, and their preferred sporting activities reflect their dynamism. Wrestling was a popular sport in many parts of the country before the arrival of the Europeans. It is still popular, as is boxing. But neither comes close to the interest in soccer, both as a participatory and spectator sport.

Checkers is popular among the lower working class, and the game is typically accompanied by friendly banter and bets (nonmonetary). The favorite way to relax is to be in the company of friends who can trade war stories about the vagaries of daily life, pass on the latest social gossip, and solve the country's problems. Nigerians in general do not seek isolation and solitude; they avoid them as much as possible.

Holidays and ceremonies are an important aspect of Nigerian culture. Both the Christian and Islamic holidays are nationally observed, as are local community festivals, some of which have been observed for centuries. Some of these local festivals were of a pagan nature but are now largely stripped of religious significance. Common ceremonies include weddings, the naming of a new child, and funerals. Such ceremonies often are lavish affairs, depending on the wealth of the celebrants. Ostentatious display of wealth is the norm, and the rich often welcome the opportunity to do so. The display is a brash, bold, in-your-face statement to the world that the celebrant is rich and is not ashamed to show it. Everyone knows what the status symbols are: the big wedding or funeral, the chauffeured car, the expensively tailored suits, and the big houses complete with servant quarters. There was even a time, during the oil boom years, when owning your own jet was added to the list.

Holidays provide a respite from what many in the populace regard as the rat race of daily life. Even the marketplaces are silent and empty on the major holidays. The favorite way to spend a holiday is to hold a feast or to attend one. Either way, holidays provide an opportunity to congregate and socialize with good friends.

Recreation and leisure in the Western sense are enjoyed by only a small percentage of the population, mainly the wealthy elite. They may play tennis or golf or go for a swim. At exclusive country clubs admission is usually based on wealth, connections, and social status. The vast majority do not have the means or the time to indulge in Western-style leisure, and the average Nigerian does not go on a road trip just to see the country or go for a hike in the woods just to commune with nature. In the cities there are plenty of nightclubs, movie houses, restaurants, and private clubs that Nigerians tend to frequent.

Balancing Tradition and Change

Modern Nigerian society is markedly different from the cultures that existed before outside influences began to take hold in the early 19th century. Nowhere is this more noticeable than in the way the center of political and economic power has shifted from local traditional rulers and their councils of senior chiefs to the educated elite, business magnates, politicians, and the military. No country wants to be an island, isolated from all outside influences. If outside influences threaten to engulf aspects of the society that are cherished by its members, however, then it is necessary to take actions to preserve them. This is true for modern Nigeria, where the trick is to modernize without sacrificing cherished traditional values.

Among the facets of daily life that were tightly controlled by the traditional rulers in precolonial times were the local marketplaces, especially the most important ones.

Among the Yoruba, for instance, the largest and most influential markets were situated in front of, or in close proximity to, the king's official residence, and they were called *oja oba* (the king's market). In some areas the official title of the local head chief was *loja* (owner of the market). Virtually all markets are now administered by local government councils, which are popularly elected bodies. Although control of the markets has shifted from the traditional rulers to the people, these rulers are still regarded in many places as the titular heads of the markets, especially in the rural areas. The changes in market administration are indicative of the changes that have occurred in Nigerian society as a result of modernization.

Other notable aspects of culture and tradition that have changed considerably over the years include religious preferences, the traditional family compound, dating behavior, language, sports, and leisure pursuits. However, the changes are not universal. As a general rule, the rate of change in the urban centers has far outpaced that in the rural areas.

Power and Influence

In precolonial days, power was heavily concentrated in the hands of local kings and their chiefs. Palace intrigues and territorial expansionism through intertribal warfare were common features of life. Again, a noted exception to this situation was the Ibos, whose egalitarian social structure did not allow for the concentration of power in any one individual or small group of rulers.

Although their power and influence have been largely curtailed by the establishment of more democratic political systems, traditional rulers still play an important role in their communities. They are regarded as the custodians of their people's legacy and identity, and they are usually at the forefront of the fight to defend traditional values against the unrelenting onslaughts of Western values. In addition, some senior local traditional rulers, especially in the north, have learned to exert influence behind the scenes. Moreover, northern Muslim traditional rulers also tend to be their community's religious leaders. This pragmatic approach has resulted in some senior traditional rulers becoming even more powerful than they would have been if they still had direct political power. But they are no longer referred to in English as *king*. The most common English term used is *traditional ruler*.

When a traditional ruler dies, the eligible candidates campaign vigorously to win ascendancy to the throne, for there is almost never an automatic line of succession. Candidates are drawn from traditional ruling families, and only the adult male members are eligible. In some areas the chieftaincy is rotated, by tradition, from one ruling family to another.

Traditional rulers are selected by a council whose members are invariably called kingmakers, consisting of the senior chiefs in the community. The selection process varies from community to community. It is worth noting that while the selection process has largely remained unchanged, the criteria for selection have changed markedly over the years. In modern times candidates with adequate formal education who have distinguished themselves in some fields are favored.

Church and Family

An important area where major changes have occurred is religion. The first religious incursions came from the north, when the Fulani under Usman Dan Fodio overran the Hausa and other northern states in the 18th century. The conquerors forcibly established Islam in the city-states that dotted the region. Usman Dan Fodio would have carried out his vow to expand his jihad (holy war) all the way down to the sea if the British, who were then establishing a presence in the south, had not put a stop to his ambition (Burns, 1963, pp. 50–51). The early European explorers and traders entering the country from the south were quickly followed by Christian missionaries. Today only a small percentage of Nigerians still adhere exclusively to indigenous animist religions. About 47% classify themselves as Muslim, 38% as Christian, and 15% as other.

Many did not find the spirituality they needed in what they considered the sterile form of worship of the European Christian sects. Some were also turned off by the perceived condescension of the Christian missionaries. Indigenous Christian sects founded and led by Nigerians soon began to proliferate, especially in the Yoruba-dominated western region. Notable sects include the Cherubim and Seraphim sect, the Apostolic Church, and the African Church. What these indigenous sects have in common is a form of worship that is more reflective of the Nigerian society: less formality in worship, singing and dancing to lively indigenous Christian music, spontaneous audience participation during worship services, and a deeper personal relationship with God. Today some of these sects are increasing their memberships at a rapid rate.

While many of the inevitable changes that have come with modernization have been welcomed by the general population, others have been perceived as socially regressive. An example of this is the way the traditional family compound has changed. In the past, as the male children in a family became adults, they were given plots of land adjacent to the family home on which to build their own houses in preparation for starting their own families; female members of the family were expected to get married and become members of their husbands' families. Today the young adults are more likely to move to the cities, first to further their education, then to work or engage in business. In the cities the relative scarcity of land and the high cost of building construction make the idea of family compounds impractical. The end result is that there has been a decline in allegiance to the larger kinship group. Urban centers continue to grow in size and population while the rural areas stagnate or shrink.

Weddings are usually major events, with members of the extended family on both sides in attendance. Alternatives for weddings include the local magistrate court, a church or mosque, or a traditional wedding. Many couples choose a combination of the traditional wedding with one of the other alternatives. Most brides take their husband's surname immediately after marriage. Divorce was at one time rare but is becoming a common occurrence due, in part, to the pressures of modern life and the decline of the family compound.

Language Evolution

Like courting behavior, linguistic patterns have changed considerably. Because of the waves of migration of the young from the rural areas to the urban centers, many of the 300-odd ethnic languages are sometimes now spoken by as few as 10,000 people. The three main languages—Hausa, Yoruba, and Ibo—are spoken by more than 65% of the population. It is telling that after 48 years of nationhood, very few Nigerians are interested in learning to speak another ethnic group's language.

Also, urban versions of the major languages have developed. For instance, *Lagosian* is the term given to the style of Yoruba spoken in Lagos, the former capital, which is in Yorubaland. *Lagosian* is a stylized version of spoken Yoruba, peppered with English words and considered by young Yorubans to be the most sophisticated version of the language. Yorubans also have a tendency to add an "o" at the end of words to add emphasis. When this gets extended to English, we have words such as "Sorry-o" shouted to someone who suffers a misfortune (Harris, 2008).

Another urban language is pidgin English, which is spoken chiefly within the lower working class. It cuts across ethnic lines, providing a means of communication among the disparate ethnic groups that live and work side by side in the large urban centers. However, the Nigerians' English (a quaint combination of Victorian-era vocabulary and grammatical and syntax structures of indigenous languages) has given rise to some concern that they will not be understood in a globalized world. Eateries are *chop houses,* street children are *urchins,* a dead or jailed robber is often said to have met his Waterloo, and the infamous Nigerian e-mail scams are committed by the "Yahoo-yahoo boys" (Harris, 2008).

Just as some markets have remained in the same location for centuries, selling many of the same traditional foodstuffs that the ancestors of modern Nigerians ate, so too have some beliefs, practices, and social norms endured the onslaught of centuries of outside influence. Aspects of the Nigerian culture that have survived colonialism and modern influences largely intact include seniority and authority relationships, social roles and status, a rigid class structure (where it existed), orientation to time, view of work, early socialization, the extended family system, and traditional festivals.

Seniority Rules

A strict system of seniority governs normal interpersonal relations. Children, for instance, are expected not to look their parents or elders directly in the eye while being scolded. This expected deference applies to any interaction where one party is significantly senior to the other. The prerogative to use first names is granted only to close friends and superiors. It is considered an insult for a younger sibling to address an older sibling casually by the first name unless they are very close in age. An age difference of just a year is enough for an older individual to expect to be addressed respectfully.

A common way to address a superior respectfully is to precede the first name with a respectful salutation. Among the Yoruba one way to do this is to precede the first name of the older sibling with *buroda* (derived from "brother") or *anti* (derived from "auntie"). Both terms have English origins, and it is not clear how they began to be used by the Yoruba as a means of conveying deference. Parents are invariably addressed using the local equivalents of the Western *Dad* or *Mom*.

This seemingly stratified ordering by age group and seniority reflects the high power distance discussed previously. But there is a weak stratified class structure per se except among the Hausa-Fulani in the northern states. The larger the age difference or seniority, the greater the amount of deference expected. Methods of greeting seniors, parents, and other adults vary among ethnic groups, but they generally involve some bowing of the head, if the greeter is male, or the bending of the knees, if the greeter is female.

In return for deference, the seniors are expected to guide, lead by example, and generally be supportive of their subordinates. Although the major criterion for determining seniority in social situations is age, there are instances where the senior in a situation is the younger person. In the workplace, for instance, a supervisor may be senior to much older workers from whom he or she may expect the deference accorded to seniors.

In business situations and among the educated, Western greeting behavior is more prevalent than traditional greeting behavior, but traditional greetings are still expected where the age or seniority gap is significant. This requirement does not extend to subordinate expatriates, who are routinely excused from traditional greeting requirements in favor of Western ones. Superior expatriates are not accorded traditional greetings either, not because they are not considered worthy, but perhaps because they know it is not expected.

Work Attitudes

The Nigerian's attitude toward work is a study in contrasts and points to a deep-seated social problem that retards economic progress. On the one hand the entrepreneurs operating firms of various sizes work prodigious hours to build and maintain their businesses. On the other hand the average salary or wage earner wants to put in as little effort as possible. Because the first group often must hire the second, a conflict of goals results. This attitude is not a manifestation of an inherent laziness; it seems to be a manifestation of a lack of faith in the system. Many workers simply do not believe that advancement on the job is based on performance. They have been conditioned to believe that favoritism, nepotism, and other unfair methods are the usual means of advancement. This attitude is further reinforced by the fact that seniority (length of service) rather than performance is a common criterion for advancement, especially in the civil service. The effect of this attitude on national productivity is not difficult to fathom.

The same deep distrust of the establishment is the reason why Nigerians of all stripes dislike paying taxes. Salary and wage earners cannot avoid having their taxes withheld, but entrepreneurs can and do work around the tax system without feeling the least bit guilty. The widespread incidence of official corruption and mismanagement is often cited as justification for cheating the government out of taxes. Quite a few entrepreneurs convince themselves that they can best use the money for the public good rather than let the bureaucrats decide how to spend it.

The role of women has not changed significantly over the years. They continue to bear the brunt of child care and household responsibilities, in addition to holding down jobs or running their own businesses. The typical Nigerian husband considers it beneath his dignity to perform household chores. There is often a clear delineation between the kinds of tasks, commercial activities, and jobs that a man and woman normally do. This delineation of tasks and activities is very much in evidence in the Nigerian marketplace, where women far outnumber men both as traders and customers. The casual observer might infer from this that the marketplace is a female domain, but this perception is inaccurate. Although women have always outnumbered men in the marketplace, the men have wielded the real power behind the scenes.

Women's Role

In this regard it was noted above that the marketplace was formerly administered by the local head chief, and more recently by elected local government councils. Both the head chieftaincy and the councils are male-dominated institutions. The true picture is that while women dominate at the retail level, men dominate at the production, wholesale, and administrative levels of the market structure. In addition, some commercial activities are perceived to be either male oriented or female oriented. Sometimes there appears to be no logical reason for the distinction other than the mere fact that the activity has been a male or female preserve for as far back as anyone can remember. For instance, butchers are almost exclusively male while fishmongers are almost exclusively female. Female-oriented activities are purposely avoided by men and male-oriented activities are often closed to women, either overtly or covertly.

Women have been active in local and national politics for years, dating from the early colonial periods, when market women rioted to protest the imposition of taxes on women. Politics in Nigeria is still largely male dominated, but women are making steady progress. The one area where women have distinguished themselves is in commerce. This is also just about the only area where society has given them wide latitude. The homes, streets, and markets of Nigeria are alive with women trading, and many women have built large-scale trading enterprises with little outside aid. In the workplace Nigerian women have as much opportunity for advancement as their counterparts in any Western country, including the United States.

Overt or covert discrimination against women persists, however, because the typical Nigerian man, in the final analysis, was not raised to regard women as equals. Polygamy is legal, but it is neither encouraged by the authorities nor tolerated by the women. It is no longer fashionable among the educated class and has been steadily declining in recent years, except in the Muslim north.

Extended Families

Furthermore, the decline of the family compound has not resulted in a parallel decline in the extended family system, which is a cherished aspect of African societies. Simply put, this concept means that near and distant relatives are embraced as full members of the nuclear family. The extended family is the African parallel of Western social welfare systems. Members of extended families take care of one another, find jobs for young members, band together to help members in times of sorrow or hardship, intervene to solve marital problems, and come together to celebrate one another's successes. When a member gets married or otherwise has a reason to celebrate, there is no need to send out formal invitations to extended family members because everyone knows they are invited and expected to attend.

In addition, Nigerians have always tended to be conscious of status differences. In the Nigerian marketplace some retailers achieve prominence and become unofficial deans of the market or leaders of the market associations. Social status is important to Nigerians, but status consciousness is not the same as supporting a rigid social stratification system. In the past, the highest status was accorded to the king, followed by his chiefs. Other people of status may include war heroes, successful traders, or great farmers. Today, as in the past, Nigerians, especially men, aspire to be recognized by their community as people of "timber and caliber," which is a popular term connoting high status and integrity. Status can be earned through education, success in business, philanthropy, exemplary character, leadership qualities, or any combination of these. However, wealth by itself does not automatically confer status. A wealthy man of dubious character and negligible philanthropy may be shunned by the community.

In modern Nigeria one of the most sought-after status symbols is a chieftaincy title, which is considered to be higher in status than an earned doctoral degree. When an individual with an earned doctorate becomes the chief of his hometown, he may use the title *Chief*, followed by the title *Dr.* in brackets (as in Chief [Dr.] XYZ). A chieftaincy is highly valued because it is an affirmation by the community that the conferee has been recognized as a community leader.

Although chieftaincies are of various ranks and importance, all conferees are eligible to use the title chief. There are separate title lines for men and women, for women also can be chiefs. Traditional rulers see two clear advantages in the modern chieftaincy system. It encourages successful citizens to be philanthropic because that is an important criterion for becoming a chief. Furthermore, it promotes the preservation of traditional culture and values, since it co-opts the conferee into the traditional structures of the community.

Ironically, the increasing popularity of chieftaincy titles has threatened the integrity of the institution. Many successful Nigerians feel incomplete without being recognized by their hometown with a chieftaincy. As a result, rumors of payoffs, favoritism, and award of chieftaincies to ineligible individuals sometimes surface.

The desire for status also carries over into the workplace. Prestigious job titles and impressive offices are prized. People are often defined by what they do for a living and how high they have risen in their profession. However, most people work simply as a means of earning a livelihood; the concept of work for its own sake is not widely valued. And, as suggested previously, evidence of large power distance is prevalent throughout the workplace. Subordinates are expected to show deference, but it is not uncommon for superiors to have strong informal relationships with some immediate subordinates. The size of the power distance is roughly proportional to the vertical distance between the two parties on the organizational chart.

Perspectives on Time

Another area that highlights the balance between change and stability is that of time. There is a common misconception among Westerners that Nigerians, like many Africans, do not pay much attention to the clock. Nigerians are quite capable of going by the clock, and they are pragmatic enough to recognize when they should do so. In senior-subordinate relationships, subordinates are expected to be prompt, and they usually show up at the appointed time. Workers, despite their attitude toward work, usually report to work on time. School children are routinely punished for arriving late to school. Nearly everyone who can afford a watch has one. Still, when possible, Nigerians are much more relaxed about time schedules than U.S. Americans.

However, certain economic and social aspects have contributed to the relaxed approach to time. In the rural areas there is usually no need to watch the clock because the pace of life is slow. In the cities severe traffic congestion and an unreliable public transportation system make it difficult to foresee how long a trip across town will take. Although city dwellers readily start off early to arrive at their jobs or schools on time, they do not feel pressed to do the same for casual social engagements. Everyone understands the difficulties of being on time, so they are not offended when appointments are missed. Just as the trader at the market patiently waits for the next customer when business is slow, so too does the Nigerian host wait patiently for the guests to arrive. Thus Western hosts should not be surprised if their Nigerian guests show up hours after the appointed time without apologizing profusely for the delay. However, when the issue is business, Nigerians will try their best to be prompt and apologize for being late.

There is a significant contrast in culture and values between the younger, college-educated generation and the older generations, which as a rule are less educated. The older generations are more conservative in their views and are alarmed that traditional values are rapidly eroding among the younger generations. The elders fully support economic and social progress, but not at the expense of cherished traditional values.

Religious Conflict

In trying to strike a balance between tradition and change, perhaps the greatest difference is between the Muslim north and the largely Christian south. Most of the values, culture, and traditional institutions in the Muslim north have survived intact; it is ultraconservative compared to the Christian south. The typical northern Muslim, regardless of age group, is not impressed by Western manners, culture, and values. Muslims are more likely to identify with the culture and values of the Islamic world. Thus a Muslim community leader in the north is more likely to address a community meeting in either Hausa or Arabic rather than English even if both he and his audience speak English fluently.

Some observers sometimes characterize the southern educated elite as trying to be more Western than Westerners. There is an element of truth in this characterization. It is not unusual, for instance, for English to be the first language of the child of an educated southern elite rather than the parents' ethnic language.

One of the greatest issues facing Nigeria is the conflict between Muslims and Christians, particularly over the imposition of Sharia or strict Islamic law in the Muslim communities; Christians are supposedly not subject to it. Penalties can be devastating, for example, having a hand removed for stealing $30. Only a few nations such as Saudi Arabia and the Sudan follow Sharia. In the year 2000, more than 2,000 people died in fighting between Christians and Muslims in the northern state of Kaduna when Sharia was first introduced; about half of this northern state is Christian, unlike most of the Muslim north. In 2002 another 220 people died in fighting between Christians and Muslims over the issue of having the Miss World Pageant in Nigeria during Ramadan. Such occurrences, while not the rule, are not rare.

The imposition of Islamic law in several states has deepened religious divisions and caused interfaith violence to erupt and separatist aspirations to take root. Another major challenge is corruption, which is estimated to have cost the country more than $400 billion and largely explains why most people in this oil-rich country live in poverty. The country also has a very active drug trade to feed the European demand for cocaine and heroin through nontraditional ports. In 2007 alone, 234 drug arrests were made at Lagos airport but according to the government, these were just grazing the surface of the country's booming trade.

Throughout this chapter, we have attempted to offer an accurate portrayal of the people and culture of Nigeria. Like a snapshot, it does not include all the detailed nuances of this highly diverse society. Still, the image of the marketplace is critical for understanding traditional and modern Nigeria. It is an appropriate metaphor for many African nations that face the gigantic task of integrating their diverse ethnic groups as they seek to modernize without losing the cherished aspects of their traditional cultures. Most of the world and especially U.S. Americans know of Nigeria because of the innovative e-mail and other Internet-based scams that are aimed at parting "gullible Americans" from their money. We hope that this chapter provides a more complete picture of the fascinating subcultures that make up this populous and oil-rich African country.

CHAPTER 20

The Israeli Kibbutzim and Moshavim

The feuding over these stark hills, ridges, and valleys south and east of Bethlehem, a 27-square-mile region that includes the Nassar farm, is emblematic of the decades-old Israeli-Palestinian conflict—a struggle rooted in land.

—Richard Boudreaux (2007, p. A1)

Israel is, in one sense, a classic cleft culture in which various ethnic and religious groups experience great difficulty living comfortably with one another (Huntington, 1996). Although most citizens are Jewish, the nation seems to be composed of tribes with radically different viewpoints and lifestyles, and 10 or more political parties vie with one another at any given time. Psychologically, it is more difficult to live in Israel than in many nations that have resolved or at least minimized their ethnic and/or ideological/religious differences, and various estimates indicate that 1 million Israelis and possibly more have either emigrated permanently or live outside of the country on a semipermanent basis. Some Israelis return to Israel for 1 month each year to serve in the military reserves and do so regularly for 30 years. However, many of these exiles will not openly admit that they are very reluctant to return to Israel permanently. Israelis would tend to view such an admission as traitorous or worse.

A perfect example of this psychological conflict and identity crisis is the comedy routine of a well-known Israeli trio, HaGashash HaChiver. An Israeli couple goes on a tour of the United States and stops to visit a relative's friends who had moved to the United States several years earlier. When the visitors ask their hosts when they plan to return to Israel, the host says: "You have caught us literally sitting on our suitcases, all packed and ready to go; we are just waiting until our son finishes college, and then we will hurry back to our beloved Israel." When the visitors inquire about the son's age, they are astonished to learn that he is "already 3 years old."

Before discussing the cultural metaphor of Israel, the kibbutz (kibbutzim in the plural) and its close cousin, the moshav (moshavim in the plural), we will provide some historical material that will help to shed some light on this small and complex nation. Although Israel was founded in 1948, its short history is as complicated as or more complicated than that of many older nations. It is difficult to identify a cultural metaphor that is expressive of the nation's basic values and, as suggested above, its conflicts. We have chosen the kibbutz because it is so intertwined with the nation's history and values, but some of these values have changed and are changing, at least in some major segments of the population.

Zionism, Types of Judaism, and the Palestinians

Israel's place in history is directly related to its geography. It is a small country lying on the eastern coast of the Mediterranean. Israel is only 260 miles from north to south and averages 60 miles from east to west, sharing borders with the Arab nations of Lebanon, Syria, Jordan, and Egypt. The country's landscape is marked by abrupt changes from mountains to plains and from fertile green areas to harsh deserts. The diversity of the scenery is immense—there are more than 15 distinct geographical regions. Israel's population is 6.6 million and 20% of its citizens are Palestinians.

The history of Israel has been characterized by uniqueness and oppression and at various times it has been ruled by Christians and Muslims. In their early history the Israelites, in opposition to the tribes surrounding them and their belief in polytheism, worshipped only one God. For this belief they suffered persecution and enslavement, thus creating the condition for their flight from Egypt under Moses' leadership. Christianity and Islam trace their origins to Judaism and this belief in only one God, and Muslims hold that Mohammed ascended to heaven from Jerusalem, while Christians maintain the same belief about Jesus, thus making the city sacred for all three religions and a source of conflict between them, particularly between Islam and Judaism. What sets Judaism apart from other major religions is its search for meaning in all aspects of life (Smith, 1958).

The opening of the Suez Canal in 1869 significantly increased the geopolitical and economic importance of the Middle East, and Jewish immigration began to grow significantly. Most of the newcomers came not for religious reasons but because of Zionism, a unique modern national and ideological movement.

Zionism, born in the last half of the 19th century, was the movement for the redemption of the Jewish people in the land of Israel. It derives its name from the word *Zion*, the traditional synonym for Jerusalem and the land of Israel. The movement gained strength because of the continued oppression and persecution of Jews in Eastern Europe and increasing disillusionment with their formal emancipation in Western Europe, which had neither put an end to discrimination nor led to the integration of Jews into their local societies. The Zionist movement aimed at attaining an internationally recognized, legally secured home for the Jewish people in their historic homeland.

Inspired by Zionist ideology, increasing numbers of Jews immigrated to Palestine at the end of the 19th century. These early pioneers drained swamps, reclaimed wastelands, founded agricultural settlements, and revived the Hebrew language for everyday use. It was at this stage that the kibbutzim became intertwined with the history of Israel, as they were founded by the Zionists seeking to create a new society for the Jewish people.

The first kibbutz was founded in 1909, and as of 2008 there were 271 of them, about 30% of which are a particular variant, the *moshav*. Although many kibbutzim are still engaged in farming, others are involved in several different types of industrial activities. Although they currently include 110,000 individuals, down from a peak of 125,000 in 1990, the influence of the kibbutzim and their democratic and egalitarian character on Israeli society is vastly disproportionate to their modest numbers.

The basic principles of kibbutz life are community ownership of all property, absolute equality of members, democratic decision making, the value of work as an end as well as a means, communal responsibility for child care, and primacy of the group over individuals. However, there are many pressures on the kibbutzim, and a variant of it, the *moshav*, allows for private ownership of some property by its members and for more individual freedom for its members; the *moshavim* now account for more than 30% of all kibbutzim. Because of economic hard times, 75% of the kibbutzim now charge for meals, 82% require payment for electricity, more than 25% have introduced salaries based on sliding scales, and more than 30% charge for health care (Leggett, 2005). In this chapter, we will describe the kibbutzim and include some final remarks about the *moshavim*.

Previously, members of the kibbutz received no pay for their work. All their essential needs—food, shelter, clothing, medical care, and education—were provided by the community. Children are raised with their peers, meals are usually eaten in the kibbutz dining hall, and members convene once a week to discuss and vote on issues. Jobs rotate among members. All positions of authority—such as those of the chief executive of the kibbutz, plant managers, and supervisors—change every 2 or 3 years. Power is not permitted to accumulate in the hands of individuals. The kibbutzim range in size from 100 to 2,000 members and are small enough so that the individual's role in the community is visible and tangible.

Because of the kibbutz's relative success as a social, innovative, political, and economic enterprise, it has attracted worldwide attention. Also, the kibbutz is a vehicle for obtaining insight into the Israeli character and the mythology from which the Israelis draw their self-identity. The perception of the kibbutz as an ideal existence stems from the initial goal of the founders: that the kibbutzim would be the core of society, rather than only an alternative.

A New Country

To understand the kibbutz fully, we need some historical background. After the Ottoman Turks were defeated in World War I, Britain took control of Palestine. Until World War II consecutive waves of immigrants arrived, and an infrastructure of towns and villages, industry, culture, and health, educational, and social systems was developed. The kibbutzim were close to their Palestinian Muslim neighbors and had fairly good relations with them at the time. Meanwhile, immediately before and during World War II, Jews in Europe suffered immensely, and the Holocaust was the most brutal form of genocide ever practiced by any culture. Renewed calls for an independent state were raised.

In 1947 the United Nations General Assembly adopted a resolution calling for the partition of the land into two states, one Jewish and one Arab; 54.5% of the land was allocated to Israel and 45.5% to the Palestinians. On May 14, 1948, the state of Israel was declared. On the very next day a combined Arab force committed to the destruction of the new state attacked. Although Israel was badly outnumbered and suffered the loss of 1% of its entire population, Israel won 6 months later. About 700,000 Palestinians became refugees, most of whom lived in U.N. refugee camps in neighboring Arab nations, camps that still exist. These refugees became a pawn of the neighboring Arab nations and Israel in the political and military wars that have occurred since 1948. Poverty is widespread and deep among the Palestinians.

In 1948 and immediately thereafter, the new state of Israel faced the daunting task of turning the Zionist ideological dreams into reality. Immigrant absorption was an immediate priority. In its first three years Israel welcomed 678,000 immigrants, many from the Arab countries, and in the process doubled its population. The nation's economic growth over the next decade was truly remarkable, especially in light of the fact that Israel has few natural resources and does not even have water in some major areas of the country. With the development of modern and sophisticated farming and irrigation techniques, orchards and groves were blooming even in the most arid regions of the country. Tens and hundreds of new towns and agricultural settlements came to life, modern industrial plants were built, roads throughout the country were paved, and a modern army was equipped virtually from scratch.

This success could not have occurred without outside help, which has continued to this day. By 1991 Israel had become the single largest recipient of charity, grants, and assistance per capita in the world, the bulk of it from the United States. The U.S. government allocates more of its foreign aid to Israel than to any other nation; Egypt receives the second-highest amount.

Continuous War

In 1956 and 1967, Egypt moved aggressively against Israel, but in both wars Israel emerged as victor. The 1967 war, also known as the Six-Day War, resulted in a devastating defeat for Egypt, Syria, and Jordan, as Israel gained possession of the Sinai Peninsula, Gaza Strip, the Golan Heights, the West Bank of the Jordan, and East Jerusalem. For the first time in 2,000 years Jerusalem was again under Jewish sovereignty. However, the country relaxed its military posture and the Egyptians and Syrians attacked again in 1973, this time on Yom Kippur, the holiest day on the Jewish calendar. Israel was caught by surprise, and although it won the war, the losses were significant. In recent years there have been *intifadas* or uprisings by the Palestinians accompanied by guerrilla warfare and suicide bombings within Israel. Extremists in the Palestinian Territory and in the neighboring Arab nations vow to destroy Israel, which makes negotiations between political leaders from Israel and the Palestinian Territory extremely difficult.

Today Israelis are in the awkward position of living in a cramped geographical area within which the Palestinians also live, but separately. In the early 1980s Israel withdrew from the Sinai and in 2006 from Gaza. Still, as Tolan (2006) points out, the Palestinians have fared even worse in terms of space, and recent Israeli proposals would provide less than 20% of the original Palestinian objective of the whole of Palestine. There are, as indicated previously, Palestinians who are Israeli citizens and they fare better than the others. Still, there is discrimination against the Palestinian citizens, whose homes in the West Bank and East Jerusalem are sometimes confiscated and bulldozed by the Israelis to make way for Jewish settlers. The Palestinian Israelis have recourse to legal remedies but their suits often fail. Also, the Israelis have built a wall separating Israel from the Palestinian Territory.

There are great tensions between these two groups, and Israel has been cited frequently by Amnesty International for violating the basic rights of the Palestinians. In the 1980s and early 1990s, there was great hope and movement toward a genuine Palestinian state by both the Palestinians and Israelis, and it looked like the Oslo Peace Accord of 1994 would be the basis of a solution to this long-festering problem. However, this hope was dashed when an ultra-Orthodox student in 1995 shot and killed Itzak Rabin, the prime minister who led this effort. An increased level of deadly conflict between the Israelis and Palestinians followed, with the Palestinians suffering much more than the Israelis.

Within Israel many Palestinian Israeli citizens report feeling uncomfortable and threatened, and there have been unprovoked attacks on them (see "When Good Men," 2002, p. 42). Even Israelis are divided on the issue of the poor treatment accorded these Palestinians, and there is a movement for Palestinian self-rule and peace in the region that many Israelis also champion. In fact, one of the great tragedies is that business people on both sides of the conflict have actively attempted to create a Mideast economic community in which trade and commerce would flourish but so far to little avail.

Religious Conflict

To complicate matters even further, Israeli citizens manifest a great variety of religious beliefs or lack thereof, which is one of the key reasons why there are so many political parties in the nation. Many Israelis are secular and even agnostic in their beliefs and do not participate actively in Judaism or attend synagogue. At the other extreme are the ultra-Orthodox Jews who demand rigid adherence to all aspects of Judaism, including strict dietary laws and the cessation of all business activities during the Sabbath. There have been instances where such Jews have demanded that females sit in the back of public buses and separately from the males. These Jews believe that religious laws and regulations should be given precedence over the legal system, at least in some instances. They refuse to serve in the military until the age of 24 and receive subsidies from the government to preserve Judaism. Other Israeli groups vary between these two extremes in terms of their adherence to tradition and change.

Jews in Israel are not differentiated by synagogue affiliations as much as by the manner in which they relate to the land and state of Israel. Because of the Jewish people's reconnection with their land and their building of a modern state, there are now five main options for defining oneself as a Jew (see also Friedman, 1989). First, more than 50% of the population is secular, and they are the ones most responsible for building the new state of Israel. For the secular Zionists, being back in the land of Israel, erecting a modern society and army, and observing Jewish holidays as national holidays all became a substitute for religious observance and faith.

The second group consists of religious Zionists, who make up about 25% of the population. These are traditional or modern Orthodox Jews who fully support the secular Zionist state but insist that it is not a substitute for the synagogue. The third group has about 5% of the population, and it also includes religious Zionists but of a more messianic bent. The members of this group believe that the rebirth of the Jewish state is the first stage in a process that will culminate with the coming of the Messiah. They form the backbone of the Gush Emunim settler movement, which acts on the belief that every inch of the land of Israel should be settled by Jews, even if it is owned by Palestinians.

The fourth group consists of the ultra-Orthodox, non-Zionist Jews, who constitute about 12% of the Jewish population. They believe that a Jewish state will be worth celebrating only after the Messiah comes and the rule of Jewish law is total. They also believe that the pinnacle of Jewish life and learning was achieved by the great 18th-, 19th-, and 20th-century *yeshivas* (Jewish schools for higher education) and rabbinic dynasties located in their native Eastern Europe. They have attempted to re-create that life in Israel and still dress in traditional dark coats and fur hats, naming their yeshivas after towns in Eastern Europe and preferring to speak Yiddish rather than Hebrew. They also refuse to celebrate Independence Day. Fifth and finally, there are the 800,000 secular Russians, among whom there are many atheists. They have come to Israel since the breakup of the Soviet Union and constitute a powerful group; so powerful, in fact, that they have created the only ethnic political party in Israel.

There are no Reform Jewish synagogues in Israel as there are in the United States and elsewhere. Members of such synagogues follow Judaic practice but in a much looser form than that found in Conservative Jewish counterparts. There is a much greater acceptance of outsiders and other points of view. Although many Jews marry non-Jews—the percentage of intermarriage in the United States exceeds 50%—members of Reform synagogues tend to do so more than other Jews.

In this chapter the focus is on the Israelis, although clearly their troubled relationship with the Palestinians at least needs to be spotlighted.

To return to our cultural metaphor, the debates about the viability of the kibbutz in a globalized economy reflect the intense debates among the various tribes of modern Israel, as we will see. Probably no other institution is as uniquely Israeli as the kibbutz, the collective farm that played a critical role in the Jewish settlement of Palestine and later in establishing and developing the Jewish state. As Lawrence Meyer (1982) describes it, "The Kibbutz is the ultimate symbol of pioneering Labor Zionism, the institutional embodiment of the dream of returning to the land, working it, receiving sustenance from it, and reviving an ancient culture in the process" (p. 327).

With this understanding of both Israel and the kibbutz's origins in mind, it is possible to see how the kibbutz reflects many of the essential features of Israeli society and the Israeli character. First, the explicit social and ideological values to which the kibbutzim adhere are also manifested in the Israeli society, although to a lesser degree. Also, the small size of both Israel and the kibbutz is related to specific behavioral outcomes, and the traumas that the Israelis have experienced have led to the evolution of a distinctive worldview and personality profile both in the kibbutzim and Israeli society. It is these points of comparison between the kibbutz and Israeli society that we will highlight.

Explicit Values

Most if not all societies possess explicit values which their members share, at least to some degree. In Israel these values are explicit in the kibbutz and, correspondingly, in Israeli society. They include democracy, egalitarianism, socialism, a group-oriented mixture of socialism and individualism, and a deep connection to the land.

One of the main features of the kibbutz is the direct and total democracy it practices. Almost all decisions and rules are voted on by all kibbutz members, and no major decision can be made by only one or a few individuals. Analogously democracy is a deep-rooted Jewish value. Having to confront many perilous situations as a group rather than as atomistic individuals, the Jews developed a self-deprecating sense of humor that deflated pretentiousness and heightened a feeling of social democracy, out of which evolved political democracy. The concepts of social and political democracy were brought to Israel by immigrants from Western countries, but mainly by the East European immigrants who founded most of the kibbutzim before the establishment of the State of Israel in 1948. The successful preservation of political democracy occurred in part because of the disproportionate influence of kibbutz members in Israeli politics.

Formally, Israel is a parliamentary democracy consisting of three branches. The executive branch (the government) is subject to the confidence of the legislative branch (the Knesset, Israel's parliament), and the judiciary branch, whose absolute independence is guaranteed by law. The 120-member Knesset is elected every 4 years. The entire country is a single electoral constituency, and Knesset members are assigned in proportion to each party's percentage of the total national vote. In this way even very small parties that succeed in gaining only 1% of the votes can be represented. This system of voting was a result of a preference for a government that would reflect not only one party but the consensus of at least half a dozen; thus rather than following one party's policy, it accepts the lowest common denominator of many. Hence the system lends greater leverage to small and even minuscule factions and interest groups, and a good amount of bargaining occurs among the parties, some of it quite dishonorable. Debates in the Knesset are sometimes vitriolic.

Bargained deals often result in undemocratic concessions to religious splinter groups, which hold the balance of power between the main political blocs. Such concessions have led to strict Sabbath laws and laws such as those prohibiting the raising and selling of pork, civil marriage, and divorce. Many of these laws are unacceptable to a large portion of the Israelis, especially to those who are secular and emphasize Judaism's culture but do not regularly participate in religious activities or attend the synagogue. Electoral reform is often spoken of, but action has been minimal until recent years, for example, with the rise of an avowedly secular party, Shinui, in the early 2000s.

Although Israel is not a theocratic state, there is a significant overlap between the government and Judaism. As indicated previously, rabbinical law prevails in such matters as marriage, divorce, and the Sabbath, and sometimes there are conflicts between the state laws and religious laws. For example, the Old Testament dictates that every seven years farmers should let their fields lie fallow. Modern Israeli farmers meet this requirement by technically selling their property to a non-Jew through the auspices of the local chief rabbi, who holds the check for one year and then destroys it. This formal practice has met the letter but not the spirit of the Old Testament since the founding of the State of Israel. However, one of Israel's two chief rabbis, Yona Metzger, changed this system by allowing local rabbis to decide whether the food raised in this seventh year was kosher, which would severely hurt a farmer who could not obtain this designation. This issue became so important that the Supreme Court ruled that, if a local rabbi decided that the food was not kosher, an additional, more flexible rabbi would be appointed (Khalil, 2007).

The Army's Role

A prominent example of how important democracy is to Israel is the status of the army within the society. Although necessity has turned the army into an essential part of Israeli life, Israel is far from being a military society. Israelis spend a large portion of their adult lives in the army, both during their compulsory 3 years and then during at least 30 days a year as reserve soldiers, but they remain very much civilians. The Israeli army is highly professional, with a basic discipline similar to that of many other armies, but it is always the first preference to train the soldier by example, not by order and punishment. Amos Elon (1971), the Israeli author, points out:

> The army is not an aristocratic institution, as it continues to be in some democratic countries. There is no deliberate attempt to break the will of recruits. It is a citizen army, and the gap between officers and men is minimal. There are few privileges of rank. The Israeli military code bluntly states that officers have no privileges whatsoever, only duties. Officers are usually addressed by their first names. There are no fancy uniforms . . . military titles are used in writing, but rarely in verbal address. The use of such titles in civilian life, after retirement is frowned upon . . . job turnover among officers is unusually rapid and the army is almost never a lifelong occupation. Most officers are weeded out soon after they reach forty. This practice has helped to prevent the establishment of a military class. (p. 252)

Hence, the Israeli army exists out of necessity, not out of choice. It combines the needs that evolve out of Israel's memories of its tragic past of persecution, the constant state of war or fear of it, and its small size with the wish to maintain a democratic society that will not be controlled by any elite. Thus far Israel has been successful in maintaining that combination.

The kibbutz is close to the ideal of an egalitarian society. The equality of all people within the small, enclosed society of the kibbutz in one of its most fundamental principles, and the periodic rotation of critical and powerful roles is one effective way of preserving egalitarianism.

Likewise, the egalitarian principle is basic to Israeli society. In Hofstede's (2001) study of cultural values across 53 nations, Israel had the second-lowest score on the dimension of power distance or the belief that inequality among members of a society should be emphasized. The more recent GLOBE study (House et al., 2004) indicates that only 5 of 61 societies promote the *practice* of lower power distance more than Israel. Still, this study also indicates that Israel scored near the middle of these 61 societies in terms of *valuing* lower power distance, a finding that accords with the objective data indicating growing inequality. In fact, Israel has become one of the most unequal societies in the world (Friedman, 2005). About 25% of Israelis live below the poverty line—one third of all Israeli children. Among people 65 and older Israel has the highest rate of poverty in the Western world. One major reason for this state of affairs is the acceptance of all Jews who wish to immigrate to Israel, many of whom come from poor and impoverished nations.

The Role of Power

Twenty centuries of oppression have made Jews skeptical of the indiscriminate use of power. Many Israelis find the mere striving for power to be objectionable. This has caused a basic disrespect for authority, which is a main characteristic of Israeli society. Also, the anarchist strain that is found in Israel has its origins in the Marxist and socialist ideas that the first pioneers espoused, for some envisioned the entire society as a network of collective agricultural and industrial associations involving a minimum degree of coercion and a high degree of voluntary, reciprocal agreements. Although this vision did not materialize, belief in egalitarianism continues to this day. Voluntarism, rather than formal authority, is highly regarded throughout Israeli society, and especially so in kibbutzim and cooperative movements. Members spend a relatively large amount of time on voluntary activities, such as serving on different committees and working extra hours. Another form of voluntarism is the high percentage of kibbutzniks who serve in the elite units in the army and as officers. It is almost a social norm to volunteer for these assignments.

Some Israelis are worried that the simultaneous growth of prosperity and inequality will lead to greater individualism and a weakening of the collective spirit (Waldman, 2007). Some kibbutzim members are selling some of their property; their small homes are now dwarfed by the mega-mansions that wealthy Israelis have built alongside of them. While the number of applicants for combat units has actually increased, the six wealthiest cities vary in the enlistment of 18-year-olds in active combat units from 37% to 60%; historically the kibbutzim have registered far higher rates of recruitment.

Furthermore, the ideological basis of the kibbutz is socialism, whose basic tenet is that the major means of production and distribution are managed and controlled by the community as a whole, and it is actively practiced in the kibbutz. Members live according to a system whereby each person works as he or she can and receives what he or she needs. Almost all needs are met by the kibbutz, including food, shelter, clothing, medical care, child care, and education. As noted previously, the cash-strapped kibbutzim are now charging their members for some of these services.

A similar type of socialism exists in the Israeli society at large. The concept of national responsibility for basic health, housing, and social welfare is a given. Even the recent influx of hundreds of thousands of Ethiopian and Soviet Jews has not led to a fundamental renewed questioning of priorities, although the costs are enormous. Education is free, and health insurance is set at a price all people can afford.

Group Versus Individual

Another key value in the kibbutz and Israeli society is the primacy of the group over the individual. From the outset of the Zionist movement until the proclamation of the state, the emphasis was consistent: the purpose of the Zionist movement was the redemption of the Jewish people. The movement, and later the state, focused on the group rather than the individual. Moreover, the early pioneers arriving in Palestine firmly believed in this socialistic approach, and they quickly realized the value of group cooperation as they attempted to make the land habitable.

Many of these pioneers were not accustomed to manual labor, were weakened by malaria and other diseases, and suffered from the hostility of the Arabs. To overcome such obstacles, making the group more important than the individual was not only desirable but absolutely necessary. This experience had the same effect on the Israeli society as a whole, although to a lesser extent.

Currently the demands made on the individual in Israel are great because of the continuing tenuous security that the nation faces, which implies that survival rests on the cohesiveness of the group. Therefore, each Israeli must sacrifice some individuality to meet these demands. Mandatory military service is required of both sexes, taxes are among the highest in the world, and Israelis can take only a small sum of money out of the country for investments elsewhere. The ultimate betrayal in the eyes of Israelis is to leave the country permanently. A similar attitude toward people who left the kibbutz to live elsewhere formerly existed. People who grew up in the kibbutz and decided to pursue a different way of life found they were rejected by the group, although this rejection was not always conscious (Bettelheim, 1969). This reaction is much less intense than it used to be, partly because a relatively large number of young people choose to leave—between a third and a half of each age group, and sometimes even more.

The relative unimportance of the individual is most apparent in encounters with the bureaucracy, which is bothersome in most nations. However, the Israeli system presents itself to Israeli citizens not as a guardian and servant but as a warder. The stories about the indifference and callousness of the Israeli bureaucracy are legion. Almost every contact with a government or public agency involves standing in line for a long time, and it is quite common to have to wait in several queues to arrange one matter. The combination of constant demands on the individual and the unpleasant bureaucracy leads to a resistance to authority that manifests itself in the disregard for whatever seems unessential to the general safety or survival. For example, cheating on taxes is a common practice.

Admittedly individualism has increased significantly in recent years. Only 2% of the workforce is now in agriculture, and most Israelis live in and around cities. Erez (1986) dramatized this movement toward individualism in a well-controlled study in which she manipulated leadership style across three Israeli organizational cultures. In the kibbutzim collectivism was the highest, and participative management was most effective in this setting. In the public sector, however, delegative management was most effective. But in the private sector, which was the most individualistic, directive leadership was most effective. Today Israel's economy is increasingly integrated with the activities of other nations and its private sector is growing rapidly. Given this situation, we can expect to see individualism increase and conflict between groups representing different varieties of individualism and collectivism to rise.

Other Values

It can be seen, then, that while the instinct for self-preservation has given rise to social discipline, there is still a widespread, almost compulsive resentment of any regimentation. The result is a peculiarly Israeli mixture of self-reliant individualism and a socialistic readiness to cooperate. This interplay between individualism and collectivism is also portrayed in Hofstede's (1991) study of 53 nations, where Israelis scored near the middle on the individualism-collectivism dimension. In the more recent GLOBE study (House et al., 2004), however, Israel ranked only 49th of 61 societies in valuing institutional collectivism, thus suggesting that Israel has moved toward more individualism since the Hofstede survey, completed in the period 1968 to 1973.

The deep connection to the land is another value held both by the kibbutz and the Israeli society, even though agriculture represents only 2.8% of Gross National Product. This connection partly stems from the knowledge that land is necessary for survival and protection, and it is partly the consequence of a rebellion against generations of Jews who had no real connection to any land. The land the Israelis inherited was poor in many ways, but the vigor of the response has been overwhelming. The people of the kibbutzim have shown their love and connection to the land through working it and making it fertile and blooming. In Israel at large the connection was shown through the attempts to settle as much of the land as possible, even in the most remote areas.

Size and Behavioral Outcomes

The kibbutz and Israel are small, both geographically and in population. As indicated previously, Israel's population is 6.6 million, 20% of whom are Palestinians, and the distances between cities are very short. Driving from the Mediterranean to the border with Jordan takes an hour and a half. Both Haifa and Jerusalem are an hour's ride from Tel Aviv. More than 80% of the population is concentrated in this small triangle. Much of the country is uninhabited desert, and it is consequently even smaller than it appears on most maps, where it is about the size of New Jersey.

As a result, both in the kibbutz and Israel, specific behavioral outcomes occur that are not common in larger organizations and societies. Relations of all kinds are much more intimate in both the kibbutz and Israel than in larger entities. In the kibbutz children spend almost all of their time with the same group of peers throughout childhood. Because there are typically only a few hundred people in the kibbutz, adult members interact with the same people almost every day. Similarly in the larger Israeli society, people tend to maintain a relatively steady and intimate group of friends, but with a relatively large number of acquaintances and contacts on the periphery.

Families also are close-knit. Small geographical distances lead to easy access and relatively frequent visits and gatherings. This sense of closeness also manifests itself in the responsibility many parents feel toward their children even when they have become adults. When parents can afford it, buying an apartment for sons or daughters when they get married is a common practice, and financial help frequently continues even after that stage. Moreover, children tend to expect this help because of the high cost of living.

Given the country's small size, maintaining anonymity and privacy requires hard work. It is extremely difficult for individuals to achieve anonymity, and intimate details about someone's past and present are common knowledge, even in the larger urban centers. There are 318 people per square kilometer in Israel whereas the comparative figure in the United States is 32. This intimacy and smallness help to establish a strong sense of community.

Furthermore, this sense of closeness leads to a heightened intensity of experience. Tragedy rarely remains isolated within the narrow confines of the family but spreads to wider social circles. In the kibbutz, with its communal dining room and similar collective arrangements for hundreds of members, this is obvious. But it is often true also in the towns, where news travels fast and immediately affects even apparent strangers. Each death in war electrifies the country as though it were the first time, drawing strangers closer to one another. At such moments, the sense of cohesion is so strong it makes the country seem more like a large village than a state.

Finally, the mass media powerfully dramatize this state of affairs, and Israelis are obsessive consumers of radio, TV, and Internet news. It seems that no people in the entire world tunes in to the news as often, as regularly, and with such fervor.

Traumas, Worldview, and Personality

The Jews have experienced proportionately more traumas than most if not all major cultural groups. Two modern traumas that have had an enormous influence on the Israeli worldview and personality, both in the kibbutz and outside of it, are the Holocaust and the continuing struggle with their Arab neighbors. More than 6 million Jews, or about one third of all Jews then living, perished in the Holocaust during World War II, and the memory of it remains vivid in Israel. Furthermore, the Holocaust had an enormous impact on the process of creating the state of Israel. As Elon (1971) points out, it

> caused the destruction of that very same Eastern European world against which the early pioneers had staged their original rebellion, but to which, nevertheless, Israel became both an outpost and heir. Because of the Holocaust there is a latent hysteria in Israeli life that accounts . . . for the prevailing sense of loneliness, a main characteristic of the Israeli personality. (p. 199)

This sense of loneliness helps to explain the obsessive suspicions, especially of outsiders, in both the kibbutz and in the society at large and the urge for self-reliance at all cost in a world that never again can be viewed as safe and secure. Clearly the Holocaust left a permanent scar on the national psychology, the tenor and content of public life, the conduct of foreign affairs, on politics, education, literature and the arts . . . the notoriously lively bustle . . . and seemingly endless vivacity of Israeli life merely serve as compensatory devices for a morbid melancholy and a vast, permeating sadness . . . It crops up unexpectedly in conversation; it is noticeable in the press, in literature, in the private rituals of people. (Elon, 1971, p. 199)

It is little wonder that the term "gripe parties" is employed to describe many social gatherings.

There are countless public and private monuments to the Holocaust. Israelis do not let themselves forget. One day a year is devoted to mourning for the victims. The public commemoration serves a compulsive need to reassert the group and demonstrate its continuing vitality. However, as the older members of society die, they are replaced by others who often do not use the Holocaust to justify questionable actions, particularly the harsh treatment of Palestinians. And the younger Israelis are focused not so much on remembering the Holocaust as on making Israel into a modern nation that is quite comfortable with different nationalities and perspectives.

Most Israelis believe wholeheartedly that the Holocaust was possible simply because the Jews had no country of their own and consequently lacked the means of resistance. To them, it is proof that the earlier Zionist theme of a need for a land was absolutely correct. The Holocaust also explains in part why Israelis are willing to endure most hardships imposed by their government with hardly any protest.

Arab hostility has helped to sharpen the outlines of the picture of utter loneliness drawn by the Holocaust. Since independence in 1948, Israelis have lived in a state of geographic and political isolation unusual in the modern world. Most countries today share common markets or at least open borders, a common language, or the same religion, but Israel does not. Also, it does not have any military, political, or economic alliance in the region. This claustrophobic isolation has given rise to a pessimism that is a main feature of the Israeli personality. It is a root cause of Israeli stubbornness and largely explains why pious reprimands from other nations have little effect.

The kibbutz, although in a completely different way and for different reasons, is also isolated in many ways from the Israeli society as a whole, choosing to keep to itself. Although egalitarian in their own society, kibbutzniks tend to view themselves as an elite, and this view about the kibbutz is also held by some Israelis. Their social systems, such as education, are almost completely separate. They also are not much influenced by criticism that comes from the outside, feeling that others who have not shared their experiences cannot judge them.

Because memories of the Holocaust are so alive, Arab threats of annihilation achieve just the opposite of their original intention. Such threats have kept Israelis wary and ready for combat long after the initial period of pioneering enthusiasm waned and have increased Israeli resolve, inventiveness, cohesion, vigor, and a nervous but fertile anxiety. It has also helped foster the sense of shared social purpose. This kind of cohesion and unity is also strong in the kibbutz, although like the rest of the Israeli society, the members of the kibbutz are the harshest critics of their own community.

Ironically, Zionism in its early period was predicated on faith in peaceful change, which is one of the major reasons why the members of the kibbutzim had good relations with their Arab neighbors. The discovery that this was nearly impossible to achieve has profoundly affected the Israeli worldview. Redemption has become intertwined with violence on a continual basis, and the resulting dissonance is now a part of this worldview, creating an outlook that is a combination of the hopeful and the tragic.

A Fatalistic Approach

The never-ending state of war has brought about an awareness of *Ein Brera*—there is no choice. This viewpoint reflects a fatalistic attitude. Zionists, the former rebels against their fate, have come to accept continual warfare. This fortress mentality was certainly not part of the original Zionist dream. Again, however, some younger Israelis are questioning the need for such constant vigilance. The young people accept their military responsibility, however, emphasizing an unreflecting, elemental urge for self-preservation for the Zionistic vision of their elders.

However, as might be expected in such a small land whose citizens are constantly living under a state of siege, claustrophobia has emerged as a major problem. This claustrophobia, combined with a sense of adventurism, drives many Israelis, including kibbutz members, to trips outside of Israel, ranging from the conventional 2-week tours in Europe to exotic trips that may last for months and years. Kibbutzim members, given the small size of the relatively confined society in which they exist, can experience this sense of claustrophobia even more intensely than other members of Israeli society. Many of the young people go off and explore the life outside, working and living for a year or two in the city. The sense of claustrophobia is in part the reason why some of them choose not to return to the kibbutz, as it is the reason why some Israelis choose not to live in Israel.

There are, naturally, additional factors besides the traumas that help to form the Israeli worldview, two of the most important being immigration and religion. Israel is rife with contradictions, such as pessimism and optimism, fatalism and determination, and so forth. One of the most salient of Israel's contradictions is the composition of its population. Many of its citizens are native born, but the rest are immigrants from across the world: Europe, Asia, Africa, North America, South America, and Australia. Even most of the native-born Israelis are only one or two generations removed from their immigrant past. Most immigrant societies face problems resulting from serving as a melting pot for widely divergent backgrounds, but in Israel there is a noticeable distinction between the Ashkenazim Jews and the Sephardim or Oriental Jews. The Ashkenazim are those Jews from Eastern and Western Europe, America, and Australia. The early Zionists and the founders of the state were almost all Ashkenazim.

The Sephardic Jews, now approximately half the population, are those who emigrated from the Islamic countries in the Middle East and North Africa and who came to Israel after it was established as a state. The Sephardic Jews had to enter a modern Western society when many of them had significantly lower literacy rates and education, were short on professional skills, and had very little money. The largely Ashkenazi establishment, unlike the newer immigrants, was socialist and secular, and it tried to transform the mostly traditional Sephardim into models of themselves. Although a large number of the Sephardim prospered in their new country, many did not. By the 1960s it became clear that a "second Israel" had emerged, and there is a real gap between the two groups. This has resulted in a campaign aimed at improving the conditions and the opportunities of the Sephardic Jews.

In a definite manner, the kibbutzim represent part of this status problem. Almost all their people are of European descent, and in many ways they represent an Ashkenazi elite, which unofficially was reluctant to have the Oriental immigrants join its ranks. However, this is slowly changing, and there is a much higher degree of acceptance of these immigrants and their offspring in the kibbutzim.

Church and State

The other major issue affecting the Israeli worldview is religion. As discussed previously, serious conflict exists between religious and secular Israelis, and it is an explosive issue that has not been resolved but merely shelved temporarily. In fact, it has been impossible even to reach an agreement in Israel over the most basic question: Who is a Jew?

Given such strong opinions and points of view, it is not surprising that these groups clash fiercely over the relationship of religion to the state. While the secular Jews want the state to be completely separate from religion, the religious Jews want them to be intertwined. Another serious clash is over the issue of the young men from the ultra-Orthodox, non-Zionist group avoiding service in the army, especially in the combat units, which is bitterly criticized by the Zionist Jews.

Most of the kibbutzim are decidedly secular, and their founders actively rebelled against everything represented by the Judaism in their native Eastern Europe. In this sense the worldview of the kibbutzim is similar to that held by the secular Jews, who constitute the largest group in the population.

Thus far we have emphasized the internal character or personality of the Israelis, but this should be related to overt behavior that Israelis manifest. The common perception is that Israelis are impolite, arrogant, and brash. A classic example of this behavior is the high-risk, macho manner in which Israelis drive. They do not keep the proper driving distance between cars because of the fear that someone will cut ahead of them. Drivers honk at each other if the car in front of them does not move within a nanosecond of the traffic light turning green. Furthermore, this arrogant behavior is exhibited in the way people cut corners in their business and personal dealings and in the impolite manner in which customers are frequently treated. This roughness of Israeli culture is a reflection of the way Israeli society works, and it is not deliberately directed at the outside but is rather a spillover from the anxiety and internal frictions. Many U.S. American Jews visiting Israel experience great difficulty relating to the Israelis because of this roughness.

In part, this behavior stems from the constant feeling of insecurity, the simultaneous need to suppress it, and the need and desire to go on about daily life and behave normally. Such contradictory needs build up great repressed tensions between the desire to admit fears and worries and the belief that it is not appropriate, helpful, or "masculine" to do so.

Personal Style

Also, this behavior is partly a result of a Mediterranean influence—Italians behave in a similar manner. Still, this is only surface behavior. Israelis are quick to react, to scold, to act impolitely, but also quick to calm down. This harshness on the outside combined with a soft inside is the reason why the image of native-born Israelis is portrayed by the *sabra*—a cactus plant with thorns on the outside but with a delicate inner part. Correspondingly there is relatively little violence in terms of rape, murder, or extortion, and the streets are safe at night.

Israelis are generally very direct people. As Joyce Starr (1991) describes it,

They are eager to get down to business, to get right to the heart of things. The little extra graces that give the Middle East its charm and style are seen by Israelis as useless decorations, impediments that only waste time. Everything is up front, without pretense, and matters are stripped down to bare essentials. Israelis almost never phrase things with "I think" or "It seems to me," the classical American way of avoiding hurting the other. They will simply attack with "you are wrong." (p. 45)

In studies of the Prisoners' Dilemma, an exercise in cooperation, Israeli men are more uncooperative at the early stages of negotiation than most men from other nationalities. *Dugry*, an Arabic term now part of the Israeli slang, reflects a similar orientation among Israelis, as it means: Don't take offense, as I am going to speak frankly and be critical, so get ready, but realize it is not personal. There are actually management programs that train Israelis to be polite to U.S. executives, who do not enjoy a particularly high rating in this area.

In addition, informality is the norm in Israel. People introduce themselves to strangers using first names. This behavior is repeated almost everywhere: First names are used between employees and employers, school children and their teachers, and soldiers and their commanders. As noted previously, Israelis seek to deemphasize status and power distance between individuals, and they have little patience for formality or protocol, for complicated rituals or procedures. Israeli men rarely wear a suit and a tie. This informality is associated with a display of openness and frankness that encourages people to speak their mind and offer their opinion freely to another without feeling that they are being presumptuous for doing so. It is also associated with the way people socialize with each other—it is common to simply drop by a friend's house without calling in advance. When someone says "Why don't you drop by," they actually mean it.

Both the directness and the informality are essential ingredients of the kibbutz way of life. Because status differences are greatly if not totally deemphasized, and all matters are subject to debate and participative decision making, the weekly meetings create a platform where opinions are usually expressed freely, including opinions about the members themselves. Informal social relationships are easy to maintain because the physical size of the kibbutz is so small. Informal clothing can actually be a source of pride.

Finally, Israel is a classic "doing" society in which action is taken proactively to control situations and overcome environmental problems. Although U.S. Americans like to think of themselves in a similar fashion, the Israelis far outshine them: They have an uncanny ability to improvise, and they are at their best when doing so. Israelis pride themselves on their ability to fashion solutions to both mundane and desperate problems as they occur. In part this ability comes from their historical experience when Jews didn't have rights, and the only way to survive was to look around for a short cut.

The more recent experience, when the new state of Israel had to overcome many difficult and complex problems in a very short time, has strengthened this ability. Brett (2001), in a negotiation study of international MBA students at Northwestern University, showed that Israelis (pragmatic individualists) are more effective than Germans (cooperative pragmatists) or the Japanese (indirect strategists) in terms of achieving favorable outcomes for themselves. However, she also points out that, even though they are hard, individualistically oriented bargainers, they achieve success in large part because they are more responsive to the needs of their negotiating adversaries, thus leaving both parties more satisfied with the outcomes than occurs with German and Japanese negotiators. Within the kibbutz, the ability to improvise was developed to its highest form. Many of the kibbutzim possessed only scant resources, and so their members learned to use all of them in inventive ways. Without such improvisations they might well have failed. Part of the fame that the modern and inventive Israeli agriculture has acquired worldwide can be related to the experiences the kibbutzim had to face.

Together with being an action-oriented society, Israel is also an intellectual society. A high percentage of people have university degrees, and the number of books bought per capita is one of the highest in the world. This trend also exists in the kibbutz, and one way kibbutzim try to maintain a high cultural level is through special cultural events, either internally produced or brought from the outside.

A Changing Culture

As organizations change, so too does the culture in which they exist. The most popular model for looking at organizational change and growth is biological or evolutionary, that is, organizations go through distinct stages. Edgar Schein (1985) has described a three-stage model, the first of which is birth and early growth, during which the main ideas and sources of inspiration come from the founders and their beliefs. These ideas serve as the "glue" holding the organization together. During the second stage, organizational midlife, there is a decline in the commitment that members make to the organization and the loss of key goals and values, which results in a crisis of identity. If an organization is to achieve the third stage, organizational maturity, and become even stronger, it needs to redefine its goals and values to meet the expectations of a changing world. In effect, the kibbutz is experiencing an organizational midlife crisis and, correspondingly, so too is Israel.

In particular, most Israelis are not willing to work the long hours that each type of kibbutz requires, although they frequently feel that their children should spend at least 3 months working on it to reconnect with the country's past. The *moshavim* have become a popular alternative simply because of the rigidity of the life that most kibbutzniks lead, and they are characterized by a lower degree of ideological fervor, the ownership of private property, and the espousal of some rights for individuals, even to the extent of benevolently accepting the fact that some members want to leave permanently.

Fortunately, the ideological rigidity of the founders of both Israel and the kibbutz has eased considerably, and the new generation is responding to a changing world. In the kibbutz the rigid rules of living with the bare essentials has changed dramatically. Life has become more prosperous, members now have all or most of the modern appliances in their own rooms, they are able to take more vacations abroad, and so on. These changes parallel those in the larger Israeli society, which has become much more materialistic.

Another dramatic shift in the kibbutzim is industrialization. Israeli agriculture has reached its upper limit, mainly because of the lack of water. By 1980 Israel was using all its replenishable water supply (Meyer, 1982). Today, many of the kibbutzim have at least one factory, employing on average 50% of the members. Some kibbutzim are engaged in tourism and such service-oriented activities as managing hotels and weddings. This transition occurred because of the belief that the future lies in industry and elsewhere rather than agriculture. Still, it has been accompanied by other changes and problems. Economic decisions in the kibbutz have become more complex, and it is harder for all members to participate in all decisions because of a lack of knowledge necessary to understand the issues. And, as noted above, many kibbutzim have suffered economically in recent years.

There is also a definite shift in the kibbutzim, and particularly the *moshavim*, from an emphasis on group effort and collective well-being to one combining such group effort with recognition of individual needs, privacy, and family. Many kibbutz members, and correspondingly many Israelis, demand more freedom for self-expression and individual preferences. One result in the kibbutz has been the decline in the practice of collective child rearing, which was once synonymous with kibbutz life and an essential part of the ideology. Another manifestation of this trend is the larger flexibility kibbutzim show toward people who want to be able to pursue the careers they freely choose.

Although the kibbutzim are still well-represented in the Israeli parliament or Knesset, their influence has declined in recent years. This is a result of the replacement of the Labor Party championed by the kibbutzim by the more conservative Likud Party as the leading political party in 1977. This was not only a signal event but a symbolic one, as it stood for the movement away from some of the traditional values of the kibbutz, particularly the emphasis on creating people who want to adhere to group norms, rather than to lead. Today the leadership of the kibbutzim is mainly setting an example, not being involved in politics.

The kibbutzim have been an absolutely essential part of Israel's struggle to survive, and their values have been a bellwether for the country. If the kibbutzim and Israeli society are to achieve and maintain the third stage of maturity, it would probably be wise to rekindle the spiritual and moral heritage of the kibbutzim throughout Israel in some form, for their basic values were and most probably are indispensable for continued growth and success. Israel, now a global leader in some areas such as high technology, is seeking to balance individual desires and national imperatives, and using the kibbutz and *moshav* as a starting point for discussion is most probably a wise course of action.

CHAPTER 21

The Italian Opera

The only form of theatrical music that is at all controversial . . . is the operatic form.

—Aaron Copland (1957, p. 133)

In March, Italian senator Paolo Amato joined placard-waving citizens furious over the removal of an iconic painting from Florence's famed Uffizi Gallery. While protesting the loan of Leonardo da Vinci's "Annunciation" to an exhibition in Japan, the senator took an unusual step: He wrapped himself in chains, looped them around a post outside the museum entrance and snapped the padlock shut. The stunt was the dramatic, even operatic conclusion to a noisy conflict raging for weeks.

—Christopher Knight (2007, p. E1)

I taly for centuries has had a well-deserved reputation for excelling in the areas of painting, music, and living life fully with a general sense of style and vivacity. Other nations excel at waging war and creating economic prosperity for most of their citizens. However, as John Peet (2005) has shown, Italy is a laggard economically, particularly in comparison to other members of the European Union (EU). Some of the causes for this poor situation include high taxes to support social programs such as early retirements from government work, a low rate of labor force participation (57% of those in the 15 to 64 age range versus 66% in Germany and 73% in Great Britain), low productivity, the continuing influence of criminal groups such as the Mafia, widespread corruption, tax evasion, a huge public debt, and inefficient government. Just in the area of tax evasion, one reliable estimate indicates that unpaid taxes are equal to 27% of Italy's gross domestic product (Kahn & Di Leo, 2007).

To make matters worse, much of the prosperity in Italy since World War II has occurred because of the dramatic success that small, entrepreneurial firms in the north have experienced. Because of the competition from China and other nations and the increasing power of large multinational corporations, such small Italian firms have faced serious problems in recent years. There are only 47 large firms in Italy, with its population of 57.3 million, while Britain with a similar population (59.4 million) has 129 (Peet, 2005).

Italy is about the size of Florida, but less fertile and less endowed with natural resources. The country is divided into two main regions. Continental Italy consists of the Alps and the northern Italian plain while Mediterranean Italy encompasses the Italian peninsula and the islands. The Apennine mountains run directly down the peninsula's center, and its northern range almost completely cuts off the north from the south. This geographical division has led to extreme regionalism, and many Italians view Italy as two separate nations: the wealthier, industrial north with its population of 37 million, and the poorer, more agrarian south with 20 million.

North and South

The political scientist Robert Putnam (1991) has empirically and convincingly demonstrated that the civic traditions of the north and south of Italy originated from different sources and that these traditions have persisted for centuries. In the 11th century King Frederico established an autocratic form of government in the south that was both politically and economically successful, but over time his successors corrupted it. The Mafia arose as the middleman between the kings and the people, which tended to reinforce corruption. In the north, by contrast, there has been a centuries-long legacy of civic involvement through guilds and democratic elections that has held corruption in check. Today the north of Italy continues to be successful, even though it is under duress because of globalization. However, the south languishes in poverty, even though millions of dollars have been devoted to eradicating it. There is even a group, the Northern League, which periodically seeks to create two separate nations in Italy. However, this movement is more quixotic than real, as the economically prosperous north depends heavily on emigration from the south for its workforce. Moreover, the north and south are closely linked to one another culturally, as noted below, even though there are regional differences.

Although Italian culture is justifiably renowned, it is in danger of extinction, as are the cultures of some other developed nations. Italy's fertility rate of 1.28 children per woman is far below replacement (2.2) and is one of the lowest in the world. Italy has the second-oldest population among nations following Japan, with 25.6% of its population over 60. Still, the population is large and these trends may reverse. Also, many people want to immigrate to Italy, and in fact there are countless numbers of illegal aliens crossing its borders daily. Thus, while Italy will probably change, its cultural base should be relatively secure, at least for the foreseeable future.

Italy has been historically victimized by overwhelming natural disasters: volcanoes, floods, famines, and earthquakes. It is a geologically young country, most of which cannot be categorized as terra firma, and there have been so many mudslides resulting in great loss of property and life both in the past and in recent years that the Italians nickname the nation "landslide country" ("Hand of God," 1998). As a result Italy exudes an aura of precariousness (Haycraft, 1985). Italians tend to accept insecurity as a fact of life. This acceptance may explain why they seem to be able to enjoy life more for the moment and why they are willing to accept events as they happen. Italians tend to feel that if something is going to happen, it will, and that not much can be done about it.

Italy's history is filled with many culturally rich and influential periods, including the Roman Empire and the Renaissance. The country emerged as a nation-state in 1861, much later than many other European nations such as France and England, as a result of national unification, known as the *risorgimento* or revival. Before unification Italians never quite identified with the many foreign bodies that conquered and ruled them. This is a possible explanation for Italians' historic contempt for the law and paying taxes. And even though Italians have been overwhelmingly influenced by foreign rule, they have managed to create a culture that is distinctly their own.

The Opera Metaphor

To understand Italy it is helpful to look at the opera, the art form that Italians invented and raised to its highest level of achievement. It is the only art form for theatrical theater in which almost all of the words are sung rather than spoken. Musical forms that have evolved from the opera include the English operettas of Gilbert and Sullivan and such U.S. musicals as *South Pacific* and *Oklahoma*. However, they have proportionately many more spoken rather than sung words in comparison to opera.

The opera represents most if not all of the major features of Italian culture. It is a metaphor for Italy itself, as it encompasses music, dramatic action, public spectacle and pageantry, and a sense of fate. It may be tragic or comic, intensely personal or flamboyantly public, with the soloists and chorus expressing themselves through language, gesture, and music and always through highly skilled acting. There is a larger-than-life aura surrounding operas, and the audience is vitally engaged in the opera itself, showing great emotion and love toward a singer whose talents are able to express the common feelings that Italians tend to share. Operas and operatic songs reflect the essence of Italian culture, and they are embraced by Italians with an emotional attachment that less dramatic peoples might find difficult to understand. Therefore we chose the opera as our metaphor to understand Italian culture.

Using this metaphor, we focus on five distinctive characteristics of the opera and demonstrate how they illustrate Italian life. These characteristics include the overture; the spectacle and pageantry itself and the manner in which opera-like activities are performed in the daily life of Italians; the use and importance of voice to express words in a musical fashion; externalization, which refers specifically to the belief that emotions and thoughts are so powerful that an individual cannot keep them inside and must express them to others; and the importance of both the chorus and the soloists, which reflects the unity of Italian culture (chorus) but also regional variations, particularly between the north and the south.

The Overture

The overture was such an influential innovation that it inspired the development of the German symphony. Essentially the orchestra sets the mood for the opera during the overture and gives some idea as to what the audience can expect to occur. Typically the overture is about 5 minutes long, and it frequently includes passages that are emotionally somber, cheerful, reflective, and so forth. At various points the overture is quick, moderate, or slow in tempo, and it employs different instruments to convey the sense of what will happen in the various acts—usually three—into which the opera is divided.

Italian culture emphasizes overtures. It stands somewhere in the middle of the continuum between U.S. low-context culture and Asian high-context culture. U.S. Americans tend to ignore the overture and get down to business rather quickly. Many Asians, on the other hand, devote a significant amount of time getting to know their negotiating counterparts before doing business with them. Italians take a midway position on the issue of personal trust and knowledge, but they do tend to convey their feelings and thoughts at least partially at the beginning of the relationship. The other party has some understanding of what will unfold, but it is only a preview that is imperfect, as the unexpected frequently occurs.

There are regional differences in the manner in which the overture is performed in daily life. In Milan, which has been influenced by its proximity to Switzerland and Germany, there is a tendency to shorten the overture. Still, it occurs, even in the smallest café, which has only a few tables and a modest menu featuring sandwiches. The sandwiches are presented in a colorful and appetizing manner, and *paninis*—invented in Italy—are frequently featured, as are various forms of coffee and espresso. There may be just a few moments of conversation with the owner, but they are pleasant, like the display of food. Department stores emphasize the same pattern of behavior, and the tendency to interact with the salesperson is much higher in Italy than in the United States.

In the south the overture can be much more involved. When looking for a pair of shoes, for example, the buyer frequently visits a shop and indicates to the owner that a mutual friend suggested he or she make a purchase there. This information leads to a pleasant conversation and perhaps the offer of a cup of espresso. Frequently no purchase is made, or expected, on the first visit. When the buyer visits a second time, the same pleasantries occur, and he or she may or may not make a purchase. However, both parties expect that a purchase will occur if the buyer visits the shop for a third time. At this time the real negotiations or opera begin. Both parties—and the audience—have a good idea about the outcome, but expressions of emotion, various uses of argument and persuasion, and different tones of voice will help to facilitate the final outcome or sale. This strengthens the general sense of living life fully and with vivacity. Still, the unexpected can and does occur, as fate is a critical element in the mix.

Pageantry and Spectacle

By and large, Italy is a land of spectacle and pageantry. Italians tend to be more animated and expressive than most other nationalities. Many Asians, for example, attempt to minimize emotions and gestures when communicating with others. Italians, particularly those in the south, enjoy communicating in an expressive manner. The level of noise in Italy tends to be high in public places, and people tend to congregate rather than to be isolated from one another. Also, some bystanders do not mind becoming part of the action. For example, a driver stopped by a police officer for excessive speed may be debating the issue while a crowd vocally supportive of the driver's plea for mercy offers comments. In one situation a bus driver argued with a mother that her son, who was obviously not a child, should pay the adult rather than the child's fare, but he also had to argue with the many other passengers who vehemently came to her defense.

Once the overture is done and the curtain goes up, the audience is greeted by a set so visually stunning that it elicits applause. This is the beginning of the pageantry and spectacle. They play such an important role in Italian life that people and things frequently tend to be judged first and foremost on their appearances.

Barzini (1964) in his classic portrait of the Italians points out that the surface of Italian life, playful yet bleak and tragic at times, is similar to what has occurred in Italian history and has many of the characteristics of a show. It is, first of all, unusually moving, entertaining, and unreservedly picturesque. Second, all of its effects are skillfully contrived and graduated to convey a certain message to, and arouse particular emotions in, the bystanders. Italians are frequently great dramatic actors, as we might expect of the creators of the opera. Current portraits of Italy in best-selling books, such as *Under the Tuscan Sun*, reaffirm Barzini's descriptions, suggesting that cultural change in Italy is occurring slowly in many areas.

The first purpose of the show is to make life acceptable and pleasant. This attitude can largely be explained by the circumstances of history, both natural and man-made, where four active volcanoes, floods, earthquakes, and continuous invasions by outsiders have created a sense of insecurity. Italians tend to make life's dull and insignificant moments exciting and significant by decorating and ritualizing them. Ugly things must be hidden; unpleasant and tragic facts are swept under the carpet whenever possible; and ordinary transactions are all embellished to make them more stimulating. This practice of embellishing everyday events was developed by a realistic group of people who tend to react cautiously because they believe that catastrophes cannot be averted but only mitigated. Italians prefer to glide elegantly over the surface of life and leave the depths unplumbed (Barzini, 1964).

The devices of spectacle do not exist simply because of the desire to deceive and bedazzle observers. Often, to put on a show becomes the only way to revolt against destiny and to face life's injustices with one of the few weapons available to a brave people, their imagination. To be powerful and rich, of course, is more desirable than to be weak and poor. However, it is frequently difficult for the Italians, both as a nation and individually, to have both power and wealth.

The Importance of *Garbo*

This eternal search for surface pleasures and distractions is accompanied by *garbo*. Although this term cannot be translated exactly, it is the finesse that Italians use to deal with situations delicately and without offense. *Garbo* turns Italian life into a work of art. Italians tend to have a public orientation, are willing to share a lot, and are used to being on stage at all times.

As this description suggests, many Italians wish to portray a certain image to those around them. This is termed *la bella figura*, which Brint (1989) has defined as the projection of "a confident, knowing, capable face to the world." Wilkinson (2007a, p. A5) expands on this definition by pointing out that *la bella figura* represents the Italian obsession that allows substance to be ignored for style; it is the art of public performance.

The importance attached to impressions is shown in the basic document of the Italian republic, the constitution. In the U.S. Declaration of Independence, the first of the self-evident truths is that all men are created equal. In the Italian constitution, the first basic principle is that "all citizens are invested with equal social dignity" (Levine, 1963). Similarly many Italians would rather sit inside their home than go outside and make a bad impression. Wilkinson (2007a) describes the extreme pursuit of *la bella figura* in Italy among those who are wealthy but not sufficiently wealthy to pay for this pursuit from their perspective as to what is expected. They have joined a club that allows them to rent either a prestigious car for an afternoon or to obtain an impressive but temporary collection of paintings on their walls for a party. While such ostentatious displays are common in other nations, Italy seems to outshine many if not all of them in this regard.

A large number of Italians pursue *la bella figura* through material possessions. Giving a good impression, not only with demeanor but with material wealth, is important. Various explanations have been offered for this tendency, including years of domination by foreign nations, overcrowding in various areas, and natural disasters. Whatever the explanation, it seems that the creation of a show through an eternal search for pleasures and distractions is an attempt to deal with the difficult realities of everyday life.

The Italian reliance on spectacle and *garbo* or turning life into a work of art helps people solve most of their problems. Spectacle and *garbo* govern public and private life and shape policy and political designs. They constitute one of the reasons why Italians have excelled in activities in which appearance is predominant. During medieval times Italian armor was the most beautiful in Europe: It was highly decorated, elegantly shaped, and well-designed but too light and thin to be used in combat. In war the Italians preferred German armor, which was ugly but more practical. Similarly, on the occasion of Hitler's visit in 1938, Rome was made to appear more modern, wealthy, and powerful with the addition of whole cardboard buildings, built like film sets.

Italians believe that "anybody can make an omelet with eggs" and that "only a true genius can make one without" (Barzini, 1964). For example, warfare during the Renaissance, as it was practiced outside of Italy, consisted of the earnest and bloody clash of vast armies. The army that killed more enemies carried the day. But in Italy warfare was an elegant and practically bloodless pantomime. Highly paid leaders of small companies of armed men staged the outward appearance of armed conflict, decorating the stage with beautiful props, flags, colored tents, horses, and plumes. The action was accompanied by suitable martial music, rolls of drums, heartening songs, and blood-chilling cries. The armies convincingly maneuvered their few men back and forth, pursued each other across vast provinces, and conquered each other's fortresses. However, victory was decided by secret negotiations and the offer of bribes. It was, after all, a civilized and entertaining way of waging war. Although this approach often left matters as undecided as warfare practiced elsewhere, it cost less in money, human lives, and suffering.

Church Ritual

The operatic pageantry of Italian life also occurs in the rituals of the Catholic Church, which still exercises considerable political and cultural power ("By Hook," 2007). Italians prize these rituals for their pageantry, spectacle, and value in fostering family celebrations more than for their religious significance. Many Italians view the church, much like the opera, as a source of drama and ritual but not authority. Although they do not attend church regularly, the church still exerts a strong cultural and social influence on their behavior, and almost all Italians identify themselves as Catholics.

Furthermore, the spectacle of Italian life is quite apparent in the area of communications. Italians tend to be skilled conversationalists. Onlookers can frequently follow the conversation from a distance because of the gestures and facial expressions that are employed to convey different emotions. Similarly one can understand many of the actions that take place in an opera without fully understanding the words the actors are singing.

Although Italians tend to be emotional and dramatic, decisions are influenced rarely by sentiments, tastes, hazards, or hopes but usually by a careful evaluation of the relative strength of the contending parties. This is one of the reasons why, when negotiating even the smallest deal, Italians prefer to look each other in the face. They read the opponent's expression to gauge his or her position and can thus decide when it's safe to increase demands, when to stand pat, and when to retreat.

Transparent deceptions are employed to give each person the feeling that he or she is a unique specimen of humanity, worthy of special consideration. In Italy, few confess to being "an average man." Instead they persuade themselves that they are "one of the gods' favored sons" (Barzini, 1964).

The operatic Italians must always project a capable face to the world. Therefore, they prefer to engage only in work that will create the image of confidence and intelligence. In Italy middle-class people tend to work when they have a profession and not while being a high school or university student. In fact, many lower level jobs such as waiting on tables are full-time in nature and not readily available to students. In the universities, students specialize immediately in their first year and do not dabble, as is the custom in the United States. However, there is no set term for most university careers—for example, 4 years devoted to undergraduate work and 3 years to law school—and so many Italians spend more time in school than their U.S. counterparts.

Business Negotiations

Pageantry and spectacle also apply to business presentations and negotiations. When presenting ideas during such negotiations, managers are expected to ensure that the aesthetics of the presentation are clear and exact; they should demonstrate a mastery of detail and language and be well organized. Polish and elegance count for a great deal. However, although pageantry is important in business presentations and negotiations, Italians also tend to expect good-faith bargaining.

Most if not all Italians feel that it is infinitely better to be rich than to seem rich. But if a man or a nation does not have the natural resources necessary to conquer and amass wealth, what is he to do? The art of appearing rich has been cultivated in Italy as nowhere else. Little provincial towns boast immense princely palaces, castles, and stately opera houses. Residents of some small coastal villages have completed elaborate paintings on the rocks that can be seen from the sea to give the illusion of wealth and prosperity. In spite of economic difficulties many Italians wear good clothes, drive shining cars, and dine at expensive restaurants. However, some of these people may have few material possessions. Decoration and embellishment are important so that the realities of economic insecurity, uncertainty, and a scantily endowed land can be changed into a spectacle of illusion.

The pageantry of Italian life is highlighted by its lack of a rigid social hierarchy. There is no permanent and rigid class structure in Italy. Rather, there are conditions on which a person is judged. These include occupation and the amount of authority a person has in that role, education, ancestry, and, in most instances, wealth. However, the major focus is placed on social behavior. The Italians term this focus *civilta* or the extent to which someone is acculturated to the norms of the area (Keefe, 1977). These norms include styles of dress, manners, and participation in the local community.

Position in the social hierarchy can be clearly judged by the amount of respect shown to a person and his or her family by the members of the community. However, respect is also associated with age and family position. Younger people will almost always show deference to older people, just as children show respect for parents.

Class consciousness and the social hierarchy will inevitably change as a result of industrialization, massive migration to urban areas, low birth rates, an aging population, and the Americanization of Italian youth. However, certain titles still command much respect. These include doctor, lawyer, and professor. Many people feel honored to speak to or have affiliations with someone holding such a title and will refer to him or her in a respectful, subordinate manner.

One final example of pageantry is appropriate, especially because it shows how some other cultures are similar to that of Italy (see Chapter 23, "The Mexican Fiesta"). Once a year in every village and city there is a celebration in honor of its patron saint (*santo patrono*) in which a parade is featured. The statue of the saint is placed in the lead position in the parade, followed by the clergy, members of the upper classes, and then all other citizens. A celebration follows in the piazza or town center. This traditional festival and parade help to strengthen the feeling of solidarity and permit people to escape from the regular routine in a dramatic, spectacular, and fun-filled manner.

The reliance on spectacle must be clearly grasped if one wants to understand the Italians. Spectacle helps people solve most of their problems and governs public and private life. It is one of the reasons why Italians have always excelled in activities in which impressions are important: architecture, decoration, landscape gardening, opera, fashions, and the cinema.

Voice

Italians tend to believe that their language is the most beautiful one in the world. Much of the beauty of the Italian language derives from the fact that it has a higher proportion of vowels to consonants than all or most languages, which gives it something of a musical effect. Italians put great emotion into their language, speaking with passion, rhythm, and changing tonality. It is the quality of the way people speak that is of utmost importance, and this bias is replicated in the opera in that there are several strikingly different types of voice registers such as the soprano and the bass. And, like the opera, the sound and cadence of the communication play a role at least equal to the content of what is said in getting the message across. Italians are often much more interested in engaging and entertaining their listeners than in conveying their thoughts accurately.

Perhaps the most difficult operatic singing is associated with bel canto, or beautiful singing, introduced by the romantic Italian composer Vincenzo Bellini. This form of music requires extraordinary exertion and ability on the part of the singers, especially when they are singing together to express their thoughts and emotions. Bel canto is so difficult a form of opera that only a limited number of singers can perform it effectively, and only Italian composers have been successful in writing music in this style.

Italians tend to talk louder than many other nationalities, but more so in the south than in the north. At meetings or informal gatherings individuals talk simultaneously and begin to raise their voices without realizing that they are doing so. Stories are told with passion, anger, and joy. The air in Italy is filled with so many voices that one must frequently talk in a loud voice to be understood, thereby increasing the total uproar.

Speech and Gesture

Oral communication in Italy is something of a show itself. Speaking is punctuated by elaborate gestures, befitting the Italian operatic tradition. Whatever the section of the country, Italians talk with their hands. Quick, agile expressive movements of the hands, arms, and shoulders contribute emphasis and sincerity to the spoken words and facial expressions. For instance, a man thinking about buying fish in a market empties his imaginary waist pocket when he is told the price and walks away without a word (Willey, 1984).

However, the gestures are not, as many believe, unrealistically exaggerated. In fact, the gestures are so realistic that they may be unapparent. The acting of the opera singers, directly derived from the Italian's natural mimicry, contributes to the erroneous impression of excessive exaggeration. In reality, Italian gestures are based on natural and instinctive movements and can therefore be understood by the inexperienced at first sight. Many of these gestures have been traced back to classical antiquity (de Jorio, 2000).

In addition to gestures, Italians sometimes employ flattery and polite lies to make life decorous and agreeable. The purpose is to reduce the turbulence of day-to-day living and to make life more acceptable. Flattery somehow makes the wariest of men feel bigger and more confident, similar to the larger-than-life opera singer. This perspective accounts for the fact that Italians sometimes make promises they know they cannot fulfill. Small lies can be justified to give pleasure, provoke emotion, or prove a point. This applies not only to the businessman who swears with soulful eyes that he will deliver his product on a certain day (and does so a month later) but also to officers of government cabinet rank. Ministers will promise an appointment, will confirm it and reconfirm it, but on the appointed day will find an excuse for evading the appointment. Casualness toward promises is part of accepted Italian behavior.

Contentious Spirits

Contentiousness is generally not far below the surface of the Italian personality. This argumentativeness surprises many because it is discordant with the usual Italian demeanor. Yet the accusatory words shouted by offended drivers, the sidewalk conversations that often sound like arguments, and the tedious monologues by Italians reciting some imagined wrong they've suffered testify to the argumentative side of Italians. For most Italians controversy is a hobby or a sport, something in which they take immense pleasure (Levine, 1963).

Italians are frequently portrayed as joyful people with an operatic song on their lips. This is not completely accurate. They can also be rather somber, just as many of their operas end in death and tragedy. This dark side is emphasized through humor, which tends to be more insulting than witty. Humor generally implies the ability to detach oneself from the object of wit. However, the Italians tend to be passionate people who easily identify with what they are talking about.

Italians make up for this type of humor by analyzing everything around them through conversation. Talking is a great pastime for most Italians, who frequently meet in cafes to discuss the latest news. Privacy is frequently lacking in any Italian community, even though Italians say they mind their own business. Unlike the reserved Englishman, an Italian is apt to tell an acquaintance of a half hour's standing all about his or her financial status, family health problems, and details of current emotional attachments. The Italian love of conversation as a pastime usually limits the chances of something secret staying so for very long, and it is this love of conversation (and voice) that is intimately related to externalization, the fourth characteristic of the opera.

Externalization

Externalization refers specifically to the fact that feelings and emotions are so overwhelming that people must express them to others. It also refers to the assumption that the event is more important than the actions of one individual, which is consistent with the Italian reaction to catastrophes and uncertainty described previously. That is, the drama is of more significance to the viewers or community than to the individuals because of its symbolism and generality. This pattern of behavior is the opposite of the Anglo-Saxon mode, which emphasizes that people should control emotions and not express them or, as the British say, keep a stiff upper lip (see Chapter 17). At Italian funerals, for instance, there is no shame in showing emotions, and both women and men cry openly and profusely to express what they feel. Similarly a major victory such as a World Cup Championship is accompanied by excessive noise, large gatherings in piazzas, and communal festivities at which everyone talks excitedly and simultaneously. Although there will be some drinking, the Italians do not tend to emphasize this aspect of victory as much as do the English and U.S. Americans.

Given their emphasis on externalization, it is generally not wise to ask Italians: How do you feel? Unlike Anglo-Saxons, who will usually give a cursory reply, the Italians may well describe in minute detail all of their aches and pains, what the doctors have prescribed for them, and even the most intimate details of any operations they have experienced.

Because of the importance of drama in everyday life, Italians value the piazza as the center of every town and village. It is the stage on which people gather to share conversation and relate experiences. The actions that take place in the piazza are minor dramas of Italian life. Most people do not schedule meetings with others, but everyone knows that gatherings will occur at regular times at the piazza, usually around noon and at 5 or 6 p.m. after work, and it is easy to see old friends and acquaintances and make new ones. This pattern of strolling (*fare la passeggiate*) is in sharp contrast to the German *spatziergang*, which occurs primarily on Sunday afternoons. Even in Italian cities where there is no central piazza, it is well-known that certain avenues serve as substitutes for them, and the village-like behavior is replicated along them. On these avenues there are many outdoor cafes at which people meet and greet one another, and sometimes there are even small open-air areas with benches that are miniature replicas of the village piazza. As might be expected, Italy has far more bars and restaurants per capita than comparable European nations: 257 inhabitants per café and restaurant in Italy versus 451 for Britain, 795 for France, 558 for Germany, and 778 for Spain (Richard, 1995). An Italian doctoral dissertation has wedded the importance of externalization with changing cultural circumstances by suggesting that the large piazza is more descriptive of southern Italy whereas bars (and the adjoining small parks) are more descriptive of northern Italy (Venezia, 1997).

The View From the Piazza

Italians tend to be great spectators of life. The show at the piazza can be so engrossing that many people spend most of their lives just looking at it. Café tables are often strategically placed in such a way that nothing of importance will escape the leisurely drinker of espresso. A remarkable result of this pastime occurred when an old woman died at her window and 3 days passed before any of her neighbors thought that something might be wrong with the immobile figure.

Foreigners are frequently impressed by the fact that many Italians seem to be doing their jobs with whole-hearted dedication and enthusiasm. This does not mean that these Italians do everything with efficiency, speed, and thoroughness. Rather, they frequently complete their jobs with visible pleasure, as if work were not man's punishment. However, when the visitors look closely at the actions of Italians, they realize that many Italians have a theatrical quality that enhances but slightly distorts their actions.

In the opera the best example of externalization is the crowd scene. There are so many crowd scenes in Italian opera that some writers have identified them as one of its essential characteristics. Members of the crowd represent the chorus for the lead singers, and there is a dynamic interplay between the lead singers and the chorus. This interplay is analogous to the behavioral dynamics in a small village in which there is a problem and everyone will comment on it in the piazza.

There is little if any private life for most Italians. Everything of importance occurs in public or is at least discussed in public. There is no word for privacy in the Italian language, and information is widely shared. There are some regional differences, as northern Italians tend to be more reserved than southern Italians. However, once a person is perceived to be a member of the community, it is assumed that he or she will externalize, at least to some degree, when in public.

Business confidentiality can be a problem in Italy. Everyone discusses secret business negotiations with friends and family as well as the press. Therefore leaks are common for most large business deals. The media in Italy are volatile and speculation-prone due to people's interest in their neighbors' activities. Hence ordinary business dramas become a theater with a large audience of spectators. Furthermore, to uncover corruption and tax evasion, the government has dramatically increased the use of wiretapping of phone conversations in recent years, which makes keeping secrets even more difficult. The Italian think tank Eurispes estimates that the government spent nearly $1.6 billion on nearly 200,000 phone call intercepts between 2000 and 2005 (Wilkinson, 2005).

La Bella Figura

We have already indicated that dressing is important to Italians, as it represents pageantry and spectacle in the form of *la bella figura*. Dressing is also an aspect of externalization, as it is an outward expression of the emotions that Italians feel and want to convey. Italians dress differently in public than at home at least in part for this reason. In the business environment clothing is well tailored and sophisticated. Through their clothing Italians externalize the feeling of confidence.

Many aspects of daily life yield endless opportunities for drama. A well-known example of externalization is Italian driving behavior. In the larger cities such as Rome and Naples, drivers blow their horns impatiently, swear at one another, gesture color-fully, and drive in a dramatic and seemingly reckless manner. To Italians, such behavior is normal because it allows them to communicate with others even when behind the wheel. Even when two drivers get out of their cars to confront one another, they rarely come to blows, although their gestures, body movements, shouts, and the torrent of words would lead the onlooker to the conclusion that blood will be spilled. The drama of the occasion is what gives Italians emotional satisfaction.

Even politics takes on a dramatic and entertaining edge, especially during elections. Television coverage of such events is said to have somewhat of a carnival atmosphere (Brint, 1989). Such an atmosphere reflects the Italian penchant for pageantry, voice, and externalization. Even though the Italian political system is changing radically, as discussed below, we can expect that such drama will continue to be part of the political process.

Feelings and emotions are expressed through direct communications and through subtleties in Italian culture. The Italian history of natural disasters and foreign invasions has created a fear of the unpredictable. This fear can be detected behind the Italians' peculiar passion for geometric patterns and symmetry that can be easily destroyed (Barzini, 1964).

Weddings are considered the highlights of life and constitute an excellent example of externalized behavior. In fact, the wedding scene is quite common in Italian opera.

Italians often save for a lifetime to marry their daughters off in an appropriate manner. Guest lists frequently include the entire village. The wedding ritual is often something of a mini-opera, beginning with the bridal party traveling in open automobiles to the church, with relatives and friends following. The success of the wedding ritual is judged by the number of cars, the size of the crowd, and the feast served. Parents are content to spend years of savings and even get into debt to correctly carry out the social obligations of a society that may, in the eyes of outsiders, be living at a subsistence level but that carries on ancient traditions with real satisfaction (Willey, 1984). Throughout the wedding ritual the emotions of sadness and joy are expressed by all. It is not uncommon for both the bride and her father to be crying while walking down the church aisle.

Family Ties

The family and the extended kinship group are the basic building blocks for externalization. Familial relationships tend to be close and emotional, and the behaviors found in the piazza are replicated within the family. The family is generally seen as one's greatest resource and protection against all troubles. For example, children are critically important, and normally everything is done for them, even to the extent of satisfying their smallest wishes. Parents often go without comforts to pamper their children and to see that they go to school and reach a higher rung on the social ladder. This attitude can be directly linked to the constant influx of foreigners who compete with the Italians for jobs, to rapidly changing governments, and to the overwhelming natural disasters that have characterized Italian history. Of course, externalization plays a part in this support, as the family views its children as public expressions of its values and lifestyles.

The Italians themselves see the irony and humor in having family relationships that seem to be excessive. A few years ago a best-selling book, *Mammoni* or "Mamma's Boys" was widely discussed in Italy, and there was even a segment about it on the U.S. TV news show, *60 Minutes*. It is estimated that 40% of Italians between the ages of 30 and 34 still live at home and males far outnumber females at a ratio of 2 to 1 (Peet, 2005). Some unmarried male Italians in this age group also live independently but come home for dinner every night. One mother ironed her son's clothes, including underwear, and shipped them to him weekly by bus; and one town was thinking about introducing a tax on unmarried males over the age of 25.

Furthermore, family connections are often extremely important for handling problems and getting ahead. Outside the family, official and legal authority is frequently considered hostile until proven otherwise. Closeness within the family is transferred to outside relationships and brings a personal edge to most social interactions. This helps to explain why Italians are so demonstrative and why men feel almost no hesitation to show affection to one another.

In the Italian business environment family contacts tend to be necessary to run a successful company. An example of the importance of contacts is the general lack of hiring policies. Hiring is usually accomplished through personal connections and recommendations. Many corporations select people not on the basis of their skills but on the basis of their relationship with employers and their families. Bond and Smith (1998) have reviewed national preferences for various selection techniques, and Italian business is distinctive in that the interview by itself is the preferred alternative, even though this method when used by itself has been shown to be the least effective of all techniques.

The family remains the center and stronghold of Italian life, despite all modern trends, where the role of each member is understood and performed as elaborately as any Italian opera. Although men are the official leaders of the families while women are subordinate to them, the reality of family life is much more complex. The main character in the family, who might be compared to the lead tenor, is the father. He is in charge of the family's general affairs. But while he holds center stage, his wife is an equally important figure, like the lead soprano. Although the Italian father is the head of the family, the mother is its heart. While yielding authority to the father, she traditionally assumes total control of the emotional realm of the family. The mother usually manages the family in a subtle, almost imperceptible way; she soothes the father's feelings while avoiding open conflicts. However, the woman of the house frequently has the last unspoken word. The factors that determine the strength of the family are placed in the hands of women. Wives engineer appropriate and convenient marriages, keep track of distant relations, and see to it that everybody does the suitable thing, not for individual happiness but for the family as a whole. The fact that women form the predominant character of Italian life can be seen through many small signs. For example, popular songs frequently highlight the role of mothers, and in some years there are more songs devoted to mothers than to romance.

Gender Issues

However, Italy tends to be a man's world. When a child is born, the proud parents tie a blue ribbon to the door for a son, but only sometimes do not-quite-so-proud parents of a girl put out a pink ribbon. The principle of male superiority is less strictly enforced in the north than in the south. For example, in some Sicilian villages, unmarried women are supposed to sit indoors during the day when unchaperoned.

Divorce and abortion have recently been legalized in Italy. Legal abortion symbolizes the loosening of individual morals and the breaking of the hold of the Catholic Church over the family. In 1974 civil divorce became legal, but it seems to be more a symbol of social independence than anything else. Not many marriages have actually ended in divorce. For example, there are only 0.8 divorces per 1,000, whereas the comparable figure in the United States is 4.8, second only to Aruba's 5.3. The number of separated couples, however, has increased significantly.

Many Italians view divorce as unacceptable because it chips away at the foundation of the family and entire clans. Others argue that it has not increased in popularity due to the fact that, without husbands, most women would be in dire financial straits. This argument is losing merit as industrialization ushers women into the workforce. Divorce, however, is still viewed seriously. If a man leaves his wife, sometimes the ex-wife will move in with the husband's family while he is ostracized, especially in the South.

The family frequently extracts everybody's first loyalty. Italians raise their children to be mutually supportive and to contribute to the family. Separation from the family is generally not desired, expected, or easily accepted. Things that create conflict in Italian families may be sources of celebration in other cultures. Some normally joyful events that may cause operatic sadness are a job promotion that necessitates moving away from the family, acceptance into a prestigious university in a foreign land, and marriage. While these experiences may be sources of personal growth for the individual, they are likely to be experienced negatively if they weaken the collective sense of family.

Many Italians view education and vocational training as secondary to the security, affection, and the sense of relatedness the family has to offer. Personal identity tends to be derived significantly from affiliation with the family, and also from one's occupation or personal success.

Family connections are extremely important for handling problems and getting ahead. Italians, not wanting to work for outsiders, often start their own family business. The business naturally adds to the solidarity of the family. Closeness within the family is transferred to some outside relationships and brings a personal edge to most social interactions. This helps to explain why Italians, like operatic singers, are emotionally passionate toward friends and extended family.

As noted previously, the north of Italy has been successful economically, and most of this success is due to the existence of small firms specializing in particular products. Moreover, these firms both compete and cooperate with one another. If one firm cannot honor an order, a neighboring firm will help it produce the product so that the commitment can be kept. At the same time firms will cut prices to move ahead of neighboring competitors. These types of close business relationships echo family and village patterns that are also expressed in the opera.

Personal Style

Sex for Italians, especially men, is seen as the essential life force (Newman, 1987). While honor is a quality all strive to achieve, virility and potency are still the basis on which many men are judged. This is confirmed by the fact that an adulterous wife is considered to be a direct reflection of her husband's manliness. For many Italians the ideal man is not necessarily intelligent or well off but rather of good character and physically and sexually strong.

Most Italians do not put much faith in what others say. Everyone is considered to be an outsider except members of the family. It is not surprising that Italians, living as they have always done with the insecurity and dangers of an unpredictable society, are among those who find their main refuge among their blood relatives. This overall fearful, suspicious attitude may be partially due to a quality on which Italians place high value, namely cleverness. Because of the constant change and struggle in the country, those who can survive through enterprise, cunning, imagination, and intelligence are held in high esteem by others around them. In other words, many Italians admire those who can create the most imaginative show. Minor deceptions, cleverly and subtly executed, are acceptable even if not necessary. Because everyone is trying to outsmart everyone else, the population as a whole is placed on the defensive. Visitors to Italy often comment on this feature of Italian life. For example, it is common practice to shortchange customers, sometimes significantly. If, however, the customer questions the transaction, the correct amount and profuse apologies will be offered immediately and without question.

While the family is the group that above all else dramatically influences the individual, another important group, the political party, can be the difference between employment and unemployment. Italians move from group to group, feeling little remorse when doing so because they tend to be skeptical of all groups besides the family. Members of groups often create powerful coalitions that are used to gain power and influence. These subcultures, groups, and parties allow for intergroup bargaining to create coordination and cooperation between all involved.

Sometimes these powerful coalitions lead to widespread corruption. The worst example of corruption in Italy's history was exposed beginning in 1992 when a company in Milan refused to give a kickback, a regular business practice that is estimated to add 15% to 20% to the final price. However, in this instance public officials vigorously pursued various leads involving large numbers of prominent people. More than 1,200 businesspeople were indicted and many served prison time, as did several Mafia members and two former prime ministers; some committed suicide in prison. While these and related activities seemed to weaken the Mafia, even more virulent and violent strains of crime became prominent in the south: the Camorra and the 'Ndranghera. There has been a backlash against this reform movement, even to the extent of limiting the use in Italian courts of evidence gathered abroad and decriminalizing convictions for false bookkeeping. Silvio Berlusconi, prime minister until 2007 and a well-known businessman, has had at least nine legal actions against him, and there are others still pending. He used his position as prime minister to change the laws so that he would be immune from prosecution.

Italians of many persuasions would like to see the emergence of a genuine two-party system that is normally associated with much less corruption. Still, the sense of family and coalition behavior is so strong in Italy that we can expect many of the behaviors described in this chapter to persist. For example, we can expect that many job referrals and recommendations will still be based at least in part on family connections, much more than in the United States.

Problematic Attitudes

Externalization also influences the management style of Italians. There is little delegation of authority or effective communication between the different levels of management in most Italian firms. Employees have little say about decisions concerning the company or about their own work. In fact, superiors are likely to cut off emotionally the ideas and suggestions of subordinates.

One final example of externalization is somewhat disturbing, as Italy is experiencing difficulty integrating the large number of legal and illegal immigrants flooding into the country, particularly those from the African nations of Morocco, Nigeria, Senegal, and Tunisia. The poorest Italians are most affected in that they feel that these immigrants are taking jobs away from them. There have been some disturbing events such as the burning of immigrants and the hanging of a 16-year-old immigrant. Many Italians are upset by this trend and, in true Italian fashion, have externalized it dramatically in newspapers, on television, and in the traditional piazza. Although such outbreaks of violence occur in many societies, these Italians are confronting the problem in an open manner involving everyone in the society, and solutions are gradually being implemented.

Externalization is directly related to the mixing and balancing of diverse elements in both the opera and in the Italian society. Because the entire community or audience is involved in the unfolding of the drama, several issues and factors come into play that involve numerous individuals. It is not an accident that the Italians have not specialized in one-person stage shows or dramas. Thus Italy, while primarily stressing individualism, is oriented toward specific aspects of group behavior or collectivism. Although individuals might not give the country as a whole much consideration, they place great importance on local and regional affiliation. This is directly related to the next characteristic of the opera, the influence of soloists and the chorus.

Chorus and Soloists

When Italians created the opera about 1600, there were no soloists. The first great Italian composer, Claudio Monteverdi, basically used different parts of the chorus to express the ideas and emotions. However, singers vied with one another for the spotlight and gradually soloists became prominent. Today it is difficult to conceive of an opera without great solo performances, which may help to explain why Monteverdi's operas are so seldom performed. This tension and balance between the chorus and soloists epitomize the struggles between regions, as each region retains certain easily perceptible characteristics among its inhabitants, even if the major division is between the wealthier north and the poorer south. Most Italians define themselves by the town where they were born. These regional differences are similar to the chorus and soloists in operas. Soloists represent the regional differences in the culture whereas the chorus is the embodiment of the overall

Italian culture. Still, in spite of the fact that there are distinct regional differences, most Italians retain the cultural characteristics described above.

The Italian word that expresses the idea of belonging first to a town, then to a region, and third to a nation is *campanilismo*, derived from *campanile*, which means "bell tower." It refers to the fact that people do not want to travel so far as to be out of sight of the piazza church steeple.

Because of industrialization northern Italians have had the benefit of a thriving economy and a relatively prosperous existence. In contrast, southern Italians, who have relied on farming for their livelihood, have tended to be poorer and less educated. The people from these two regions are similar in that they love life and enjoy creating the illusion of a show. However, the difference between the people from these two regions is that southerners tend to cling to past ways of living whereas many northerners look toward the future.

To most northerners, wealth is the way to ensure the defense and prosperity of the family and close friends over the long term. Northerners are perpetually trying to acquire wealth in its various forms. They want a job, a good job, and then a better job. They also want the scientific and technical knowledge that will assure them better paid employment and advancement. Many southerners, on the other hand, want above all to be obeyed, admired, respected, and envied. They want wealth, too, but frequently as an instrument to influence people. Southerners are preoccupied with commanding the respect of the audience throughout the many operas of life. Many southerners, be they wealthy or poor, want the gratitude of powerful friends and relatives, the fear of their enemies, and the respect of everybody. Southerners seek wealth as a means of commanding obedience and respect from others (Barzini, 1964).

A Culture of Death

In the south the culture centers around death. The Italians worry that the operatic drama of life will fall apart when a family member dies. The experience of dying and the fear of not being able to react properly to such an event has forced southerners to create a complex strategy to enable them to face and conquer death (Willey, 1984). The strategy encompasses hundreds of beliefs, customs, and rituals. The purpose of these rituals is to reestablish contact between the living and the dead because, according to southern beliefs, the family includes both the living and the dead. Each family member has reciprocal rights and duties. The dead have to protect the living and the living have to keep alive the memory of the dead, and people tend to attach great value to these obligations. Thus, in the early 1980s a parish priest in one Calabrian village thought he would discourage the long local funeral processions by levying a special fee per kilometer on the family of the deceased. His bishop promptly transferred him to another part of Italy (Willey, 1984).

This distinction between north and south is easily found in Italian operas, which tend to depict the characteristics of one region or the other. For example, *Cavalleria*

Rusticana by Pietro Mascagni represents a classic southern opera with its emphasis on humble folk, suspicion, and revenge. In contrast, Giuseppe Verdi's *Tosca* emphasizes the north, as it takes place in elegant urban settings in which important personages display more subtle and complex feelings, although no less passionate (and perhaps even more so) than those found among the southerners.

Education has been a major mechanism for bringing about the modernization of Italy and minimizing the regional differences. By diffusing the national culture through the teaching of one Italian language and by raising literacy levels, Italians have stressed cultural unity and deemphasized regionalism to a greater degree than in the past.

Different Ways of Doing Business

Although many of the differences between "the two Italys" have been defused and minimized through time, business dealings differ in the two regions. Working with people from northern Italy, a low-context subculture that emphasizes written rules and agreements, is much like dealing with U.S. Americans or Germans. To negotiate effectively with northern Italians essentially means communicating with them in a straightforward sophisticated manner. Social talk should be kept to a minimum to get down to business. When negotiating with people from southern Italy, a high-context subculture in which oral communication and subtle nuances are stressed, visitors must spend time establishing rapport with their counterparts. Long-term relationships are important to southern Italians and trust must be built up before business dealings become truly effective.

In spite of these subtle differences, most Italians use a collaborative style of negotiating in that they will continue a dialogue until everyone's needs are met. Part of this style is due to the emotional nature of many Italians. Communication involves much more than words. The emotional nature of such Italians enables them to see past the words spoken to the emotions felt. This insight helps them empathize and understand the needs of those with whom they negotiate. Such aggressive yet emotional Italians want to please everyone, including themselves. Therefore, the collaborative style of negotiating is common throughout Italy.

Similar to the operatic importance of both the chorus and soloists, the regional differences combined with the overall Italian culture helped Italy gain economic strength during the 1970s and 1980s. This strength resided in the proliferation of small-scale commercial and industrial enterprises, usually family run. People whose families used to be southern farmers brought their deep-rooted traditions to the north. Some of these traditions and skills enabled many southerners to move their families from farming into the commercial market. The tradition of keeping ownership within the family, combined with the skills of managing a diverse product mix and willingness to work long hours, enabled the new entrepreneurs to compete effectively with their European neighbors. However, as noted earlier, in the modern era of globalization, such small firms face increasing competition from China and elsewhere and do not enjoy the scale of operations of large multinational corporations, which are becoming more and more common throughout the world.

Italy is facing major economic problems, and today its economy is only about 80% of the size of that of Great Britain, which it had overtaken in 1987; the populations of these two nations are very nearly equal in size. Some governments in villages and even large towns do not have the money to manage critical services. For example, in 2007 Naples experienced a garbage strike during which the high and extensive piles of garbage created a major health crisis. To maintain a leading role in the EU, Italy must change this situation.

Chorus and Culture

The characteristics of the chorus and soloists can also be applied to other aspects of Italian culture. Italians tend to be individualistic people, yet they place importance on the group. As we have seen, the family is the primary group that influences individuals. Whenever important decisions are made, the individual usually consults with family members to get their opinion and evaluation of the situation. Although the family's opinion is important, the final evaluation is made by the individual. This is similar to the relationship between the chorus and soloists in the opera. The chorus frequently gives the soloists the facts and opinions about the drama unfolding on stage, yet the soloists are the ones who dramatically decide how to handle the peril or situation, even when such decisions lead to disaster and tragedy.

The influence of the group is also felt in the business environment. In business meetings people will externalize their feelings and opinions about a subject. They will listen to everyone's ideas and freely give their own opinions. Business meetings tend to be productive in Italy due to the openness displayed by everyone. However, like the chorus and soloist, the decisions coming out of a meeting are frequently made by one or two dominant or domineering people. In spite of the influence of the group, Italians tend to be aggressive and materialistic individualists due to their bias toward spectacle and externalization.

Furthermore, the vocal sections of the chorus are directly comparable to the many regional variations of the Italian language. Although each section is important to the harmony of the chorus, each has its own melody. The Italian history of invasions has produced many dialects, each the product of the particular regional invader. The language of Italy consists of both local dialects and Italian, a derivative of the Tuscan language. Although linguistic variations are disappearing to some extent due to the homogenizing effects of television and telecommunications, Italians still cling to the notion of having their own regional dialect.

In the opera there is inherent inequity in the amount of fame given to each of the soloists. The same is true in the Italian economy. Today poverty still exists in some areas, particularly in the south. While Italian industry has been successful since the end of World War II, success is not shared equally by everyone. Because northern companies believe there is a difference in the northern and southern work ethic, they are reluctant to locate branches in the south. The large migration of workers from the south to the north has left behind people who are not willing to give up customs and traditions that have been followed since early European civilization. Still, although the Italian culture consists of many regional subcultures, the modernization of Italy has begun to unify these subcultures. The Italian culture is continuously changing to one where the melody of the regional soloists blends together into a harmonic chorus.

In this chapter we have not explicitly treated the Italian orientation toward time, space, and the other culture-general concepts (see Chapter 1). As the discussion of the piazza suggests, Italians tend to be more polychronic than monochronic, performing many activities simultaneously. In Geert Hofstede's (1991) study of the cultural values of 53 countries, Italy clusters with those countries emphasizing a large power distance between groups in society, that is, Italians tend to accept the fact that some groups are and perhaps should be more powerful than others, and they act accordingly. The GLOBE study (House et al., 2004) confirms this feature of Italian life in the daily practices that take place. Italians also try to avoid risk and uncertainty in everyday life, preferring friends over strangers and familiar over new or strange situations, as their behavior in the piazza and externalization would confirm. Although the family and kinship are important, Italians tend to cluster with those countries that are more accepting of individualism and aggressive, materialistic behavior, all of which reflect their externalized bias.

This, then, is Italy. It is still a grand and larger-than-life society whose citizens love pageantry and spectacle, emphasize a range of voices in everyday life, externalize emotions and feelings, and feel a commitment to the town and region of the country in which they were born. Italians have had a difficult history. The institutions have changed, rulers have come and gone, and people have to survive—and do—thanks to their personal, unofficial relationships. In Italy life is the theater, each act carefully played out for everyone to witness. From this perspective the opera is not only helpful but possibly essential for understanding Italian behavior and culture.

CHAPTER 22

Belgian Lace

It is clear that while the diversity of the cultures alive in Belgium may sometimes be a source of friction, this same diversity can and must also be above all a source of spiritual and material enrichment. For this very reason, and because Belgium is the common ground where two great European cultures meet, we must continue to be what we have always been in the past: Pioneers in the construction of a united Europe.

—The late King Baudouin, 1986
Christmas radio/TV message

Belgium is a small country slightly larger than the state of Maryland, with just over 10.3 million people. The distance between its farthest points is 175 miles. Given its size and population, the casual observer should be careful not to underestimate its high degree of importance and interrelatedness in Europe and the world. The bitter tensions between the two major ethnic groups, the Wallonians (French-speaking) and Flemings (Dutch-speaking), are considerable, and the manner in which they address and resolve them may well reflect in many ways the ethnic and national cultural tensions in the European Union (EU) considered as one entity. But just as the gifted lacemaker takes fine threads from many spindles and weaves them into a fabric with a beautiful but strong pattern, so too history, geographic position, and religion have woven the Belgian culture into a diverse and complicated design, similar to all of Europe. To understand Belgium and even some of the tensions within the EU, it is helpful to describe it in terms of the metaphor of lace.

Before doing so, however, let us explore its turbulent and fascinating history. For hundreds of years the people of Belgium, both the Flemish (of Dutch ancestry) in the north and the Wallonians (of French ancestry) in the south, were ruled by a number of other nations. Over the last 500 years Austria, Spain, France, and the Netherlands ruled, at one time or another, the area that is now Belgium. Although there were indigenous nationalistic movements that sought to create the nation of Belgium, it was finally established in 1830 thanks to the machinations of Lord Palmerston, arguably the most influential foreign affairs minister that Britain has ever produced. He wanted to create a buffer zone between the Germans and the French in the event of war between these two nations and in the process to provide England with some breathing space before she was forced to enter any conflict. Lord Palmerston's instincts were unerringly correct, as Belgium was the site of some of the bloodiest battles in both World War I and II, and without this buffer zone it is conceivable and even probable that Germany would have overwhelmed both the French and the English (Tuchman, 1962).

The Flemish and Wallonian cultures, existing side by side, have given birth to a sense of Belgian pride. At multinational or world-class competitions, a Belgian is a Belgian and the entire country unites to watch the performance, regardless of the side of the language border on which the participant or team originated (Schrank, 2008).

Wallonian Versus Flemish

Still, it cannot be denied that there are conflicts and tensions between the Wallonians and the Flemish. Unlike many other cultural groups living in one nation or area who escalate these differences into bloody confrontations and war— for example, the fighting between the Catholics and Protestants in Northern Ireland or that among the Croatians, Bosnians, and Serbians in the former Yugoslavia—the Wallonians and the Flemish have patiently addressed their differences in a systematic way since 1968. Today Belgium is a federation of states in which the central government has relinquished almost all of its power except in such areas as finance, defense, and international relations. There are, in fact, a bewildering number of parliaments: One is for the nation in its entirety; one for Brussels (an international city in the middle of Flemish-speaking Flanders, in which 80% of the residents nevertheless speak French); one for Wallonia; one for Flanders; one for the tiny German-speaking community; and one even for French speakers regardless of location.

Following national elections in June 2007, the country was without a federal government for a record 195 days because the prime minister-elect failed several times to form a coalition government that represented all interests. In 2006, when a French-language television program was interrupted with a spoof news flash announcing that the Flemish parliament had declared independence, the king had fled and parliament had been dissolved, it was widely believed.

Admittedly, this solution to ethnic and regional tensions is complex and has also led to some bewildering outcomes. In Brussels, the capital, some streets receive only

Flemish-language cable TV, while adjoining streets only French. Belgium's embassies abroad have three separate commercial attachés. Although this solution to governing amid ethnic tension may seem contrived and unduly complicated, it is far preferable than the alternative of war and destruction. As *The Economist* aptly pointed out ("Belgium: Fading Away," 1992), Belgium "deserves a clap" for avoiding the mindless ethnic wars and destruction from which other nations so grievously suffer. Protests are relatively civil and even tinged with some humor. One Belgian posted the country for sale on eBay but the listing was pulled after attracting a bid of $13 million. Flemish protesters slashed signs with names in both Dutch and French and paraded coffins symbolizing Belgium's demise.

Still, there is a serious movement among some prominent Wallonians to argue for national disintegration so that the Flemish region can unite with the Netherlands and possibly parts of France and Germany ("Could Flanders," 1997). This has caused concern in other European nations where there are also secessionist movements (e.g., Spain). Whatever the eventual outcome, Belgium serves as a useful model that other nations—and even groupings of nations such as the EU—with diverse cultural groups can study and at least partially follow to avoid such alternatives, and this model closely resembles our metaphor of Belgian lace.

History of Lace

Lace first appeared in the late 15th century in Flanders and Italy. It was originally used as a modest ornament for undergarments. By 1600, lace had become a fabric of utmost luxury and a key article of trade and commerce. Nobility from all over Europe began to enhance their clothing with lappets and cuffs of lace. Dresses were soon adorned with lace overlays and shawls. Beyond clothing, lace began to drape tables, windows, and bed linens. Lace was a sign of social status and wealth. As a lace-making center, Flanders established itself as a prominent industrial region within Europe.

Bobbin lace, first introduced in Flanders, requires the weaver to manage hundreds of bobbins in creating a single piece of lace. Systematically and rhythmically the threads from the 10 to 20 sets of bobbins are braided around brass pins skillfully affixed to a cardboard pattern to form a geometric or pictorial design. The design incorporates a delicate balance of lace and space that is best exhibited when the lace is placed on a dark background fabric such as deep-colored velvet or satin. During the Renaissance, lace-making emerged as both an industry and an art form, as the following passage confirms:

> Form became more important than color; instead of intricate effects of limitless expansion, designers stressed clarity, symmetry, and stability. In the decorative arts, the new spirit was most noticeably expressed in sharp distinctions between pattern, background, and frame. No minor art form embodied Renaissance ideals more fully than lace. (Benton, 1970, Vol. 13, p. 85)

Traditionally, the best lace makers were cloistered nuns. Emperor Charles V decreed that lace-making should be a compulsory skill for girls in convents throughout Flanders during the Renaissance. To create a square foot of lace with an intricate design may require hundreds or even thousands of hours. Legend has it that some nuns would spend their entire lives producing lace for the pope's robes. Although lace is no longer a predominant industry in Belgium, and machine-produced lace is the norm, the Belgians still take great pride in this traditional art form; most homes have at least one piece of handwoven lace on a table or displayed in a frame.

Many Belgians can hold up a piece of lace and tell if it was handwoven or machine made, what the relative retail value of the piece should be, and the region of the country where the piece was produced. The texture and pattern of the lace have regional characteristics. For instance, in Brugge and Turnhout the texture is delicate and the patterns are ornate, whereas in St. Hubert the thread is coarse and the design simpler. Hence lace from Brugge might typically be used as a table runner between meals, and lace from St. Hubert might adorn the edge of the tablecloth used during a meal.

We selected the famous Belgian lace as a metaphor for this country because it encapsulates the beauty of this land of contrasts where different cultures coexist in relative harmony, where individualism is compatible with the control of very strong family values, and where cooperation and harmony win out over competition and the people are quietly proud of their traditions and heritage.

A Land of Contrasts

The beauty of lace can be fully appreciated only when the contrast of woven cloth and dark background are recognized within the overall form. Just as lace interweaves contrasting elements of a complex design into a single structure, so too there are many contrasting elements interwoven in Belgium, a small country that includes three recognized languages, two major cultures, three separate official regions, three flags, two major economic centers with separate industries, and three unique geographic landscapes. The starkest contrasts for Belgium are linguistic and political. In addition, other areas of contrast include the following:

- The struggle between individualism and the need to belong to a group, conform to societal rules, and have strong family ties
- The dichotomy between working and social welfare
- The stark difference between urban and rural areas
- The desire for art and beauty, but also practicality

Still, just as a lace maker integrates contrasting designs in the masterpiece, the Belgian government attempts to integrate separate cultures into a unified republic, even when there are three separate parliaments in each of the three major regions.

A Land of Three Languages

Linguistically the Belgian government recognizes three native languages through regulations regarding the conduct of education and commerce within the areas populated primarily by native speakers. In Flanders, the northern portion of the country bordering the Netherlands, business and education are primarily conducted in Dutch, the native language of the Flemish people, who represent about 58% of the Belgian population. In Wallonia, the southern portion of the country bordering France, business and education are primarily conducted in French, the native language of the Wallonian people, who constitute about 31% of the Belgian population. In Limburg, a very small area bordering Germany with about 1% of the population, German is the native language. To be impartial and economically viable, merchants, educators, government officials, and others in the capital city of Brussels conduct business in Dutch, French, or English, at the mutual preference of the participants.

Brussels is at the crossroads of Europe, and many multinational firms have established European offices there in recent years because of the unification of the European Community in 1992. At any given time about 5% of the population of Belgium is foreigners.

In Belgian schools, children are first required to study the native language of the region in which they live for at least 8 years. They are also required to study a second language, which could be Dutch, French, English, or German, for at least 4 years. It is not uncommon to meet a Belgian, usually a Fleming, who is proficient in each of the four languages as well as others. However, Belgians are proud of their native languages and cultures. If two headstrong Belgians, a Fleming and a Wallonian, have to communicate, they will usually defer to English because the Wallonian will probably not be able or want to speak Dutch and the Fleming will not condescend to speak French to a Wallonian if it can be avoided.

In everyday life the impact of these linguistic differences varies significantly and may not be encountered at all by those who reside deep within Flanders or Wallonia. By contrast, Belgians living closer to the language border or in Brussels must deal with language differences daily. For instance, new regulations in some municipalities near NATO headquarters require that public land can be sold only to Dutch speakers or those who show a willingness to speak the language. One man reported that police failed to respond to his French-language request to investigate a disturbance. Belgians tend to appreciate the use of their languages and will generally warm up to outsiders more quickly if an attempt is made to learn at least one of them. Although many Belgians deal with the language differences every day, it is in poor taste for a foreigner to point out the differences or to wonder aloud why the differences cannot be resolved through diplomacy.

The oldest Catholic university in the world, the University of Leuven, located 20 miles east of Brussels in Flanders, illustrates the consequences of the language conflict. For hundreds of years the university served both Flemish and Wallonian students and scholars, using French as the official language on campus. After World War II linguistic tensions began to mount, causing tensions between student groups and sporadic scuffles in cafés around the university. For more than 20 years the hostility evolved to the point that the university was divided into two campuses: one in Leuven (Flanders) and one on the other side of the language border in Wallonia. This separation divided not only students and faculty but also resources such as the library collection. Books were literally divided by title, with those in the first half of the alphabet staying in Leuven and those in the second half going to the new campus. Conducting research now necessitates visiting two campuses, but this cumbersome and inefficient solution is preferable to a long-standing linguistic conflict.

Region Versus Nation

As an extension of the dual languages and cultures of the Flemish and Wallonians, the country has three prominent flags: the tricolor of black, yellow, and red, representing the Belgian country; the Flemish lion; and the Wallonian rooster. Most Belgians identify more emotionally with the flag of their respective cultures than with the tricolor flag of the nation. The Flemish lion is the symbol of the Flemish resistance to French domination over hundreds of years.

While regional and ethnic identification is clearly important, Belgians do identify themselves as Belgian when outside the nation and come together on critical social issues. For example, 300,000 Belgians—as a proportion of the total population, the equivalent of 7.7 million U.S. Americans—rallied against the seemingly careless and perhaps corrupt manner in which the judiciary handled a chilling case involving a pedophile ring that had killed four young girls.

Belgium is a constitutional monarchy, divided into three republics: Flanders, Wallonia, and Brussels. Each republic has its own representatives in the central government or national parliament and its own divisions of the national political parties, which are also divided into subparties by language. Each subparty operates autonomously yet generally follows the same political agenda as the parent party.

The two largest political parties are the Christian People's Party, whose members are primarily Flemish and Catholic, and the Socialist Party, whose members are less strict Catholics and non-Catholics and more likely to be Wallonian. These two parties are also responsible for publishing most of the major newspapers in Belgium (five of seven) and supervising social facilities for youth, sports confederations, care of the aged, hospitals, savings banks, and unions. Much of the workforce is unionized. The successful administration of the trade unions and other social programs is heavily dependent on social stability. Due to the extensive influence of these two political parties in the life of the average Belgian, other political parties have not gained a significant foothold. Although all Belgians are required to vote in every election, they are free to vote for candidates of any political group.

However, a right-wing, anti-immigration, secessionist political party, the Vlaams Blok, created in 1978, started gaining popularity in the 1990s. On November 14, 2004, the party changed its name to *Vlaams Belang* (Flemish Interest). In several European nations (e.g., France and Spain) similar political parties have been campaigning against the free entry of immigrants into their nations.

Individual Versus Group

Furthermore, Belgians tend to be very individualistic. Hofstede's (1991) cross-cultural study ranked Belgians as eighth out of 53 nations on this dimension. It is considered a major accomplishment for young adults to become established in their own houses, apart from the family. Hard work is also valued, as it is seen as providing an incentive for individuals to get away from the home. Still, even when the young leave the immediate family, they do not venture far geographically, usually living within 30 miles of their parents' home. This practice creates a strong family/community tie, essentially enhancing the family orientation of the Belgians.

Belgian individualism is also displayed in their preference for owning their homes. Despite the great expense, about 65% of Belgians own the place where they live, residences that range from condominiums to suburban-country homes. A person's home is his castle, and Belgians tend to take great pride in the cleanliness of their homes. Early in the morning it is common to see people sweeping their porches and walkways. While a U.S. American might use a garden hose to wash down a sidewalk, a Belgian would typically use a bucket and a scrub brush. This preference may be traceable to the historical shortage and high cost of fresh water. Belgians also tend to take short baths; use soapy water to wash the dishes and then wipe the soap off with a towel, rather than rinsing them with water; and drink soda water, juice, beer, or wine rather than tap water or anything made with tap water.

Just as each region of Belgium has a distinctive style and texture to its lace, so too are condominiums and row houses often organized and decorated distinctively, reflecting the tastes of their owners. Unlike the case of England's traditional houses (see Chapter 17), the interior layout of a Belgian row house cannot be assumed from its exterior appearance. In Flanders it is common for extended families to gather together to construct the exterior brick walls of a new or remodeled home.

While Belgians tend to be individualistic, their family orientation is strong. The family plays a central role in daily life. Family activities and meals are highly cherished. Most Belgian children go home after a long day at school, do their homework, eat dinner, help with chores, and spend time with their families. Family time often includes reading or watching television. On TV they can see programs broadcast in French, Dutch, German, and English, usually with subtitles in a second language. Because the state owns the Belgian television stations, there are few commercials.

Family Life

On the weekends many Belgian children participate in scouting programs. Sunday lunch is usually the biggest meal of the week. The meal will normally take place at the same relative's house every Sunday, and family from the immediate area will all attend. If a husband and wife live close to both in-laws, they will frequently rotate Sundays between the two families.

Mothers often work. Out of a total workforce of 4.4 million, 1.9 million are women; however, women typically are found in traditionally "non-male" occupations such as nursing and secretarial jobs. Few have completed a university education or occupy top management positions. Day care is heavily subsidized by the state.

Families often spend weekends and vacations together, traveling to the Belgian coast for short stays and to Spain or the Riviera for longer stays. In the warm months camping in the Ardennes (mountains in southeast Belgium) is a frequent pastime.

When a couple marries, the woman takes the name of the man and connects it with her maiden name, so that if she was a Smith and married a Jones, her married name would be Mrs. Jones-Smith. Especially in Wallonia, it is not uncommon for a couple to live together until the woman becomes pregnant with their first child, at which time the couple will revert to Catholic tradition and get married in church. Due to the Catholic influence divorce is rare and only recognized after years of separation. However, the economy, more than the Catholic Church, has influenced the size of families. The socially correct family size is two children, and a family has received a perfect gift from God if the two children are a boy and a girl.

In the family it is typical for the father and the eldest son to be the ultimate decision makers. Mothers usually administer discipline and rule household matters. Since about 1970, much has been done to emancipate women through legislation and political decision making. Although legal equality has been mostly achieved, it will take time to eradicate the convictions and prejudices that have been nurtured for so many years. Moreover, the gender-based segregation of half of the country's elementary classes (primarily in parochial schools) reinforces the traditional view of women's role (Verleyen, 1987, pp. 189–192) although this is changing.

Aside from family, over their lifetimes Belgians will have only a few truly close friendships. Unlike the United States, where people tend to be quick to establish friendships, the Belgians generally require a great deal of time before relationships mature. Most people with whom Belgians interact regularly will remain simply acquaintances, although they may have known each other for years (see also Chapter 12, "The German Symphony").

Consensus and Compromise

Belgians normally hold their right to privacy and their own opinions sacred. As lace makers have a range of motions with which they pull from their bobbins and weave their threads around the brass pins, so too the Belgians seem to have similar ranges in their varied activities. Still, rarely does a Belgian stray far from the group by taking an extreme position. Belgium is generally a country of consensus, compromise, and cooperation. However, eccentric artists are revered as the embodiment of the eternal voice of protest, as long as they are not hostile or confrontational. For the individual, however, exclusion from the group can entail loneliness, pain, and more often than not, material poverty (Verleyen, 1987, p. 119).

Overarching the individualism of each Belgian is a set of common political rules that have governed public policy and conduct for decades. This set includes the following:

- The monarchy and Brussels are untouchable; in other words, all else could change, but there should still be a monarchy and Brussels should still be the capital.
- European unification and the EU are to be advocated without any reservation. As noted above, however, the political party, Vlaams Belang, now advocates the breakup of the EU.
- The authority of NATO and the American ally may be the subject of frequent criticism, but they also enjoy the privilege of inviolability.
- Prosperity is to be shared and not to be used as a justification for confrontation or class struggle.
- Labor strikes should not be allowed to jeopardize the existing economic order.
- Privacy and personal liberty are not to be impeded in any serious ways (Verleyen, 1987, pp. 147–150).

The varied types of occupations and pastimes enjoyed by the Belgian population present yet another area of contrast. Belgium has a long tradition of agricultural activity. The central area around Brussels has rich soil and flat terrain conducive to farming, yet only 2% of the workforce is agricultural and only 1% of the Gross National Product comes from farming, while the population is increasingly concentrated in urban areas and engaged in industrial and service activities. Nonetheless, Belgians vigorously defend and subsidize their traditional agricultural base. Although Belgium has the 17th-highest population density among nations at 337.5 per square kilometer, urban planning has helped to ensure that most urban areas are pleasant.

Industry and Business

In the past few years a great debate has raged over national subsidies to the mining and steel industries located particularly in the Wallonian region. Many believe there is no prospect of these industries ever regaining their former status or profitability. Many assert that governmental support should be rechanneled to help expanding industries where Belgium can compete in the EU and worldwide.

The contrasting fortunes of the north and south of Belgium have undergone a dramatic change over the last century. In 1830 French was the official language of the

Belgian kingdom. The nobility spoke French, government proceedings and church services were conducted in French, and the wealthiest citizens, Wallonians, spoke French. Therefore French came to be perceived as the cultured language. Conversely, the Flemish were perceived as the uncultured and poorer members of Belgian society. Some Flemish families remain embittered over stories that their ancestors fighting as foot soldiers in World War I died simply because they could not understand the commands of their French-speaking superiors. As fortunes shifted from the coal and steel industries in Wallonia to the shipping and international trade industries of Flanders, the Flemish found themselves in a position to demand recognition of their language and cultural heritage.

Over the last decade Flanders has continued to prosper, while most of Wallonia has been devastated by unemployment averaging over 15% and by depression. Today, as *The New York Times* observes in the Belgium country profile, old factories dominate the gray landscape where once "a kind of provincial snobbery was polished to a fine sheen" ("Belgium," 2009).

Belgians must attend school from age 6 to 18. Programs are rigorous, and most children have at least 2 hours of homework every night after 8 hours of school. To pass from one level of education to the next, each student must take an exam. The results of the exams determine what schools the student may attend and what vocation the student may undertake. As in Germany, students begin studying for a vocation in their early teens. By the time they come out of secondary school, most begin internships in their chosen vocations or go straight into college. Belgians are highly educated and skilled craftspeople. Thus, the high rates of unemployment are disconcerting and depressing for many of them.

Immigration Issues

In general, Belgians are hard-working people, just like lace makers, who may work nearly 1,000 hours on an intricate piece of lace that measures only 1 square foot, investing not only time but pride in the quality of their work. Because of the great pride that they take in their work and craftsmanship, regardless of how depressing the prospects of long-term unemployment might be, many Belgians would prefer to draw unemployment over doing menial work below their level of training or skill. Therefore immigrants from third-world countries are viewed as necessary to correct a temporary labor imbalance.

Typically immigrants are brought in from North Africa and Turkey to work in the mines and other menial jobs that Belgians reject. These immigrants are given access to many of the social programs such as medicine and government welfare, but they are ineligible to vote and are rarely allowed to become citizens. Although Belgians tend not to warmly welcome third-world foreigners, antiracism legislation makes discriminatory public expressions against immigrants punishable by law. Furthermore, although the Belgians feel that third-world immigrants are only temporarily needed, long-range forecasts show that the Belgian population is shrinking because of low birth rates and emigration while the immigrant population is increasing. As mentioned earlier, social tradition has led most Belgian families to have two or fewer children despite the fact that the Catholic-influenced legal system prohibits abortion and most forms of birth control.

Overall, Belgians believe that people have a need and desire to work. In an effort to minimize unemployment there has been an effort to reduce the workweek, thus allowing a few more workers into industry. Retirement age is already set at 60 years, and all Belgians have at least 4 weeks of annual vacation paid by the government. The call to reduce working hours further, and even radically, is related to a strong belief in trade union circles that the redistribution of available work will help force down unemployment rates (Verleyen, 1987, p. 231).

Today, most Belgians are forward thinking in their economic activities and are moving heavily into service and diplomatic industries in response to their native skills and worldwide economic demands.

Urban Versus Rural

Another sharp and abrupt contrast is between urban and rural settings. The Belgian countryside is green and lush due to the abundant rainfall and high humidity throughout the year. As one drives through Belgium, long stretches of rolling green hills with little white farmhouses can be seen. The older farmhouses have the barn on the first floor and the family lives upstairs; the farmer with his large workhorse plowing his field is a frequent sight. At times like these, it is hard to imagine that Belgium ever entered the 20th century.

The cities typically look like medieval fortresses, many with a wall surrounding them, interrupting the gentle slope from urban to suburban to rural. There are no U.S.-style suburbs in Belgium. An aerial view at night would show roughly circular areas of dense lights surrounded by thin networks of lights along major highways and a few spots of light in rural areas. This view might resemble the contrasting lace made around Brussels, which has densely woven patches that are delicately connected by threads to other densely woven patches.

Inside the cities, one finds further contrasts between the modern industrial facilities and the grandeur of the older architecture. In Brussels, the Atomium, a large structure built during a recent world's fair, resembles a giant silver atom. It has elevators that take visitors up for a view around the city. The NATO buildings, among others, are modern and efficient. By contrast, the Brussels town square, the Grand Place, is surrounded by cobblestone streets and buildings with ornately carved facades, and it draws the visitor back to the city's origins in the 16th century.

In recent times, there has been an active movement by more than two dozen groups to reenact life in the Middle Ages. It is so strong that, "juvenile delinquents in Flanders have been allowed to atone for their misdeeds by making the Chaucerian pilgrimage to Santiago de Compostela, in northern Spain, 1,250 miles, on foot, carrying backpacks and accompanied by a guard" (Bilefsky, 2007a). According to Herman Konings, a Belgian behavioral psychologist who studies national trends, the medieval craze can be attributed to a strong desire to return to a more glorious past. The country, divided between Dutch-speaking Flanders in the north and French-speaking Wallonia in the south, is experiencing anxiety about its identity.

Beauty Versus Practicality

A final area of contrast in Belgian culture is that between their love for art and beauty and their sense of practicality. Obviously most lace is beautiful and artistic, but it is also designed to adorn clothing and linens rather than to be simply admired. The artistic history of the country is evidenced by the prominence of museums, theaters, and artistic venues. In Belgium there seems to be a museum for everything. The largest is the open-air museum at Bokrijk. Rural buildings from various centuries have been excavated from all over the Belgian area and painstakingly rebuilt here. Many retirees dress up in the authentic costumes of the times and display the tools and crafts of the era from which the buildings come.

The Rijksmuseum in Antwerp has one of the finest collections of Flemish art in the world. Flemish art was the world standard from the 16th through the early 18th centuries. Exemplified by Brueghel, the van Eycks, and Rubens, Flemish art started with the vivid colors and figures of the Italian Renaissance but made the people less elongated, plumper, and more realistic. Typical signs that a painting comes from the Flemish era are rosy red cheeks and the presence of lace on the clothing in portraits or on the table in a still life.

Inside many of the oldest churches in Belgium, one can find wonderful works of art, most with no apparent security systems attached. For example, it is not uncommon to locate a work by Rubens, van Eyck, or van Dyck hanging in an obscure village church where the doors are never locked. Some of these churches appear old and worn on the outside, but inside they have beautifully carved vestries and pedestals for the priests to stand on while preaching. Hung around the inside walls of the church are artworks representing the 12 Stations of the Cross. Most of the churches were constructed so that the lighting is primarily natural and from candles. The art here was created to inspire the worshipper rather than simply to be admired, so the concept that someone might steal these works from the churches is absurd to the average Belgian. The oldest churches are all Catholic, each having a statue of the Virgin Mary with blazing little white candles all around her. When Belgians, even non-practicing Catholics, are troubled by work or family matters, one will frequently hear them tell how many candles they have lit for Mary.

The study of artistic history is strongly supported by the school system. School children flood the museums throughout the school year for lectures and sightseeing. Early lessons stress the importance of Belgian artists, nearly to the exclusion of others. This emphasis on art is initiated at a young age and is carried forward throughout most Belgians' lives. The people of Belgium do seem to have painting "in their blood" and seem to like hearing themselves described as a nation of painters. Belgium is proud of its artistic past, and not without reason.

Another example of Belgian practicality is the site where the Battle of the Bulge was waged in World War II. This battle was a major victory for the Allied forces and has been a major tourist attraction for many years. The site is marked by a large circular monument with pillars listing the names of the dead in that battle, resembling a modern Stonehenge. The monument is surrounded by rolling hills that, during tourist season, are meticulously covered with rows of white crosses to mark the resting places of the dead. After tourist season (September through April), the crosses are removed and cows are pastured on the hills.

Control

Lace production requires strict and complex controls to avoid needless flaws. In addition, there must be balance between the spindles and the thread and between the patterns in the cloth. This emphasis on controls and balance is similarly found in the Belgian preference for controlling behavior, interacting with friends and colleagues in familiar situations and surroundings, and leading balanced lives. In Hofstede's (2001) cross-cultural study of 53 nations, Belgium ranked sixth in terms of avoiding uncertainty. There are several areas where Belgians tend to exhibit this high degree of control and uncertainty avoidance, including lifestyle, transportation, social conventions, rules and procedures, and stress management. Belgians are regulated and suspicious of anyone who would change their adopted schedule or routine. Most Belgians have a deep-rooted attachment to what is called a "3 × 8" time schedule: 8 hours for work, 8 hours for play, and 8 hours for sleep. Modern management proposals to change the organization of work so as to increase industrial output are frequently opposed. Many who own small businesses work longer hours, but they also tend to live over their shops so they are never far from home and family. Most shops and businesses are closed on Sunday, although since the Catholic Church changed its policy to allow parishioners to attend High Mass on Saturday evening, more Belgian businesses have Sunday hours.

Transportation

To avoid uncertainty, the political and linguistic borders are well-defined and mapped. Savvy travelers can usually determine which region they are in by the first reference on the city and street signs. However, this can all be very confusing to the visitor, as a sign may list two or more names for the same place, for example, Antwerpen (Dutch), Anvers (French), and Antwerp (English). In the city most maps and signs will say Antwerpen. However, if the map consulted came from Wallonia or France, the city would be marked Anvers.

Belgium has several large ports, including Antwerp, which is the second-largest port in Europe. To accommodate the huge amount of trade from its port cities, Belgium has designed the most immense transportation system in the world for a country its size. Like the threads in a piece of lace, the highways, railways, and canals in this system connect every city and village in the country, allowing goods to flow efficiently all over Belgium and into the rest of Europe.

Major roads are constructed in a hub-and-spoke fashion. There are usually roads encircling larger cities with spokes extending out toward other major cities. These highways have many signs and directions in various languages designed to accommodate industry and travelers. As one drives along the spokes from one city to another, the name of the road will change halfway between the cities. Should one take the Mechelen road from Brussels to get to Mechelen, halfway along the route the name of the road will change to the Brussels road, indicating that the traveler is now closer to Mechelen than Brussels. Changing road signs can severely complicate the trip of inexperienced travelers, especially when the languages change. However, this practice is orderly to the Belgian mind, just as the pattern and texture of a piece of lace is set by its origin.

As one drives over the cobblestone roads of the inner city and the occasional dirt roads in the country, there are no stop signs. It is assumed at every intersection that the driver to the right has the right of way, no matter what the circumstances. Therefore, technically at every intersection drivers need look only to the right before proceeding, and Belgians will sometimes grow quite impatient with travelers who stop to look in all directions before proceeding.

One of the most interesting jobs keeping some Belgians away from the unemployment office is that of road worker. Many of the city streets are paved with cobblestones. To maintain these streets, road crews can be seen working on a small area at the side of the road, digging up the cobblestones, placing them in a pile, cleaning them carefully, and then relaying them. However, the stones are not simply replaced in a haphazard fashion. Rather, they are laid in an orderly pattern similar to the manner in which lace is arranged, often with some artistic, geometric design. Hence this job, although seemingly menial, appeals to the Belgian sense of industry, beauty, and practicality.

Social Rules

In social settings, Belgians are generally formal in their interactions with others and employ a number of explicit rules. For instance, when greeting others, the surname is the proper way to address all but close friends. The casual manner in which first names are used in many countries would probably offend most Belgians. This formality is common in business and social settings, even with neighbors and acquaintances.

Formal titles are used in many business and social settings in lieu of the full name; for example, *Monsieur* or *Mijneer* for a man, *Madame* or *Mevrow* for a married woman, and *Mademoiselle* or *Jevrow* for a single woman. It is important not to use

Monsieur or *Madame* with a Flemish speaker; instead, Mr., Ms., Miss, or Mrs. is the preferred form. Titles are also used in many subordinate-superior relationships, such as student to professor or employee to boss. Both the French and Dutch languages incorporate formal and informal manners in which a person can be addressed. Informal address is used when addressing a child, but adults generally need to know one another well before communicating in this way. Among younger people and in business with English-speaking partners there is, however, much less formality and it is common to use first names.

Greetings are also an extremely important part of the social ritual. Handshakes are the norm unless the two parties are good friends. A person will typically shake hands with everyone during the greeting and again before leaving the meeting or event. Belgians will often lament if they do not have a chance to shake hands before parting. In addition, women greeting either men or women in an informal manner will generally kiss three times on alternating cheeks.

Belgians are conscious of social rituals such as gift giving. Gifts are given at any type of social visit, and not to do so is seen as rude. However, gifts are normally not a part of business relationships. A typical gift when one is invited to dinner is flowers (except chrysanthemums, which signify death), a small box of chocolates, or an unusual fruit such as a pineapple. If the visit extends to a number of days, it is common practice to offer a gift for each day spent. Acknowledgments and thanks are made in writing for any gift received, even birthday and holiday cards. In Belgium, New Year's cards are sent in lieu of Christmas cards.

Almost all social gatherings are somewhat formal. When alcoholic beverages are served, it is customary to wait until everyone is served and then, with some ceremony, to raise one's glass to each person in turn, catching his or her eyes and wishing him or her *gezondhiet, santé*, or the like. Likewise, it is customary to wait for everyone at the table to be served before eating. Just before taking the first bite of food, everyone says *smaaklijk* or *bon appétit* or, in other words, wishes everyone else a tasty meal. Invitations are frequently sent to dinner guests, even to close friends, and a formal four-course meal is normally prepared. Afternoon tea can also be a formal event.

People, including close friends and family, seldom drop by another's house without first telephoning. Neighbors also observe this protocol, maintaining a high level of formality when interacting. All this formality, however, does not keep Belgians from visiting each other. They normally love to socialize; they just like to plan ahead when someone is coming. Their plans often include going to the bakery on the corner to get a special dessert to serve with tea, which may be traditional tea or herbal tea.

When one enters a Belgian home, there is generally an unheated entryway with stone or linoleum flooring where coats are hung and wet shoes are kept. By contrast, the living area of the house is generally warm and cozy. Thoughts of putting on a cold coat and wet shoes may explain why guests stay so long on winter nights.

Belgians tend to require certainty and privacy in social settings as well as in everyday life. In Belgium almost every door has a lock. In a home or a business this means that even cabinets and closets can be locked. In business settings, it is appropriate to knock on a door and wait for an answer and also to keep doors closed in the office. A home-owner may need 50 keys to guarantee privacy and security. In the older buildings of some universities, there is a light above the professor's closed office door. When a student wants to see the professor, he or she must ring a bell, in response to which the professor activates the light's color to indicate whether the student should come in (green), wait a few minutes (yellow), or go away (red). While this may seem extreme, it appeals to the Belgian sense of practicality and, in fact, the same system is used in some northern European nations to some extent.

Public Gatherings

Certainty is also highly desirable in social agreements. For Belgians, their word is their bond and promises must be kept. However, beneath all the formal rituals and procedures, the Belgian people tend to be quite friendly, laid back, and good-natured. To see this side of their personality, one can go to a local pub and observe the closeness of the relationships and abundant consumption of beer. Beer is not only a cherished drink but a source of pride for Belgians. Most Belgian cities have at least one brewery, and the smell of the hops hangs in the damp winter air. Belgians will make beer out of most any grain and wine out of most any fruit.

Another place where Belgians are sure to congregate is at the local *frituur* or french-fry maker's shop. In most neighborhoods it is a place to escape from the cold weather and to learn all the latest gossip. Belgian french fries, *frits* (pronounced "freets"), are thickly sliced and fried once at low heat long enough to soften them. Then they are allowed to cool completely before being fried again in very hot lard. The result is a crispy outside with a soft inside, much like the bread in the bakeries. Frits are served in a paper cone with a choice of 20 or more toppings, including stew meat (usually horse meat with a beer gravy), curry sauce, and "sauce American," which is ketchup. They are as popular in Belgium as hamburgers are in America.

Before lace makers arrange the spindles and thread to begin weaving a piece of lace, they first design the pattern on cardboard and carefully punch brass pins in the card-board pattern to guide the thread during weaving. Similarly one manifestation of high uncertainty avoidance is that Belgians have regulations and customs that act as brass pins to guide behavior in most situations. They tend to dislike the government and police officers but to love order, so they tolerate the intrusion. Dealing with government offices is tedious. There are specific papers and experts for everything. First one must figure out what paper to fill out for the needed service, and then the trick is to figure out what building and what window has the expert who can process the paperwork.

A foreigner moving to Belgium has 30 days to register and get a resident visa from the city that governs the area where the foreigner will be living. Then the visa must be renewed every 3 months, and a new one must be issued whenever the foreigner changes addresses. Citizens of the EU are not required to carry visas, but most

Belgians carry personal cards or business cards with their family name, home address, and telephone number, which they can exchange or leave with a note on the back if they missed someone on whom they came to call.

As noted earlier, the complex federated governing structure of the nation causes confusion, especially for the newly arrived foreigner. The city of Brussels, one of the two capitals of the European Parliament, is technically not equivalent to the Brussels region, although geographically they are the same. It is understandable why some visitors call Brussels the capital of confusion (Waxman, 1993).

Living With Stress

Hofstede (2001) demonstrated a strong correlation between high uncertainty avoidance and high anxiety and stress. This characteristic is manifested in Belgians in several ways. One is that they are a "doing" society: busy all day long, typically rising early. Many Belgian sayings refer to the judicious use of time, such as "Time is money"; "Make hay while the sun shines"; and "Time heals all, so get back to business" (Verleyen, 1987, p. 55). They try to prepare for everything so that nothing can go wrong. Much of this feeling may be attributed to their geographical location and historical lack of control over invading armies. As noted earlier, modern Belgium has been a country since 1830. The North Sea on the west is Belgium's only natural boundary. Therefore any emperor conquering the continent generally started in Belgium (the Low Countries) and moved on from there, as did Bismarck in the 19th and Hitler in the 20th century.

The religious orientation of the Belgians also contributes to their need to be constantly busy. Catholic doctrine places high importance on good works and personal industry in this life to assure a comfortable station in the next. About 75% of Belgians are Catholic. Historically, Belgium produced more nuns and priests per capita than any other nation in the world. Although religious intensity has declined since World War II and most Belgians do not attend church regularly, a majority are still educated in Catholic schools.

Because Catholicism is the official religion of Belgium, Catholic parishes, priests, and schools are heavily subsidized by the state. Freedom of religion is guaranteed in the Belgian constitution, and most major religions can be found somewhere in Brussels. But for most Belgians, religions other than Catholic are for foreigners.

One result of the great effort that Belgians make to follow tradition and avoid uncertainty is that they frequently have minor breakdowns. It's not uncommon to call on them for an appointment and be put off, because today they are having a *zenuw inspanning*, a Dutch phrase translated as a nervous strain. Typically this means they overslept, overexerted themselves, or generally aren't prepared to face the world for a while.

When they feel a breakdown coming on, Belgians will call a doctor who will routinely prescribe tranquilizers and bed rest. In Belgium medicine is socialized and easily obtainable, although quality is often questionable. The health care infrastructure is immense, given the size of the population. Home visits by doctors and nurses are common, and paramedic support is widespread.

Cooperation and Harmony

The lace maker's primary thoughts while weaving a strand of lace must always focus on the balance and symmetry of the design. A careful balance is maintained to ensure that neither the lace nor the linen overpowers the other. The overall effect should be one of harmony, with no stray ends or spaces to mar the unity of the structure. Each movement of the lace maker may be quick and concise but always within a small range of motion. There are no extreme movements.

Furthermore, lace itself is a subtle and unobtrusive material. Traditional lace, being all white or ivory, is used to neutralize the bold colors of background material, to soften the hardness of wooden tables, or to balance an object visually. The carefully balanced nature of lace with its neutralizing and non-extreme tone is reflected in the Belgian approach to the world, especially in continuity, statesmanship, and modesty.

Belgians exert a great deal of energy maintaining continuity, the status quo, balance, and a set standard of living for all. They normally work to ensure that their internal cultural and linguistic differences do not disrupt daily life, squelching conflicts by remaining neutral or by reaching mutual consensus among the involved parties. Throughout its existence Belgium has had linguistic tensions of varying intensity. Yet these tensions have never mounted to violence or bloodshed. When necessary, concessions were made and tempers were defused so that tensions would be carefully balanced. In lace making, this means the threads are held tightly to reduce both the slack and the potential for unnecessary gaps or holes that might weaken the durability and quality of the cloth.

The Middle Road

Although Belgians can do little to control the weather in their country, the moderate climate seems to reflect their approach to life: the middle road between extremes. Due to the gulf streams off the coast, the climate is mild, ranging from about 35°F in the winter to about 75°F in the heat of the summer. The weather is frequently cloudy and/or rainy, with little snow and few fully sunny days. However, a winter day with clouds is preferable to a cloudless day because the clouds tend to insulate the land.

The middle road between extremes is also where Belgian managers can usually be found. Subordinates will refer to managers using formal titles and last names; however, managers will seldom stand out because of appearance, conspicuous consumption, or attitudes that might separate management from line workers, as frequently happens in the United States. Again, to Belgians wealth is to be shared rather than flaunted.

Furthermore, the Belgian balancing and peace-making approach to life is replicated in their competitive (or not-so-competitive) nature. Belgians tend to be cooperative to a greater extent than they are competitive. Their social and economic structure was designed in a fashion that actually lessens the competitive drive. For instance, the socialized system for health and welfare provides full health care and monetary subsistence for all citizens. Furthermore, there is no time limit on eligibility for unemployment, as long as the person appears at the appointed time each week at the unemployment (labor bureau) office and regularly applies for openings in her or his stated occupation. The appointed time changes each week to prevent working around the appointment. There are, however, proportional benefits with an upper limit.

Recently there has been discussion of limiting the unemployment allowance to 5 years, but as long as the government can afford it, Belgians will probably not limit this subsidy. Most citizens feel it is a social right to have a certain standard of living regardless of a person's ability to find a job for which he or she is suited. Needless to say, the unemployment subsidy, social security/health care system, and other socialized services in Belgium are largely responsible for one of the highest tax rates in Western Europe. Although Belgian workers tend to be highly skilled and hard working, thus creating a situation in which there is high productivity per worker, the tax rates are so high that the cost of labor almost offsets the productivity.

Another factor of Belgian society that induces a less competitive nature is that the salary earned in many jobs does not vary to reflect effort and performance. For instance, a professor who instructs classes, devotes an inordinate amount of time to research, and publishes regularly will not be paid any more than a professor who maintains a much more relaxed lifestyle. Other than personal achievement, which is not necessarily highly valued, there is little reward for working harder than the average professor or coworker. There is a greater emphasis on cooperation between colleagues rather than climbing over one another to get to the top.

Born Negotiators

Furthermore, Belgians share a great tradition of statesmanship and the ability to negotiate on behalf of others. Peter Paul Rubens, the Flemish painter, was a legendary statesman whose greatest efforts were in fostering international cooperation. He was hired by royalty and wealthy merchants to paint portraits of others. While he had the subject seated for long periods of time, he was paid to put forth the philosophy or argument of his benefactor. For example, he was sent by Philip IV, king of Spain, to paint Charles I, king of England, all the while negotiating a trade agreement between the two nations. This is one reason why Rubens painted about 1,600 royal portraits.

Modern Belgians have carried forward this tradition. The cooperative nature of the Belgians is again apparent in their approach to the European Community and the world. Belgium's linguistic and cultural diversity made it an ideal model for the "EU 92" negotiations. Due in part to this unique status, Brussels was selected as one of the two capitals of the EU. In 1967, for many of the same reasons, NATO moved its headquarters from Paris to Brussels. Belgians are also joiners and there is an association for every kind of need. Due to Belgium's cultural traditions and unqualified acceptance of these international organizations, its balanced nature, and its central location, Brussels has become one of the most important international areas in the world.

In addition, international corporations seeking expansion into European markets have flooded the Belgian borders over the past 30 years. About 40% of Belgian industries are foreign owned. In an interesting twist, the maker of the King of Beers in America, Anheuser Busch (Budweiser and Bud light), agreed in 2008 to a $52 billion takeover by the Belgian brewer InBev SA (Stella Artois, Beck's and Bass) in a merger that will create the world's largest brewery. The Belgian government offers international groups major tax incentives to bring central offices or high-tech activities into the country. Despite generous wages, high manufacturing costs, and exorbitant personal tax rates, companies have sought out manufacturing facilities in Belgium to take advantage of the transportation infrastructure and friendly economic climate. Thus, another element that the Belgians seemingly balance effortlessly is the internationalization of their small country while maintaining their own identity.

A Proud People

Finally, although there are obviously many reasons for the Belgians to be proud and boastful, people outside the country are rarely aware of Belgian accomplishments. For instance, many readers may not have been aware of the importance of Belgium to lace and vice versa before reading this chapter. Also, it is not generally known that Belgians produce some of the best beer and chocolates in the world. Belgium has the 16th-highest per capita gross domestic product among the world's nations ($34,210).

Despite Belgian prominence in international circles, many people are still unfamiliar with the country. Most people struggle to name prominent Belgians after fictional detective Hercule Poirot and comic strip hero Tintin! The reason underlying this lack of international recognition seems to be primarily the result of the Belgian approach to life. Belgians are not generally boastful, tending not to expound about their capabilities and accomplishments. In contrast, they act much like a small piece of lace on the collar of a garment; they bring out the best qualities of those with whom they work without being flashy or calling attention to themselves. Some view this approach as an inferiority complex of sorts, but this idea is contradicted by the definite pride that Belgians feel toward different characteristics of their country. Belgians do not feel they are less important or worthy of praise, as an inferiority complex would suggest; in fact, quite the opposite. Belgians tend to believe that the country can and will be an integral player in the EU and the world. The typical Belgian emphasizes the role of coordinator and peacemaker among international markets, sometimes sacrificing publicity and worldwide recognition.

In short, Belgium is as complex as its renowned lace, and its citizens have learned to meet the challenges posed by the interactions that must occur among its cultural and linguistic groups. While this country has been internationally recognized as an independent state for less than 200 years, its past has been rich. Its future is bright but complicated, just like the lace for which the nation is well known. Belgium adds character and richness to the EU, just as a small piece of lace adds beauty and depth to a well-crafted garment.

PART VII

Torn National Cultures

Some nations have experienced major assaults on the core values of their cultures. That is, they have been torn from the cultural roots that have nourished them for decades if not centuries. We provide two outstanding examples of such nations in this part of the book, Mexico and Russia. As indicated throughout the book, each nation can fit into more than one category or part of the book, and this is true of Russia and Mexico. But if readers desire to plumb the depths of what a torn culture really means, they can benefit from the analysis of these two nations.

CHAPTER 23

The Mexican Fiesta

The solitary Mexican loves fiestas and public gatherings. Any occasion for getting together will serve, any pretext to stop the flow of time and commemorate men and events with festivals and ceremonies. We are a ritual people, and this characteristic enriches both our understanding and our sensibilities, which are equally soft and alert. The art of the fiesta has been debased almost everywhere else, but not in Mexico. There are a few places in the world where it is possible to take part in a spectacle like our great religious fiestas with their violent primary colors, their bizarre costumes and dances, their fireworks and ceremonies, and their inexhaustible welter of surprises: the fruit, candy, toys, and other objects sold on these days in the plazas and open-air markets.

—Octavio Paz (1961, p. 47)

Mexico represents a classic torn culture, that is, one torn from its roots by invaders and external factors such as widespread famine that devastate it (Huntington, 1996). In Mexico's case this process has occurred several times, most recently beginning about 1990 as this nation's leaders began the process of making it a major player in the global economy. There are three distinctive cultures in Mexico: the Indian culture, the Spanish culture, and the *mestizo* culture of mixed Indian and Spanish ancestry. About 60% of the population of 105 million is *mestizo* and 30% pure Indian (Amerindian) or predominantly Indian. The Indians are descendants of the Mayan and Aztec empires. Finally, about 9% are of Spanish ancestry. As might be expected, there is implicit but imperfect ordering of these three groups in terms of status: Spanish, *mestizo*, and Indian.

Sometimes this ordering leads to conflicts. For example, some business firms have fallen on hard times because the Spanish and *mestizo* owners and managers have found it difficult to work together, and the Indians have experienced many forms of discrimination. In more recent years there have been attempts to alleviate the plight of the impoverished Indian groups, and the move toward globalization has helped to decrease the historical antipathy of the Spanish for the *mestizo*.

Still, the threefold distinction is a critical part of Mexican culture and is celebrated at the well-known Plaza of Three Cultures in Mexico City where Hernando Cortes, the conqueror of the Indians, built a Spanish church in 1521. The church was built on and around the site of an ancient pyramid structure constructed by the Indians. A marker at the church succinctly describes the three-partition culture: "On August 13, 1521, heroically defended by Cuanhtemoc, Tlateloko fell into the hands of Hernando Cortes. It was neither a triumph nor a defeat; it was the painful birth of the *mestizo* nation that is Mexico today."

When talking about Mexico, it is important to distinguish between the northern states bordering the United States and the nine states of the south and southeast, which account for nearly a quarter of the population. These nine states tend to be more rural and poorer than the rest of the country. Over time, however, all of Mexico has advanced in terms of many measures. For example, since 1960 the number of years that the average Mexican child spends in school has increased from 2.6 to nearly eight (Reid, 2006).

Mexico, like its fiestas, is a complex blend of reality, tradition, art, and people. The population is growing at a 1.3% annual rate. Mexico City, the capital, has a population of more than 19 million and is one of the largest cities in the world.

Most Mexicans tend to identify with their Indian or Spanish heritage. Spanish is the official language of Mexico although English is understood by many in large urban areas. As many as 100 Indian languages are still spoken in parts of Mexico. Mexico is about three times the size of Texas or one fifth the size of the United States.

Mexico is several thousand years old and still influenced significantly by its past. For example, plans to build a subway line under the capital's main plaza were canceled when it was discovered that construction would destroy hidden remains of the Aztec empire. The past is celebrated in numerous statues, in streets named after past— including pre-Hispanic—heroes and historic dates, and in the calendar. For example, the entire month of September is devoted to ceremonies commemorating Mexico's independence from Spain. This devotion to the past is also revealed in the fact that the Mexican government spends more of its budget on anthropological research of its past than does any other country.

Even the official past remains of current interest. News a half-century old is highlighted in interviews and columns. Symbols are sometimes taken from pre-historic times. A 1978 monument to police officers and firefighters who died in the line of fire was a statue of Coatlicue, goddess of the earth and fertility, with a fallen Aztec warrior at her feet. Important figures of Mexican history have been divided into good and evil and personify concepts such as heroism, nationalism, and revolutionary ideals or cowardice, treason, greed, and repression.

Thus, the past is important for understanding Mexican thought and action. As such, a brief review of Mexican history is essential to a comprehensive understanding of the Mexicans and their culture.

Historical Background

When Cortes arrived in 1519, Mexico was populated by hundreds of indigenous tribes including the Aztecs, the Incas, and the Mayas. Some of these represented advanced societies. It took only 2 years for Cortes to conquer the indigenous people in the name of Spain. Although most of the indigenous people were wiped out by the conquest and European diseases, those remaining were instructed by the Spanish in the Catholic faith during the colonial era. The silver mines and the people of Mexico were exploited for riches sent back to Spain.

After the victory over Spain in the War of Independence in 1810, Mexico began a period of independence and turmoil. By 1853 half of Mexico's territory was acquired by the United States, mostly in the Mexican war. This period featured unstable leadership and exploitation of the poor peons by the elite landholders, who made the peons virtual slaves on the hacienda farms. One notable leader during this time was Benito Juarez, famous in that he was pure Zapotec Indian.

When the Mexican revolution ended in 1920, presidential elections began. The government attempted to break up the Catholic Church, and a constitution was drawn separating church and state. The church was not allowed to hold land and priests could not vote, although religion remained inseparably woven throughout the life of the Mexican. Land holdings of the church and large land holders were taken away and given to the peasants and indigenous people under the land reform measures.

Catholicism in recent years has decreased in popularity, going from nearly 100% of the population in 1960 to about 89% today. Protestantism, particularly of the evangelical variety, has correspondingly risen from nearly zero to 6%. Some of the reasons for its appeal to the poor include its firm stance against alcohol and its more open and less hierarchical approach. Anti-Catholic films highlighting the faults of the clergy have become popular. *El crimen del Padre Amaro* has become one of the most popular films in Mexican history; it relates the true tale of a 19th-century priest who violated his vows of celibacy and paid for an abortion that resulted in the death of a poor young woman.

In spite of the beneficial outcomes of the revolution, many problems still remain today including poverty, unemployment, a high birth rate, an inadequate educational system, inefficient industry and agriculture, a high foreign debt, and a variable economy lacking financing capabilities. In some ways the economic situation has improved markedly over the last 50 years, in large part because of the discovery of oil. Also, the North American Free Trade Agreement (NAFTA) helped to increase trade between Mexico and the United States, which buys nearly 90% of Mexico's exports. NAFTA, established in 1994, also includes Canada, and it is controversial, as it has benefited the larger Mexican firms much more than the smaller firms.

Real wages for most Mexicans are actually lower today than in 1993, but this outcome is in large measure attributable to the peso crisis and devaluation in 1994 and 1995. Still, the poverty rate of 40% is 10% better than in 1980, even though the absolute number of those in poverty has increased by more than 20 million owing to population growth. About 15% of the workforce is in agriculture, 30% in industry, and 55% in services. There is a huge informal or black market economy that is estimated to be 50% of the total, a figure similar to that found in many Latin American nations.

One political party, the Partido Revolucionario Institucional (PRI), dominated politics for 72 years. In 1988 a near-loss by the PRI marked a turning point in Mexican politics, and viable opposition parties became a reality, culminating in the PRI's defeat in the national elections in 2000. The PRI had experienced several scandals in recent years and it no longer has the unquestioned authority to operate as it wishes. Its leaders had attempted to maintain an aura of gravitas and imperial mystery, but now they must compete against one another and members of opposition parties in U.S.-style elections.

Although this chapter generalizes about the culture of Mexico, in reality the culture is not uniform. There are five more or less distinct regions (Kras, 1989). The northern border region has been influenced by its proximity to the United States and the presence of many foreign-owned businesses under the *maquiladora* program, in which foreign businesses ship parts into Mexico, use Mexican labor to assemble the products, and ship the products back out of Mexico free of tariffs. The northern people tend to be more aggressive and independent than their fellow Mexicans.

The central region is characterized by more traditional, conservative, autocratic, family-owned businesses. In the southeast region, people are more relaxed. There are more plantations and less industry. Business is paternalistic and autocratic. There is also a large indigenous population in the region. The capital, Mexico City, represents the fourth region, and it boasts about 40% of Mexico's Gross National Product. Although it is overcrowded, it is very much a modern, cosmopolitan city. The fifth region consists of the areas along the Gulf of Mexico, which are rich in oil.

The mixture of the indigenous population and the Spanish population is nearly complete, resulting in what Mexicans call *mestizos*. Despite the tragedy inflicted by the conquistadors, the Catholic faith helped the indigenous population assimilate into the Spanish culture. The traditional indigenous religions have blended with Catholicism, resulting in a uniquely Mexican culture that is reflected in unique religious beliefs and practices. The mixing of these two cultures (Indian and Spanish) can also be seen in the distinctive architecture, the language, and many other aspects of people's lives.

For example, during the fiesta of the Day of the Dead, people flock to the graveyards to put flowers and food on the tombs of their relatives. These symbolic gestures combine the Catholic concept and the ancient Indian concept of the afterlife. One day each year people dress up like peasants and take their children to the cathedral to be blessed. In front of the church are booths where children get their pictures taken. In the backdrop of these booths and in the children's costumes are visible signs of the indigenous backgrounds of Mexicans.

The Mexican Fiesta

Although there is more to life in Mexico than the fiesta, many aspects of the fiesta can be related to Mexico to more fully understand its culture. According to the Mexican Department of Tourism, the number of official fiestas varies yearly between 500 and 600. Fiestas are held for a variety of reasons, including the celebration of historical events such as Mexican Independence Day; reinforcing the traditions of individual towns, cities, and states; honoring Catholic saints and practices; paying homage to special foods and crops found in a particular region; and recognizing birthdays, baptisms, weddings, and graduations. As such, the fiesta is an appropriate metaphor through which to view Mexico.

One broad-based and partially inaccurate stereotype is represented by the popular tourist ads having the tagline "Come to the land of the fiesta." Such ads feature seductive girls dressed in sequins and ribbons dancing with beautiful boys to a background of romantic guitar music. This fiesta represents that of the movies, tourists, and perhaps a big town project. The real Mexican fiesta is the country fiesta—the more remote and difficult it is to get to the fiesta, the more attractive.

The pure pleasures of the simple life can be experienced at a fiesta: indigenous, native dances in beautiful costumes and people dancing happily about. The fiesta gives Mexicans the opportunity to enliven and richly enhance their existence: They can see people from their neighboring towns, sell a few ornaments, party, and enjoy the camaraderie of friends and family members. Even the most modest fiesta produces a particular mood, full of humor and energy. The fiestas are a blend of beautiful costumes, language, bustling activity, and aimless walking around.

In the tropical coastal towns, rural fiestas are especially festive. Although fiestas have humor, they are not good-humored or consciously funny. If they are dramatic, it is unplanned. The participants may not understand fully the ceremonies and rituals, even when they have been involved with the fiesta for years. A fiesta cannot be likened to a big block party. It is irrational, mysterious, and moving.

Generally, the state of Oaxaca has wonderful fiestas with beautiful regional costumes, flower arrangements, and tall *castillos* (towers) of native fireworks. The people of Oaxaca are predominantly Indians who share good looks, grace, calm, and an un-Indian worldliness. They are quite interested in new people and new ways. The state of Chiapas, the scene of revolutionary activities in recent years, has pagan ceremonies that are alluring and frightening. Other regions only eat, sing, dance, and drink, leaving religious meanings behind. This is especially true of the tropics, which are devoted to the present. Given the diversity of fiestas, each of them must be experienced, not simply observed.

However, in more recent years at least some of the traditions found in fiestas have grown weak. Even the Fiesta of the Dead on November 1 is being altered. This fiesta highlights death as transformation rather than as the end of life. In this way the importance of family history is stressed. But the U.S. celebration of Halloween is being grafted onto the Day of the Dead, and in the process many traditions have been discarded ("Mexico, Haunted," 1999, p. 36).

Four important aspects of Mexican culture are revealed in the fiesta. First, the primary focus of the Mexicans is on people, and the fiesta offers them a chance to be with and enjoy the company of their family, friends, and community. Second, religion is important throughout the Mexican's life, and the abundant religious fiestas are a chance for Mexicans to communicate with God. Third, experiencing the present is important to Mexicans, and the fiesta is an opportunity to do this. Last, within the social order Mexicans find freedom, and within the social order of the fiestas Mexicans exhibit this freedom.

Primary Focus on People

People are of paramount importance to the Mexicans. A fiesta is a time to enjoy the company of family and friends. It also forms a bond between all Mexicans, uniting them into a common people. Mexico is collectivist, and there is a tight social framework in which members of the in-group, particularly those in positions of authority, are expected to look after all of its members in exchange for their loyalty. Gabrielidis and his research team (Gabrielidis, Stephan, Ybarra, Dos Santos-Pearson, & Villareal, 1997) found that the collectivistic Mexicans display concern for others and use accommodation and collaboration more than the individualistic U.S. Americans do. As a whole Mexico is an authority ranking or vertical collectivistic but paternalistic culture in which there is a good amount of social distance between superiors and subordinates. Paternalism is manifested in various ways, such as the *aguinaldo* or Christmas bonus, which can be as much as three months of a person's salary. But in some areas, such as those bordering the United States, individualism is becoming stronger.

The basic building block in the society is the immediate family. In more rural areas several generations will live together in the same house. Children are encouraged to stay dependent on the family and not leave home. The family spends most of its nonworking hours together and the children go everywhere with the parents. Babysitters are rarely if ever used, except by a minority of the population that is rich enough to afford servants. The family is an informal welfare system in Mexico. If a cousin cannot afford to feed the children, the rest of the family will provide the groceries. Mexican families tend to regard homelessness among its members very negatively and will do as much as possible to eliminate it.

The fiesta is a chance to enjoy the company of the family. Some fiestas even extend the family. Godparents (*madrina* or *padrino*) are selected for baptisms and other important festive ceremonies in a child's life. It is an honor and a responsibility to be chosen as a godparent. A child can have different godparents for each ceremony, with the new ones adding to the existing ones rather than replacing them. Because these godparents will be present for every important occasion in the child's life, this practice actually extends the family to include the godparents and their families. *Compradazco* connotes coparenthood shared by the parents and the godparents. A Mexican friend likened *compradazco* to a network of invisible lines connecting many houses or families in Mexico together into one big family.

Friends are also important in Mexico. Once a person becomes a friend, he or she is incorporated into the vast family network and becomes one of the trusted members of the in-group. Friends almost become a part of the family, as in the case of *compradazco*. Friends of friends also become part of this network. Friends are often called by familial titles such as brother, sister, and cousin.

Relationships Rule

Mexicans are ruled much more by relationships than by abstract concepts. Mexicans will interrupt whatever they are doing at the sight of a friend or a relative. Diminutive endings such as *ito* or *ita,* which mean small, are often tacked onto the ends of names or words to suggest affection, for example, *Silvita* instead of *Silvia.* These endings are also used for minimizing problems and saving face.

Mexicans tend to hire relatives and friends over strangers, no matter what the qualifications or achievements of each, although this is less true in the more modern businesses. In Hofstede's (2001) study of 53 nations, Mexicans were found to have a high need for uncertainty avoidance, indicating that they feel threatened by uncertain situations and strangers and try to avoid them. Therefore, knowing someone before hiring or doing business with him or her is important. The old adage "It's not what you know, but who you know" is taken literally. If business must be done with strangers, much time is spent getting to know the person before any deals are made. Mexicans are not loyal to an organization, but they are committed to the people in the organization.

Considering the central role the family plays in Mexican life, foreign executives must take the concerns of the employee's family seriously. For example, they might show interest and concern when there is illness in a family and an absence results. Understanding family ties is an excellent way for foreign managers to explain subordinates' behavior and to develop good relations with them.

Mexicans tend to view success in terms not of achievement but of affiliation. It is important to be someone who is important to other people. Achievement is, however, more important in larger, more modern companies. The uniqueness of each person as an individual is highly valued, even though these individuals are an integral part of a collectivist culture. Mexicans refer to *alma* (the soul) or *espiritu* (the spirit) to describe this inner individual.

It is important to protect this soul or dignity or honor. Managers should rarely, if ever, criticize subordinates in front of their friends or family. Praise is important to Mexicans. Mexicans tend to be consensus seekers and will lie to avoid hurt feelings and confrontation. Avoiding the placement of blame on anyone is important. In Spanish, people do not break things, things just break, *se lo rompió*. Lies and truth are not absolute. Mexicans will frequently say what you want to hear to avoid confrontation and the loss of someone's dignity, even if what is said is not the complete truth. When asked directions, Mexicans will sometimes give a false answer rather than say they do not know. When they beg one's pardon, they say "pretend it never happened, señor."

Michael Agar, a cross-cultural anthropologist who was asked to facilitate the relations between U.S. Americans and Mexicans involved in a joint venture, recalls vividly how the Mexicans perceived the U.S. Americans. After working with the U.S. businesspeople all day, the Mexican executives went out for a drink, and one of them said *el capo* as he waved an imaginary torero's cape before the bull or, metaphorically, the U.S. Americans, and the other Mexicans laughed uproariously. The Mexicans perceived themselves to be too polite and sophisticated to tell the U.S. Americans to their face that their behavior was boorish. However, this politeness was a major reason that the joint venture eventually failed, for the U.S. Americans became exasperated when the Mexicans indicated that it was "no problem" meeting a specified deadline, only to fail to do so repeatedly. The U.S. Americans bluntly told the Mexicans that they had lied and the Mexicans were insulted. Given the communication and negotiation styles of U.S. Americans and Mexicans, such outcomes are frequent.

Mexicans tend to view humans as a mixture of good and evil. Possibly because of this perspective and the gray area between truth and lies, one can encounter occasional obstacles where someone may need to be rewarded to gain cooperation. Analogously, in the fiesta one often dons a costume and a mask. This mask protects one's dignity. It allows an escape from reality without the loss of dignity. Octavio Paz (1961) even argues that the Mexican is "a person who shuts himself away to protect himself; his face is a mask and so is his smile" (p. 29). From Paz's perspective, Mexicans tend to mask painful realities, hiding more than they reveal.

Communication Style

Nonverbal communication is important in most if not all cultures. In Mexico, gestures, facial expressions, glances, posture, and clothing all reveal important facets of the culture. Greetings and salutations take a long time and are often full of handshakes, hugs, kisses, and pats on the back. There is more physical contact between members of the same sex in Mexico than is common in the United States. Men greet each other with an *abrazo* (embrace) whereas women may kiss on the cheek. Overall, Mexicans employ more physical closeness and smaller interpersonal distances than their counterparts in the United States.

Thus, U.S. Americans may withdraw from Mexicans, communicating emotional or social distance. Mexicans may seem too overbearing to U.S. Americans when in fact they are not. Rhythms are also different between the two cultures. Mexicans tend to use more of the trunk of the body while North Americans use their head and neck. Mexicans use their hands extensively in self-expression. Use of the hands helps illustrate and emphasize what a person is saying.

Neat clothing and appearance are also important because they show respect. In Mexico, clothing and jewelry are representative of the country's great ethnic diversity. The type of dress worn indicates a person's region or ethnic background. In hats alone, the diversity is wonderful. Upper-class Mexicans distinguish their status by wearing fine clothes, expensive jewelry, and fancy hairstyles. The concern with being respectable or *decente* is revealed in clothing.

Conversation is an art in Mexico, and the manner in which things are said is as important as what is said. The speaker beats around the bush in a dramatic and flowery style and often repeats things. Allusions and double meanings are a delight to Mexicans. Cultures differ in the importance they place on words to convey information (Hall & Hall, 1990). U.S. Americans and Europeans place great emphasis on words. In Mexico, context is more important, and Mexicans may view a statement as honest in one situation but rude in another. Who says something and how it is said are also vital to understanding what is meant.

Meetings often become social events. The participants simply enjoy each other's company. There is a lot of emotion and passion in any conversation. When a Mexican man courts a woman, for instance, he typically produces flowery praises of her beauty. Similarly in the fiesta, extreme joys and sorrows are exhibited.

The language often says more than it means. For example, *mi casa es su casa* literally means "my house is your house," but in reality it means "you are welcome in my house." When Mexicans are pleased to meet you, they say they are *encantado*, that is, they are enchanted to meet you.

Education and Training

By and large, the emphasis in the schools has been on theoretical concepts and social competence. Intellectual pursuits such as philosophy, science, and the arts are important, whereas practical applications are largely ignored. This practice often leads to problems in business when subordinates are charged with implementing a plan but have insufficient skills to do so. Thus international managers must realize the importance of training their subordinates, especially given their tendency to be agreeable and not to question authority.

In recent years there has been an increase in the importance of practical training. For example, Monterey Tech (ITESM) produces a large number of first-rate business and engineering students, and some of its programs such as the MBA program are ranked among the best in the world. This university also provides a year of remedial work for the undergraduates who are admitted on a probationary basis but fail to measure up to its tough entrance standards.

The Emphasis on Religion

The significance and number of religious fiestas demonstrate the importance of religion to life in Mexico. The religious fiesta is a chance to communicate with God. In addition, some fiestas that are nonreligious in purpose have religious features in them.

An important fiesta is Carnival, which is celebrated before Lent. Another celebration, the feast day of Our Lady of Guadalupe, is one of the more important religious holidays: The country celebrates the anniversary of the day when the Virgin Mary is said to have appeared on the hill of Tepeyac (now part of Mexico City). On the 9 nights before Christmas Mexicans act out Mary and Joseph's search for lodging in Bethlehem. Every night after these ceremonies or *posadas,* the children play the *piñata* game. Holy Week, the week before Easter, is celebrated with fiestas. In addition, each city or village has an annual fiesta for its patron saint. During the religious fiestas Mexicans pray and light candles in the churches.

There are also family fiestas or fiestas celebrating the traditional passages in the life of a person, and they include baptism, first communion, and weddings. The whole family, including the godparents and their families, will be present for these fiestas, which will frequently begin with a church service during which Holy Communion takes place.

Even some fiestas that are not religious in purpose have aspects of religion in them. There is a fiesta, *quinceañera,* for 15-year-old girls similar to the sweet-sixteen party or coming-out party in the United States. It is social in origin but includes a service in the church with communion, a multi-course meal, and dancing, preferably to a large band's music. This is one of the most important fiestas in a girl's life and the girls wear expensive, elaborate dresses for the occasion, while the boys accompanying them are in suits and ties.

What God Wills

Mexicans tend to believe that life is beyond their control. What God wills, happens. The common expression *si Dios nos presta el tiempo* means "if God lends us the time." The expression *ni modo* means "no way" or "tough luck." Many Mexicans generally consider nature to be dominant over people. When problems are encountered, Mexicans will tend to modify themselves rather than try to modify the environment. For this reason class lines are not easily broken and problems are often minimized. Similarly, the mask in the fiesta is a means of modifying the self to deal with reality. But the mask also embellishes reality and adds a creative dimension to life that is missing in a typical U.S. American cocktail party and even during official holidays such as July 4th.

Because Mexicans tend to believe life is beyond their control, they tend to share a fatalistic attitude espousing the same belief. They also exhibit a fascination and even an obsession with death. The newspapers are full of gory stories about death. This obsession is reflected in their fiestas. The Day of the Dead, also called All Souls Day, is celebrated with a fiesta to honor and remember the dead. Food and flowers are brought to the tombs of deceased relatives. Some people even set a place at the table for the dead. Candy skulls with a person's name on them are purchased and eaten or given as gifts. To Mexicans this cheerful celebration is for both the living and the dead.

Mexicans tend to believe that death is only part of a larger picture. This belief may stem both from the indigenous religions and Catholicism. In the circular Aztec calendar stone, the center contains the sun god with two snakes. Supposedly human sacrifices were made on this stone to pay back the gods so that the continuous regeneration of life could continue. Christianity also views death as a transition between two lives. Although some have argued that to the modern Mexican death is no longer a transition, Mexicans still tend to view death as an integral and inseparable part of life. This belief in communion with the dead is an outgrowth of the knowledge that the past is not dead. Similarly fiestas are also a time of death and rebirth for Mexicans in that the fiesta revitalizes and strengthens them. Thus fiestas are a source of continual energy and renewal for the Mexican people.

The bullfight, which is clearly a ritual emphasizing bravery in the face of death rather than just a sport, is often a part of fiestas (see Chapter 30, "The Spanish Bullfight"). The death dance of the bullfight is an expression of the Mexican preoccupation with death. Also, bullfighting is an exhibition of *machismo*, in that the *torero* must show that he is not afraid of death. At some fiestas there are other opportunities to show courage in the face of death, as when men hold onto the end of long poles rotating like a Ferris wheel either vertically or horizontally around a tall center pole.

Effects of *Machismo*

Machismo is an important concept in Mexico. Manliness and virility are greatly admired, and perhaps even more admired is authority. Although there are now genuine competing political parties, strongmen such as trade union bosses are still the most powerful personages in the majority of Mexican towns (Sullivan, 2002). It is important for men to exhibit masculinity and stoicism at all times and to show that they are not afraid of death. Male children are more exalted than their female counterparts. Boys are expected to act like little men and are told to look after the family when the father is away. One day each year families take their children to the cathedral to be blessed. When they have their pictures taken outside the cathedral, they draw mustaches on the boys to make them look like men.

A Mexican male's insecurity is manifested by his fear of betrayal by women. One generic anthropological explanation suggests that because Mexican mixed bloodedness began with the mating of Spanish men and Indian women, male-female relationships were based on the concepts of betrayal by women and conquest by men. The macho man must protect himself against betrayal, just as the conqueror could not fully trust the conquered.

The male worship of the female ideal is illustrated frequently by their adoration of the Virgin and their adulation of their mothers. The wife, however, is sometimes viewed as a sexual object. A husband who is faithful and affectionate shows vulnerability and weakness. Mistresses let the man conquer before he is betrayed. Wives, resentful of their husbands, are sometimes too attentive to their sons. Sons, in turn, view their mothers through the feminine ideal, yet tend to adopt their fathers' ways once married.

Women, like men, spend most of their time with members of their own sex. At social gatherings they stay close to their husbands or mingle with other women. Rarely do they spend significant time with other men. A female seeking a career runs into great pressures. The traditional role of mother and homemaker makes career choices a difficult matter. Women, however, remain central to the family, reliable in a society where illegitimate children, broken homes, and absentee fathers are common.

While they seek security, Mexicans can be introspective. Fiestas provide an emotional outlet from their solitude and self-restraint. They provide a release that is not found elsewhere.

Experiencing the Present

Although Mexicans are very much influenced by the past, they live for the present. The fiesta allows them to stop and enjoy the moment. It is an escape from work and poverty. The fiesta is a release of the soul.

The Mexican concept of time is relaxed and there is always plenty of it. This is called polychronic time, meaning more than one thing goes on at a time (Hall & Hall, 1990). Mexicans call it *la hora Mexicana* (Mexican time). Interruptions and tardiness are common and deadlines are flexible. Absence after a long weekend is institutionalized in *San Lunes* or Saint Monday, and this is accepted as an explanation for absence. Many projects are never seen to completion. People are expected to be late to social occasions, and hosts actually plan on this eventuality. Arriving on time is considered rude. Mexicans are dismayed over North American invitations that state in advance what time a party will be over.

Even though Mexican bus or taxi drivers may have paying passengers, they will sometimes stop to give a friend a ride to an unscheduled destination. This view of time is also exhibited in communication, which is often spontaneous and simultaneous, and ideas flow from many different directions. What some term the *mañana syndrome* stems not from Mexican laziness or inefficiency but rather from a different philosophy of time. Disasters are considered unavoidable, *ni modo*.

However, sometimes other factors, such as the importance of authority, override the cultural expression of time. One U.S. trainer on his first trip to Mexico experienced this phenomenon. He was ready to start his class, but not one of the 50 managerial trainees was in the room. Wanting to be polite, he waited 30 minutes before starting, at which time about half of the class was present. When he complained to his Mexican superior about the lax approach of the trainees, the Mexican startled him by saying: It is your fault. The explanation was that the trainer should start class on time, even if no one was in the room, and everyone would be there forthwith because he was the figure of authority. He did so in the second class, and within 10 minutes all the trainees were in their seats. In all subsequent classes the managerial trainees were in their seats when class began.

There is more pressure to be on time and to produce in larger, more modern companies. Strict adherence to time schedules is important for modern companies and U.S. executives must be patient with the Mexican attitude toward time. They must plan for time slippage and the unanticipated.

Finding Happiness

Mexico is a country of extremes, both in terms of geography and rigidity of the social class structure, which often makes a Mexican's life hard. However, Mexicans frequently know how to find happiness in life. They have developed the ability to find a silver lining in any cloud. The fiesta is a time for Mexicans to find joy, and the mask worn at the fiesta allows them to do this.

Mexicans have a plethora of expressions for their displeasure with work. They say *el trabajo embrutece* or "work brutalizes" and *La ociosidad es la madre de una vida padre* or "idleness is the mother of the good life." Even the word for business, *negocio*, connotes the negative of leisure, that is, work is completed only for the purpose of obtaining leisure (Fisher, 1988). Although many Mexicans work hard, they value leisure time spent with their family and friends more highly than work. Again, they tend to be more motivated by affiliation than by achievement, and their emphasis is on being rather than doing. Work is done only because it has to be done, a concept in stark contrast to the Protestant work ethic, which holds that work is good in and of itself. Mexicans need a balance between work and leisure, which requires them to leave some jobs until "*mañana*." For this reason foreigners stereotype Mexico as the land of *mañana*.

One reason for the de-emphasis on work is that success in life is frequently based not on achievement but on personal relations. Career advancement often occurs because of personal contacts rather than achievements. Hard work does not necessarily lead to success or prosperity, and many Mexicans work hard all their lives only to live in dire poverty.

There is no strict dichotomy between work and leisure. Mexicans like to have friends and relatives working with them to make work more fun. Many businesses are run by the family and they hire family members and friends before they hire strangers.

Mexicans tend to deemphasize the future, and little long-term planning occurs except in the larger and more competitive firms. Each day is taken as it comes. This is due in part to the unstable economy. Follow-through on projects is often difficult to obtain. Concepts such as quality control are uncommon in all but the largest, most modern companies.

Freedom Within the Social Order

Mexican society is hierarchical (vertical collectivism), and everyone has a role within the society. Mexico has a high score on Hofstede's (2001) power distance dimension, meaning that the society accepts without question an unequal distribution of wealth and authority; the GLOBE study (House et al., 2004) confirms this finding. Thus the distinction between the two forms of the word *you* in Spanish is noteworthy. *Tu* is informal and used between friends, and *usted* is formal and used to show respect to superiors or elders.

In the smaller villages, there are specialized roles within the community. This hierarchy can be clearly observed in the patriarchal family structure in Mexico. The father is the authority figure, and traditionally the mother's role is to love her children and husband and to lead a life of self-sacrifice for them. Children are taught to obey their father and mother. Just as the parents take care of the children, so do the more affluent members of a family take care of the poorer members. When a family holds a fiesta, the wealthier members are expected to help pay for a large part of it. It is common for Mexicans working in the United States to send money home to help pay for a younger sister's schooling, coming-out party, and many other things. Such aid is in the billions of dollars and now surpasses all government aid. Similarly it is expected that Mexicans in the United States will send money to less affluent relatives to help pay for the family fiestas.

At work, there is a rigid hierarchy where the subordinates traditionally do not question the boss. A problem typical of authoritarian administrations is that Mexican subordinates tend to withhold negative information, even if important, and convey only good news. Everyone knows his or her role and conforms to it. Similarly the government is bureaucratic, the Roman Catholic Church is hierarchical, and schools are traditionally structured in an authoritarian manner so that questioning is not encouraged. This stratification began before the coming of the Spaniards. Today, absolute authority within the hierarchy is beginning to break down, especially in the cities, as larger businesses move toward more participative management styles.

Traditional Versus Modern

As this discussion implies, there are conflicts between traditional values and modern life. In one family the older brother had started a promising business and hired the next oldest brother to be his chief salesperson, and the latter complained to the father that his older brother was not paying him the promised salary but a much lower one.

When the father asked the older brother about the matter, he was incensed and indicated that he would indeed pay his brother all that he owed him—to the detriment of investment in the growing business—but that he would immediately fire him, regardless of the impact on the family. When the father communicated this information to the younger brother, he withdrew his request for the promised salary until the business became more stable.

Status is important to the Mexicans. Impressive academic degrees or titles command respect. Title usage reinforces the sense of hierarchy within Mexican society. At low levels of the bureaucracy the use of *licienciado* for a university graduate is common. It requires wearing a suit and tie and implies influence. The head of an office is not *licienciado* So and So but *el licienciado,* as if he were the only one. *Maestro* is often used either for plumbers, painters, or carpenters or for senior officials teaching at the university.

Members of the elite were formerly drawn from purer Spanish backgrounds and were proud of their light skin and Spanish heritage. Although there are now elite from the indigenous and *mestizo* populations, there is still a preference for lighter skin and eyes. For the most part, being Indian is defined culturally rather than racially in Mexico. A person is considered Indian only if he or she speaks the language and follows the culture of the Indian. In the revolution of 1910 the indigenous heritage of Mexico was glorified and even celebrated in an attempt to erase the stigma applied to Indians.

Structure and security form a curious amalgam with freedom in Mexico. Everyone knows his or her role and will usually not violate it or challenge another's role; people are careful not to trammel another's dignity. Yet there is not as much pressure for conformity as there is in less hierarchical societies. Mexicans have the freedom to do what they want as long as they do not affront others or go beyond their role in society. Jobs offer security to Mexicans and they consequently enjoy freedom at work. As previously mentioned, Mexicans feel free to be absent from work on Monday, with the only excuse necessary being that it is "Saint Monday." Children feel secure and warm in the family structure, yet because they are so loved and honored, they have more freedom to do what they want to do.

Mexicans as individuals may be freer from guilt and obligation than U.S. Americans. It is not the individual that is guilty; the group shares the blame. The obligation is not the individual's, but the group's. Also, the Catholic religion allows Mexicans to be free from sin by asking God's forgiveness for their sins in confession. Again, friends and family usually take care of those in their group who are in need.

Chaos in Structure

In a large cathedral in Mexico City, this freedom and even chaos within the structure and order could be clearly seen. Although the Roman Catholic Church is hierarchical and structured, what went on inside this cathedral was not. People came in to pray independently and left at their leisure without any reference to leadership from the priest. This was true of the main chapel as well as the many side chapels. A woman knelt beside a priest at the front of the church, apparently having her confession heard completely independently of everything else in the cathedral. The chairs, seemingly well used, were in disarray instead of in neat rows. Within the church's hierarchy there is security and freedom.

The fiesta is a chance to observe this freedom and even chaos. Sometimes participants will become inebriated and exhibit rowdiness and other related behaviors. In some ways fiestas are a celebration of chaos. Roles, hierarchies, norms, and laws vanish, for example, the poor dress up like the richer members of society. Everyone dresses up in gaudy costumes and becomes someone they are not. During the fiesta it is almost as if people are free to choose which mask they wear. Although they indulge in freedom and chaos during the fiesta, by and large they operate within accepted norms of behavior.

Mexicans easily identify with the song of longing written by Jorge Negrete, a Mexican who became homesick in Cuba and penned the following well-known words:

> Mexico beautiful and beloved
> If I die far from you
> Tell me I am sleeping
> And bring me back here

Mexicans have changed over the years and will continue to do so. In the cities and larger corporations, Western concepts such as achievement and rigid schedules are becoming more common. Larger corporations are less autocratic and paternalistic than the smaller businesses. This may be due in part to the fact that many of the larger corporations are owned by foreigners. Although these changes are visible, they depict only a small part of the culture of Mexico. For the most part Mexico is composed of smaller, more traditional businesses, with much more traditional culture. Throughout the years the fiesta has been, and will remain, an integral part of the Mexican culture. And the fiesta will endure as a metaphor describing the importance Mexicans place on people, religion, experiencing the present, and the freedom that exists within the social order.

CHAPTER 24

The Russian Ballet

*Russia is an exceptional place. In the 20th century, over a single lifetime—
70 years—it saw three civilizations. Each of the first two was rejected by its
successor, forcing people to renounce their convictions. You can imagine the
chaos of ideas and beliefs in their hearts.*

—Edvard Radzinsky (2006, p. A10)

Russia represents a classic case of a torn nation (Huntington, 1996), as it has
been severed from its social, economic, and cultural roots not just once but
four times in its tumultuous and long history, and three of these pivotal periods
have occurred within the past 70 years. The first period began in the 12th century
when the Tartars or Mongols from central Asia invaded and brutally subjugated the
country for 400 years. In the early 18th century Peter the Great initiated the second
period when he began to westernize Russia while deemphasizing its Asian influences.
He defeated his chief enemy, the Swedes; created the Russian navy to facilitate trade in
the landlocked nation; and constructed St. Petersburg (later Leningrad) out of
swampland to connect Russia by land with the West. However, prominent Russians
emphasize their exceptionalism created by their Slavic heritage and ethnicity, and
they portray Russia as neither Asian nor European ("Russian Exceptionalism," 1996).

In the third period, starting in 1917, the Russians moved away from rule by the czar
to communism, and the results were disastrous socially, economically, and culturally.
Millions of people were imprisoned and died; Russia lost more soldiers and civilians
in World War II than all other nations combined (an estimated 50 million of the
100 million); the economy became extremely weak after World War II as the
Russian government poured most of its resources into military and atomic weapons;
priceless Christian Orthodox churches were destroyed; and religious worship was

banned. Ironically the Western press after World War II tended to portray Russia as a vibrant economy, but part of this image was based on the "Iron Curtain" that stood between the West and Russia. Good information was difficult to obtain.

Thus, just a few years prior to the formal dissolution of the Soviet Union in 1992, the West was ignorant of many of the realities of Russian life. During these years former President Mikhail Gorbachev led the movement toward Western capitalism, and two new words entered the English language: *glasnost* or openness and *perestroika* or restructuring.

A tumultuous fourth period in the nation's history, starting in 1992, was initially marked by optimism and hope. Elections involving candidates with widely different philosophies and agendas became possible and a free press encouraged debate. Economic reforms were supposedly based on Western capitalism, but they failed miserably. For example, to make Russians into capitalists, the government gave Russian citizens small shares of stock in formerly nationalized companies that were privatized as part of the largest privatization effort ever undertaken; at least 6,000 large and midsize firms were involved in this experiment; by comparison only about 8,000 firms worldwide were privatized in the 1980s. However, most of these firms failed to meet the exacting competitive standards of early-stage capitalism, at least partially because of the antiquated technology used in many industries, and most of the shares became worthless. Simultaneously, a small group of businessmen gained control, frequently in a questionable manner, over the most profitable parts of Russian industry, and they purchased the stock certificates of profitable companies for a pittance from ordinary citizens who were unaware of their value. Corruption and crime became rampant. Some U.S. businesspeople seeking to do business in Russia in the early 1990s were shocked at the open manner in which bribes were sought.

In 2000 Vladimir Putin became president. Largely because of Russia's oil and gas preserves, the economy began to improve. However, there is now a movement to take away many of the rights such as genuine political elections and a free press that the creation of the new Russia in 1992 encouraged. In 1988 under former President Gorbachev about 5% of the national political leadership was *siloviki* (power guys) or those with career military and security backgrounds such as in the former KGB (the Russian spy agency). By 2003 this figure zoomed to nearly 60% (Lichfield, 2004). A related analysis indicated that a quarter of Russia's senior bureaucrats are *siloviki;* the proportion increases to three quarters if people simply affiliated with the security forces are included ("Russia Under Putin," 2007). Although Putin relinquished the presidency in 2008, he was elected to the newly created and supposedly largely ceremonial post of prime minister and there is a widespread perception that he is directing Russian government and his hand-picked successor.

There is concern that the Russian economy will experience an economic bubble if oil and gas prices fall significantly. Russia does not have any leading multinational firms that can compete globally, and its smaller businesses tend to be inefficient. Furthermore, there is a general consensus that the Russian state is performing very poorly in many areas such as public safety, health, corruption, and the security of property rights when compared to other nations (McFaul & Stoner-Weiss, 2008). Life expectancy at 59 years for males is the lowest among all developed nations.

After the fact, it is easy to see how mistakes occurred when Russia as a torn nation gyrated quickly from one civilization to another, especially since 1917. This chapter focuses on the culture of Russia, with the recognition that culture is influenced by, and influences, social infrastructure and economic practices.

We focus on Russia, with its population of 143 million and a land mass that is almost double that of the United States. There is now a Commonwealth of Independent States consisting of 12 states, which was established to reformulate the Soviet Union without the economic or political costs of union, but it has virtually disintegrated even though politicians are seeking to revive it. Russia's population is about equal to that of the other states in the commonwealth, the largest of which is Ukraine with a population of 48 million. Moscow is located in the western region of the nation, and St. Petersburg is situated to the northeast of Moscow.

An Apt Metaphor

Russian ballet provides an apt metaphor for Russia for several reasons. Russia has excelled at producing ballet of the highest and arguably unsurpassed quality, as noted below. While Russians did not create ballet, they enriched this art form immeasurably. Also, it seems appropriate that a torn nation should be represented by a cultural metaphor that did not originate there. Furthermore, since 1992 and until recently Russian ballet experienced political intrigue, lack of funding, and some low levels of performance, thus mirroring the nation's troubles. At the current time, however, there are attempts to reinvigorate Russian ballet, the major ballet companies are now performing frequently outside of Russia, and the famous Bolshoi Theatre has undergone a major renovation.

To better understand Russian culture, we will take a look at three important characteristics: echelons, theatrics and realism, and the Russian soul. These elements also apply to our metaphor and help to distinguish this art form from its counterparts in other lands. With a quality larger than life, Russian ballet represents the complexities of Russian culture—the grand expression of aristocracy and the gentle beauty of a countryside.

Russian ballet means many things to many people. Some consider it the type of instruction given at the famous Bolshoi and Kirov academies; others identify Russian ballet with specific eras and performances. Those who witnessed the sumptuous performances of Diaghilev's *Ballet Russe* in the early 20th century surely identified his productions as "the Russian ballet." And audiences familiar with specific works such as *Swan Lake* and *Sleeping Beauty* proclaim these magnificent dance programs as Russian ballet. Essentially, all of these definitions are correct. As we will soon learn, the words *Russian ballet* evoke feelings of tremendous joy in anyone who is familiar with the finesse and style of the assured dancers associated with any part of this art form.

A brief background on the evolution of ballet in Russia gives us insight as to why the ballet remains so important to Russians today and explains why Russian ballet enjoys its excellent reputation throughout the world. Ballet in its modern form originated from court dancing in Italy during the Renaissance period. Although many U.S. Americans today think of ballet as a feminine occupation, in the 17th century male dancers dominated the spectacles for the audience of Italian nobility. Ballet emerged in the next century as a pantomime dance to the fascination of wealthier Europeans. In France, royalty participated in this form of entertainment by assuming starring roles. As it gained popularity throughout Europe, ballet performances became theatrical and were enjoyed by the public at opera houses. Like Louis XIV, who established an academy in France to improve the art, Peter the Great during the 17th century encouraged the development of social dance in an attempt to increase Russian awareness of the outside world. His determination to transform Russia into a powerful empire was achieved through modernization using Western Europe as a model. Russians have always had a great love of dancing, and when ballet was introduced in Russia in the 17th century, they responded enthusiastically to it. The monarchy placed emphasis on the arts by paying salaries for dancers, who became known as "artists of his Imperial Majesty."

During Peter the Great's reign, one of the most significant changes in Russian ballet was the elimination of cumbersome robes to permit freer movement of dancers. In 1736 the city of St. Petersburg, also the capital of Russia at that time, became the home of the Imperial Russian Ballet. Two years later, Empress Anna, who was also fond of ballet, sponsored the first school at her Winter Palace with the goal of developing professional Russian dancers.

A Flourishing Art

With the assistance of French and Italian choreographers, ballet flourished in the 18th century in St. Petersburg, Moscow, and Warsaw. Perhaps the most influential of those who staged ballet compositions was French-born Marius Petipa, who in the late 1800s created such masterpieces as *Swan Lake, Sleeping Beauty*, and *The Nutcracker*. Among the legacies of the Russian ballet is its tradition of brilliant composers, such as Tchaikovsky, who collaborated with Petipa to create the music of those well-known ballets, and Prokofiev, who wrote the stirring music that accompanies the ballet of *Romeo and Juliet*.

Although ballet lost much of its vitality in Western Europe during the 19th century, Russia preserved the tradition and elegance of the art form for the rest of the world. In 1909 Sergei Diaghilev revived the splendor of classical ballet with his *Ballet Russe*, or "Russian ballet" for Parisian audiences. His company was hailed as the "most exciting artistic force in Europe for the next twenty years" (Clarke & Crisp, 1976, p. 27). Sadly, Diaghilev's extraordinary *Ballet Russe* did not survive his death in 1929, due to lack of either a school to develop new talent or a permanent theater to call home.

After the revolution in 1917, the "artists of His Imperial Majesty" in the former St. Petersburg adapted to the socialist way of life as did the rest of Russian citizens. Russian dancers became state employees of state-owned companies. The world-renowned Maryinsky Theater in St. Petersburg became the famous Kirov Theater in Leningrad (the former St. Petersburg). Likewise, the Petrovsky Theater, home of the Imperial Ballet in Moscow, was renamed Bolshoi, signifying that which is big or grand.

In fact, ballet became more popular than ever as the government discovered that ballet could express the problems and ideals of a socialist state. Soviet promotion of socialist realism emphasized all types of artistic creativity, and citizens were encouraged to attend cultural events.

To match the public fervor of ballet, an infrastructure was developed for its expansion. A. Y. Vaganova, a legendary ballerina from the old Imperial Ballet, was called on to create a training program for dancers, and it eventually produced the finest dancers in the world. She was responsible for much of the physical richness of Russian ballet, and her method of teaching is still widely used throughout Russia. Other maestros such as George Balanchine and Mikhail Fokine inspired eager students including the magnificent Anna Pavlova, Vaslav Nijinsky, and Galina Ulanova. More than 40 new ballet companies were formed, and theaters were built throughout the Soviet Union. Gifted female pupils hoping to become prima ballerinas were provided with free academic and artistic support. Soviet teachers also searched for boys with strength, agility, and stamina who wished to pursue careers as members of a ballet troupe. Even today, a retired Russian male dancer usually retains his special status as *premier danseur*.

In spite of recent difficulties, Russian ballet serves as the standard for the rest of the world. This measurement is based on its fine traditions of classical ballet and intense national commitment to preserving and improving the art form.

Echelons of the Ballet

Clear echelons of status and privilege exist within the ballet company. At the very top is the prima ballerina, the prominent star of the troupe. Years of schooling and practice alone do not set one dancer apart from other dancers. The principal dancer must possess a natural gift of virtuosity and inner beauty. As is true in other nations around the globe, Russian citizens revere such accomplished individuals as national heroes. Although she is the most famous personality within her industry, the ballerina does not necessarily receive the largest salary. However, her status as No. 1 dancer allows the ballerina much greater privilege than almost any payment could. Similarly the producer, choreographer, and conductor are held in high esteem.

The next level within the ballet company consists of the director, set designer, and costume designer, revered not only for the titles they hold but also for their personal reputations. Then there are those who report to the director—specifically the production, stage, company, and wardrobe managers; musical arranger/director; set design assistant; makeup director; light designer; and dance captains. Many of these individuals possess the ambition and potential to become experts or even members of the elite. Furthermore, there is the corps de ballet or the members of the ballet company who perform dances. They constitute by far the largest of the groups. Finally, supporting musicians and various technicians, who specialize in everything from lighting to makeup, are at the bottom of the hierarchy. They sustain the other classes or groups both within and outside of our metaphor.

During the Soviet era the government provided ample financial support for ballet, and the echelons received rewards commensurate with the rank ordering described above. In the Soviet era privileges were selectively awarded and were primarily given to members of the Communist Party, about 20% of the total population. Many prominent members of ballet companies were in the Communist Party; even if not members of the Communist Party, they received ample rewards because of their association with ballet. Hence being a member of the ballet was associated with privileges that were denied to most citizens. While the echelons still exist in the ballet, the rewards are fewer, given the inability of the government to fund adequately. However, some new funding has been found from worldwide tours that have made Russian ballet much more accessible to non-Russians. Members of the Bolshoi have even performed in Las Vegas.

As the discussion of the four periods or eras of this torn nation implies, Russia historically has been an autocratic nation in which the few at the top ruled with an iron hand and echelons of power and prestige were created in accordance with this perspective. For example, even though Russia under communism proclaimed the doctrine "From each according to his ability, to each according to his needs," 20% of the population were members of the Communist Party and received preferential treatment. Currently there is some concern that Russia could easily revert to its former ways and behaviors, especially if most citizens begin to feel that the costs of capitalism outweigh the benefits. In contrast, the basic meaning of the echelons in Russian ballet is that outstanding, honest, and open performance is rewarded. It is easy to see when a dancer performs at a superior level, and the common citizen proudly identifies with such performance and culture. But the wide popularity that Prime Minister Putin enjoys, even though many of the rights of Russian citizens have been taken away in recent years, suggests that having a military and security background leads to favorable treatment.

Russians tend to possess a strong sense of dignity and composure. Even though the lower classes may never experience the thrill of a performance, Russians share a great pride in the reputation of their form of ballet as both enchanting magic and a symbol of magnificence. The very Westernness of ballet reflects Russian dignity and superiority, regardless of class status. Historically the most prestigious of dance companies was the Bolshoi Ballet, which tours from the Bolshoi Theater in Moscow throughout the world. The Moscow Academic Choreographic Institute, better known as the Bolshoi School, is formally affiliated with the ballet company and is located adjacent to the theater. Less famous but sharing an equally superb tradition of dancers, teachers, and choreographers is the Kirov Ballet Company, which in recent years has outshone the Bolshoi. This company and its school descend directly from Empress Anna's Imperial School and ballet company and still make their home in the same building in St. Petersburg today.

Drama and Realism

Audiences throughout the world recognize Russian ballet as the most spectacular form of dance. No other art form requires such a combination of elegance and simplicity—Russian ballet is distinctive from ballet of other national origins due to its singular innovations, which were developed from classical ballet foundations. While ballet stagnated for many years in the West, Russia along with the rest of the Soviet Union cultivated its theatrical resources to develop unimpeachable superiority over all other national ballets. Only the pageantry of the Italian opera rivals the brilliance of the Russian ballet, and only the Spanish bullfight compares with its drama. The pursuit of perfection in Russian ballet has become an inspiration to thousands of audiences and a permanent part of Russian culture and pride.

Audiences feel the vitality of the corps de ballet during the most dramatic moments of the performance. Throughout the ballet, dancers seek audience approval by delivering increasingly lofty performances. Sometimes the audience will respond to a dance so positively that a dancer will step out of character and take a bow, something that rarely happens in the West. In the final scene the dancers pour their remaining ounces of energy into the dance to make an unforgettably grand impression on every last spectator. Each member of the company strives to be remembered in the minds of the audience as a glorious image of movement. In response, the audience expresses approval in the form of applause, floral gifts, and ovations. The dancers offer their performances, again and again, in return for audience gratification.

Like the dancer who makes a personal attempt to appeal to people in the audience, the ordinary Russian citizen tries to make an impression on those who are in a position to help. The most efficient way to accomplish anything is through personal favors. Rather than jeopardizing careers by assuming risk or showing initiative, some Russians will avoid taking responsibility or other action that may be construed as controversial, especially when jobs are scarce.

Similarly the interchange between the dancer and the audience is symbolic of the basic exchange between total strangers in mundane matters. Whether the dancer gives pleasure to those who watch and is rewarded with faithful appreciation from the audience or two citizens or firms barter with one another, each participant offers the other one something that the other desires.

Just as some Mexicans tend to use the fiesta to escape a mundane existence, so too do Russians place great value on the theatrical component of the performing arts. Films, plays, and the ballet offer a chance for diversion. The ultimate cultural fantasy involves a melodramatic storyline surrounded with extravagant scenery and fancy costumes. The greater the pull of the heartstrings, the more Russians tend to love it; this sentimentality is part of the rare self-indulgence of the culture. Like the ballet company, which relies on theatrics to please the audience, the Russian shopper uses drama to persuade the clerk to decrease the price of an item. An indisputable part of Russian culture is the custom of bargaining and negotiating to beat the system. Similarly at an early age students learn to supply the "right" answer in the classroom. Sharing of answers is tolerated, and children become adept at skirting around authority with minimal confrontation.

As might be expected in an autocratic and authority ranking culture, favoritism and corruption have been and continue to be used extensively to get around the system. Disobeying laws is a part of Russian culture, even if for mere trifles. It is little wonder that firms must have their own private police forces. Friends and connections are important, and many Russians have developed complex networks of relationships that allow them to survive in an unpredictable world. A great deal of effort is required to create and maintain these relationships, in the hopes that benefits will result. Russians are continual gift givers, hoping to influence future generosity on the part of others. Such behavior is not likely to change, even with the qualified acceptance of competitive markets in which there is a level playing field.

Russians initially tend to take extreme views and offer few concessions when negotiating. If a foreign opponent makes a concession, it is perceived as a weakness. Unlike U.S. negotiators, Russians have little authority to make on-the-spot decisions and prefer not to make decisions spontaneously. Chess is a national pastime, and much of the relative strength of the many regional and world champions from Russia comes from their ability to think ahead. Players rely on strategic planning, whether offensive or defensive, to consider potential repercussions, similar to what the Russians do when negotiating.

The Vodka Pastime

Furthermore, many consider drinking vodka a national pastime. Drunkenness is accepted as a socially approved method of entertainment and escape that is helpful in coping with years of suffering and hardship. Russians respect the need to drink thoughtfully or sorrowfully. At the same time they join together in triumphant discovery—singing, laughing, and forgetting time. Especially in rural areas, drinking is a part of any type of festivity. Despite severe shortages, Russians manage to make a memorable feast out of nothing. Kitchen tables are the most likely setting for sharing conversation and vodka with friends. Throughout this culture there is a tenderness, and even an affection, shown by sober people to those who have overindulged. Everyone seems to understand the Russian need to consume vodka, although alcoholism is a widespread problem that has been linked to many social ills, including the premature deaths of Russian males at an average age of 59.

Another important difference is Russian patience in working out compromises. Whereas time is a precious resource for American negotiators, deadlines often go ignored in Russian negotiations. Budgets and production schedules have traditionally been difficult to meet, and the prevailing attitude has been "why bother?" Within the bureaucracy, it is apparent that citizens are used to tolerating long delays and complacency.

Compromise is not always necessary for achieving goals; compassionate appeals may be equally effective. Yale Richmond (1992) describes the time that he arrived at his hotel in Moscow hungry, after a long plane flight, only to learn from the woman in charge of guest services that the currency exchange office was closed for the day. He was traveling alone, did not have any rubles, and therefore would be unable to pay for dinner. When asked, she had no suggestions to offer. Instead of retreating, Richmond chatted with this complete stranger about his trip, the weather, his family, and her children, and eventually he returned to the subject of his hunger. This time she reached into her own purse and lent him some rubles until the next day. Although he did not expect to bargain with the woman, he approached her as another human being, and she responded with kindness. This is a common experience in Russia.

Realistic Depictions

Haskell (1963, 1968) makes several points about the role of realism in Russian ballet. First, Russian choreography tends to emphasize realistic interpretations through the expressions of characters, as noted in the part of Albrecht in *Giselle*. Whether Nureyev or Vikulov played the role, it was played with such vivacity that the transition from self-centeredness to remorsefulness over his lover's suffering was genuinely evident to the viewer. Besides individualistic realism, the spectacle is subordinated to the human values in the story. In Russian versions of *Sleeping Beauty*, the scene in which Princess Aurora falls into her slumber is choreographed with more than 100 dancers on stage. Even in the midst of a Russian crowd, the audience can sense a real interaction among the principal dancers. Western productions of the same scene tend to be vacuous, usually employing only Princess Aurora and four suitors at most.

Another Russian contribution to the world of ballet is the reliance on experts for accurate set and costume design. Leon Bakst, famous for creating lavish sets and costumes, was the first to tie his depictions to historical information relative to the era in which a ballet took place. Thus Russian costumes have always been able to boast a thread of truth, and sets on the Russian stage have likewise lent credibility to the ambience of the ballet.

It is well known that idealism is important to most Russians. During Communist rule, atheism became the official policy on religious affiliation as the party ideology replaced the church. Official pressures made it risky for most of society to openly participate in religion, which led to an underground system of hidden beliefs. Since the adoption of a 1990 law allowing religious freedom, many Russians have renewed their interest in the traditional values of the Orthodox Church. In 1991, for the first time in more than 70 years, the Orthodox Christmas Day (which does not coincide with December 25) was proclaimed a national holiday in Russia.

Realism also plays a role in everyday life in Russia, as people attempt to survive daily obstacles. Connections, or personal contacts, are easily the most valued perquisite for the privileged. Knowing someone who is able to influence an outcome is an accepted way of life in Russia. For example, pulling strings is an effective way to arrange the swap of apartments between strangers and acquaintances.

Connections can also be serious business, as in the case of a doctor willing to recommend a surgeon to a patient who wishes to circumvent the bureaucratic health system. (It is generally acknowledged that the public health system totally collapsed in the early 2000s and only recently has the government focused its attention on this area, with some good results.) The ultimate accomplishment of parents is to secure better educational opportunities for their children; special arrangements can be made outside of the formal admission system of a school. Regardless of the type of favor asked, it is understood that a return favor may be redeemed at a later date.

Another dramatic tool borrowed from a ballet performance is the use of *vranya*, or bluffing, and it is consistent with Russian historical norms. *Vranya* is best characterized by the statement: "You know I'm lying, and I know that you know and you know that I know, but I go ahead with a straight face and you nod seriously." It is publicly distasteful for most Russians to admit that anything has gone wrong. Social ills have been repressed in an attempt to maintain the image of Russian superiority.

Personal Qualities

From a foreign perspective, Russians have been characterized as serious and unexpressive, betraying no emotion in public. They hold a reputation for hardiness and stoicism. Often this external toughness comes across as coarse indifference or pushy discourtesy to the visitor. In the brusque surliness of service people and glum faces of crowds full of impassive stares, foreigners find the gruff, cold impersonality they expect. But this exterior is *maskirovannoye*, a false front. The stereotypical demeanor is part of the drama of Russian culture. Underneath the mask is a raw humanness that could not be more real.

Ballet is much more than a dance on stage; it is a marvel of choreography, music, costumes, and lighting that create a distinct mood for each audience. Choreographers express ideas through every dancer's movements, whether presenting a precise theme or leaving an audience free to use its imagination. Composers write scores to accompany the beautiful technique of ballet, and orchestral conductors lead musicians to perform scores at the proper tempo, rhythm, and volume for dancers. Wardrobe designers and lighting experts use special knowledge of their crafts for ballet productions, and visual artists also contribute to the mood by designing sets to provide the proper backdrop for each dance. Finally, trained athletic dancers combine flexibility, strength, and balance to perform a series of dances that often tell a story for spectators of all ages.

A final consideration of this characteristic of Russian culture is the seeming contradiction of drama and realism. On the surface, these principles appear to be opposites. Herein lies one of the major dichotomies within the culture. Theatrics project a contrived atmosphere, while realism is the rejection of visionary ideals. One of the more puzzling aspects of Russian culture is the illusion of disparity. Russians are categorized as both lazy and hardworking; they want to have money, yet they despise it. Russians are depicted as uninterested in building relationships (Adler with Gundersen, 2007) and at the same time as "warm and helpful" after new interpersonal relationships are formed (Richmond, 1992, p. 3). Russians seemingly want to get money rather than earn it. Russian people do not usually respect those who simply work hard, and yet they are willing to make sacrifices themselves. Denied religious freedom for more than 70 years, Russians still possess faith and hope. Russians tend to be dreamy and pragmatic; they place great value on the work of writers, artists, and dancers. Like members of other cultures, Russians are faced with choices about priorities.

Conflicting roles exist between public and private relationships. In their public roles, Russians are characterized as careful, cagey, and passive; in their private lives, they are depicted as honest and direct. Foreigners describe Russians as suffering, unruly, and stoic; they are also known as cheerful, generous, obedient, and hospitable. Like other nationalities they can be publicly pompous and privately unpretentious, caring or unkind. Perhaps the great novelist Fyodor Dostoyevsky described it best as "half saint and half savage." With friends, Russians feel free to pour out woes to one another and are not burdened by the need to disguise the realities of disappointment or pain.

The Russian Soul

From the 12th century until the time of Peter the Great, and then from 1917 to 1992 during the Communist era, Russians were isolated from the rest of the world. Two events, the Tartar (Mongol) invasion and the Ottoman conquest of Constantinople, abruptly halted commercial, religious, and cultural exchanges with other nations and limited Russia's development for centuries. Russia had relatively little need for trade because it was self-sufficient in agriculture; most of its interaction with other nations related to land disputes. From the Russian perspective, territorial expansion resulted from victories over foreign invaders. At one time or another, Teutonic knights, Lithuanians, Poles, Swedes, French, Germans, and Asian groups invaded Russia, which has no natural border defenses. On all sides, fear of invasion by neighboring adversaries led to a Russian preoccupation with state security. A tradition of citizens serving the state, whether peasant or nobleman, evolved from the suspicion of the hostile powers that surrounded Russia. In subsequent centuries Russian rulers purposely limited foreign influence on the empire, and as a result, later attempts to introduce new methods and technologies were met with skepticism by the people. Putin, for example, although he and former President George W. Bush were friendly on a personal level, reacted very negatively to Bush's plan to station nuclear weapons in Poland, which borders Russia. Russia's military incursion into Georgia in 2008, which is seeking to join NATO, seems to be partially based on the fact that Georgia was a part of the Soviet Union.

Today Russian distrust is still prevalent, as indicated by their feelings about these Western influences. Also, Russia is concerned about the rising importance of China, with which it shares borders. As we might expect, Russians want to be independent of foreign influence, and this is refracted in the concept of the Russian soul, which basically reflects the desire for distance so that the culture can be preserved, as it has been for so long. Reinforcement of the idea that distance is essential for survival is a characteristic of Russians, even today. As a people, Russians are considered self-reliant, strong-willed, and full of inner resources.

Although it may seem that so many years of tsarist regimes and the iron fist of communism have hindered the outward use of such resources, the argument is well supported that, in fact, individual perseverance and creativity have thrived in spite of absolute rule. Undoubtedly many Russians have been entrepreneurial since 1992, but the inadequacy of the infrastructure such as effective laws and banks and the increased intrusion of the Russian government in the private sector have stymied their efforts.

Each member of a Russian ballet troupe must endure laborious training on a daily basis and possess unfailing discipline to perfect the art. Classes, rehearsals, and performances compose an exhausting routine for Russian dancers; this cycle requires a dancer's complete attention on the world of dancing with little time for much else. This distance or separation from everyday life allows dancers to put all of their energy into the dance. Orthopedic surgeons, masseurs, and medicine have become part of the reality of contemporary ballet; physical demands often require dancing through bodily aches and pains. Relying on Russian self-determination, each dancer is able to achieve total concentration, fueled by the passionate intensity of the Russian soul. The result is unmatched excellence, both in technical performance and artistry.

The Russian soul is a mixture of intense feeling; it is emotion, sentiment, and sensitivity combined. It is triumph over seemingly insurmountable troubles or sadness arising from the discovery of stark truth. Over the years much has been written about the magic and mystic nature of the Russian soul. What is known as something mysterious seems to boil down to this: At the very core of individual Russian expression is the enigmatic principle or soul that places absolute value on decency, respect, honesty, and moral goodness. Part inherited and part learned, the stony resolve of the Russian people is unlike any other, and it can be interchanged with human compassion in the blink of an eye. Creative, electrifying, and daring, the Russian soul is the source of inspiration within the dancer.

There seems to be a strong relationship between the severe Russian winter and the Russian soul. To keep warm, the entire family would sleep together on top of the *pech* (Russian fireplace and stove), and to survive under such harsh conditions, it was necessary for everyone to work closely together. A strong in-group orientation developed in which people expressed emotions openly and automatically helped others; it was a classic generic culture of community sharing. For example, the U.S. coffee break is designed to "help people get through a day" or "wake up," but the Russian tea break is really about continually reconnecting emotionally with others (Ries, 1994).

A Contact Culture

Probably because of the harsh winters, which were instrumental in facilitating the defeat of both Napoleon and Hitler, Russians developed a contact culture reflecting the Russian soul. A contact culture is emotionally expressive and encourages people to communicate warmth, closeness, and availability for conversation. Intense eye contact, frequent touching, close physical distance, and smiling are emphasized (Andersen, 1994). And, given such a positive view of winter, it is little wonder that Russian art emphasizes winter scenes and activities. For example, many of the grand buildings in St. Petersburg and Moscow were designed and colored to look their best in the snow, the pure white of the fresh snow contrasting with the pastel colors of the buildings' exteriors.

The severe Russian climate is also responsible for both the resourcefulness and the strength of Russian people. For centuries, peasants have endured long winters of bitter cold temperatures, brutal winds, and frozen ground while longing for spring. Springtime inevitably brings the considerable work of quickly tilling soil and planting seeds so that crops may be harvested before the first frost less than 5 months later. Unpredictable weather and danger of crop failure are of great concern, yet Russians have proven able to endure hardship. Unable or in many cases unwilling to take risks with imported technologies, peasants rely on familiar if primitive tools and techniques. They are masters of patience and resolve, recognizing that they do not have control over everything in life. Yale Richmond (1992) explains,

> In contrast to Americans, most Russians have lived, until recently, much as their ancestors did before them—in small villages, distant and isolated, their freedom of movement restricted, and without the comforts and labor-saving tools provided by modern society. (p. 9)

In more recent years there has been an increased acceptance of modern technology, but productivity still lags when compared to other developed nations.

The geography contributes to feelings of powerlessness; the mountainous terrain and desolate barren spaces outside of the cities are under continual permafrost. Russia is one of the least densely populated nations in the world. There are only eight inhabitants per square kilometer.

Intimacy appears to be more pronounced among Russian families than elsewhere in the world. Accommodations are typically crowded, with multigenerational families living in small spaces. Although this is changing, extended families have resided together in cramped situations under relentless scrutiny for decades, usually within apartments. Beds serve as couches in sitting areas, and bathroom facilities are shared among many within the apartment. Privacy is an oddity; no Russian word exists to describe isolation in positive terms.

Throughout the arts, the Russian soul is the unmistakable power that compels the writer, actor, painter, or dancer to aspire to greatness. A potent force, the Russian soul shaped the insightful expression of Dostoyevsky's writings. His literature revealed the extent of responsibility that Russians feel for the afflictions and pain of the rest of the world. Company and conversations begin to matter as soon as one steps into a Russian's personal life. Among friends and family, Russians become the wonderful, flowing, emotional people of Leo Tolstoy novels, sharing humor and sorrows and confidences. They enter a simple intimacy in which individuals are less self-centered than in the West. Russians tend to pour out their hearts in total commitment to friendships and become easygoing, affectionate, and tender.

Furthermore, intellectuals look for a spiritual compensation to escape the boredom of everyday life, whether in ballet, novels, or poems. (There are, to be sure, exceptions to this generalization, for example, the night life of St. Petersburg is lively.) Russian culture reveres poetry for its wit, courage, and creativity. Poets such as Alexander Pushkin are regarded as heroes for daring to use metaphor to conceal meanings in light of censorship.

Many Russians like to think that they have a monopoly on virtue, whether it refers to family loyalty, a sense of duty, or love of nature. Emotionality is considered a positive attribute, and some of the most convincing arguments appeal to the irrational element of human existence. Guilt is also a factor in emotionality. Similarly the presence of guilt helps the dancer to feel more deeply. Russians feel that they can communicate emotions through their eyes, conveying love, hate, or passion. Whether expressing unbounded joy or woeful melancholy, the Russian dancer performs with an inner dynamism that members of other cultures find difficult to duplicate.

Just as Russian workers in the city or country survive harsh winters and hot summers, children also learn to adapt to hardship. Parental indulgence is a national cult, and toys teach meaningful lessons. Russian-made toys tend to break easily, and when children experience the disappointment or anger of a new or favorite toy falling apart, they learn that they have little control over their world. Russian children learn to accept without too much bitterness the fact that their world is not perfect. Perhaps this is why Russian ballet students are able to adjust without bitterness to grueling schedules of dance instruction, traditional classroom subjects, and physical workouts that build strength and flexibility.

Shame and Guilt

At all levels of society, the device of public shaming is central to the Russian system. There is a collective responsibility for children, exhibited through overprotectiveness. Russian babies are traditionally wrapped in *kosinkas*, a tight mummy-like wrapping that allows little movement. Russian culture reasons that swaddling is necessary for the first 9 months of life to protect babies from hurting themselves. Babies are unswaddled only for nursing and bathing, at which time they are showered with love and affection. Babies learn to develop complete trust in their mothers, who return their trust with complete and uncritical love. Thus infants are unable to fully explore their environment and must rely on only their eyes or their cries to express emotions.

During the period of swaddling, babies experience repeated feelings of security and anger, which are replaced with freedom and happiness when the *kosinka* is removed from time to time. Besides being totally dependent on their parents, Russian babies continually experience polarized emotions in their infancy. As a result, Russian children grow up with the dual beliefs of powerlessness and guilt for personal actions. Like the special relationship between Japanese mothers and their sons, which results in the creation of *amae* or dependency (see Chapter 3), long-lasting psychological dependency is generally emphasized in Russian culture.

Without a doubt, Russian society has a pronounced collective emphasis. Szalay (1993) found that U.S. Americans make almost twice as many associations with the word *individual* compared to Russians, whereas Russians make almost three times as many associations with the word *people* compared to U.S. Americans. Young children are taught how to behave and relate to others at an early age. Respect and deference for authority figures is a norm in every classroom. Landon Pearson (1990) points out that "A respectful and, indeed, slightly fearful attitude to adult authority is inculcated into Soviet children from the moment they set foot into an educational establishment" (p. 111). Prime Minister Putin and his followers have reinforced this message in recent years, as exemplified in the adage that all should follow a strong leader without question, just as migratory birds follow one bird. From childhood on, Russians acquire an acute sense of place and propriety, of what is acceptable and what is not. They generally know what they can get away with and what should not be attempted. Student discussion is limited and individual thought is not encouraged. Pupils learn that supplying "official" answers ensures good grades, which can lead to admission to higher education and a good job after graduation.

Educational institutions are designed to shape young minds using structured discipline and group dynamics. Much learning is based on memory drills and the modeling of rigid prescriptions. Enormous peer pressure is exerted over children to achieve stringent expectations. Russian children learn early the futility of challenging authority; parents are often chastised by teachers and other officials for the failure of offspring to meet social norms.

One of the more remarkable new universities reflects this modeling of strict behavior, but in both a new and traditional way reflective of the era of Russian royalty. The New University of the Humanities emphasizes the teaching of proper manners, particularly within the Institute for Noble Young Ladies, modeled on an imperial-era boarding school that trained the daughters of wealthy merchants and noblemen in the tsar's court. Other educational groups are also stressing etiquette, as it is critical in business, particularly international business. Many of the new rich are sending their children to these schools, as are many others who have difficulty paying the tuition (see Williams, 1998).

Formal ballet productions bring together dancing, music, design, and an audience. With the exception of solos by individual dancers, the ballet troupe dances together as one large body. Even these solos are precisely staged by choreographers to balance the entire performance. All members of the company are mutually dependent on each other, and individual needs, rights, and desires are subordinated to the whole. It is the choreographer's job to integrate the dancer's body with the music score, the possibilities of the stage, and the mood to be conveyed. Choreographers must "explore and expand the dancer's capabilities" while ensuring that the components of the ballet production, as well as the dancers, are all in sync (Clarke & Crisp, 1976, p. 38). As a result, Russian dancers think of themselves as members of a company, and "people's artists" rather than as individual dancers.

A Group Ethic

The group ethic can be traced to the *mir*, or the small village cooperative that made all farming decisions for the community, even through the early 1900s. Not only did the *mir* choose which crops to plant and when, it also collected taxes and resolved conflicts. As Richmond (1992) notes, "Its authority, moreover, extended beyond land matters. It also disciplined members, intervened in family disputes, settled issues which affected the community as a whole, and otherwise regulated the affairs of its self-contained and isolated agricultural world" (p. 15). The *mir* consisted of household heads who discussed issues and reached group consensus about courses of action. Decisions based on the "collective will" were binding on every household.

Even before the 1917 revolution, Russian tsars approved of this type of informal local government because such units were capable of controlling the vast populace. The *mir* enforced tax collection and military conscription as well as other programs that influenced the lives of the peasants. In describing the *mir*, Richmond (1992) aptly concludes, "Because the *mir* affected so many people, it played a major role in forming the Soviet character" (p. 15). It is interesting to note that popular usage of the word *mir* refers to the definition of peace.

A few examples of collective behavior include communal seating in restaurants and group participation in attending cultural and sports events. Physical contact such as jostling or pushing among total strangers is typical in crowds, as is holding hands among friends of the same sex. Older Russians generally are more than happy to offer unsolicited advice, even to foreigners, and it is seldom considered impolite to drop in on friends unannounced (Richmond, 1992, pp. 17–20).

The Globe study (House et al., 2004) identified Russia as part of its Eastern European cluster, which scores high on gender egalitarianism, in-group collectiveness, and assertiveness, but low on performance orientation, future orientation or deferring gratification to achieve long-term goals, and the acceptance of risk or uncertainty avoidance. These findings add support to the profile of Russia in terms of its cultural metaphor, the Russian ballet.

In sum, like the ballet, Russian culture can be unpredictable and enigmatic. Russian culture today fuses together the earnest elements of emotion, spirituality, and survival once found in Old Russia. Echelons within Russian society, drama and realism, and the Russian soul as expressed in Russian ballet are characteristics that have made the Russian people culturally unique. Although Russia may privatize its economy significantly and become more Western in its economic orientation—and the jury is still out on this matter—the characteristics of the Russian ballet described in this chapter should be of help in understanding its underlying culture for years to come.

PART VIII

The Base Culture and Its Diffusion Across Borders (Clusters of Nations)

The Example of China

n the preceding chapters the basic unit of analysis was each nation and its cultural metaphor. However, migration has played a critical role in cultural transmission. For example, the English culture has had a significant influence on the national cultures of Australia, Canada, the United States, and New Zealand.

In this part of the book we focus on the base culture of the Chinese through a discussion of China's Great Wall and cross-cultural paradox. We then examine how this base culture influenced the overseas or expatriate Chinese living in various nations through an analysis of the Chinese family altar. Finally, we look at the creation of the Singaporean national culture in a nation that is predominantly Chinese but also includes significant representation of other ethnic and cultural groups. The development of the Singaporean hawker centers, our cultural metaphor for Singapore, is a distinctive expression of this nation and its culture.

A useful discussion among readers of this book could center on other base cultures and their influence around the world. Several researchers have shown that there are cultural clusters of nations considered in such terms as shared religions, languages, and geography. In Chapter 1 we have described one of the most prominent clustering studies, the GLOBE study (House et al., 2004). The reader may want to review these clusters.

It should be emphasized that the concept of country clusters is increasingly important in the modern world. Although we have a global economy, most nations tend to trade most heavily with their neighbors, for example, the NAFTA Agreement among the United States, Canada, and Mexico. But bold experiments are taking place, such as the European Union's attempt to include a wide diversity of national cultures within its scope, not to mention the World Trade Organization whose membership includes 156 of the world's 220 nations. Thus both regionalization as exemplified by NAFTA and globalization as exemplified by the World Trade Organization are simultaneously occurring, as are bilateral agreements between two nations. In such a complex world understanding both the cultural metaphors for each nation and the clusters in which they exist is not only helpful but necessary.

CHAPTER 25

China's Great Wall and Cross-Cultural Paradox

Funeral spending has increased with incomes in recent years, making the industry among China's 10 most profitable, with many urban Chinese spending several years' worth of disposable income on funerals, government figures show.

—Mark Magnier (2007, p. A1)

Donning a rough cotton white-and-red headdress, I took a place in the line of her descendants, trudging in her funeral procession and kowtowing to her casket as we wound our way through her lakeside township to a mountainside burial site. And as I did, I was struck by the enduring power of ancestors in the lives of the Chinese. Spirituality and authority emanate from a church or temple; they reside with the elderly and one's family tree. People don't disgrace themselves in the eyes of God; they do so in the eyes of their forebears.

—John Pomfret, a *Washington Post* writer married
to a Chinese woman describing the funeral
of the wife's grandmother (2003, p. B2)

China is undergoing massive changes that began in the late 1970s, at which time a new Communist government stressing a mixture of communism and free enterprise replaced the disastrous Mao Tse Tung government. Mao and his associates ridiculed and decimated the middle class and attempted to stamp out Confucianism with its emphasis on traditions, relationships, and education. Starting in the late 1970s, the government began sequentially to emphasize free enterprise and infrastructure development in various regions of the nation. As a result, the economy grew rapidly and China's Gross National Product supposedly will surpass that of the United States by 2025.

While the standard of living rose, however, so did economic inequality. Regional governments own the land and began to move farmers off of it, with the result that China is now undergoing the largest transition from a rural to an urban population in history. The International Monetary Fund ranks China's economy on a per capita basis 109th among all nations. Although its exports have increased very significantly, nearly 60% of this output is generated by non-Chinese-owned companies, and 89% of China's high-tech exports follow this pattern. Soon China will surpass the United States as the No. 1 emitter of greenhouse gases; it leads all other nations in destroying the ozone layer; 70% of the country's lakes and rivers are polluted; and half of the population does not have access to clean drinking water.

China's demographics are very problematic as a result of the one-child-per-family rule. The number of children born to a Chinese woman stands at 1.8, below the replacement level for population stability of 2.1. This is occurring at a time when life expectancy has increased dramatically, from 35 in 1949 to 73 years. Thus its population is aging and most of its citizens are not protected by either a government or a private pension system (see Pomfret, 2008).

Still, visiting modern-day China is a shock for anyone who has not been there since 1978, at which point the major mode of transportation was the bicycle. People resided primarily in the countryside and they were poor, and the Mao ideology emphasized the importance of the iron rice bowl from which all would eat, even though at least 50 million starved because of his ill-conceived policies. Today China is an automobile-clogged, fast-moving nation that is consuming vast amounts of resources such as steel and oil. China's population of 1.35 billion is the largest among the world's nations. There is a great stress on consumerism and satisfying one's own needs and desires, even at the expense of short-changing aging parents, long revered in the Chinese culture. Public shaming of grown children who have abandoned their parents is common, to the extent that they have been sent to jail (Magnier, 2006).

Still, there is a great emphasis on the past and tradition, as our introductory quotes suggest. Yu Dan, a relatively unknown professor of communications, has become famous as the author of a best-selling book on Confucius, and there are now schools for children stressing Confucian thought (Fan, 2007; Ni, 2007).

Attempting to describe and explain any cultural group is a daunting task, and this is doubly true of the Chinese, who have been and continue to be stereotyped in various guises, such as the shifty-eyed Chinese, the unfathomable Chinese, and so forth. Even the Chinese have difficulty explaining themselves, in part because they tend to be high-context communicators; they also have faced exclusion and discrimination in various nations and tend to play things close to the vest. There are countless books and articles on the Chinese, but most of them focus on "do's and taboos" or obvious facets of Chinese culture, such as that it is an authority ranking and high-context culture and that the color red is a sign of good fortune. However, as we will see in "The Singapore Hawker Centers" (Chapter 27), even this generalization about high-context communication must be qualified.

In 1999, Tony Fang published a masterful book on Chinese business negotiating style that provided the kind of insightful analysis that has been sorely missing from the literature. Fang is a native of China who was a business negotiator for Chinese firms for 20 years, after which he completed his doctorate at Linkoeping University in Sweden. These experiences have provided him with a unique understanding of both Chinese and Western perspectives. His book revolves around a paradox, namely that Western negotiators have consistently described Chinese negotiators as both sincere and deceptive. How can this be?

To understand the Chinese, it is helpful to focus on both the paradox that Fang (1999) has identified and a cultural metaphor unique to China, its Great Wall. While we use the visible Great Wall as our cultural metaphor, we employ it to understand the invisible Great Wall of Chinese values, attitudes, and behaviors.

Paradox is central to Chinese thought. Essentially a paradox represents the concept of interdependent opposites or the existence of two opposite states of reality simultaneously. For example, *many* and *few* combine to mean *how much* in the Chinese language, and *inside* and *outside* together mean *everywhere*.

As in previous chapters, each subdivision of this chapter represents a dimension or aspect of the cultural metaphor. In this chapter, we use the three related explanations that Fang (1999) has proposed in answering his paradox as the distinctive features of our cultural metaphor: China's long, tortuous, and complex history; the influence of Confucian thought on Chinese behavior; and the Chinese perspective on war and the marketplace, particularly as enunciated in the writings of Sun Tzu and applied to the marketplace in terms of six principles of Sun Tzu (McNeilly, 1996).

The Great Wall: Long, Tortuous, and Complex History

Anyone who has ever visited the Great Wall of China tends to be overwhelmed by its scope, magnitude, and majesty. The Wall is 7,200 kilometers or 4,474 miles in length. It is frequently said that the first Emperor of China, Qin Shi Huang, began the construction project in 221 BCE by restoring the ruins of older walls and linking them with new construction; the result was a 3,000-mile wall. The intention underlying the construction was to ward off marauding warriors and tribes. Frequently marauders settled near the Wall after being repulsed, traded with the Chinese, and in many cases, intermarried with them and eventually were allowed to enter China. Unfortunately, however, the first emperor was repressive, harsh, and autocratic, even to the extent of outlawing Confucianism (see below). As in the case of many such autocrats and individuals seeking to demonstrate that they are superior to others, Emperor Qin also constructed a Great Tomb for himself, on which 700,000 workers labored for 34 years. The well-known 7,000 terra-cotta warriors found in his tomb are a popular tourist attraction as well as a stark reminder of the futility of such efforts; the Qin Dynasty fell in a peasant revolt in 209 BCE, a year after the emperor's death.

In that same year, the Han dynasty arose. Today, about 85% of the Chinese are classified as Han, although China is home to about 400 ethnic groups (some as small as a few thousand in number) and 56 spoken languages. China is much more culturally integrated because of this Han association than most nations, as evidenced by the fact that there are only about 400 Chinese family names. It is easy for the Chinese to categorize one another because of such uniformity.

The Han conquered central Asia, extended the Wall for another 300 miles into the Gobi desert, and provided protection for the legendary Silk Road linking traders from distant points to one another. It was impossible for traders to reach their destination without passing through the Great Wall. The Han also added beacon towers to the Wall and used a sophisticated system of smoke signals to warn of impending invaders and their approximate numbers.

The Ming dynasty, which arose in 1368 CE, extended the Wall to its current length and is responsible for the structure we see today although it is based on earlier work. Ming walls were bigger, more ornate, and more impressive. During this dynasty China was at its height of power, and the nation was increasingly referenced as "the Middle Kingdom," that is, the center of the Earth. Geographic maps were drawn to show that China was the center of the Earth. There was good reason for this perception, as China was far ahead of the West during this period.

As this discussion suggests, China has experienced a long, tortuous, and complex history extending over thousands of years. There have been periods of tranquility, sometimes lasting hundreds of years, but there have also been periods of extreme social unrest. Throughout these eras the Wall has served the Chinese well and preserved their culture.

U.S. Americans who are at all familiar with Europeans quickly notice how they have a grasp of their respective nation's history that far exceeds the time perspective of many if not most U.S. Americans; Europeans talk easily about 10th-century kings and queens, for example, such as King John III or King John IX. In the case of the Chinese, the time horizon is much, much longer. It is frequently said that U.S. strategic planners think in terms of 3-year or 5-year plans whereas Chinese strategic planners focus on plans extending 100 or more years into the future. This long-term perspective is associated with tradition and ancestors, as many firms are controlled by one family and frequently by the patriarch in it. In Asia, with its emphasis on the family and the past, two thirds of publicly traded companies are controlled by a single shareholder, compared with just 3% in the United States, according to the World Bank (McBride, 2003). China follows this pattern.

Furthermore, China exemplifies a torn culture, having been ripped from its cultural roots several times. Since 1900 China has experienced a dizzying array of governmental arrangements, including the disastrous Mao Communist era starting in 1949, which effectively blocked China's entry into the global economic market. In the late 1970s a newer group of more enlightened Chinese Communist rulers emerged who fostered China's integration with the global marketplace. As indicated previously, they introduced economic development in phases, developing specific areas of the nation before others to use scarce resources effectively. This approach, in sharp contrast to the approach used in Russia (see Chapter 24), has proved to be successful, and development economists uniformly praise it (see Stigler, 2002).

A visitor to Shanghai—which seems to combine the best of U.S., Asian, and European architecture—is immediately impressed by the skyline, and this positive impression is constantly reinforced when talking with governmental officials and business executives, who exhibit a cultural uniformity of purpose. Shanghai is frequently called "the city of the future," and it and other Chinese cities are being built in the manner that guided the Great Wall's construction: patience, hard work, a long time horizon, and a systematic approach integrating major cities and regions.

Confucianism and Taoism

It is impossible to understand China's culture without understanding Confucian thought. Confucius died in 479 BCE, frustrated that he was never able to become a major adviser to any of China's regional rulers. However, his ideas took hold in large part because of the vicious warfare that was occurring between these rulers, and his emphasis on forming and permanently maintaining tradition deliberately in accordance with China's idealized past period of Great Harmony had wide appeal.

Confucian thought is encapsulated in five terms, the first of which is *Jen*, meaning human-heartedness or the simultaneous feeling of humanity toward others and respect for oneself (see Smith, 1991). Second, *Chuntzu*, "the Superior Man," is someone fully adequate and poised to accommodate others as much as possible rather than to acquire all he can selfishly. At all times he should be sincere with others, which accounts for the sincerity of Chinese negotiators that Fang (1999) highlights. Third is *Li* or propriety, which refers to the way things should be done. For example, the Five Relationships involve the appropriate conduct that should occur between father and son, older and younger siblings, husband and wife, older and younger friend, and ruler and subject. The Five Relationships are central in the Confucian system and clearly encourage the creation of an authority ranking culture.

More specifically, *Li* reflects the importance that Confucius attached to the family, and the idea evolved from the earlier era of ancestor worship to include the concepts of filial piety and veneration of age. From the Chinese perspective a person exists only in relation to others, particularly but not exclusively members of the family. There is a second meaning of *Li*, ritual, and it is envisioned as encompassing all of a person's activities during his or her whole life.

Te is the fourth term, and it refers to the power by which men should be ruled, not by force but by moral example. Finally, there is *Wen*, which accords a place of prominence to the arts as a means of peace and as an instrument of moral education.

The device for developing tradition deliberately was extreme social sensitivity, and this is still reflected in the Chinese concept of face. As discussed in "The Japanese Garden" (Chapter 3), face is an unwritten set of rules by which people in society cooperate to avoid unduly damaging one another's prestige and self-respect. In bargaining, for example, the winner should allow the loser some minor tactical reward, especially when all the observers can ascertain the identity of the vanquished. If a father's business becomes bankrupt and he dies suddenly, frequently his sons will work for years to pay off the debt to maintain face for the family. When an individual loses face, he or she tends to adopt a stony or blank expression as if nothing has happened; generally face is forfeited through loss of self-control or a display of frustration and anger (Bonavia, 1989, pp. 73–74).

Of course, individuals in all cultures want to protect their face or honor. As a general rule, people do not enjoy losing, being humiliated, or experiencing any related outcome. However, the Chinese added a unique dimension to the concept of face, as it is incumbent on everyone in a transaction not only to preserve face but to give face, saving themselves and everyone else from embarrassment. This is how a Superior Man should behave.

Some experts have argued that Confucianism is not a traditional religion, as it has no concept of a personal god and only an amorphous concept of heaven or a shadowy netherworld in which ancestors live and help to guide the living. Thus the Chinese tend to focus on this world and not the next. David Bonavia (1989) aptly captures the essence of being Chinese in the following description, although he was primarily discussing modern China:

> The most determining feature of the Chinese people's attitude to the world around them is their total commitment to life as it is. . . . In this world view, all human activity—religion, sex, war—consists of functional acts aimed at achieving something. Only the arts are considered to have intrinsic value, and they are chiefly reflections of the real world or an imagined world, not abstract patterns. . . . Action must have a purpose, the Chinese feel; there is nothing ennobling about pain, and death is an infernal nuisance. . . . Most Chinese, seeing a Hindu holy man stick a knife through his cheeks or walk on coals, would conclude that he was either a fool or a fraud. . . . The central concept of Chinese society is functionality. (pp. 56–57)

Taoism is usually considered the primary religion among the Chinese, while Confucianism relates to the manner in which an individual is to behave in society. Taoism was supposedly created by Lao Tzu, who was born in 604 BCE, and its main tenets were outlined in one small book, *Tao Te Ching* or "The Way and the Power." *Tao*, or the way, has three overlapping meanings, the first of which is that it can be known only through mystical insight. In the second sense *tao* refers to the ordering principle behind the universe or all life, and it represents the rhythm and driving force of nature. The third sense of *tao* is the way man should order his life to be in balance with the universe.

The Nature of Power

Power, the second part of the title of the book, refers to the belief that a person gains power by leading a life that is in harmony with the dictates of the universe. This approach continues to shape the Chinese character and is manifested in the desire to achieve a state of serenity and grace. The basic approach to life that is consonant with the universe is *wu wei* or creative quietude. Paradoxically, it is simultaneous action and relaxation and letting behavior flow spontaneously. Creative quietude is never forcing or straining but, rather, seeking the empty spaces in life and nature and moving quietly and without confrontation through the avenues of least resistance.

Taoists reject all forms of self-assertiveness and competition and seek a union with nature and simplicity of life more than materialistic possessions. There is an extreme aversion to violence that verges on pacifism. Taoism also includes the traditional Chinese symbols of yin and yang, which connote that there are no clear dichotomies, but rather that all values and concepts are relative to the mind that entertains them, that is, values and concepts are by definition paradoxes. Thinking in yin-yang terms means analyzing the universe into pairs of interacting opposites, such as shadowed and bright, decaying and growing, moonlit and sunlit, earthly and heavenly, and feminine and masculine. Whether something is classified as yin or yang depends not on its intrinsic nature but on the roles it plays in relation to other things, which is consistent with the relation-based system of the Chinese. Clayre (1985) expands on these ideas as follows:

> In relation to Heaven man may be classified as yin, but when paired with Earth he would be seen as yang. Heaven itself is the supreme embodiment of the yang aspects of the cosmos: ethereal, bright, active, generative, initiatory and masculine. Earth is seen as deeply yin: solid, dark, cool, quiescent, growth-sustaining, responsive and feminine. (p. 201)

Men and women are not seen as exclusively yang or yin: Each has a predominance of one aspect or the other, and the balance within them and between them may change. The relation of the two elements is always changing, a continuous cycle where each may dominate in responsive sequence. This idea may have evolved from the annual cycle of growth and decay found in agricultural China and in the rotation of day and night and seasons of the year.

Taoists traditionally had little empathy for Confucians, whom they viewed as pompous and ritualistic. A more fundamental difference is summarized by the

Chinese themselves, who say that "Confucius roams within society, Lao Tzu wanders beyond." However, in many ways Taoism, Confucianism, and Buddhism are complementary, and it is commonly said that they are "the Three Faiths in One." Thus many Chinese accept Confucianism as a guide to daily living, have recourse to Taoist practitioners for ritual purifications and exorcisms, and employ Buddhist priests for funerals.

We have sketched some of the major concepts and values that make up the Chinese value system, and there are wide divergences from it in many situations. Still, this value system has not been stamped out or radically changed for thousands of years, as Mao Tse Tung and others would have liked. Rather, it has evolved slowly over time. Ambrose King (as quoted in Kotkin, 1993) has pointed out that many of the 60 million expatriate or overseas Chinese have developed a culture of "rationalistic traditionalism, a combination of traditional filial and group virtues with a pragmatism shaped by the conditions of a new competitive environment" (p. 177). Like the Italian Catholics who attend church only irregularly, these expatriate Chinese and even those living within the confines of China are more culturally than spiritually influenced by their Three Faiths.

Many Languages

In a similar manner the Chinese culture is closely linked to language. People in China's various regions are actually a mix of diverse ethnic origins who speak mutually unintelligible dialects. In southeastern China, where Cantonese is the primary dialect, it is difficult if not impossible to understand the dialect of northern China, which is Mandarin. Spoken Mandarin is very formal, whereas Cantonese includes many more slang words and is spoken more loudly, even to the extent that two people expressing love to one another can sound to outsiders as if they are quarreling (Pierson, 2006). Remarkably, all dialects of spoken Chinese can be written in the same way and understood by anyone, regardless of the pronunciation of words. Whereas the English language employs only 26 letters to represent all spoken sounds and written words, the Chinese language is a storehouse with more than 50,000 different symbols or logograms, each standing for a different word. Amazing feats of memory are required to master such a language, which tends to reinforce cultural values. This generalization is also valid for the Japanese language, which was derived from the Chinese (see De Mente, 1990).

As this discussion suggests, English is a low-context language that is direct and understandable, and this has facilitated the use of English worldwide, especially in business. However, both the Chinese language and the Japanese language are high context, and the writer must choose words carefully, as there are complex meanings that can be inferred from the context in which they are spoken or written. Norman Fu, a well-known Chinese journalist with the *China Times*, once remarked that he much preferred to write in English than in Chinese because of this fact (see Chen, 2001).

When a visitor treks along part of the Great Wall, it is easy to imagine not only the repulsion of invaders but also the long and complex development of the Chinese perspective, thought systems, and language. Like the Wall, Confucianism and Taoism did not happen overnight, and although the Chinese adopt other perspectives such as Buddhism, they retain the best of their own systems. Chinese thought is conservative, but it does allow for change, a topic to be explored in depth in the next chapter.

Sun Tzu, War, and the Marketplace

Around 400 BCE, one of the greatest, if not the greatest, treatises on war and the manner in which it should be conducted successfully was written by Sun Tzu, a Chinese general. It is impossible to do justice to this great work in a short summary, and readers should form their own judgment by reading the original, which is reprinted fully in McNeilly's (1996) book. McNeilly derives six principles from Sun Tzu's work and applies them specifically to business and the marketplace. In this section we describe these six principles and relate them to the paradox that Fang (1999) identified, namely that Chinese negotiators are both sincere and deceptive simultaneously.

As both Fang (1999) and Chen (2001) point out, Sun Tzu's book should have really been called *The Art of Peace*, as the very first principle that Sun Tzu postulates is that war should be avoided if at all possible, as it necessarily results in the destruction of resources and people. McNeilly (1996) phrases this principle as: Win all without fighting, that is, capture the market without destroying it. In the airline industry, for example, about half of the strategic and tactical moves that companies make are price reductions, which tend to result in vicious price wars and frequently excessive losses for all companies involved (Smith, Grimm, & Gannon, 1992). Part of Cisco's success, by comparison, was based on acquiring competitors through stock swaps that greatly benefited all parties. To alleviate anxiety among employees in the newly purchased subsidiary, Cisco also established the principle that the CEO of Cisco and the CEO of the subsidiary would need to agree before a major downsizing of the subsidiary's workforce would occur.

The other five principles are: avoiding strength and attacking weakness or striking where competitors least expect it; deception and foreknowledge or maximizing the power of market information; speed and preparation or moving swiftly to overcome your competitors; shaping your opponent or employing strategy to master the competition; and character-based leadership or providing effective leadership in turbulent times. In addition to Sun Tzu, an anonymous writer also penned 36 famous stratagems somewhere between 1368 and 1600, for example, clamor in the East but attack in the West. These tend to be consistent with Sun Tzu's principles and help to enrich the Chinese perspective.

Fang (1999) profiles the differences in Chinese and Western thought by comparing *The Art of War* to Karl von Clausewitz's (1832/1989) classic *On War*. Both generals agree that war should be used for political purposes. However, whereas Sun Tzu emphasizes winning without fighting and related principles, von Clausewitz promotes the use of maximum force to overwhelm the enemy. World Wars I and II and their extreme devastation of populations, land, and cities are examples of the use of von Clausewitz's principles.

Fang (1999) proposes that there is an invisible Great Wall separating Western and Chinese negotiators; he also points out that many other Asian negotiators have been heavily influenced by the Chinese and exhibit many of the same patterns of behavior and thought. While Chinese negotiators are hospitable to Western negotiators and take them as a matter of courtesy to visit the Great Wall, they protect their invisible fortresses as much as possible.

Three Forces

For Fang (1999), three forces have molded the Chinese personality or negotiating style. The first is China's long history and particularly its recent history since 1949, when Mao came to power. The Chinese are acutely aware of the damage that Mao inflicted on the nation and want to avoid it in the future, especially as it may affect them. About 1990 two Carnegie Mellon professors completed a video on business in China, but many of the interviewed executives demanded that their faces be unidentifiable on the videotape because of fear that their interviews would come back to haunt them. Such fears are justified, as having worked for a Western company in China in the 1920s was grounds for arrest under Mao after 1949. Capitalism and "capitalist roaders" or sympathizers were personified as intrinsically evil. Thus Chinese negotiators in many ways tend to be conservative, although business executives in China today typically do not mask their identities as they did in 1990.

Fang's (1999) second force is the Chinese perception of the marketplace. Chinese negotiators have been heavily influenced by Sun Tzu, and they view the marketplace as a place of war in which the goal is to win all without fighting. They employ all sorts of ruses, subterfuges, strategies, and stratagems (including the famous 36) in their interactions with Western negotiators, whom they frequently regard as marauders seeking to penetrate the invisible Great Wall. U.S. negotiators, for example, get down to business as soon as possible and quickly establish an opening position from which they bargain. Chinese negotiators, however, tend to focus on establishing a relationship, trying to understand the other party, and allowing a position to be established naturally over a long period of time. They are accustomed to haggling for long periods, which tends to bother many U.S. negotiators.

Fang's (1999) third force is the Chinese personality or negotiating style, which has been molded in large part by Confucianism and its concept of the Superior Man who is sincere. Thus Western negotiators confront a conservative negotiator who is well prepared, kind, sensitive to the needs of the other party, but deceptive.

China is a large nation, about equal in size to the United States, and its citizens constitute about one out of every five people in the world. As Western and Chinese negotiators learn more about one another, it is most probable that their perspectives will begin to overlap. Still, the long, convoluted, and complex history of China, plus its simultaneous emphasis on both deception and sincerity, will continue to support the invisible Great Wall that must be patiently addressed if it is to be scaled. It is not an impossible task, as Western companies have learned. After years of complaining that Western investment in China was not profitable, economic reports suggest that many Western companies are now profiting handsomely from their investments. Still, the message of this experience, and the tripartite model that Fang has developed to explain the Chinese perspective, suggest that the task will be long, involving, but ultimately rewarding both in personal relationships and market success.

CHAPTER 26

The Chinese Family Altar

One of the most famous sites on the campus of Tianjin University . . . is a stone engraved with a copy of the very first diploma. . . . Next to the date on the diploma, 1900, the graduate's name is printed, along with the names of his father, grandfather, and great-grandfather. . . . The diploma speaks volumes about the strength of family tradition in Chinese society. Like any individual in Chinese society, the graduate exists primarily in the context of his family. His achievements belong to them all.

—Ming-Jer Chen (2001, p. 19)

When Westerners visit a traditional Buddhist temple in Asia, they see different images of the Buddha, such as the Reclining Buddha, the Sitting Buddha, and so forth. However, in a Chinese Buddhist temple these images are supplemented by a formidable group of statues that seem to represent warriors who are massive in size and strength. These fierce-looking warriors frequently have rough-looking beards and mustaches, and they carry large swords. It is a surprise for many Westerners, then, to learn that these warriors are actually representations of important people within extended kinship and family groups and that they actually lived in Chinese villages 1,000 or 2,000 years ago. The position of honor that these warriors have in the temple is but one demonstration of the importance that the Chinese place on the family and kinship group, even to the extent of allowing them to be situated alongside the Buddha images.

Throughout this book, we have given prominent attention to the nation as our unit of analysis. However, as many studies such as the GLOBE study (House et al., 2004; see Chapter 1) have demonstrated, it is possible to cluster countries into groups that are similar to one another in terms of language, religion, and geographical closeness. Nevertheless, we have found it difficult to identify metaphors that are appropriate for groups of nations.

The Chinese, however, represent one ethnic group for which one metaphor is appropriate, regardless of the nation in which they reside. This does not seem to be true for other ethnic groups. Thus Joel Kotkin (1993) has described the major ethnic groups—the expatriate or overseas Chinese, the Japanese, the Jews, the English, and the Asian Indians—who have achieved remarkable economic success outside of their homelands. However, the metaphors for the non-Chinese groups as described in this book—the Japanese garden, the kibbutzim and *moshavim*, the traditional British house, and the dance of Shiva, respectively—do not apply fully when members of these groups become expatriates.

The Importance of Family

The importance of the family and kinship group cannot be overestimated for the expatriate Chinese, and the metaphor we have selected for them—the family altar—reflects this fact. One Taiwanese diplomat who lived in several countries had a family altar in Taiwan but, when traveling, carried drawings of his ancestors to whom he and his children prayed; Amy Tan's (1991) best-selling novel, *The Kitchen God's Wife*, includes the concept in its title; many expatriate Chinese travel to their original homeland so that they can visit the burial places of their ancestors at least once in their lives because in Confucianism the living relatives have the obligation to tend such places on a regular basis if at all possible or at least to visit them; and the Chinese family name is traditionally given before the personal name, unlike the Western practice of mentioning the personal name first. Furthermore, many Chinese long to return to their native villages for burial alongside their ancestors. These examples—and countless others could be cited—indicate the prominent place that family and kinship groups hold in Chinese society, whether at home or abroad.

Of course, some may argue that other ethnic groups are also family oriented. However, Chinese society is neither individualistic nor collectivistic as are other family-based ethnic groups; instead, it is based on relations (Bond, 1986). Confucians, for example, feel that individuals have roles they must fulfill but, in doing so, their individualism can enlarge and enrich them for the greater good of the family and kinship group. From the Chinese perspective, people exist only in relationship to others (Chen, 2001). The relation-based system of the Chinese in which roles are differentiated is quite different from the more collectivistic and non-differentiated system of the Japanese, which helps to explain why the Japanese seem to favor the development of corporations such as Sony whereas the Chinese seem to favor family-centered businesses. As indicated in the previous chapter, 25% of the publicly traded companies in Asia are controlled by a single shareholder, very often the founder, but the comparable figure in the United States is just 3%. Of course, many of these are Chinese family businesses (McBride, 2003).

Chen (2001) indicates that the correct term is not *family businesses* but *business families* uniting generations. Similarly Fang (1999) points out that these business families operate like insurance companies focusing on building and preserving family assets. Nakane (1973) has noted the following:

> In the Japanese system all members of the household are in one group under the head, with no specific rights according to the status of individuals within the family. The Japanese family system differs from that of the Chinese system, where family ethics are always based on relationships between particular individuals such as father and son, brothers and sisters, parents and child, husband and wife, while in Japan they are always based on the collective group, i.e., members of a household, not on the relationship between individuals. (p. 14)

Hence the Chinese tend to take seriously their relations with others, whether inside the kinship group or outside of it. They have separate words for older brother, younger sister, and each aunt or uncle from each side of the family, and they tend to make life-long friends who are clearly demarcated from acquaintances. Still, like many other high-context ethnic groups, the Chinese usually spend a long time getting to know people before doing business with them. Other unique characteristics of the Chinese family system are described below.

Furthermore, like the Japanese and other group-oriented cultures, the Chinese tend to experience difficulty differentiating the individual from the group. As noted previously, there is no equivalent word for *privacy* in Chinese and some other languages that stress the importance of the group rather than the individual, such as the Japanese and the Arabic languages. In fact, the word *I* has negative connotations associated with it in the Chinese and Japanese languages, as do many of the other characters that are used in conjunction with it.

Chinese culture is closely associated with Confucianism, Taoism (pronounced Dowism) and, to a lesser extent, Buddhism (see Chapter 25). We will describe the Chinese altar and its three dimensions and assume that readers are familiar with these three religions. The specific aspects of the altar are: roundness, symbolizing the continuity and structural completeness of the family; harmony within the family and the broader society; and fluidity or the capacity to change while maintaining solid traditions. We will then briefly describe the activities of the 60 million expatriate Chinese and the countries in which they have been such an important economic driving force.

The Expatriate Chinese

Today there are more than 60 million expatriate Chinese, and many of them have been spectacularly successful in businesses throughout the world when compared to members of many other cultural groups. Initially these emigrants from China did not come from mainstream Chinese society, which was at times insular and suspicious of foreigners whose cultures they generally considered to be inferior to that of the Chinese. Rather, they originated from the peripheral southern regions of China where control and subservience to the emperor was less rigid.

In some of the countries in which these expatriate Chinese settled, such as Thailand and the United States, they have integrated themselves effectively and have even married non-Chinese in increasingly large numbers. In other countries such as Malaysia and Indonesia, the Chinese experience discrimination and resentment, and they tend to be less effectively integrated with the other cultural groups. Since the 1960s, there have been sporadic violent riots directed against the Chinese and their properties in Malaysia and Indonesia in which thousands of Chinese were killed and injured and their properties destroyed.

In the late 1990s the Indonesian government assumed control of major Chinese companies valued in the billions after a violent period of Chinese-focused riots; at the time 5% of the population was Chinese, but they controlled 70% of the nation's resources. Ironically, most of this wealth was controlled by only a small percentage of Chinese families who partnered with the government to expand business. These Chinese families were well-positioned through *guanxi* and knew how to take advantage of business opportunities.

In Malaysia a Chinese cannot be the chief executive officer of a company seeking government contracts. To get around this restriction, the Chinese put a native Malay into this position while retaining most of the power; in recent years the Malaysian Chinese have begun to withdraw into protected communities and Chinese-only private schools.

Some of this discrimination seems to be religious in nature, for example, Muslims in Malaysia and Indonesia versus Confucians. Whatever its causes, it has divided such countries into Chinese and non-Chinese groups that experience difficulty in developing harmonious relations.

Taiwan, a country of 22.7 million people, is an interesting example of the success that the expatriate Chinese have wrought. Taiwan was established as a separate country after World War II when the Communists drove Chiang Kai-shek and his followers out of China. Its rulers realized that the most important resource of Taiwan was the Confucian dynamic of its people, and so laws were established that fostered entrepreneurship. Taiwanese businesses are generally small and family-dominated, and they have been so successful that the country is now one of the richest in the world.

Hong Kong, 95% of whose citizens are Chinese, has a population of 7.1 million people. Although it occupies only a small area of 412 square miles, more than 120 million visitors were flying into Hong Kong each year prior to the takeover of Hong Kong by China and today the same pattern continues; most visitors to Hong Kong are there for the purposes of either business or shopping. Business became so successful that executives were forced to build "vertical factories" housed in tall buildings, each factory built on top of the other. In recent years some of these factories have been shut down, as the work has been transferred to China.

In Singapore, with its population of 4 million, there are three ethnic groups: Chinese (77%), Malays (15%), and Indians (6%). Like Taiwan and Hong Kong, this country has been resoundingly successful, in large part because of the herculean efforts of the Chinese, as the country was extremely poor until recent years. These three countries have been so successful that they, along with Korea and Thailand, are nicknamed the five tigers of Asia.

Thailand has a population of 64 million with two main ethnic groups: native Thai (75%) and Chinese (14%). These two groups are quite comfortable with one another, and as indicated previously, intermarriage is common. While the Chinese play their traditional role as businesspeople, particularly small entrepreneurs, the Thais tend to control the government, the military, and the banks. Ironically, the Thais are continually debating the merits of "Taiwanizing Thailand," as many of them prefer to keep their long-established and traditional ways of doing business and enjoying life, ways that are frequently incompatible with modern approaches.

The expatriate Chinese are influential in most if not all of the countries in which they reside. Unlike many expatriate Japanese, who tend to separate themselves from the societies in which they reside and eventually want to return to Japan, the expatriate Chinese seek integration in the countries in which they reside, but they stay in contact with one another through their far-flung families and the activities of the Chinese "spacemen" who roam the world seeking business opportunities.

In short, the expatriate Chinese are similar to the traditional Chinese in the sense that the family is the basic social unit through which all are united in a relation-based system. Roundness is emphasized by these families, who form a complex but informal network throughout the world, as are harmony and fluidity. The Chinese, regardless of their country of residence, tend to exhibit conservative and high-context behavior. While some Chinese may have moved away from the practice of honoring ancestors at the family altar, they typically accept the importance that is attached to the need for roundness, harmony, and fluidity. In this sense the family altar is not only an appropriate metaphor for the Chinese but also one that clearly illustrates the major values of this ethnic group, no matter where they settle permanently.

Roundness

The Chinese family altar is the cornerstone of family life for Chinese in many parts of the world. The altar is the "ties that bind" a dispersed family and serves as a focal point for viewing an extended family as including the living, the dead, and those as yet unborn. Although it may be thought of as a traditional and ancient aspect of Chinese society, and even as an anachronism in present-day mainland China, the Chinese altar is helpful in providing insight into the values, attitudes, and behaviors of the Chinese today, wherever they may live.

The physical presence of the altar represents the family as a well-knit integrated unit. A house with two altars contains two families; a household with no altar usually considers itself part of another family, and its members will go to the house in which the altar resides for important rituals. In many parts of the world the Chinese tend to live in family or kinship compounds that have several houses, and frequently each of them contains a family altar. Typically the altar stands in the central room of the house, opposite the principal door. Such altars are meant to be publicly seen: Many are visible from the street, and guests usually sit in front of them. There is normally an incense pot on the left side of the altar because this is the position of honor for any of the animistic gods that the Chinese revere; the ancestral tablets are placed on the right side alongside their incense pot. Most of the altars have a backdrop depicting some of the more popular deities. The newer, more prestigious backdrops are painted on glass in glittering bright colors.

Varying amounts of religious odds and ends, including written charms and souvenirs of visits to temples, also decorate the altar, and frequently a Buddha image also stands on it. The ancestral tablets affirm the idealized picture of the Chinese family as a patrilineal kinship group and commemorate its immediate patrilineal forebears and their wives. The family may also have a larger ancestral hall housing the tablets of more distant forebears. In many Chinese restaurants around the world there is a Chinese altar, frequently by the cash register, which possibly symbolizes the functional and earth-focused perspective of the Chinese.

Many families worship at the family altar regularly and sometimes every day. A representative of the family, usually one of the older women, burns three sticks of incense. One stick goes into the ancestor's incense pot, another stick is positioned in the incense pot reserved for the gods that the family members revere, and a third stick is placed outside of the door to welcome the gods and ancestors. As might be expected, the older members of the family tend to worship more frequently than the younger members.

Furthermore, the family tends to commemorate the death days of their closer ancestors or those whom they have personally known. On such occasions the living members of the family serve its deceased members full meals complete with bowls and chopsticks, rice and noodles, and some of their favorite dishes; food is also provided for the animistic gods in many instances. In this context people treat ancestors almost as if they were still living kin. Family members typically eat the food after the ancestors have supposedly eaten, that is, once the incense sticks have burned out. Ancestors also receive food offerings during major festivals for the gods; domestic worship of the gods tends to occur on the first and fifteenth of every lunar month.

Roundness, the first of the three characteristics of the family altar, stands for the continuity and structural completeness of the family; it symbolizes that the family is the basic, distinct, and enduring feature of the Chinese culture. The family altar as such represents the critical point of continuity between the natural world and the supernatural world. As indicated previously, Confucianism did not develop a

Westernized conception of heaven, but it holds that the dead exist in a shadowy netherworld and can communicate with and directly influence living relatives. For the Chinese, there is no sharp demarcation between birth and death. Rather, all humankind is considered to be part of an organic system; in yin-yang fashion, individuals are continually being born into it and processed in it, but they do not experience final separation from it or death in the Westernized sense. To the Chinese, the system, life, and time itself are circular, and all are united to one another because of this circularity. This perspective is consistent with both Buddhist and Taoist beliefs.

According to this rounded viewpoint, what is crucial is that the Chinese have descendants who can worship them and provide them with food and sacrifices; under such conditions they will lead contented lives in the netherworld. All of the efforts of the living are directed toward obtaining a rounded family in which the descent line is preserved. Thus roundness suggests a unity of the family circle that is related to the structural ideal of flawless Chinese patriarchy. Male children (especially the oldest son), who are obligated to perform ceremonial and ancestral rites, are preferred over female children. Hence roundness also suggests that men should have wives who can bear sons.

The Role of Women

There are some problems with this idealization of roundness. Some women cannot bear children or do not bear sons. Under classical Chinese law, failure to bear a son was grounds for divorce. Some children die before they are of an age to produce offspring, and historically it was common for them to be buried unceremoniously and without mourning. Sometimes the father does not live to an old age, and his family does not receive the honor accorded to those who have achieved such a distinction. However, the premature death of parents does not preclude their participation in the family, as they can be enshrined in ancestral tablets; the premature death of a husband does not weaken his wife's claim to a position in the descent line; and the failure to have children is correctable by adoption.

Such a patriarchal approach has led to inequality of the sexes. Daughters occupy an inferior position to sons in the family, as do wives to husbands, and the mother assumes a position of relative equality with her husband and authority over her adult sons only when she is old, at which point she is accorded a position of honor and prestige. In such a context, most activities are segregated by sex.

In previous eras polygamy was common among the Chinese, and it can still be found in isolated instances in rural areas. This practice complicated the familial structure, but the first wife was usually accorded the most honor and authority. In the modern world it is typical for a Chinese man to marry only one woman, although the practice of having a "minor wife" or mistress is common. Because of the emphasis on the family, the Chinese businessman frequently works alongside his wife, or she will handle office matters while he is in charge of external relations. This pattern has limited the growth of many Chinese businesses, as only the wife and other family members are given positions of authority. Sometimes the husband and wife rarely see each other because of his external activities, but this is acceptable behavior provided that the husband is working hard on behalf of the family.

One Chinese family living in Thailand is typical in that every Sunday the wife of the oldest son hosts all family members at her home, including the younger sons and their families. Even though some of the brothers are not very competent, the eldest son and chief executive officer of the family business allows them to keep their positions as vice presidents and heads of the family's factories, but non-family members actually are the key decision makers in their factories. However, these non-family members participate in the Sunday activities irregularly. Because they are excluded from becoming full members of the familial in-group and can never aspire to become owners of the factories, it is common for them to quit after a few years and establish their own businesses. Frequently the family provides the financial support to start the new enterprise for such loyal employees, thus strengthening the family-based networks of long-term contacts.

Looking at the Long Term

This rounded perspective leads the Chinese to take a long-term perspective on problems and issues that they face. Unlike U.S. Americans, most of whom feel that a long-term plan involves a projection into the future of 3 to 5 years, the Chinese tend to be comfortable with plans that are expressed in 10, 20, and even 100-year increments. Geert Hofstede (2001), whose 53-nation study of cultural values is highlighted in this book, has expanded his four-dimensional framework into five dimensions because of recent research on the Chinese. He calls this dimension the Confucian dynamic, and it reflects a long-term rather than a short-term orientation; it also resembles in some ways the Protestant ethic in that deferred gratification of individual needs is accepted for the purpose of long-term success that helps the family.

Louis Kraar (1994), commenting on the causes of the success of the overseas or expatriate Chinese, begins his analysis by emphasizing this Confucian ethic:

> Regardless of where they live or how rich they are, the Overseas Chinese share an abiding belief in hard work, strong family ties, frugality, and education. Yes, this same constellation of virtues defines much of what Westerners have labeled the Protestant Ethic, but for the Overseas Chinese these attributes are not musty relics from their culture's past but compelling rules to live by. (pp. 92–93)

This rounded conception of familial relationships is distinctively Chinese, and it encompasses members of the family over space and time. The family rather than the individual has been the basic unit of social organization among the Chinese since the time of the Duke of Chou in the 12th century BCE. He established the system of *bao-jia* or family/kinship rule and divided society into units of 10, 100, 1,000, and 10,000 families arranged by neighborhoods and districts. Each unit chose a leader from its ranks who was responsible for the behavior and welfare of all of the families in that jurisdiction, and this leader in turn reported directly to the leader of the next *bao-jia* unit. If a man committed a crime, the head of his own household would be held accountable, followed by the head of his 10-family unit, the 100-family unit, and so on. As a result, minor crimes were rarely reported beyond the smallest unit, which had the authority to settle such matters. In one way or another, modern Chinese follow this pattern of collective responsibility, although practices vary widely by the nations in which they reside.

Furthermore, the concept of roundness helps to explain the well-known Chinese practice of *guanxi* (relationships or connections). As indicated in the previous chapter, from the Chinese perspective a person exists only in relation to others. The lifeblood of a Chinese company is *guanxi*. Penetrating the layers of *guanxi* is like peeling an onion. First come connections between people and ancestors; then between people from the same village; then between members of the family; and finally between the family and close associates who can be trusted, such as the competent executives who are not family members but who actually run the factories. All of these relationships are considered to be continual in nature, and obligations go far beyond what can be put into a contract. For example, a Chinese businessman will help household servants establish their own small business after they have served the family faithfully for several years; he will also expect Western colleagues to help his son gain admission to a prestigious Western university after they have worked together on a joint venture for some years.

Historically, the Chinese, like many other high-context cultures, are less concerned with what is written in the contract than with the actions that people take to meet their obligations as they emphasize *guanxi*. For example, the U.S. Chamber of Thailand spent 2 years identifying a U.S. partner for a Chinese Thai businessman who was seeking to establish a joint venture. At long last all of the principal parties were to gather for dinner in Bangkok to finalize the arrangements. However, the Chinese Thai businessman angrily stormed out of the room when the U.S. business-man showed up with a lawyer and a written contract, thus putting an end to the potential joint venture. To the high-context Chinese, this was an affront of the highest order. In recent years, especially since China joined the World Trade Organization and must accept the sanctity of contracts, many Chinese firms now accept the inevitability of strict contracts. Still, *guanxi* is such an ingrained value that it continues to influence Chinese behavior.

The Chinese also tend to bypass the use of banks because of this system of *guanxi*. In fact, banks were unknown in China until the 20th century. Throughout the world the Chinese establish communal investment and credit associations. A group of friends, relatives, and colleagues pool their money to form a mutual fund from which each member may take turns borrowing, and the Chinese rely on the personal close-ness of the group to inhibit fraud or default. Sometimes these associations will lend money to other Chinese who cannot afford to contribute to the association. Furthermore, the Chinese are well known for lending their personal funds to their relatives in other nations, for example, Taiwanese investment in Malaysia, Thailand, and China. It was Taiwanese investment in China starting in the late 1970s that largely jump-started the Chinese economy.

The largest Chinese families even establish private social welfare agencies that are open to anyone from the family. For example, the Chinese Lee family—there is also a Korean Lee family—operates a social welfare agency in a large building in New York's Chinatown that helps any Lee of Chinese origin obtain employment or solve any crisis such as dealing with immigration officials who may be trying to deport him and his family.

The Modern Businessman

The conception of the modern Chinese businessman is also related to the concept of roundness. Many Chinese businessmen fly around the world so frequently that they have been caricatured as "spacemen." They meet regularly with business associates in other nations who are united to one another through *guanxi*. If they are working with non-Chinese businessmen, they spend a great deal of time getting to know them so that the non-contractual nature of their relationships is given prominence.

Still, it must be stressed that *guanxi* has severe limitations. It is developed slowly and excludes many individuals who might provide new business opportunities. *Guanxi* also is at odds with the strong movement throughout the world toward contract-based business transactions championed by many nations and particularly the World Trade Organization. In the past, Chinese firms have found it difficult to become global in scope because of the use of *guanxi* at the expense of contractual relationships, and many of them are now emphasizing the importance of contracts and law.

Harmony

As described in Chapter 3, "The Japanese Garden," the need for harmony and its development as a cultural ideal seem to have originated in the activities surrounding wet rice farming, and this generalization is also valid for China. Given the need to cooperate closely to provide subsistence for all, the Chinese accorded primacy to harmony rather than to other ideals, such as the Western conceptions of life, liberty, and the pursuit of individual happiness. When disharmony infects the relations in the family, supernatural explanations for it are frequently offered, and the support of ancestors and gods is the natural way to reestablish harmony.

Roundness is a necessary but not sufficient condition for the second characteristic of the family altar, harmony. Ideally the harmonious family is one in which there are few if any quarrels, financial problems, or illnesses. The most common prayer among many devout Chinese is for harmony, and it is frequently printed on door frames, charms, wedding cakes, and the walls of homes. It is a prayer that is directed to both ancestors and gods. And, as indicated above, the members of the living family offer food and rituals in return for the harmony that they are seeking.

People make all kinds of requests at the family altar, for example, helping a child pass an examination, curing the sick, and obtaining employment. Although the requests are not always met, devotees often feel renewed hope and comfort, in large part because of the rituals that unite deceased and living family members through the medium of the family altar. Even the skeptical carry out the rituals—just in case.

While the Chinese seek to achieve harmony, their worldview is not nearly as comforting as that of the Christian, who can aspire to a tranquil heaven that is under the supervision of a benign and personal God. The shadowy netherworld represents a degree of uncertainty that the Christian does not have to take into account. It may be for this reason that the Chinese tend to believe in luck and fate and that they have little if any control over events. In a Chinese Buddhist temple, for example, worshippers can purchase a set of fortune sticks which they throw on the ground and then rearrange before consulting a book that predicts their future in terms of this rearrangement. Similarly many Chinese do not use car seatbelts because they believe that nature will take its normal course, no matter what they do.

Also, although other cultural groups like to gamble, many Chinese clearly are attracted to it, seemingly in part because of this sense of uncertainty, luck, and fate. The major sense of security and harmony comes from the rounded family and the all-encompassing family altar. Many times gambling and games of chance reinforce both harmony and roundness. For example, large Asian families, each of which may include 30 or more individuals and several generations, go to restaurants for several hours of convivial conversation and relaxation, during which time they leisurely play cards and other games of chance. Similarly it is not unusual for a Chinese family of 30 or more members, including several generations and close family servants, to go to the seashore for 4 or 5 days and spend most of the time inside an air-conditioned house talking to one another, drinking, eating sumptuous meals, and playing games of chance in which all participate. Sometimes no one ventures to the beach because of the enjoyment that family members experience interacting with one another under such circumstances.

Fluidity

The third characteristic of the family altar is fluidity or the capacity to change while maintaining solid traditions. It reflects the relation-oriented approach of the Chinese, which allows them to be individualistic as long as they meet their many obligations to various family members, including the deceased.

Thus, although the Chinese tend to be conservative, they are frequently innovative and entrepreneurial. The number of Chinese inventions and scientific breakthroughs is startling. Joseph Needham and his Associates have written a famous multivolume series of books, *Science and Civilization in China*, that describes these inventions and breakthroughs in detail. These include the first suspension bridge, fishing reels, rudders for ships, hang gliders, parachutes, fireworks, lacquer, wallpaper, paper, armor made from paper, wheelbarrows, and the first design for a steam engine. Similarly, as Kotkin (1993) and many others have described, the Chinese have been successful in establishing business enterprises, and they are disproportionately successful and entrepreneurial compared to most if not all other ethnic groups. This balanced emphasis on conservatism and innovation clearly reflects the concept of fluidity.

In traditional Chinese religion there is a huge pantheon of gods and goddesses, most of whom are the heroes of Chinese myths, legend, and history and deified either by imperial order or popular choice. Some communities have cult followings that have evolved around a particular historical figure who supposedly protects and guides the town or who may have worked a miracle in it. The reputed powers of the best-known deities have been affirmed by generations of the Chinese over the centuries. However, devotees believe that the power of a deity tends to decline with age and eventually loses its efficacy completely. The Chinese generally pray only to gods who have answered at least some of their petitions. When the decline in power occurs, the devotees begin to worship a new god. When Neil Armstrong walked on the moon in 1969, many elderly Chinese stopped praying to it because of this supposed loss of efficacy. Thus there is fluidity, as change does occur but devotion at the family altar continues.

The history of China can be interpreted in terms of the three characteristics of the family altar. Han, the largest cultural group, includes about 85% of the population. However, given the 1.35 billion people in China, it is not surprising that there are about 400 ethnic groups ranging in size from millions of people to only a few thousand, and most of them have and have had harmonious relations with one another. Almost all of Chinese history can be divided into periods defined by dynasties. Families or rulers would occupy the throne until another would take the power away. History was viewed as endless cycles of renewal and decline, starting with the first dynasty in approximately 1953 BCE and not ending until the retirement of the last emperor in 1911 CE. Each ruler was considered to have a mandate bestowed on him by the gods to rule fairly and wisely. When an heir became corrupt or lazy, rebellions would break out and a new emperor would surface. Typically he would introduce reforms that were consistent with the underlying cultural values of roundness, harmony, and fluidity.

The fact that China was the most developed civilization in the world for a significant portion of this long time period may also be attributed to the emphasis on these three characteristics. As a broad generalization, the dynastic cycle helps to explain, through the mechanism of periodic renewal, how the Chinese were able to retain a consistent and continuous pattern of government for thousands of years (Major, 1989, pp. 46–47). Thus this cycle exemplifies all three characteristics of the family altar: Roundness or continuity and structural completeness of the family and family-based nation; harmony over thousands of years marred only by periodic revolutions; and fluidity or maintaining the past while accepting change.

CHAPTER 27

The Singapore Hawker Centers

Some [hawkers] were such excellent cooks that they became great tourist attractions. A few became millionaires who drove to work in their Mercedes Benz and employed waiters. It was the enterprise, drive and talent of such people that made Singapore.

—Lee Kuan Yew (2000)

Hawker centers are one of Singapore's most unique and notable social icons, an important part of the Singaporean way of life . . . and have helped to put the country on the world map.

—Lee Yuen Hee, CEO of the
National Environment Agency (2007)

First impressions of Singapore among first-time visitors tend to be positive, and it is easy to understand why. The taxi from the airport to the hotel cruises along miles of flower-bedecked streets and buildings, some of which emphasize traditional Asian architecture while others are strikingly modern; the taxicab driver slows down when a bell automatically sounds, indicating that he is driving over the 50 mph speed limit; and everyone is courteous and efficient.

If the visitor arrives around noon at Raffles Place, the business center of Singapore, the sun is normally blazing. People begin to emerge from the skyscrapers one after another. Most of them are the businesspeople who represent Singapore's high-powered finance industry; they are going to take their lunch break. Like many cities of the world, the crowd includes a mixture of various cultural and ethnic groups. Asians might be Chinese, Japanese, or Thai; others may be Malay, Indian, or Indonesian. Caucasians could be U.S. American, European, or Australian. Everyone is headed for one of the most unique and emblematic eating places in Singapore, the hawker center. Regardless of ethnic groupings or residential status—citizens, permanent residents, or expatriates—many people go to the hawker center to find something they like during their limited lunch break.

The hawker centers are venues where one can try a wide variety of traditional ethnic foods, the latest fad in beverages, or even simple fruits. These centers are clean, and the personnel serving the food are, like the cabdriver, efficient and courteous. Each center is a collection of at least 20 food stalls at the same location and is casual or informal in motif, presentation of food, and seating arrangements. The hawker centers are such a unique aspect of Singapore's culture that the National Environment Agency commissioned a book entitled *Singapore Hawker Centers—People, Places, Food* in 2007 and followed up the launch with a traveling exhibition in 2008 (Kong, 2007).

The hawker centers constitute a remarkable parallel to the features of modern Singapore—an integration of variety, efficiency, tremendous energy, safety, and a symbiosis of the old and new. This island nation of 4.3 million was poor 50 years ago, and there were some ethnic issues involving the Chinese (76% of the population), Malays (14%), and Indians (9%). Some of these issues related directly to the lack of education and infrastructure, and the crowded conditions only exacerbated them. Singapore's population density is 6,729 per square kilometer whereas the United States' is 32. Nevertheless, today Singapore is a leading banking and finance center; its workforce is highly educated; and government and business emphasize the development of cutting-edge technologies and industries. As a result, the gross domestic product per person is $24,840, 29th in the world; the comparable figure in the United States is $39,430 at 9th in the world. Singapore has become one of the world's most prosperous countries and is the world's busiest port. *From Third World to First: The Singapore Story* (1965–2000) is the apt title of the memoir of Singapore's legendary leader, Lee Kuan Yew (2000), who helped bring about this remarkable change. The broad elements of the story are described below.

Origins of the Hawker Centers

Numerous hawker stalls were a familiar sight in Singapore's streets and alleys in the 1950s and 1960s. They were not much different from the hawker stalls that visitors to many developing nations see: colorful, arranged in a disorderly pattern, inviting visitors to use them, but unconsciously sending the message *caveat emptor*. While there's little argument that these hawkers brought a certain charm to the social and cultural landscape of Singapore, many stall holders plied their trade under less than desirable conditions. More often than not, hawkers operated under unhygienic conditions, contending with a lack of piped water and inadequate facilities to prepare and cook their food. To compound the problems, the authorities and wider population had to deal with the indiscriminate disposal of wastes into drains. Over a period of time, these conditions caused considerable pollution in the island's water courses, endangering both public health and the environment.

As a result, the government developed a plan to build designated areas for hawkers. These areas, which we now know as hawker centers, were to have a complete infrastructure to support hawkers' day-to-day operations. An island-wide census in 1968 and 1969 was the first step in the systematic phasing out of old-style hawkers. The census registered a total of 18,000 street hawkers, who were then issued temporary hawking licenses. The exercise effectively curbed illegal hawking in Singapore, as all hawkers now needed to follow strict regulations or face financial penalties as well as the inability to operate their businesses.

In 1970, the government began to relocate street hawkers. The new hawker centers were equipped with proper facilities for food preparation and cooking and were supported by clean and efficient drainage systems as well as proper sanitation facilities. By February 1986, all street hawkers were completely relocated in hawker centers. The process, although long and difficult, was ultimately completed.

In 2000, there were 139 hawker centers owned by the Ministry of the Environment (ENV), the Housing and Development Board (HDB), and the Jurong Town Corporation (JTC). Collectively, these hawker centers hold a total of 17,331 occupied stalls; 10,333 are market stalls and 6,998 are food stalls for preparing and serving cooked food. The environment and public health agencies have been largely responsible for transforming the hawking from a public health disaster to a world-renowned icon that is uniquely Singaporean and celebrates its ethnic diversity and foodie culture. The 2007 book on the hawker centers referred to earlier reports that there were 113 centers with more than 6,000 cooked food stalls.

All stall holders in hawker centers are licensed by the ENV. They are required by law to ensure that food prepared in their stalls is safe for public consumption. Stall holders are expected to be vigilant in adhering to proper food and personal hygiene practices and to ensure that their stalls are free from pests and other infestation.

Launched by ENV in April 1994, the Stall Ownership Scheme enables stall holders to own their stalls. Under the scheme, stalls in ENV and HDB hawker centers are sold to incumbent stall holders on a 20-year lease at a discount. Those declining to buy their stalls can surrender them and opt for cash grants or else they can continue to rent their stalls at a revised rate. Stallholders buying their stalls were allowed to sublet them. As a result, many of these stalls are open for many hours. These stall holders were also permitted to sell their stalls. Some did and, consequently, more innovative and enterprising players joined the trade.

ENV launched the Stall Assignment Scheme in February 1998 to allow stall holders to assign their stalls to operators who are willing to pay rent as assessed by the Chief Valuer. The assignees can sublet the stall or get someone to run the stall on their behalf. Legal restrictions apply in all cases, but there are more restrictions for stalls subsidized by the government.

As an extension to the Stall Assignment Scheme, stall holders currently paying subsidized rent may opt to assign the stall to themselves. This effectively means that they can opt to pay rental rates as assessed by the Chief Valuer and can operate the stall free of the more onerous restrictions imposed on subsidized stalls. The assignees are granted a 3-year tenancy agreement with an option for renewal if the hawker center is not affected by redevelopment. If the stalls in the hawker center are offered for sale under the Stall Ownership Scheme, the assignees may also buy the stalls at the assessed values. Whether subsidized or not, stalls must follow strict guidelines on food preparation, cleanliness, and polite treatment of customers.

Singapore's History

Just as the hawker centers faced many turbulent changes, Singapore has also experienced struggles and challenges to reach its current status as a prosperous cosmopolitan society. Modern Singapore began in 1819 when Sir Stamford Raffles claimed this Malay island, at the time nothing more than a large fishing village, as a regional base for the British East India Company; it flourished in large part because of its central location at the crossroads of Asia. In 1832 Singapore, along with Malacca and Penang (now part of Malaysia), were established as the Straits Settlements under the control of British India, and in 1867 the settlements became a separate British crown colony after the Singapore merchants started agitating against British India rule.

During World War II the Japanese occupied the nation after the largest surrender of British-led forces in history and renamed it *Syonan-to* or "Light of the Island" in Japanese. In 1946 Singapore (including Penang and Malacca) became a British crown colony with a civil administration headed by a governor, while Penang and Malacca became a part of the Federation of Malaya in 1948. The Communists attempted to take over both Singapore and Malaya in 1948, resulting in a state of emergency for 12 years. After a few more tempestuous years Singapore became a self-governing state. Lee Kuan Yew, a legendary figure in the nation's history, became prime minister.

At least in part because of the continuing Communist threat, Malaysia was formed in 1963; it included Singapore, Sarawak, and North Borneo. However, the merger was shaky from the beginning with continual race riots and allegations of unfair treatment. In 1965 Singapore became a sovereign, democratic, and independent nation much to the anguish of Lee Kuan Yew, widely considered as the father of modern Singapore, who had always believed in the merger of the territories.

Now on its own, Singapore initiated a massive industrialization project. After the shock of two oil crises, in 1979 the government initiated a program of economic restructuring emphasizing technology and computer education and offering financial incentives to industrial enterprises.

Almost immediately, the Confucian ideals of the Chinese were employed as public policy. The Confucian philosophy argues that before an individual can be expected to perform as an ideal and sincere person (see Chapter 25), he or she must be satisfied with basic necessities, clothes, food, and a place to live. Lee Kuan Yew, the first prime minister, endorsed a program of public housing because of his Confucian ideals. It proved to be enormously successful, and today the vast majority of Singaporeans live in such housing. To encourage home ownership, Singaporeans were allowed to use their Central Provident Funds savings (similar to a pension fund) to pay for the apartments.

There are at least five features of the hawker centers that help us to profile Singapore. These are ethnic diversity but unity, efficiency, the power of women, safety, and synthesizing traditional and modern values. We discuss each briefly.

Ethnic Diversity But Unity

As discussed above, customers can find any cuisine—of course, Chinese, but also Malay, Peranakan, Indian (both north and south), and international food (Japanese and Western)—in a typical hawker center. The stalls are close together, and their owners frequently sit and talk with one another. Singaporeans enjoy visiting their friends and neighbors from other ethnic communities and celebrating their food and festivals. Chinese noodles, Malay satay, and Indian curry are all part of the hawker center experience.

The Singapore population includes three major ethnic groups: Chinese, Indian, and Malay. The government encourages citizens to intermingle and understand one another in a variety of ways, for example, by ensuring that public housing includes all ethnic groups. There is a strict allocation of apartments determined by the proportion of each ethnic group in Singapore, which is maintained over time. In like manner, the government determines the distribution of hawker stalls on the percentage of each ethnic group in the population. There is one section for each group in each hawker center, with the largest number of stalls naturally serving Chinese food.

Some of the older hawker centers share a communal water supply located in the center; hence they automatically learn how to behave as responsible shop owners and cooperate with others who have different skin colors, customs, or cultures. To survive in competition with other hawker centers, the stall holders work together to maintain their hawker center, making it clean and efficient enough to attract more customers. This is exactly a miniature of the Singaporean ethos: all races making a supreme effort and sharing values beyond ethnic boundaries to survive in the face of international competition.

To shape the destiny of such a multiethnic country, the Singaporean government has embarked on a program called Shared Values to avoid conflicts among races. First discussed in 1988, Shared Values was to be a blueprint for the development of a national ideology that Singaporeans of all races and faiths could comfortably accept. They also have a pledge that captures the idea that Singaporeans can overcome any difficulties if they come together as one. It was written in 1966 during a time of great turmoil and uncertainty shortly after independence.

The aim of this national ideology was to create a Singaporean identity by incorporating the relevant parts of citizens' various cultural heritages as well as the attitudes and values that have helped them survive as a nation. There was also an anti-Western bias in this formulation because the government wanted to safeguard against "undesirable" values such as an excessive emphasis on individualism and self-centeredness at the expense of group obligations. This concept of a national ideology had already been developed by some of Singapore's neighbors. For example, Indonesia had drawn up the *Pancasila* or five principles, a set of common beliefs to unite its peoples.

The five Shared Values are:

Nation before community and society before self.

Family as the basic unit of society.

Community support and respect for the individual.

Consensus, not conflict.

Racial and religious harmony.

In accordance with these Shared Values, Singaporeans are actively encouraged to understand different points of view. In business discussions Singaporeans try to avoid conflicts and strive to solve problems with compromise. They are genuine collectivists. In the GLOBE study (House et al., 2004), Singapore was ranked with the Confucian Asia cluster and ranked high in both institutional and in-group collectivism.

In many ways Singaporeans are competitive collectivists. Although this is a paradox, it is a reality and is expressed in a local term, *Kiasu-ness*. Mr. Kiasu originated as a cartoon character and is roughly equivalent to "the ugly American," cutting into lines, heaping food on his plate at the expense of others in a cafeteria line, and loudly insisting on fast and efficient service before others are handled. Singaporeans are aware of this paradox, which one would normally associate with individualism, and any long-term visitor will probably experience its manifestation. The norm, however, is to follow the rules, be efficient but courteous, and be sensitive to the needs of others.

In addition to *Kiasu-ness*, Singaporeans have other terms that reflect a peculiar brand of English called "Singlish" and mix English with common phrases from Chinese dialects and Malay. Needless to say, as in other countries (e.g., Nigeria, India) with a colonial past, die-hard Anglophiles bemoan the loss of the Queen's English but "Singlish" makes for efficient communication.

Military training also contributes to cooperation among different races. British forces remained in Singapore after independence but were withdrawn by 1971 and Singapore built up its own military. The Singaporean military includes an equal number of professional and conscripted soldiers because of the small population. This training is compulsory for men, and they serve periodically until they are in their 40s. During hard training their cohesion or friendship across racial boundaries becomes deeper.

Singapore's educational system also focuses on cooperation among the different races. The government encourages students to mingle with other races to strengthen ties between ethnic groups. Because Singapore possesses no natural resources, no major industries, and a limited workforce, the united power of its citizens is an important concept if Singapore is to survive and prosper in international competition.

Efficiency

When customers order food in a hawker center, they must follow one well-known but unspoken rule: Be quick and efficient. At lunchtime each hawker center is teeming with prospective diners. It is a major time for stall owners to transact business, and they naturally like to take as many orders as possible. Likewise, other customers want to receive an order as soon as possible and enjoy their lunch. When a stall owner's eyes meet yours, he or she slightly jerks the chin upward, signaling that it is your turn to order. You have to immediately say loudly: "One curry chicken packet." This is the correct way to order food in the hawker centers in Singapore: Succinct orders are the rule. All meaningless conversation is left out, for instance, "How are you doing today? I would like a curry chicken for takeout."

This tendency to make conversation short and efficient is typical of Singaporeans. They are not interested in niceties, and they prefer a quick yes or no answer so that their time is not wasted. Normally it is sufficient to say either yes or no on the telephone when a person asks if it possible to obtain a loan, to purchase a ticket to an event, or something similar. Singaporeans are paradoxically polite, but they become annoyed when such interactions are not efficient.

This efficiency is also demonstrated by the cutting-edge technology found in Singapore, which employs numerous advanced technologies for its finance and transportation systems. We have already talked about the automatic bell in taxis indicating that the speed limit has been exceeded. We should also emphasize that these taxis are a convenient form of transportation because of availability and reasonable price. A central satellite monitoring system links all taxies; when you book a taxi over the automated phone system, the nearest taxi is hailed automatically by the system and dispatched to your precise location. Public buses use the satellite system, too. When public buses approach traffic lights during morning and evening rush hours, the lights are controlled to turn green immediately.

There is a downside to this excessive emphasis on work, efficiency, and productivity. Singaporeans tend to be involved in their work and deemphasize leisure. The birth rate is only 1.4 babies per woman, much lower than the 2.1 necessary to maintain a steady population and by 2030 it is estimated that one third of Singapore's population will be over 60 years of age. Married couples routinely point out that they do not have sufficient time for leisure and do not even see their spouses on some days. Of course, U.S. Americans voice a similar complaint, expressed in the acronym, DINS: double income, no sex. Unmarried Singaporeans frequently say that they have no time for dating. In true Singaporean fashion the government created a department, the Social Development Unit (called facetiously "single, desperate, and ugly"), to encourage college graduates to marry. SDU sponsors events featuring speed dating (see Chapter 16, "American Football") and zodiac dating.

The Power of Women

A distinctive feature of the hawker center is the ratio of men and women who work there. Almost the same number of men and women work in a hawker center in similar positions and with equally long hours. Women can be seen frequently occupying the upper hierarchy of the power structure—many owner-operators of the hawker stalls are women who deal with issues like accounting, food preparation, and sales.

Because human resources are comparatively scarce in Singapore, the government encourages equal opportunity in the workplace. In fact, women are among the most powerful influence brokers in Singapore. This is remarkable in view of the fact that in most other Asian countries, women rarely have high positions in politics and business. However, Singaporean women operate successfully in these areas. This is reflected in the high scores for Singapore on gender egalitarianism in the GLOBE study (House et al., 2004). Singapore's mean score (3.70) is closer to that of the high-scoring Eastern (3.84) and Nordic European (3.71) cluster nations than that of its fellow Confucian Asia cluster nations (3.18).

Women have become CEOs of major companies, ambassadors, and leaders in all walks of life. However, this is not to say that Singaporean women do not face the challenges of pay differentials with men, the glass ceiling (especially in the civil service), and challenges of work-family balance. The government is very cognizant of these challenges and responds with public policy initiatives such as flex time, job retraining, and family care leave to help families manage the work-family balance issues.

Safety

As we have seen, first-time visitors to Singapore tend to have a positive reaction, but they also observe the many rules and regulations governing Singaporean life. The standing joke is that Singapore is a "fine" nation, for everything you do is fined, including the failure to flush public toilets. Some youth tend to feel stymied because of the large number of rules. Hence it was not surprising that some of them began urinating in elevators, a practice that was soon outlawed. They then began carrying water bottles with them and, when caught by inspectors, indicated that water had spilled. Soon the inspectors developed a test that could determine immediately whether the offending pool was indeed urine.

Although this example is humorous, others are less so, and one involved a teenage American who was arrested and caned for spray-painting automobiles at night. There was a public outcry, and former President Bill Clinton became involved. Although the number of strikes or hits was decreased as a result, the caning still took place. There are fines for littering (first-time offenders are fined 1,000 Singapore dollars and second-time offenders could end up not only paying double that fine but cleaning up litter wearing bright jackets with an audience to boot) and the death penalty is standard for trafficking in drugs ranging from 15g of heroin to 1.2 kg of opium. The import, sale, and possession of chewing gum are banned because of the high cost and difficulty of removing chewing gum from public places.

Arguments can be made for and against the Singaporean system, which has tended to loosen somewhat in recent years. However, in accordance with the yin-yang tradition of looking at each situation before making a value judgment, we must take into account Singapore's tortuous history, its need to align the goals of individuals and the state closely if economic success for all is to be attained, and its crowded conditions. Unlike the situation in so many other modern nations, there are no homeless in Singapore, and the population is well-off in terms of education, standard of living, and the ability to withstand major difficulties.

The government makes use of Singapore's Internal Security Act (or ISA), which lets officials detain people without trial, although extremely judicious use of this legislation has been the norm in the past. Basically this nation is small enough to control tightly. This is a major reason Singaporeans are law abiding. Paradoxically, however, prostitution is openly countenanced in the better sections of the city, which is a contrast to the power of women as discussed above. Singapore is clearly an authority ranking culture in which there is a high degree of collectivism but also a relatively high degree of power distance.

Furthermore, this strict control seems to have had an undesirable side effect in that citizens have become more comfortable being followers than entrepreneurs. This may well be reflected in the low risk acceptance or tolerance (or high uncertainty avoidance) score (5.31) that emerged in the GLOBE study (House et al., 2004). Compared to Hong Kong, another Chinese-dominant society, Singapore doesn't have many major local businesses. Citizens look to a paternalistic government as an all-providing authority; they delegate decision making to the government and expect to be well cared for in exchange.

The hawker centers are open for long hours, and many of them remain open 24 hours a day, 7 days a week. This is reflected in the Singaporean work ethic: The average Singaporean has been taught from birth to be hardworking and diligent. The Singaporean government abhors safety violations and is concerned about assuring the public that hawker centers are safe places. The ENV even started a 24-hour service to attend to complaints from members of the public. Enforcement officers respond quickly to illegal hawking complaints as well as environmental health complaints such as noise pollution, illegal dumping, smoking in prohibited areas, and so on. As might be expected, Singapore has a low crime rate compared to neighboring nations.

Synthesizing Traditional and New Values

The hawker centers sell traditional fare from the Southeast Asian region as well as newer Western food. It is common in hawker centers throughout Singapore to see traditional Chinese herbal tea being sold next to fried chicken. Old and new are mixed without any sense of incongruity. Similarly Singapore is attempting to maintain its traditional values while pursuing efficiency and practicality. An example of this emphasis on traditions is the celebration of festivals. All three ethnic groups (Chinese, Malays, and Indian) celebrate their major festivals, thus reinforcing traditional values and heritages. The government has officially decreed that all such ethnic celebrations are national holidays. Family and close friends celebrate most festivals, and the traditions of the culture are passed to the young through such communal celebrations. Neighboring Malaysia does the same (see Chapter 18).

Similarly, as Singapore prepares for the challenges of the 21st century, it is increasingly looking at blending the old and the new in its educational systems. Nothing is given more importance in Singapore than the preparation of its young for the workforce of tomorrow. The main challenge has been keeping in touch with Singapore's cultural heritage (be it Chinese, Malay, or Indian) and the cosmopolitan ways that a global marketplace is demanding. Also, Singapore is discarding the old-style educational tenets of rote learning and memorization, holdovers from the British colonial past, and a rigid emphasis on mathematics and the sciences, for a gradual broadening of the curriculum geared toward broad and creative thinking appropriate for a globalized world. In this globalized world, Singapore's critical role as a trading center between East and West continues into the age of telecommunications and it has one of the world's busiest ports and one of the world's leading airports.

On an individual level, parents of children still motivate and pressure their children toward constant improvement, but the previous emphasis on the strictly academic has been broadened to include some new areas, for example, extra classes promoting computer skills, confidence building, and an appreciation for the arts.

This, then, is Singapore. It is a unique country, and there are some who question whether its approach would work in other situations, for example, large nations, countries with even more ethnic diversity than Singapore, and so on. Regardless, Singapore's achievements are significant, and all of us can benefit from studying this traditional but modern nation that is clearly a major player in the global economy.

PART IX

India, Shiva, and Diversity

I ndia is a fascinating nation in which there is a great amount of diversity such as ethnic, religious, and geographic diversity. While arguably India possesses more contrasts and diversity than any other nation, as visitors to this nation attest, it also has a surprising degree of unity. Although there are 16 official languages, one of them, English, is used by many if not most Indians and is the language of business and trade. Similarly, more than 80% of the population is Hindu, although other religions also have large memberships, sometimes numbering in the millions.

Given both unity and diversity, we present two chapters on India. The first looks at the dominant culture, Hinduism, as manifested in our cultural metaphor, the Dance of Shiva. The second chapter, "India: A Kaleidoscope of Diversity," describes the various types of diversity. This chapter serves as a model for understanding many nations, as less than 10% of the 220 nations in the world are monocultural.

CHAPTER 28

India

The Dance of Shiva

Sex may drive the soap operas in America. But in India, what really moves the dishwashing liquids are serials based on ancient myths of Indian gods.

—Jonathan Karp and Michael Williams (1998, p. A1)

ndia is the second-largest country in the world with a population of more than 1 billion (after China, with 1.3 billion residents); it is about one third the size of the United States. India is the world's most populous democracy. It became independent in 1947 after the British ceded control. At the same time neighboring Pakistan was also created. India is a poor nation, but on many measures it has achieved substantial success since 1947. Since 2003, it has been identified as a member of the BRIC club (Brazil, Russia, India, and China), nations that are expected to become wealthier by 2050 than most of the leading economic powers of today. Life expectancy has increased from 32 to an average of 65 years (for men and women) and adult literacy is 61%. Although Gross National Product per person has increased in recent years, it is still only $640, compared to $39,430 in the United States and $1,470 in China. This nation also has the largest number of college-educated scientists and computer specialists in the world and a middle class that is estimated to include 100 to 150 million people, but 53% of the population lives on less than $1 per day. India has surpassed such nations as Pakistan on most measures of economic and social success, but its success is limited when compared to that of China. A major reason for India's limited success is its dramatic increase in population without corresponding growth in resources.

In 1991 *The Economist* published an influential "Survey of India"; a picture of a caged tiger was on the cover, with the title "caged" (Crook, 1991). The basic message was that India was too bureaucratic and centralized. Manmohan Singh, India's finance minister at the time, and others were influenced by this negative depiction and began efforts to privatize the economy. These efforts are continuing but are retarded by infighting among political parties and interest groups.

Religious diversity is a major feature of India, and it is fitting that our image of, and cultural metaphor for, this country should be based on religion. As Swami Vivekananda, one of India's foremost thinkers, so succinctly stated, "Each nation has a theme in life. In India religious life forms the central theme, the keynote of the whole music of the nation."

For 2,000 years of its history, India was almost completely Hindu, but for the last millennium or more, Indian culture has been a synthesis of different racial, religious, and linguistic influences. Hinduism itself has undergone many changes owing to the impact of other faiths. It is, therefore, incorrect to contend that Indian culture is solely a Hindu culture, although Hindus represent 81% of the population.

To begin to understand India, we must start with Hindu traditions. The overwhelming majority of Indians are still tradition oriented, and changes in their culture and society cannot be understood without reference to that tradition. In this chapter we examine Hindu culture using the "Dance of Shiva" as our organizing principle.

Shiva's Dance

There are numerous deities or gods in the Hindu religion, each being a different manifestation of one Supreme Being. The most important gods are Brahma (the Creator), Vishnu (the Preserver), and Shiva (the Destroyer). Among the greatest names and appearances of Shiva is Nataraja, Lord of the Dancers. The Dance of Shiva has been described as the "clearest image of the activity of God which any art or religion can boast of" (Coomaraswamy, 1924/1969, p. 56), and it also reflects the cyclical nature of Hindu philosophy. Through this metaphor we will begin to explore Indian culture and society.

Among Hindus dancing is regarded as the most ancient and important of the arts. Legend even attributes the creation of the world to dance: Brahma's three steps created earth, space, and sky. Every aspect of nature—man, bird, beast, insect, trees, wind, waves, stars—displays a dance pattern, collectively called the Daily Dance (*dainic nrtya*). But nature is inert and cannot dance until Shiva wills it; he holds the sacred drum, the *damaru*, whose soundings set the rhythms that beat throughout the universe. Shiva is like a master conductor and the Daily Dance is the response of all creation to his rhythmic force.

Shiva is seen as the first dancer, a deity who dances simply as an expression of his exuberant personality (Banerji, 1983, p. 43). His dance cannot be performed by anyone else, as Shiva dances out the creation and existence of the world. However, just as the mortal dancer gets tired, Shiva too periodically lapses into inactivity. The cosmos becomes chaos, and destruction follows the period of creation. This concept of the Dance of Shiva is innate in Eastern ideas of movement and history; it is continuous and both constructive and destructive at the same time (Gopal & Dadachanji, 1951).

The Dance of Shiva represents both the conception of world processes as a supreme being's pastime or amusement (*lila*) and the very nature of that blessed being, which is beyond the realm of purpose or understanding (Coomaraswamy, 1924/1969). The dance symbolizes the five main activities of the supreme being: creation and development (*srishti*); preservation and support (*sthiti*); change and destruction (*samhara*); shrouding, symbolism, illusion, and giving rest (*tirobhava*); and release, salvation, and grace (*anugraha*).

Considered separately, these are the activities of the deities Brahma, Vishnu, Rudra, Mahesvara, and Sadavisa, respectively. Taken together, the cycle of activity illustrated by the Dance of Shiva encapsulates Hinduism as the main driving force of Indian society. The idea of cycles is a common thread in traditional Indian philosophy and is the theme that will run through our discussion of its culture.

In January 2003, Martin Gannon and his wife took a car trip from New Delhi to Agra to see the wondrous Taj Mahal and Red Fort. Leaving at 6:30 a.m., they saw incredible diversity along this main highway, with two lanes in each direction. The poor with their deeply lined faces were huddled by fires alongside the highway; there were no sidewalks, so people walked on the highway, as did animals of all types, including monkeys, water buffalos, cows, and so on. There was even a caravan of camels. Under these conditions travel was very slow, averaging 15 to 20 miles an hour, and many vans and cars had a sign, "honk your horn when passing," which resulted in a constant din. All of these discomforts disappeared at the first sight of the magnificent Taj Mahal. This experience, repeated several times in other parts of India in one form or another, confirmed that the choice of the Dance of Shiva was appropriate for understanding India.

Indian Culture: Early History

Two distinctive variants of basic Indian culture spring from the people's Dravidian and Aryan ethnic origins. The Dravidians probably came to India from the eastern Mediterranean coast, forming the highly developed Indus Valley civilization 3,000 years before Christ. About 1500 BCE this civilization fell into decline, and its people migrated to the southern part of the Indian subcontinent. At about the same time Aryans arrived in India from Persia, settling almost all of the Indo-Gangetic Plain. Today 72% of the population is of Aryan origin, while Dravidians account for 25%. The remaining 3% is made up of a myriad of other groups including Mongoloids. India's most populous cities, all ranking among the 40 largest in the world, are Mumbai (18 million) to the west, Kolkata (13 million) to the east, Delhi (12 million) to the north, and Chennai (5 million) and Bangalore (5 million) to the south.

India's history reflects the cycles of chaos and harmony epitomized by the Dance of Shiva. Time after time India has recovered from episodes that would have ended the existence of any other nation. In fact, Shiva's son, Ganesh, is the symbol of good arising from adversity. According to the legend, Parvati, the consort of Shiva, would spend hours bathing, dressing, and adorning herself. This often meant that Shiva was kept waiting, so Parvati set their son, Ganesh, on guard to prevent Shiva from bursting in on her unannounced and catching her in a state of unreadiness. One day Shiva was so frustrated by Ganesh's actions that he cut off the child's head. Distraught, Parvati completely withdrew from her lord, and Shiva realized he would have to restore the child to her if he was to win her back. He resolved to use the first available head he could find, which happened to be that of a baby elephant. The boy regained his life and now had the added advantage of the elephant's wisdom. Similarly, India's past and present contributions to art, science, and the spiritual world are immense, despite periods of turmoil and apparent anarchy.

The estrangement of north and south India, illustrated by the debate over language, has historical roots that reach far into the past. The south has enjoyed calm and relative tranquility throughout most of its history, whereas the north has been subjected to a series of foreign invasions, often on a grand scale. Consequently, the northern culture is more a product of a mixed heritage. Among the most significant modifying influences in the north were the various Muslim invasions, beginning about 1000 CE. As a result of these invasions, the administrative structure of northern India was repeatedly destroyed, society often deprived of leadership, and religious faith shaken.

Muslim Rule

Muslim rule of north India began early in the 13th century and lasted until the middle of the 19th century. Muslim rulers were harsh on Hindus with the notable exception of the great Mughal emperor Akbar, who married a Hindu princess and attempted to foster the development of a new faith, *Din-i-Ilahi* that blended the best of the leading religions. He also fostered tolerance for all religions and routinely appointed Hindus to high positions and received Jesuit priests in his court. However, it is against Muslim beliefs to worship any idol or image of God, so most of the invaders destroyed many thousands of Hindu temples and replaced them with mosques. A discriminatory tax was imposed on non-Islamic subjects, and Hindus were given low-level positions if they were employed at all.

Forceful conversion of Hindus to the Islamic faith was widely carried out. Hindus were turned into second-class citizens in their own land and almost never shown the beautiful side of Islam. The confrontation between two virtually incompatible religious systems led to implacable mutual hatred between their respective followers. The echoes of this conflict resound even today. For instance, in 1992 more than 200,000 Hindus stormed and destroyed a 450-year-old Muslim mosque erected by the Mughals to replace a Hindu temple on the site that marks the birthplace of the Hindu god, Rama, and hundreds of people were killed in the ensuing bedlam. Such instances are relatively common. In 2002, a group of Hindus were returning by train from a pilgrimage to the same trouble spot when they were reportedly attacked by a Muslim mob in Gujarat, a Western Indian state. This set off a full-scale massacre in retaliation, which left nearly 2,000 dead, mostly Muslims.

Unlike the north, southern India enjoyed a stable, almost uninterrupted regime of Hindu kingdoms until 1646, when the Mughal rulers succeeded in conquering and unifying all India. The Muslim Mughal empire began to disintegrate during the 18th century, with independent regional kingdoms springing up everywhere. The influence of the British East India Company rose as Britain ousted rival Western colonial powers in the south. During those days of weakness plundering invaders came from Persia and Afghanistan. North India entered a state of anarchy from which it did not emerge until the British gradually extended their control, leading to the establishment of the British *Raj* (Rule) in the 19th century.

The British Raj

The British government instituted direct rule over India in 1858 following the Sepoy (Indian) mutiny. Many Indians think of this event as the first war of independence. The Sepoys, Indian soldiers in British employ, mutinied over a rumor that animal fat was being used in the cartridges they had to bite to load their rifles. Hindus heard it was beef fat, while Muslims heard it was pig fat, thereby violating the taboos of both. British troops barely put down the insurrection; the British garrison at Kanpur, with its women and children, was slaughtered, and Kanpur became a rallying cry for British vengeance (Arden, 1990).

Early expressions of nationalism first crystallized in the Indian National Congress in 1885 and the All-India Muslim League in 1906. Following the infamous massacre of more than 400 unarmed demonstrators at Amritsar by General Dyer's troops in 1919, Indian leaders put aside their previous faith and hope in the good intentions of the British Empire. Inspired by M. K. Gandhi, the Indian National Congress began a program of peaceful noncooperation (*satyagraha*) with British rule. Tragically, just months after India's independence was finally granted, Mahatma (Great Soul) Gandhi was killed by a Hindu extremist who had denounced him as an appeaser of the Muslims. However, Mahatma Gandhi's legacy lived on in the civil rights movement in the United States, Nelson Mandela's fight against apartheid in South Africa, and Lech Walesa's Solidarity movement in Poland. It continues to inspire all those who live under tyranny.

The British granted independence in 1947, but the Raj was partitioned into a largely Hindu India and a Muslim Pakistan. Overnight partition created great communal strife, and 12 million refugees moved across the new India-Pakistan border from 1946 to 1947, Hindus into India and Muslims into Pakistan. More than 200,000 people were killed in the accompanying riots, giving the world a lasting image of a modern India seemingly at war with itself. In fact, except for times of crisis, India has managed to accommodate and contain the destructive forces latent in group differences. But it is also true that today the political and social compromises that have permitted the country to deal with its diversity are under extreme pressure.

As India's central authorities have faced threats of secession, caste warfare, and sectarian violence, they have sometimes adopted stern measures. The army has been called out more frequently in recent years to restore order to the country's troubled provinces than to defend the country from external threats.

Modern Leaders

Jawaharlal Nehru, the head of the Congress party, became the first prime minister of India in 1947. He was unwaveringly loyal to the basic concepts of freedom, democracy, socialism, world peace, and international cooperation and emerged as an eloquent statesman for the world's nonaligned, less developed nations. Two years after Nehru's death in 1964, his daughter, Indira Gandhi (no relation to Mahatma Gandhi), succeeded to her father's office. Mrs. Gandhi struggled to modernize India and make it an economic power, but she may have lacked her father's devotion to "the spirit of man." She invoked the emergency provisions of the constitution in 1975 and suspended civil liberties, citing the need to address some of the nation's persistent problems "on a war footing." When elections were called in 1977, the Indian people expressed their resentment against the methods of "the Emergency" and voted Gandhi out of office. After full democracy was restored, an apparently chastened Indira Gandhi returned to power in 1979, where she remained until she was assassinated 5 years later.

Rajiv Gandhi, Indira's son, became prime minister on her death but amid claims of widespread corruption in the government, his Congress (I) Party lost a general election. The succeeding government was short-lived, unable to sustain a parliamentary majority for their policies. During the next campaign Rajiv Gandhi was also assassinated, and the Congress (I) Party was swept back to power on a huge sympathy vote.

It appears that the Nehru-Gandhi dynasty, which has dominated India's modern political system, is still a force to reckon with in Indian politics. Rajiv Gandhi's widow, the Italian-born Sonia Gandhi, is the president of the Congress Party and was identified by *Forbes* magazine as the sixth most powerful woman in the world in 2007. The family's journey—their sequence of evolution, power, death, return, creation, destruction, and, perhaps, ultimate salvation—epitomizes the actions of the Dance of Shiva and the cyclical nature of Hindu philosophy.

Sonia Gandhi's rise to her position of power followed the 2004 shocking defeat of the ruling Bhartiya Janata Party (BJP), a conservative Hindu-centric party, which had risen in stature and won major elections. The BJP party leadership was responsible for continuing and building on the reforms that jump-started the Indian economy in the 1990s and led to India's emergence as an Asian superpower. However, people at the lower end of the socioeconomic ladder did not experience the benefits of the reforms and voted out the BJP in favor of the Congress Party and its allies.

Cyclical Hindu Philosophy

The Indian perspective on life tends to differ most sharply from that of Europe and the United States in the value that it accords to the discipline of philosophy (Coomaraswamy, 1924/1969, p. 2). In Europe and America the study of philosophy tends to be regarded as an end in itself—some kind of mental gymnastics—and as such it seems of little importance to the ordinary man or woman. In India philosophy tends to overlap with religion, and it is regarded as the key to life itself, clarifying its essential meaning and the way to attain spiritual goals. Outside of India, philosophy and religion pursued distinct and different paths, which may have crossed but never merged (Munshi, 1965, p. 133). In India it is not always possible to differentiate between the two.

In Hindu philosophy, the world is considered illusory, like a dream, the result of God's *lila* (amusement). According to one interpretation, *Bharata Varsha*, the ancient name of India, literally means "land of the actors" (Lannoy, 1971, p. 286). In an illusionary world, people cannot achieve true happiness through the mere physical enjoyment of wealth or material possessions. The only happiness worth seeking is permanent spiritual happiness as distinguished from these fleeting pleasures. Absolute happiness can result only from liberation from worldly involvement through spiritual enlightenment. Life is a journey in search of *mukti* (salvation), and the seeker, if he or she withstands all the perils of the road, is rewarded by exultation beyond human experience or perception (*moksha*). In the same way that the Dance of Shiva leads the cosmos through a journey, Hindu philosophy directs each individual along a path.

There are basically four paths or ways that lead to the ideal state: intense devotion or love of God (*bhakti yoga*), selfless work or service (*karma yoga*), philosophy or knowledge of self (*jnana yoga*), and meditation or psychological exercise (*raja yoga*). The four ways are not exclusive, and people may choose among or combine them according to the dictates of temperament and circumstance. Whatever path is followed, every Hindu is aware of the difficulty of reaching the ideal state in a single lifetime. This is the point at which the concept of reincarnation, or the cycle of lives, becomes important.

Individual souls *(jivas)* enter the world mysteriously; by God's power, certainly, but how and for what purpose is not fully explainable (Smith, 1958, p. 100). *Jivas* begin as the souls of the simplest forms of life, but they do not vanish with the death of their original bodies. Rather, they simply move to a new body or form. The transmigration of souls takes an individual *jiva* through a series of complex bodies until a human one is achieved. At this point the ascent of physical forms ends, and the soul begins its path to *mukti*. This gives an abiding sense of purpose to the Hindu life, a god to be actively sought and patiently awaited through the cycles of many lives.

The doctrine of reincarnation corresponds to a fact that everyone may have noticed: the varying age of the souls of people, irrespective of the age of the body. Some people remain irresponsible, self-assertive, uncontrolled, and inept to their last days; others are serious, friendly, self-controlled, and talented from their youth onward. According to Hindu philosophy, each person comes equipped with a highly personalized unconscious, characterized by a particular mix of three fundamental qualities: *sattva* (clarity, light), *rajas* (passion, desire), and *tamas* (dullness, darkness). Their relative strength differs from one person to another but, in the Hindu idea of destiny, the unconscious has an innate tendency to strive toward clarity and light (Kakar, 1978).

The birth of a person into a particular niche in life and the relative mix of the three fundamental qualities in an individual are determined by the balance of the right and wrong actions of his or her soul through its previous cycles. The rate of progress of the soul through this endless cycle of birth, life, and death—the soul's karma—depends on the deeds and decisions made in each lifetime. One way of mapping the probable karma of an individual is to consult astrological charts at the time of his or her birth, and this is an important tradition in Indian society.

The Dance of Shiva portrays the world's endless cycle of creation, existence, destruction, and re-creation, and Hindu philosophy depicts the endless cycle of the soul through birth, life, death, and reincarnation. We will now turn to examining the cycle of individual life within that greater series of lifetimes.

The Cycle of Life

According to Hindu philosophy a person passes through four stages of life, the first of which is that of a student. The prime responsibility in life during this stage is to learn. Besides knowledge, the student is supposed to develop a strong character and good habits and emerge equipped to produce a good and effective life.

The second stage, beginning with marriage, is that of a householder. During this stage, human energy turns outward and is expressed on three fronts: family, vocation, and community. The needs of pleasure are satisfied through the family, needs of duty through exercising the social responsibilities of citizenship, and the needs of success through employment.

The third stage of life is retirement, signifying withdrawal from social obligations. This is the time for people to begin their true education: to discover who they are and what life is all about. It is a time to read, think, ponder over life's meaning, and discover and live by a philosophy. At this stage people need to transcend the senses and dwell in harmony with the timeless reality that underlies the dream of life in this natural world.

The Hindu concept of retirement is exemplified in a story told by a traveler in India (Arden, 1990). The traveler saw a white-bearded man seated on a blanket, writing in a notebook. The man looked up and smiled as the traveler walked past. "Are you a Buddhist?" the traveler asked, to which the man shook his head. "A Hindu? A Muslim?" Again, he shook his head and replied "Does it matter? I am a man." The traveler asked what the man was writing. "The truth," he said, "only the truth."

The final stage is one of *sannyasin*, defined by the Bhagavad-Gita as "one who neither hates nor loves anything." In this stage people achieve *mukti* (salvation) and are living only because the time to make the final ascent has not come. When they finally depart from this world, freedom from the cycle of life and death is attained.

People can pass through the four stages of life in a single lifetime or stay at each stage for many lifetimes. Even Buddha is reputed to have passed through several hundred lives. Progress is determined in the light of the activities and inclination of the person at each stage of life. For example, Indian religion is replete with rituals, the primary purpose of which is to receive the blessings of God. Each ceremony involves the singing of religious songs (*bhajans*) and discourses by priests and other religious persons (*satsang*). The sincerity with which people indulge in these activities and apply the tenets of the philosophy in their practical life determines their progress through the cycle of life and death. A person may expound philosophy at great length, go to the temple every day, and offer alms to saints and the poor, yet indulge in all sorts of vices. These contradictions in life are resolved on death by karma, which dictates that on reincarnation, each person will receive rewards or punishment for their accumulated good and bad deeds.

The Hindu desire for positive outcomes of daily activities, resulting in positive karma, leads us to a brief discussion of the importance of astrology. With so much at stake, most Hindus in India consult the stars, if not on a daily basis, then at least on important occasions. Matching the horoscopes of a bride and groom is as much a part of planning a marriage as choosing the flower arrangements. It is routine for Indians to consult the stars about the best day to close on a house or sign an important contract. Indian astrology is one of the oldest systems in the world and differs from Western astrology in that it is based on the actual constellations as seen in the sky instead of divisions in the plane of the ecliptic. When it was revealed that an astrologer helped former President Ronald Reagan's wife, Nancy, set her schedule, U.S. Americans hooted with derision. In contrast, India leaders routinely consider the charts, local television channels devote airtime to discussing political and personal astrological predictions, and all major newspapers and magazines run daily predictions or articles based on astrology. It is reported that New York based financial astrologer, Ashok Motiani, predicted the 2008 sub-prime crisis in the United States. In an interview with CNBC TV 18, a sister station of NBC's CNBC in India, he also predicted the fortunes of the Indian stock market, the SENSEX index. In general, he is bullish on both India and the United States based on his study of the stars.

Like Hindu philosophy, the Indian concept of time is cyclical, characterized by origination, duration, and disappearance ad infinitum. This is reflected in the dramatic structure of a traditional Sanskrit play, typically based on the themes of separation and reunion and tending to end as they begin. Various devices are used—the dream, the trance, the premonition, and the flashback—to disrupt the linearity of time and make the action recoil on itself (Lannoy, 1971, p. 54). Similarly, the Dance of Shiva is a repetitive cycle of creation, existence, and destruction; constant change within a period of time, but ultimately time itself, is irrelevant.

In an attempt to neutralize the anguish of impermanence and change, the carved religious images that every village home possesses are made of permanent materials, such as clay or metal. This also reveals the functional role of the image in a materially restricted environment. The practice of religion at home is one of the main reasons Hinduism was able to survive the invasion of foreign powers over the centuries. Just as religion is important to the family, so, too, the family plays a dominant role in Indian society.

The Family Cycle

Most Indians grow up in an extended family, a form of family organization in which brothers remain together after marriage and bring their wives into their parents' household or compound of homes. Recent migration to cities and towns in search of economic opportunities has contributed to the weakening of many traditions, including that of extended families. In this section we describe family traditions that exist most strongly in the India of about 500 million people, which continues to be only marginally affected by industrialization. While weakened in some parts of society, many aspects of the family cycle are still important to all.

The preference for a son when a child is born is as old as Indian society. A son guarantees the continuation of the generations, and he will perform the last rites after his parents' death. This ensures a peaceful departure of the soul to its next existence in the ongoing cycle of life. The word *putra*, son, literally means "he who protects from going to hell." In contrast, a daughter has negligible ritual significance. She is normally an unmitigated expense, someone who will never contribute to the family income and who on marriage will take away a considerable part of her family's fortune as her dowry. Although formally abolished by the Dowry Prohibition Act of 1961, the institution of dowry is still widespread in India, but it is becoming increasingly fashionable among educated Indians not to indulge in the practice. It is also the case that a number of embittered wives have misused the provisions of the act to seek revenge on their husbands and their husbands' families, causing women activists to be alarmed at the possibility that hard won anti-dowry laws could be brought into disrepute.

A striking reflection of this gender preference is the continued masculinization of the Indian population, particularly in the north. There are 1,000 males to 930 females. The main reasons for this ratio are the higher mortality rate of female children and the tendency to limit family size, once there are a sufficient number of sons. Also, the recent availability of sex determination tests has allowed women to ensure that their first-born is a boy, as they can abort unwanted female children. It should be noted, however, that Indian legislation enacted in 1994 and amended in 2002 (Pre-Natal Diagnostics Techniques Regulation and Prevention of Misuse Act) bans sex determination tests.

Nurturing Children

Just as the Dance of Shiva represents preservation, caretaking, and support, parents tend to nurture their children with great care. A Hindu child grows up in the security of the extended family and has few contacts with other groups until it is time for school. Although the mother is chiefly responsible for the care of the child, there is also close contact with other females and mother-surrogates, and this continues for much longer than in many other cultures. The strong ties of home life do not conflict with the Hindu belief in the liberty of the soul removed from worldly concerns. Love of family is not merely a purpose in itself but a way to the final goal of life. Love will not yield the rewards of *mukti* (salvation) when it remains self-centered; that is why the Hindu try to diffuse their love over sons, daughters, guests, and neighbors (Munshi, 1965, p. 115).

Children in India are considered sacred, a manifestation of God, but if the Hindu ideal is a high degree of infant indulgence, reality is somewhat different in the poorer areas of India. Here, there are typically many young children under one roof, and 1 in 10 will die in infancy, so babies are not regarded as extraordinary creatures. Except for the first-born son, children tend to be taken for granted. This is reinforced by the belief in rebirth; as an individual is not born once and once only, he or she cannot be regarded as a unique event. The mother has probably witnessed the birth of several babies and may have seen some die, too. When her child cries, falls sick, or is accidentally hurt, she is not beset with feelings of intense guilt. A mother's work may be long and hard, both in the home and in the fields, so she is unable to give her child undivided attention.

Even as the Dance of Shiva leads the world through the joys of existence, an element of chaos is inherent in the world's Daily Dance. Similarly, nature in India has been full of threats to a child's safety: famine, disease, and chronic civil disorder. As a rule, until modern times more than half of all deaths befell children in their first year of life. As the nation got a grip on its affairs and as campaigns against diseases such as malaria and smallpox took hold, mortality rates fell. The cultural importance of children is derived from the need to carry on the cycle of life. This continued importance is reflected in statistics showing that while death rates since 1921 have fallen, birth rates have declined much more slowly. One of India's great assets is its population of children, which exceeds the total population of the United States.

Government attempts to regulate the birth rate have become synonymous with sterilization programs. Resentment against coerced sterilization in India helped to defeat Indira Gandhi's government in 1977. As a result of the political fallout, birth control was set back as a popular cause. Middle-class Indians, influenced by education and the desire for an improved standard of living, are increasingly adopting family planning methods. However, when the formidable psychic barrier of traditional Hindu beliefs in the life cycle is considered, it seems clear that rapid population growth will continue in the poorer, rural areas.

An Indian father is frequently remote, aloof, and a much feared disciplinary figure, just as Shiva is distant from the world he nurtures, but there are also special bonds between father and son, and the relationship is one of mutual dependence. A son must obey his father unquestioningly, pay him respect, and offer complete support in every need, both in life and after death. The father owes his son support, a good education, the best possible marital arrangement, and inheritance of property. One Indian proverb reads, "A son should be treated as a prince for 5 years; as a slave for 10 years; but from his 16th birthday, as a friend."

The Status of Women

Very early in life the son learns that women are lower in status than men. The position of women in this hierarchical society means that they must constantly be making demands and pleading with superiors for one thing or another. The son soon develops an attitude of superiority. A female's authority can seldom be absolute, except for the unchallengeable position that the senior grandmother may inherit. A son finds out that anger may be productive; violent outbursts of anger are often effective if directed against someone of uncertain status. Similarly the destructive powers of the Dance of Shiva are effective in creating new opportunities and patterns. We underscore the point here that we are referring mainly to those millions of Indians who live in the rural areas and are marginally affected by industrialization. We recognize that Indian society is undergoing some tremendous changes, especially in the urban areas, following the liberalization of the Indian economy in the early 1990s.

The relative position of men and women is clear in Indian society, and the question of competitive equality is not customarily considered. The Hindu marriage emphasizes identity, not equality. Generally, women are thought to have younger souls, and therefore they are nearer to the world than men and inferior to them. Girls are trained to be submissive and docile, to fulfill culturally designated feminine roles. The ideal of womanhood in Indian tradition is one of chastity, purity, gentle tenderness, self-effacement, self-sacrifice, and singular faithfulness. Throughout history, Indian women have had dual status: As a wife, a woman seduces her husband away from his work and spiritual duties, but as a mother, she is revered.

Among the crosses women have had to bear in Indian society are female infanticide, child marriage, *purdah* (feminine modesty and seclusion), marital mistreatment, and the low status of widows. Until the mid-19th century, the voluntary immolation of the widow on her husband's funeral pyre (*sati*) was not uncommon; the widow believed her act would cleanse her family of the sins of three generations. Poor families are more likely to be fearful of not being able to scrape together enough money to find their daughters husbands and may resort to killing infant girls.

The Dance of Shiva is not destined to lead to joy throughout the world, and if the corresponding experience of humanity includes some unhappiness for women in society, that is simply the way things are. However, many activists attribute the rise in sexual harassment against women to the influence of Western culture. Police often treat such crimes as minor incidents and there is a tendency to suggest that women invite sexual harassment by the way that they dress. At the same time, India selected Pratibha Patil as the first woman president (a largely symbolic position in a parliamentary democracy) in 2007. The country has also produced many successful women entrepreneurs and political leaders.

Marriage and Family

A man's worth and recognition of his identity are intimately bound up in the reputation of his family. Lifestyle and actions are rarely seen as the product of individual effort but are interpreted in light of family circumstance and reputation in the wider society. Individual identity and merit are enhanced if people have the good fortune to belong to a large, harmonious, and close-knit family, which helps to safeguard children's upbringing and to advance them in life. The family contributes to decisions that affect an individual's future, maximizes the number of connections necessary to secure a job or other favors, comes to aid in times of crisis, and generally mediates an individual's experience with the outside world. For these reasons the character of the respective families weighs heavily in the consideration of marriage proposals.

Arranged marriage is still the norm in India. Advertisements regularly appear in European and U.S. newspapers and on the Internet for the purpose of identifying potential candidates. The Western concept of romantic love arises from the Western concept of personality and, ultimately, from the un-Indian concept of equality of the sexes. In addition, the concept of life as an illusion makes the idea of a loveless marriage easier to understand.

Marriages are traditionally for a lifetime, as divorce is considered socially disgraceful although this is changing in the urban areas of the country. There is even a Web site, Secondshaadi.com, catering to the needs of divorcees, a phenomenon that would have been unthinkable a decade ago when a divorcee would have been a social outcast and treated like a misfit in society. One major reason for the rapid increase in divorce rates is the growing number of working women in urban areas who have independent incomes and are refusing to submit to their husbands' traditional views of the role of men and women within a marriage.

In the case of child marriage, the girl lives at her parent's home until she is about 15 or 16 years of age, after which she moves to the home of her husband's family. A newly arrived daughter-in-law is sometimes subjected to varying forms of humiliation until she becomes pregnant. This treatment originated historically from the urgent need to ensure the early birth of a son in times of low life expectancy. Also, the size of the dowry that a girl brings with her can determine how she is treated or mistreated in her husband's home. The husband's family may keep making demands on her for additional support from her family, and if it is not forthcoming, she may be tortured or even burned alive, although the outcry against such treatment seems to have diminished such illegal practices.

The restricted life of women in the conservative atmosphere of India does not prevent them from developing a strong sense of self-respect. Their ultimate role is to preserve unity and continuity in the chain of life, and there is pride and dignity in their sense of identity with the family and their roles as wife and mother. Indian society seems to have given women, rather than men, resilience and vitality under the difficult circumstances of life in that country. Ultimately, all respond to the Dance of Shiva, and whether that brings great joy or unhappiness to the current life is irrelevant compared to the ongoing search for salvation, or *mukti*.

Since the beginning of time, dancing has been a rite performed by both men and women; Shiva and his wife, Parvati, are often depicted in ancient sculptures as one composite figure, half male and half female (*Ardhanarishwara*). Typical figurines of Shiva are four-armed, with broad masculine shoulders and curving womanly hips. Similarly both genders contribute to Indian society today. In this century, Indian women have undergone a social revolution more far-reaching and radical than that of men. While this process has been going on, women have attained positions of distinction in public and professional life.

In summary, it can be seen that the extended family unit is still a strong feature of Indian society. Just as the Dance of Shiva wills all nature to respond to its rhythm, so, too, each member of the family fulfills a role dictated by family tradition.

The Cycle of Social Interaction

A sense of duty (*dharma*) is the social cement in India; it holds the individual and society together. *Dharma* is a concept that is wider than the Western idea of duty, as it includes the totality of social, ethical, and spiritual harmony (Lannoy, 1971, p. 217). Dharma consists of three categories: universal principles of harmony (*sanatana dharma*), relative ethical systems varying by social class (*varnashrama dharma*), and personal moral conduct (*svadharma*). Among the prime traditional virtues are: generosity and selflessness, truthfulness, restraint from greed, and respect for elders. These principles are consistent with a virtual global idea of righteousness. Hinduism has progressed through India's moments of crisis by lifting repeatedly the banner of the highest ideals. The image of the Dance of Shiva is strongly evoked by the following passage from the *Bhagavad-Gita:*

> Whenever the *dharma* decays, and when that which is not *dharma* prevails, then I manifest myself. For the protection of the good, for the destruction of the evil, for the firm establishment of the national righteousness, I am born again and again.

The oldest source of ethical ideas is the *Mahabharata*, or Great Epic (of Bharata), the first version of which appeared between the 7th and 6th centuries BCE. It is a huge composite poem of 90,000 couplets in 18 books, which traces the rivalry between two families involved in an unrelenting war. The story is interrupted by numerous episodes, fables, moral tales, and long political and ethical discourses, all of which serve to illustrate the illusory nature of the world and encourage the reader to strive for God. This sacred book, a repository of Hindu beliefs and customs, is based on the assumption that *dharma* is paramount in the affairs of society. The epics took at least 1,000 years to compose and are still the most widely read and respected religious writings of the Hindus. The most popular and influential part of the epics is the *Bhagavad-Gita* (Song of the Blessed One), a book Gandhi once said "described the duel that perpetually went on in the hearts of mankind."

A recent European traveler in India gave the following illustration of the power of *dharma*. Sitting precariously among local people on top of a bus during a long journey, a traveler was astounded when a sudden shower of money fell into the dusty road behind them. An Indian alongside the traveler began shouting and pounding on the roof of the bus for the driver to stop. Some way down the road the bus pulled over, and the man rushed away. All the passengers disembarked and waited for the Indian to return, laughing at his comic misfortune and manic disappearance. Eventually the man reappeared, clutching a big handful of notes, including Western money. It was then that the traveler realized his own wallet was gone from his back pocket; it had come loose and blown away from the top of the bus, scattering the equivalent of a year's income for the average native. The Indian, a total stranger, had run back and convinced the poverty-stricken locals to hand over the money they were ecstatically gathering from their fields. The traveler began to thank his new friend for his troubles, but with the comment, "It was my duty," the Indian declined to take any reward. This is a comment that many travelers hear when similar incidents occur.

Power and Purity

Hindus generally believe that social conflict, oppression, and unrest do not stem from social organizations but originate in the nonadherence to *dharma* by those in positions of power. Their actions have created the cycle of disharmony. Hindus see a quarrel as a drama with three actors—two contestants and a peacemaker—and it is not the protagonists but the peacemaker who is seen as the victor in the dispute, as it is he or she who has restored harmony (Lannoy, 1971, p. 198).

Individuals who head institutions are believed to be the sole repositories of the virtues and vices of the institution. Traditionally social reform movements focused not so much on abolishing hierarchical organizations or rejecting the values on which they are based but on removing or changing the individuals holding positions of authority in them (Kakar, 1978). For example, during the declining years of both the Mughal and British Indian empires, the ruling classes enjoyed lives of luxury and extravagance in India. Conspicuous consumption by the aristocratic elites at the expense of the productive classes is with us still in the India of the early 21st century. The identity of the elites may have changed, but their attitude remains.

The issues behind the social and political ferment in India today are not rooted primarily in economic deprivation and frustration, although these make the mix more volatile (Narayana & Kantner, 1992, p. 2). Rather, there is a widespread feeling that the institutions on which the society was founded no longer work.

The tragic recourse to mob violence by religious followers at different times in the country's history is a contradiction that astounds casual observers of India. How can such terrible things happen in a country where everyone believes in harmony and awaits the ultimate consequences of good and bad deeds in reincarnation? Hindus believe that *sila*, character or behavior, has its roots in the depths of the mind rather than in the heat of the action (Lannoy, 1971, p. 295). Because all worldly acts are transient or part of the illusion of life, they can have no decisive moral significance. Within the Dance of Shiva, destruction exists as strongly as creation and preservation; so it is with India.

This is not to say that violence is condoned by Hindu faith; just the opposite is the case. However, Hindus avoid the theological use of the terms *good* and *evil* and prefer to speak of knowledge and ignorance, *vidya* and *avidya*. Destructive acts done by people who are ignorant are not regarded as sins, but those acts committed by people aware of their responsibilities are counted against them in their seeking of *mukti*.

Bathing in the holy water of the Ganges is believed to wash away all of a person's sins and is required of all Hindus at least once in their lives. Indians tend to synthesize or integrate with nature because they assume this is the natural relationship of human beings with the world, unlike others who may exploit the physical environment for their own purposes. But the belief in the spiritual purity of the Ganges is so strong that government attempts to clean up the badly polluted waters have little chance of being effective. Many people simply do not accept that anything can spoil the Ganges' perfection. As a consequence, rotting carcasses of both animals and partly cremated people are a common sight along the river banks. The image of death among life, decomposition next to creation, and pollution mixed with purity is evocative of the Dance of Shiva.

Caste Tensions

Another pervasive social dimension in India is the caste system (*Varna*). Discrimination based on this system is now officially outlawed but is still a source of constant tension. Following the assumed natural law that an individual soul is born into its own appropriate environment, Hindus assume an individual belongs to a caste by birth. There are four main castes, each of which contributes to society in specific ways: *Brahmans* as priests, teachers, and intellectuals; *Kshatriyas* as warriors, police, and administrators; *Vaishyas* as skilled craftspeople and farmers; and *Shudras* as artisans and workers. Each of these natural classes has its appropriate honors and duties, but as privilege has entered the scale, with top castes profiting at the expense of those lower down, the whole system has begun to disintegrate.

Below the system is a group that has been termed the *panchama* (the fifth *varna*) or the Untouchables (now called *Dalits*), who lie outside of the major activities of society. *Dalit* is an ancient word from Marathi (a Western Indian language), which may be defined as "ground" or "broken in pieces." According to the Dalit Freedom Network, it refers to people who have deliberately been broken or ground down by those above them. Members of this group are engaged in work that is considered socially undesirable and unclean. Clay cups were historically used by restaurants and other establishments to serve Dalits and these were supposed to be destroyed so that no upper caste customer would ever use it and risk contamination by a Dalit's "uncleanness." Untouchability, as it exists today, is often described as a perversion of the original caste system.

Within each caste or group there are numerous subcastes, or *jati*, which influence the immediacy of all daily social relations, including work. About 3,000 *jati* exist, and they are further divisible into about 30,000 sub-*jati*, with unwritten codes governing the relationships between *jatis*. Friendships with members of the same *jati* tend to be closer and more informal than those with members of other *jatis*. As a general rule, a person's name provides information not only about his *jati* but also about the region of the country from which the person's family originated. For example, Gupta is a family name from the trading class, although many have gone into the professions, especially teaching. Most Guptas come from the north Indian states of Haryana and Uttar Pradesh and also from West Bengal, an eastern state.

The *jati*'s values, beliefs, and prejudices become part of each individual's psyche or conscience. The internalized *jati* norms define the right actions or *dharma* for an individual—he or she feels good or loved when living up to these rules and guilty when transgressing them.

When society was divided strictly by caste, there was no attempt to realize a competitive equality, and within each caste all interests were regarded as identical. But that also meant equality of opportunity for all within the caste: Every individual was allowed to develop the experience and skills which he or she needed to succeed at the caste's defined role. The castes were self-governing, which ensured that people were tried and judged by their peers. Central authorities viewed crimes committed by upper caste members more severely than those committed by the lower caste. As it was simply not possible to move outside of the caste, all possibility of social ambition, with its accompanying tension, was avoided. This suits the Hindu belief in harmony. The comprehensiveness of the caste system, together with holistic *dharma*, contributed to the stability that prevailed among the vast mass of people for much of India's history. Preservation of order, interspaced with disorder, is a characteristic of the Dance of Shiva.

The Untouchables

The worst fate of the caste system falls on the Untouchables. This caste has come to be the symbol of India's own brand of human injustice, victims of a system that kept people alive in squalor. Of course, social hierarchy is universal, found not only among the Hindus but also among the Muslims, Christians, Sikhs, Jains, and Jews (Srinivas, 1980). In addition, pollution taboos are prevalent in all civilizations, including the most advanced and modern, and eliminating dirt is an attempt to introduce order into the environment (Lannoy, 1971, p. 146). But Hindu society pays exceptional attention to the idea of purity and pollution, and historically this has resulted in the virtual ostracism of the Untouchables from the rest of society.

By way of historical explanation, Hindus believed that proximity to the contaminating factor constitutes a permanent pollution, which is both collective and hereditary. Therefore, they had a dread of being polluted by members of society who were specialists in the elimination of impurity. Hindu society was more conscious of grading social groups according to their degree of purity than of dividing castes precisely into occupations. The Untouchables—traditionally society's cleaners, butchers, and the like—were at the bottom of the Hindu hierarchy because they were considered irrevocably "unclean." A similar caste system was developed in Japan, and it too has been outlawed, although its effects are still being felt.

While India's traditional social structure was based on institutionalized inequality, today the government, and supposedly the nation, is committed to social equality. Beginning with Mahatma Gandhi, public figures have tried to reform the attitudes of Indian society toward the Untouchables. Gandhi named them Harijan, literally "children of God." The entire caste system was declared illegal by the constitution, and today, Untouchables are guaranteed 22.5% of government jobs as compensation for traditional disfavor. These policies have met with some success, but such a deep-rooted prejudice cannot be eliminated by a mere stroke of the pen. As recently as 2008, terrible atrocities (including torture and murder) were reported against Dalits who dared fall in love with a member of the upper caste.

The ambiguity of caste in occupational terms is another wedge by which the lower castes push their way upward on the scale. However, ambiguity is not so great as to render the system inoperative. Violations of caste norms, such as intercaste marriage, still evoke responses of barbaric ferocity. Educated Indians look on such incidents as throwbacks to the inhumanity of feudal times and believe that they must be dealt with sternly by the authorities. But efforts to create greater equality of opportunity for members of the traditionally disadvantaged castes meet stiff resistance from these same ranks (Narayana & Kantner, 1992, p. 5). There have been times when activists from the upper castes immolated themselves outside government offices to protest against any kind of accommodation through "affirmative action."

Attempts at Reform

Additional reforms remain problematic, as recent history suggests. In 1990 the government introduced policies reserving 27% of jobs in federal and state jobs for these castes and for Christian and Muslim groups that were socially backward. In protest, dozens of upper-caste students burned themselves to death. The upper caste Brahmins, a mere 5.5% of the population, have traditionally run government departments, but the struggle for jobs in India is so intense that the students saw themselves as victims of injustice, not as historical oppressors. The prime minister at the time, V. P. Singh, was forced to resign when the government's coalition partner withdrew its support of the government, mainly over caste reform issues.

The Harijan quickly realized their ability to assert their democratic rights as equal citizens through organized political activity. The effect of politicization of caste in modern times has made it clear that power is becoming ascendant over status. In 2008, Mayawati (who goes by one name), a Dalit leader, was elected as the chief minister of the state of Uttar Pradesh, which is India's most populous state and home to several former prime ministers. She is reported to be eyeing the biggest prize of them all—the prime ministership of India. Her meteoric rise, achieved through an unlikely coalition of supporters from opposite ends of the caste spectrum (Brahmins and Dalits), has truly shaken up Indian politics (Chu, 2008).

Modern education also acts as a solvent of caste barriers. The government has been playing its part by offering financial incentives to people who marry members of the lowest Hindu castes. These factors hold out the best hope for the disappearance of caste over the longer term.

The hierarchical principle of social organization goes beyond caste and has been central to the conservatism of Indian tradition. Among the criteria for ordering are age and gender. Elders have more formal authority than younger people and, as we have already related, men have greater authority than women. Many times women are not involved in social functions or conversations and are required to cover their heads in front of elders or mature guests. Most relationships are hierarchical in structure, characterized by almost maternal nurturing on the part of the superior and by filial respect and compliance on the part of the subordinate. The ordering of social behavior extends to every institution in Indian life, including the workplace, which we will examine shortly.

It is clear that the traditional social structure of India is undergoing change and reform. This is consistent with the evolutionary aspects of the Dance of Shiva. But any change requires the destruction of old ways, and pressure is beginning to build within the old system. It may be that before those changes are complete, Shiva will rest, and chaos will rule for a time. Or perhaps a new rhythm is beginning for the dance of the 21st century, and the Daily Dance of Indian society will quietly adjust in response.

The Work and Recreation (Rejuvenation) Cycle

There are several different perspectives on the importance of work: to earn a living; to satisfy the worldly interests of accomplishments, power, and status; and to fulfill the desire to create and care for the family. An aspect considered more important in India is that work enables, prepares, and progresses the individual through the cycle of life toward the ultimate aim of achieving *mukti*. The Indian approach to work is best defined by the *Bhagavad-Gita:*

> Both renunciation and practice of work lead to the highest bliss.
> Of these two, the practice of work is better than its renunciation.

Work was originally prescribed as duty (*dharma*) without any concern for material outcomes. Castes were occupational clusters, each discharging their roles and, in turn, being maintained by the overall system. But meeting obligations to one's relatives, friends, and even strangers and maintaining relationships constituted the ethos of the system. Even with the rapid expansion of industrial activity in the 20th century, requiring large-scale importation of Western technology and work forms, Western work values have been only partially internalized by Indians. Today, with government-mandated affirmative action, it is not unlikely that someone from a higher caste may work for someone from a lower class. Many Indians have developed a state of mind that allows them to put aside caste prejudice in the workplace but, on returning home, to conduct all their social activities strictly according to caste norms.

In some parts of the country, many Brahmins are feeling left out of the economy's rapid expansion. In the southern state of Tamil Nadu, for instance, 70% of government jobs are reserved for people from the lower castes and other historically disadvantaged groups. It is the Brahmins who now face ridicule and are locked out of professional careers in that state.

We have already seen how family life develops an acute sense of dependence in people, which serves to fortify the participative and collective nature of society. Similarly most Indian organizations have numerous overlapping in-groups, with highly personalized relationships between the members of each group. Members cooperate, make sacrifices for the common good, and generally protect each other's interests. But in-groups often interfere with the functioning of formally designated sections, departments, and divisions, and their presence can lead to factionalism and intense power plays within an organization. Just as disorder within order is a characteristic of the Dance of Shiva, so too incompetence is often overlooked because work performance is more relationship oriented than contractual in nature. Competent people may be respected but not included in a group unless they possess the group characteristics.

Loyalty to Group

Family, relatives, caste members, and people who speak the same language or belong to the same religion may form in-groups. There are typically regionally oriented subgroups, formed on the basis of states, districts, towns, and villages from which people's families originated. Within the group, Indians are very informal and friendly.

Geert Hofstede's (1980a) attitudinal survey of the cultural differences among some 53 countries is especially helpful in the case of India, which tends to cluster with those countries where there is a high degree of uncertainty avoidance. Indians tend to work with lifelong friends and colleagues and minimize risk-taking behavior although this is changing in modern India. Interestingly, India and the United States are tied in the recent GLOBE (House et al., 2004) rankings of uncertainty avoidance at 4.15 (Switzerland was highest at 5.37). This orientation is consistent with the Hindu philosophy of life as an illusion, Indians' preoccupation with astrology, and their resignation to karma. India also falls with those countries characterized by large power distances and this is consistent across both Hofstede's study and the recent GLOBE study. However, India ranks 21st of the 53 nations on individualism and is fairly high on the GLOBE collectivism rankings, especially with regard to in-group collectivism which reflects close ties with family members and respect for authority. Finally, India has a high score on masculinity, which is consistent with the emphasis on male-domination in Hinduism. This is also borne out by the low scores on gender egalitarianism in the GLOBE study. Generally the values described by Hofstede and the GLOBE studies reflect a historical continuity and resilience of the Indian social system, despite the onslaught of foreign invasions, colonial rulers, and economic dislocations.

The hierarchical principle continues to be a source of stagnation in modern Indian institutions. Younger people have a limited say or no say in decision making. Persistent critical questioning or confrontations on issues necessary to effect change simply do not occur. Any conflict between intellectual conviction and karma manifests itself in a vague sense of helplessness and impotent rage. Gradually the younger workers resign themselves to waiting until they become seniors in their own right, free to enjoy the delayed gratification that age brings with it in Indian society. The apparent lack of control and ambition displayed by workers is similar to the resignation of the world to the will of Shiva. The world responds to Shiva's rhythm, captive of its pace, and is unable to influence it.

Bureaucracy and Corruption

The importance of honoring family and *jati* bonds leads to nepotism, dishonesty, and corruption in the commercial world. These are irrelevant abstract concepts in India; guilt and anxiety are aroused only when individual actions go against the principle of primacy of relationships, not when foreign standards of ethics and efficiency are breached. This gives rise to legendary tales of corrupt officials, which are widely shared among travelers and businessmen who have spent time in India.

Indian organizations have been shaped by colonial experiences that have bureaucratized them and polarized the positions of the rulers (managers) and the ruled (workers). As a consequence, the role of the manager tends to be viewed as that of an order giver or autocrat. As a general rule, American managers perceive their role to be that of problem solver or facilitator and attempt to involve subordinates in routine decisions (Adler with Gundersen, 2007). In a much used cross-cultural training videotape, a U.S. manager attempted to use this style with his Indian subordinate, who wondered about his superior's competence and held him in some contempt for not being autocratic. The clear implication is that U.S. managers must act more authoritatively in India than in the United States. However, in India's Silicon Valley, Bangalore, where managers are trying very hard to recruit and retain talented engineers and information technology professionals, human resources is adopting some interesting practices that reverse the traditional hierarchical relationship between managers and employees. In one company, the best-performing employees of the month are treated to a special breakfast cooked by their bosses in appreciation of their efforts.

In the area of rejuvenation and recreation, one of its sources for the Indian people is the many religious festivals held throughout the year. These festivals are usually associated with agricultural cycles or the rich mythology of India's past. In some regions community festivals involve the active participation not only of Hindus but of members of other religions, too. Family bonds are recurrently emphasized and strengthened through the joint celebration of religious festivals.

The Indian sense of fun and play is given free rein during the festivities, which often include riddles, contests of strength, role reversals, and rebellious acts. Just as the Dance of Shiva is an expression of his joy and exuberance, festivals give Hindus an opportunity to express their feelings of devotion and happiness.

Gods and the God-like

Memorials to the grand line of India's "modern gods"—Mahatma Gandhi, Jawaharlal Nehru, and now Indira Gandhi—are as much the objects of pilgrimage as any temple or festival. Indira Gandhi was killed by her own trusted Sikh bodyguards, just 5 months after she ordered the storming of the Golden Temple at Amritsar by the Indian army to dislodge Sikh rebels. Her home in New Delhi is now a museum and shrine visited by thousands daily. The spot in her garden where she was gunned down is bracketed by two soldiers; her bullet-riddled sari is on display inside. Crowds gather before these, many weeping.

Another example is that of N. T. Rama Rao, a former movie star and chief minister of the state of Andhra Pradesh from 1983 to 1989. Rama Rao acted in leading roles in more than 320 films with mythical, historical, and folkloric themes. Among the masses, Rama Rao was associated with the qualities of the gods he played, and when he gave up his movie career to establish a new political party, he was immediately voted into office. This tradition has continued with many members of both "Bollywood" and the regional film industries being voted into office almost as an extension of their reel life roles. The fact that his party's radical Hindu fundamentalist policies sometimes caused strife within society is not inconsistent with the concurrently constructive and destructive nature of the Dance of Shiva.

The favorite pastime of Indians is watching movies, either at movie theaters or by renting videos from the shops that have sprung up all over the country, and today India's "Bollywood" is the largest producer of films in the world, releasing upward of

800 movies per year and these days, Hollywood studios come calling on a regular basis for financing and distribution tie-ups. Movies that draw on images and symbols from traditional themes are dominant in popular Indian culture. They incorporate but go beyond the familiar repertoire of plots from traditional theater. Films appeal to an audience so diverse that they transcend social and spatial categories.

The language and values from popular movies have begun to influence Indian ideas of the good life and the ideology of social, family, and romantic relationships. Robert Stoller's (1975) definition of fantasy, the "protector from reality, concealer of truth, restorer of tranquility, enemy of fear and sadness, and cleanser of the soul" (p. 55), includes terms that are equally attributable to the illusory nature of the Dance of Shiva, and it is easy to understand why films play such a major role in Indian recreation and rejuvenation.

India is the heart of Asia, and Hinduism is a convenient name for the nexus of Indian thought. It has taken 1,000 to 1,500 years to describe a single rhythm of its great pulsation, as described by the *Mahabharata*, or Great Epic. By invoking the image and meaning of the Dance of Shiva, and drawing parallels between this legendary act of a Hindu deity and many of the main influences of traditional Indian life, we have attempted to communicate the essence of India's society in this chapter.

It is not always possible to identify a nicely logical or easily understandable basis for many of the contradictions and paradoxes that exist in Indian society, just as it is difficult to explain the existence of racism, sexism, and other forms of intolerance and injustice in Western countries. In India the philosophy of life and the mental structure of its people come not from a study of books but from tradition (Munshi, 1965, p. 148). However much foreign civilization and new aspirations might have affected the people of India, the spiritual nutrient of Hindu philosophy has not dried up or decayed (Munshi, 1965, p. 148); within this tradition the role of the Dance of Shiva is accepted by all Hindus (Coomaraswamy, 1924/1969):

> Shiva rises from his rapture and, dancing, sends through inert matter pulsing waves of awakening sound. Suddenly, matter also dances, appearing as a brilliance around him. Dancing, Shiva sustains the world's diverse phenomena, its creation and existence. And, in the fullness of time, still dancing, he destroys all forms—everything disintegrates, apparently into nothingness, and is given new rest. Then, out of the thin vapor, matter and life are created again. Shiva's dance scatters the darkness of illusion (*lila*), burns the thread of causality (*karma*), stamps out evil (*avidya*), showers grace, and lovingly plunges the soul into the ocean of bliss (*ananda*). (p. 66)

India will continue to experience the range of good and bad, happiness and despair, creation and destruction. Through it all, its people will continue their journey toward *moksha*, salvation from the worldly concerns of mankind. Hindu philosophy is the key to understanding India and how a nation of such diversity manages to bear its immense burdens while its people seem undeterred and filled with inner peace and religious devotion.

And through it all, Shiva dances on.

CHAPTER 29

India

A Kaleidoscope of Diversity

Studying India is like looking into a kaleidoscope which is turning constantly. India is complex. It is colorful. It offers multiple evolving patterns which are contained within one framework.

—From a 10-part radio
documentary series, *Passages to India*,
produced by Julian Crandall Hollick, 1986–1989

The diversity in the Indian culture is not merely in its ethnic or racial composition. It is in every walk of its life. Starting with the geographical features, climatic conditions, and the vast regional and intraregional differences one can go on to religion, customs, attitudes, practices, language, food habits, dress, art, music, theatre and notice that no two regions are alike in these matters.

—Sreenivasarao Subanna, writing in
allempires.com, an online history community, April 5, 2007

India is often described as one of the world's most culturally and geographically diverse nations. It is also a land of great contrasts where you can expect to see Mercedes Benzes and BMWs vying for road space with ox carts and motorcycles, especially in the major urban centers. High-rise residential complexes for the moneyed upper classes rise like so many lotuses from the swamps. Slums develop around these complexes and they exist to house domestic servants who take care of the needs of the apartment dwellers. According to Forbes.com, Mumbai will soon see the completion of the world's first billion-dollar home, a 27-story skyscraper in the downtown area with a cost nearing $2 billion for the family of one of the richest men in the world, Mukesh Ambani of Reliance industries. Writing in a 2008 *Harvard Business Review* post online, world-renowned executive coach, Marshall Goldsmith, observes, "In the cities, I saw a longing for extreme opulence—countless ads with rich people living lavish lives—next to the reality of extreme poverty—countless shanties with poor people living harsh lives."

In 2006, 60,000 fans of an aging but iconic regional movie star, Rajkumar, went on a rampage in India's hi-tech hub, Bangalore, after he died of a sudden heart attack. They set fire to autos and buses and attacked Microsoft and other software firms, bringing India's fastest-growing city to a virtual standstill. The actor was seen as a defender of a regional cultural identity, which was perceived as being increasingly under siege. Some wondered whether the two Indias, representing prosperity and poverty, were drifting apart.

Once famous for holy men in sackcloth and ashes who renounced the world, and also known for its elephants and sacred cows, today India is equally as well known for having one of the largest pools of educated scientists and engineers in the world and it is a base for R&D for leading multinationals. This is a country where millions of Hindus, Muslims, Sikhs, Christians, and Buddhists have lived in relative harmony for several decades. Thomas Friedman (2005), *The New York Times* columnist, pointed out that the reason for this state of affairs is to be found in India's multiethnic, pluralistic, free-market democracy. In 2004, India experienced an election event unprecedented in human history: A nation of more than 1 billion people went to the polls in large numbers and were willing to elect a party headed by a foreign-born (Italian) Catholic political leader (Sonia Gandhi), who peacefully and firmly made way for a Sikh (Manmohan Singh) to be sworn in as prime minister by a Muslim (President Abdul Kalam). This took place in a country that is more than 80% Hindu.

The kaleidoscope is an appropriate metaphor for modern India. It presents an ever changing set of colorful images not only to foreigners who flock to do business there and the nonresident Indians who return every year for quick visits but also to the people living in the country. This is the result of India's unmatched diversity of cultures and its dramatic transformation from a third-world country to an Asian superpower in the 21st century. As Chhokar (2008) suggests, "What may be termed as the culture of India today is the outcome of, or merely the current stage in, a process of evolution of a continually living and changing culture" (p. 972). Thus, using the kaleidoscope metaphor, we will explore the diversity that lies within India. The kaleidoscopic areas we discuss include the variety of religious and culture celebrations, the distinctive festivals and feasts held throughout the year; modern India's call centers and cell phones; and the changing game of cricket, which has recently imported cheerleaders from the United States.

The Kaleidoscope of Religions and Cultural Celebrations

Religion and language separate the people of India far more than ethnic background or geography. India is the birthplace of four major religions, Hinduism, Buddhism, Jainism, and Sikhism. Although 81% of the Indian population is Hindu, sizable numbers belong to other religious groups: Muslim (14%), Christian (3%), Sikh (1.86%), Buddhist, Jain, and aboriginal animist. Hindus live throughout India, with smaller concentrations at the southern, northeastern, and northwestern extremities. The minority populations (the term is relative, as Muslims alone number more than 140 million, the third-largest concentration of Muslims after Indonesia and Pakistan) actively resist being dissolved into a Hindu melting pot. Muslims are in the majority in Kashmir, and Sikhs are concentrated in Punjab.

Hinduism, Islam, Christianity, and Sikhism are all associated with specific languages in which their original scripts were written: Sanskrit, Arabic, Latin, and Gurumukhi, respectively. The first three are not spoken languages in India, and the fourth in a modernized form is a state language of Punjab. In addition to Arabic, the Muslims in India evolved a language of their own, Urdu, and a number of other regional languages. There are at least 300 known languages in India, 24 of which have at least 1 million speakers each.

However, India has had a long tradition of embracing other religions throughout its history. Christianity came to India as early as the middle of the first century CE. Jewish people have lived in India for centuries and Islam came to India in the 13th century and its influence lasted for several hundred years. Through extensive trading relationships over the centuries, India has been able to absorb characteristics of foreign cultures and also to share the best of her culture with other nations of the world in an early version of globalization. Thus, not only has India witnessed the birth of several world religions but has also absorbed and integrated other faiths. "India: The Dance of Shiva" (Chapter 28) presents the tenets of Hinduism in great detail and in this chapter we highlight some of the other great religions that were born in India.

Buddhism

Buddhism began in India with the life of Siddhartha Gautama, a prince from a small kingdom in the foothills of the Himalayas who grew up in luxury but soon left home in search of the meaning of existence. He finally found enlightenment under a tree in the forests of Gaya (the modern state of Bihar in northern India) and achieved the knowledge that he later expressed as the Four Noble Truths: all of life is suffering; the cause of suffering is desire; the end of desire leads to the end of suffering; and the means to end desire is a path of discipline and meditation. Gautama was now the Buddha, or the awakened one, and he spent the remainder of his life traveling about northeast India converting large numbers of disciples.

The great emperor Ashoka of the Maurya dynasty was largely responsible for spreading Buddhism outside of India's borders in the 3rd century BCE and dispatched his own son, a monk, to spread Buddhism to Sri Lanka. Ashoka is said to have converted to Buddhism after the battle of Kalinga where his men killed more than 100,000 people. He was filled with great remorse and gave up war and violence.

Buddhism today has about 350 million adherents all over the world and is reputedly the fourth-largest religion in the world, practiced mainly in Asia. The Dalai Lama, who is considered to be a reincarnation of a great Buddhist master, is exiled from Tibet and has been living in Dharamsala, India, since the Chinese takeover of Tibet in 1951.

Jainism

Jainism, another religion that was born in India, attained major status in the time of Mahavira, who was born into a princely family in 599 BCE and abandoned his aristocratic surroundings when he was 30 years of age in favor of a more ascetic life. Jainism, which has nearly 4 million followers today, teaches that the way to liberation is to live a life of nonviolence and renunciation. Jains believe that animals, plants, and human beings have living souls and that each of these souls has equal value and should be treated with respect. The three jewels or the guiding principles of Jainism are right belief, right knowledge, and right conduct.

Jains are strict vegetarians and believe that they have to live in a way that minimizes the use of the world's resources. This strict practice of vegetarianism has created some interesting problems with nonvegetarian tenants of high-rise apartment complexes in Mumbai. Some critics have attached the label of *apartheid* to the intolerance of nonvegetarian food in these upscale, Jain-dominated housing complexes, where prospective buyers have been grilled about their food habits, charged exorbitant prices, or banned outright from owning property.

Sikhism

Sikhism, yet another religion that was born in India, is considered to be the youngest of the world religions, founded about 500 years ago by Guru Nanak, who was born in 1469. Nanak believed that we are all one, created by one Creator of all creation. He taught his followers to align with no other religion and to respect all religions. Today, Sikhism's followers are in the millions (23 million at last count) and most of them live in the Indian State of Punjab, with several hundred thousand more living in North America, especially Canada.

They are highly visible because of their distinctive turbans, which caused some of them in the United States to be mistaken for members of the Taliban after the September 11, 2001, terrorist attacks in New York and Washington, D.C. In one incident in Mesa, Arizona, a peaceful Sikh gentleman, Balbir Singh Sodhi, was killed in a hate crime, prompting the Sikh community to explain their religion to their American counterparts.

Sikhs form 1.86% of India's population but 8% of the Indian Army. By contrast, Muslims are 14% of the population and 2% of the army. The Prime Minister of India is Manmohan Singh, a Sikh who is the first person in that highest of offices to hold a doctorate. He is widely respected for being the architect of the economic reforms of the early 1990s, which set India on the path to economic prosperity. The ethnic backgrounds of India's presidents and prime ministers over the years highlight not only the country's amazing diversity but also the secular nature of its constitution.

Zoroastrianism (Parsis)

It is also important to mention the small community of Parsis, whose numbers belie their extraordinary contribution to Indian industry. Members of the Parsi community, who originally migrated from Iran, are followers of one of the oldest religions in the world, Zoroastrianism. Many Parsis escaped persecution in Persia (Iran) and settled down in India between the 6th and 7th centuries. Their numbers (a little more than one hundred thousand) are dwindling because they guard their culture and customs zealously. Most of them live in Mumbai, which is one of the most cosmopolitan cities in the country.

The Parsis believe in the existence of one invisible God. They also believe that there is a continuous war between the good forces (forces of light) and the evil forces (forces of darkness). The good forces will win if people do good deeds, think positively, and speak well. God is represented in their temples through fire, which symbolizes light. Parsis were favored by the British because of their knowledge of English, as well as their trading skills and facility with global trade.

The first steel plant in India was built by a Parsi, the industrialist J. R. D. Tata, over a hundred years ago in Jamshedpur as a model workplace. Tata's heirs are among the wealthiest industrialists in India today and they manage the widely diversified Tata conglomerate, which rivals industrial dynasties in other parts of the world. In 2008, Tata Motors unveiled the world's cheapest car, the Nano, and took control of iconic brands like the Jaguar and Land Rover from Ford Motors. Other famous Parsis include Zubin Mehta, the renowned conductor; Homi Jehangir Bhabha, the founder of India's nuclear program; Rohinton Mistry, the award-winning novelist who lives in Canada; and Adi Godrej of the Godrej family, which controls another of India's leading billion-dollar conglomerates.

Despite all this diversity, as M. C. Chagla, a renowned former Indian jurist, diplomat, and Cabinet minister observed, there is something to the idea of Indian-ness:

> There is an Indianness and an Indian ethos, which has been brought about by the commission and intercourse between the many races and the many communities that have lived in this land for centuries. There is a heritage which has devolved on us from our Aryan forefathers. There is an Indian tradition which overrides all the minor differences which may superficially seem to contradict the unity. (quoted in Iyer, 2000)

Images of Festivals and Feasts

Every turn of the kaleidoscope reveals a different festival in all its myriad colors. These festivals are celebrated in different parts of this vast country. Sometimes, it feels as if there is a festival every day of the year. Religious raptures, possessions, and trances are common during Indian seasonal festivals. This is a structured and, in some cases, highly formalized phenomenon that enriches the consciousness of the individual and the group. There are also the attendant dangers, however, that the mood can degenerate into a hysterical mob psychology, which Indian history has witnessed many times. Festivals allow the discharge of intense emotion, which is otherwise submerged in a network of reciprocity and caste relations, but they also can be used to reestablish order.

Religious Festivals

The birthday of Lord Ganesh is one of the most popular festivals in India and often involves 10 days of feasting, fund raising, and performances of music and dance, with music blaring from loud speakers on every street corner, especially in the city of Mumbai. The festival gained currency during the Independence Movement in India as a way of uniting various castes and communities against British rule. Massive processions through the streets feature large and small clay idols of Lord Ganesh, and the final goal is to immerse these idols in the sea. Millions of citizens patiently tolerate the ensuing traffic jams for several days every year in August as the monsoon rains finally taper off and the festival season begins in earnest.

Another festival that is celebrated throughout India is Deepawali or Diwali. It celebrates the victory of Lord Rama over the 10-headed evil demon king, Ravana, and his return from exile with his devoted wife and brother. Diwali ushers in spring cleaning, and it includes decorating the home with lights ranging from beautiful earthen lamps to colorful lanterns swaying in the breeze. The celebration also involves setting off all kinds of firecrackers and visiting family and friends with plates of mouthwatering sweet treats and gifts.

A festival that aptly captures the kaleidoscope metaphor is the festival of Holi, which is celebrated all over India at the advent of spring. One of the most colorful festivals in India, Holi involves people spraying each other joyously with colored water or powder in all hues of the rainbow to celebrate the vibrant colors of spring. Markets are filled with vats of yellow, blue, pink, green, and red powders several days before the festival, and old and young get into the spirit of this colorful celebration.

Holi is also marked by India's own version of Guy Fawkes Day (see Chapter 17, "The Traditional British House") with the lighting of bonfires across the country and burning effigies of Holika to commemorate one of the many legends associated with the festival. The legend holds that King Hiranyakashap was displeased with his son Prahalad, who worshipped Lord Visnu and thus was not his father's greatest admirer. The king ordered his sister, Holika, to sit in a blazing fire with Prahalad on her lap. Although

Holika had been granted a boon that protected her from being burnt, that blessing did not work when she entered the fire with her nephew in her lap and she burned to death. Lord Vishnu, however, protected his young devotee, Prahalad, who survived.

Janamashtami or the birthday of Lord Krishna is celebrated with the popular Dahi-Handi ceremony, which is the most visible aspect of the festival and has become synonymous with it. In his childhood, a naughty Lord Krishna and his friends raided the houses of their neighbors in search of milk and butter and merrily helped themselves to the caskets, which had been hung from the ceilings to prevent domestic animals from spoiling the precious contents. They accomplished this by building human pyramids.

During the modern-day Dahi-Handi ceremony, a large earthenware pot is filled with milk, curds, butter, and honeyed fruits and is suspended from a height between 20 and 40 feet. In cities like Mumbai, these pots are sometimes suspended between tall buildings. Several young men work collaboratively to claim the prize. They build a human pyramid by standing on each other's shoulders until the pyramid is high enough to enable the person at the top to reach the pot, break it, and take the contents. Prize money, which is usually tied to the rope by which the pot is suspended, is shared among those who participate in the pyramid building.

Muslims in India celebrate the various Eids (celebration or festival) and are greeted by their non-Muslim fellow citizens in a display of religious unity that generally permeates festivals of all religions in this secular country. The main Eid is the Eid-ul-Fitr, which marks the end of a period of fasting during Ramadan. Eid is celebrated with feasts and the exchanging of food and gifts with neighbors and family. Many give food, clothing, blankets, and books to orphanages and the poor. This Eid is also a time when religious and political leaders throughout the nation stress the need to remain united across various religious beliefs.

Christians in India celebrate Easter and Christmas with their Christian counterparts all over the world. Many non-Christians get into the Christmas spirit and pay a visit to Santa Claus in the department stores in the shopping districts of major cities and decorate their homes with colorful lights. Some even attend midnight Mass with their Christian neighbors and friends.

Buddha Purnima or the birthday of Buddha is the most sacred day in the Buddhist calendar and is celebrated all over the world by the faithful. By a strange three-fold coincidence, the day commemorates the birthday, the enlightenment, and the attainment of Nirvana by Lord Buddha. Buddha Purnima is a holiday in most Southeast Asian nations, especially India, Sri Lanka, Laos, Cambodia, Thailand, and Myanmar. The main Jain festival is Mahavir Jayanti, which is the birthday of Mahavira. It is marked more by quiet reflection and prayer than any demonstration of outward gaiety.

Nanak Jayanti or Guru Nanak's birthday is widely celebrated by the Sikhs in India. His birth anniversary is marked by prayer readings and processions, especially in Amritsar in Punjab State and Patna in the State of Bihar. The Sikhs also celebrate Baisaki or Vaisaki with great gusto, for it marks the beginning of the new harvest season. Much feasting and dancing ensues after prayers are offered at the *gurudwaras* or Sikh temples.

Indian Cuisine

No discussion of festivals and feasts is complete without a mention of the variety of Indian cuisines. India is widely considered to possess one of the most diverse cuisines in the world and has more regional variations than the countries of Europe. Food plays a very important role in Indian culture and is an integral part of weddings, festivals, and day-to-day living. Wedding feasts in India are unlike any other in the world, catering to hundreds of guests and influenced by regional tastes and practices. Frequently a wedding celebration unfolds over several days of feasting and rejoicing. A wedding feast in southern India (and other parts of the country) may be served on banana leaves with guests sitting cross-legged on the floor while a small army of servers doles out delicious items with the precision of an assembly line. The colorful arrangement of food on each guest's banana leaf changes with each pass by the rapidly moving servers through the line, resembling a food kaleidoscope.

A modern Indian wedding feast is often a marvelous integration of international and national cuisines. Thai, Mexican, Chinese, and Italian food stations are interspersed with various regional Indian cuisines. Multiple courses include appetizers, chutneys, dips, and desserts. Depending on the preferences and origins of the bride and groom, both vegetarian and nonvegetarian dishes may be served. The changing patterns on the plate remain a constant feature of these multicourse wedding feasts. If the traditional wedding feast menus from the different regions of the country were displayed on plates, they would look like a kaleidoscope of food, ever changing and forming interesting combinations of rice, lentils, vegetable dishes, yogurt, sweets, and sometimes meat, fish, and chicken.

Cell Phones, Call Centers, and Curriculum

Among the many images that emerge with each turn of the kaleidoscope of modern India is the cell phone, which has permeated every walk of life. Cell phones are simply everywhere you go. Shashi Tharoor (2007), a U.N. undersecretary general for communication and public information and author of a new book, *The Elephant, the Tiger, and the Cell Phone*, observes, "One of my favorite photographs of India shows a *sadhu* right out of central casting—naked body, long matted hair and beard, ash-smeared forehead—chatting away on a mobile phone" (p. 377).

The mobile phone revolution in India is nothing short of remarkable in a country that had millions on waiting lists for landlines right up until the 1990s. When

Rajnandini Pillai left India in the late 1980s, it was a privilege to have a landline. Returning to India in 2005 after a 9-year absence, she was struck by the proliferation of cell phones as the primary means of communication, not only among young people but also service providers like the electrician and the taxi driver. It was particularly mortifying, as she was armed with a cell phone purchased in the United States, which did not have international call access.

In 2006, India overtook China in the number of new telephone subscribers per month and it is now the fastest-growing cell phone market in the world followed by China and the United States. In 2008, the number of new telephone subscribers reached 8.5 million per month. Low cell phone charges, competition in the telecommunications industry, and lifetime unlimited incoming calls for a low one-time fee have made the cell phone accessible to people at the lower end of the economic hierarchy. It is quite common to see street vendors (vegetable sellers), launderers, and tea stall owners taking orders on the cell phone. In fact, observing his fellow countrymen and their mobile phone craze, one Indian exclaimed half in exasperation, "They won't have food on the table but they will have their cell phone!"

In a December 2007 National Public Radio segment, Laura Sydell, who had spent time in India, was amazed to find that cell phones were ubiquitous there. She reported that according to *The Economist*'s intelligence unit, only about 6 million people had access to broadband, but more than 200 million people owned cell phones. As a direct consequence of this, most Indians use their cell phones for double duty: communication and entertainment. They call and text each other (as is common in most parts of the world), and they also download Bollywood movies and Mp3 files, a popular pastime in this movie- and music-mad culture. It was not uncommon, according to Sydell, for cell phone users to have huge flash memory cards to download movies; the idea that it was not routine for U.S. Americans to have FM radio on their cell phones was incomprehensible to most Indians. In recent years, a few film schools dedicated to teaching the art of making movies for the mobile phone have sprung up, and one even partnered with Nokia to hold a competition for such movies. Cell phone companies today are using incentives to make major inroads into rural India where 70% of the more than 1.1 billion people still live.

Call Center Employment

Yet another image associated with modern India is the call center. After India's independence from Britain in 1947, Hindi, a north Indian language, was proclaimed the national language. Hindi was spoken at that time by about 25% of the people, but it was the single most prevalent language in the country. The south had virtually no Hindi speakers, and the southern people opposed starting the process of learning an "alien" language. As a result of these objections, the original constitution recognized no fewer than 15 official languages (3 more were added in 1992), and English was designated an "additional official language" as a compromise. Today, Hindi is the native tongue of about 30% of all Indians, but English is often the language of national communication, spoken throughout the country.

This facility with spoken English has been partly responsible for the enormous growth of call centers in India. Today, it is almost impossible to call the customer service number for a major U.S. corporation and not be struck by the fact that you are talking to someone from India. To keep up with demand from the United States and the United Kingdom, accent training classes have mushroomed in the cities where most of the call centers are located.

Despite these attempts at erasing the lilting Indian accent, most Westerners can recount at least one or more experiences of dealing with customer support delivered with an unmistakable Indian accent all those many miles away. Call centers are part of the industry called Business Process Outsourcing (BPOs), which was created when Western companies discovered the large pool of English-speaking educated workers in India who could handle customer service at a fraction of the cost in the United States or the United Kingdom. This has given India a competitive edge among other Asian nations.

In the beginning, many young Indians wanted to work in call centers because of the attractive salaries. Call center employees have tough schedules because they work at night in India when it is daytime in the United States. They usually party together and their jobs motivate them to learn American pop culture and pronunciation. Many even adopt a dual identity (e.g., Susan or John by night and Savitri or Venkateshwar by day). However, the job is stressful, and employee turnover is high. One call center owner told Rajnandini Pillai that he had to constantly work at keeping the employees motivated by changing his BPO service offerings so that the work was interesting and challenging to them. To this end, although he had started out with a customer service call center operation a few years ago, he had now evolved to a medical transcription services business, having outsourced the initial operation to other English-speaking countries such as the Philippines.

In 2005, Chetan Bhagat wrote a fictional account of call center life titled *One Night @ the Call Center: A Novel*, which chronicled the mundane lives of several call center workers. It became a best seller in India. The next area of growth in Indian BPO is legal process outsourcing. Thousands of English-speaking lawyers in India are employed by U.S. companies and law firms to do legal work at a small fraction of the cost of using U.S. legal assistance (Lakshmi, 2008).

Call center employment has changed the dynamics between members of different generations within the family. Young people are able to afford the same brand names as their counterparts in the West, can splurge on high-end restaurants and night clubs, and can enjoy a lifestyle that was beyond the reach of their parents. They are able to move from job to job, motivated almost entirely by salary considerations, and it is quite common for some to hold four jobs in the space of 2 years. Often, the BPO workers are seen as avaricious and individualistic, an anomaly in a largely collectivistic society that espouses the philosophy of work as an end in itself. After all, the essence of the

Bhagvad Gita, which is the most famous text of ancient India and exerts considerable influence to this day, can be summed up as "your business is with the deed, and not the result" (Chhokar, 2008, p. 978).

BPO jobs have often changed perceptions of the female child in the family. Instead of considering her a burden for whom a massive dowry has to be amassed, in some families parents rely on their daughters, smart upwardly mobile call center employees, to help finance the education of their younger brothers. The growth in demand for information technology (IT) services from India has provided bright young English-speaking college-educated women some remarkable opportunities for career development. However, India faces a potential shortage of 500,000 professional employees in the IT sector in the next few years, according to the National Association of Software and Service Companies (NASSCOM), a trade group.

The highly competitive environment has forced the BPO companies to adopt many elements of U.S. business culture. Managers constantly monitor the performance of their employees, offering them rewards and perks for exceptional performance and working hard to keep them motivated with company-paid dinners and contests. Because many employees spend hours dealing with U.S. Americans and their credit cards, they are reported to be quite comfortable with the idea of credit card debt (Kalita, 2005). Cars, refrigerators, and vacations are paid off in equal monthly installments (EMIs) deducted from every paycheck for years to come. This creates another division between the older and younger generations because the former did not grow up in an India where financing for education, housing, and consumer goods was freely available. In the past, most parents admonished their children to spend within their means and never get into debt. Today, it is a source of conflict between the older and younger generations.

The Down Side

Parents of call center employees also complain that young people no longer accompany them to religious functions and weddings and instead observe U.S. holidays like Labor Day and Christmas (Kalita, 2005). In a rapidly changing India, young people are making much more money than their parents, are less likely to follow traditional dating practices, and want the same things as their Western counterparts. In an interesting compromise, however, many IT professionals are finding a balance between the traditional arranged marriage and the Western idea of marrying for love. They go online to Indian matrimony Web sites, exchange e-mails and photographs, and if they find that they have a lot in common with each other, they introduce the two sets of parents, who seal the match.

As the kaleidoscope turns, so does the image of the call center employee. Today, the reputation of a call center job has taken a nose dive and call center employees are often perceived by their peers to have limited educational opportunities. A traditional call center job with the long hours (usually at night) poses widely reported health risks such as multiple personality and bipolar disorder and excessive stress. Often such jobs subject the employee to verbal abuse from unhappy overseas customers, and they involve tedious work. However, the maturing Indian economy is no longer only about call centers handling back office operations for U.S. companies. It is more about R&D centers springing up with a view to turning India into a global hub for companies like Microsoft, GE, and Motorola.

Education Highs and Lows

India is becoming known as a country with a highly educated workforce and a very strong scientific community. Indeed, large pools of educated workers and professionals have been crucial to India's ascent in both manufacturing and services. As a legacy of the strategic vision of the first Indian prime minister, Jawaharlal Nehru, India has a vast pool of engineers and scientists. The Indian Institutes of Technology (IITs) have drawn high praise from Microsoft founder Bill Gates and other U.S. business leaders for the whiz kids that they produce. These whiz kids then go on to contribute in a substantial way to companies like Microsoft, Intel, and Cisco systems.

"This is IIT Bombay. Put Harvard, MIT, and Princeton together, and you begin to get an idea of the status of this school in India" observed Lesley Stahl, co-anchor on CBS's *60 Minutes,* when she profiled the Institutes for a segment of the highly rated U.S. television program. There are seven IITs in India and competition for admission is fierce, as only a few thousand students are admitted each year. Entrance is determined by rankings in a nationwide examination.

Japan has recently discovered that it can learn a lot from the Indian model because many Japanese see India as an ascendant education superpower. Japanese bookstores are filled with titles like "Extreme Indian Arithmetic Drills" and the "Unknown Secrets of the Indians." At the Little Angels English Academy and Kindergarten in Tokyo, students are taught by South Asian teachers, and classroom posters depict animals out of Indian stories (Fackler, 2008). This is not the first time that Asian powers have turned to the Indian education model. Several centuries ago, India was home to one of the world's first universities, the famous Nalanda University, which was founded by Buddhist monks in 427 CE and played host to more than 10,000 students of various religious traditions. It was destroyed three times by invaders but rebuilt only twice. Asian scholars who travelled to India to study at Nalanda have left vivid descriptions of the university. In December 2006, India and a consortium of other nations announced plans to rebuild the university as a postgraduate research university.

Turn the kaleidoscope again and a different picture of India's education system emerges. Underfunded schools, a massive bureaucracy, outdated teaching methods, poorly compensated teachers, and financial pressures contribute to a public education crisis of vast proportions. Private school education is often beyond the means of most families. Government statistics indicate that about 25% of students make it past eighth grade, and 15% get to high school. About 200 million start high school but only about 7% graduate. The adult literacy rate is 61% compared to 99% in the United States. Only one in five job seekers has had any vocational training, and about 14 million people are added every year to the labor market. India needs to invest in education and devise a strategy to find employment for these people. Otherwise, these challenges will greatly inhibit India's ability to compete as a global power.

With foreign investment and job opportunities exploding, India may have a shortage of trained professionals because of the poor educational infrastructure. India graduates about 900 engineering PhDs per year, which is simply not enough to meet the growing staff requirements at Indian universities. Efforts are under way to increase the number of private colleges and universities and ease rules for foreign universities that may want to establish branches in India. Furthermore, leading Indian companies such as Infosys have achieved nothing short of a miracle by relying not on India's education system but on their own resources to recruit, train, and develop their workforces.

A recent report co-authored by Vivek Wadhwa, a former Indian CEO and executive in residence at Duke University, entitled "How the Disciple Became the Guru," found that Indian companies are picking up best practices from foreign companies outsourcing to India and perfecting those techniques (Wadhwa, de Vitton, & Gereffi, 2008). According to Wadhwa, "Indian industry has developed a surrogate educational system that can take workers with weak educational backgrounds and then turn them into world class R & D specialists."

The Changing Image of Cricket

When we discussed our metaphors for India with a focus group of Indian professionals, many suggested that the apt metaphor for India would be the game of cricket. The unofficial national sport of India is cricket although the official sport is field hockey. According to official cricket sources, in simple terms, cricket is a team sport involving two teams of 11 players each. Although the game play and rules are very different, the basic concept of cricket is quite similar to that of baseball. Teams bat in successive innings and attempt to score runs, while the opposing team fields and attempts to bring an end to the batting team's inning. After each team has batted an equal number of innings (either one or two, depending on conditions chosen before the game), the team with the most runs wins. A formal game of cricket can last anywhere from an afternoon to several days.

Test matches last for 5 days in an extended version of the one-day game. As in many other sporting events such as World Cup football, test matches between rival cricketing nations like India and Pakistan arouse a great deal of nationalistic fervor. When test matches are played, work in government offices almost comes to a standstill, as employees listen to the broadcasts. In the past when the only means of obtaining the scores was through a radio broadcast, it was quite common to see people walking around with their ears practically glued to transistor radios, which looked like curious appendages attached to one side of the head. Today, giant screens are set up at street corners and text messages are sent on mobile phones. Work still comes to a standstill during crucial international test matches.

In 2007, India clinched the inaugural Twenty20 cricket World Cup after a thrilling five-run victory over archrival Pakistan in the final in Johannesburg, South Africa. It was India's first major triumph since winning the 60 overs-a-side World Cup in 1983. The team was given a parade through the streets of Mumbai, which practically shut down the city as thousands of fans lined the streets and the stadium to honor their cricket heroes.

Cricket was introduced to India by the British. Some accounts suggest that the first game was played in 1721 by British sailors on shore leave. The Parsis were the first Indians to play the game with the Europeans and participated in many tournaments. Soon, the Hindus and Muslims fielded their own teams and played in the famous Bombay Quadrangular tournaments. The game went through many different formats and is now the most popular sport in India. Cricketers are almost on par with movie stars in the degree of attention that their personal lives receive. Not surprisingly, many of them date and marry Bollywood stars and some even graduate to starring in Bollywood blockbusters.

The game can also be seen as slow moving and dull, however, by countries that do not share the enthusiasm of the few cricketing nations. The Indian Premier League was formed with the goal of changing these perceptions abroad and attracting a new and youthful fan base at home with a mixture of sportsmanship and entertainment. The Premier League version of the sport evolved into a 3-hour match play with a shortened format in 2008, and teams chose colorful names—for example, the Bangalore Royal Challengers and the Kolkata Knight Riders. Entertainment came in the form of the Washington Redskin cheerleaders.

As Emily Wax (2008), writing in the *Washington Post,* put it, "It can all be seen as a metaphor for India itself, which is growing younger, hipper and more willing to take chances, awash in cash as its economy expands at 9 percent per year" (p. A01). Of course, as is inevitable there was a stir across conservative, Hindu-dominated India where traditional women usually wear saris or *salwar kameezes,* which cover most of their bodies. Some TV pundits argued, however, that the dance sequences in Bollywood films, featuring scantily clad gyrating women, were far worse than the spectacle of the foreign cheerleaders in their go-go boots, yellow shorts, and bikini tops. Cricket purists also criticized the shortened format of the hallowed game but by all accounts, the new formats resulted in a financial success for the promoters. Sony reportedly signed a $1 billion deal for exclusive rights to film and photograph Indian Premier League games over the next 10 years (Wax, 2008).

Images of the Future

The dictionary has a technical definition of the kaleidoscope as a tube-shaped optical instrument that is rotated to produce a succession of symmetrical designs by means of mirrors reflecting the constantly changing patterns made by bits of colored glass at one end of the tube. It also describes a constantly changing set of colors or a series of changing phases or events. We have used the kaleidoscope as a metaphor to describe the diversity that is integral to India and the dynamic elements of its modern culture. It seems as if with every turn, one sees a pattern that is different and one is never sure whether one is likely to see the same configuration again.

On the one hand, there is an economic boom that is unprecedented in the country's history. On the other, the crumbling infrastructure threatens to derail its economic agenda. On the key access road to the famous electronics city in Bangalore (dubbed India's Silicon Valley), cars, motorcycles, taxis, rickshaws, dogs, and cows compete for space and create a traffic jam that even seasoned Los Angeles commuters would balk at. The same goes for Mumbai where negotiating the roads can seem like a Darwinian challenge where only the fittest survive. Horns blare constantly and evoke the same response as car alarms in New York City; they are routinely ignored. Highways, airports, the power grid, and major bridges need to be upgraded before they crumble. Spending on major infrastructure such as ports, power plants, and highways is expected to top $60 billion annually over the next few years but there is a tremendous shortage of skilled labor in a country of 1.1 billion people. This is causing major delays and cost overruns in almost half the projects that are under way (Bellman & Range, 2008).

Asia's largest slum is on the route from Mumbai airport to downtown with its fine hotels and gleaming corporate headquarters. Some experts have argued that India should follow China, which launched a major infrastructure upgrade over a decade ago that is paying dividends today. In the 1960s, India had a higher per capita gross domestic product (GDP) than China. Today at $640, India's GDP is less than half of China's ($1,470). Growth in modern India has taken place in a chaotic, often unplanned, entrepreneurial, noisy, democratic way despite the bureaucratic obstacles presented by the government. The global financial meltdown of 2008 is expected to affect both India and China in the foreseeable future, belying the hope that the two Asian economic powers would be able to sustain the world economy in the context of a downturn in North America and Europe. For all the talk about decoupling, it appears that the global economy is still very interdependent although Indian exports, at 22% of the country's GDP, are less vulnerable to a global recession than China's exports, at 37% of its GDP.

Another threat is the growing Naxalite movement (Maoist insurgents who despise the landowning and business classes and seek to overthrow the state), which is operating on the periphery of some major industrial states. Prime Minister Singh has acknowledged that the Naxal groups are targeting all aspects of economic activity and that we "cannot rest in peace until we have eliminated this virus." The movement has been in existence for more than 40 years but today members operate in 30% of India, up from 9% in 2002. More than 1,400 Indians were killed in Naxal violence in 2007 (Kripalani, 2008).

However, as Edward Luce (2008) contends in his book *In Spite of the Gods: The Rise of Modern India,* although India's challenges may be Herculean, it also has colossal advantages. He goes on to describe how the Indian economy will benefit from a demographic dividend in the coming years with the proportion of dependents to workers falling from 60% of the population to 50%. India also has built-in institutional advantages such an independent judiciary and a free media and has accomplished "a high growth rate without the tools of an autocratic state" (Luce, 2008, p. 352). After independence, many predicted that India's linguistic, religious, and social divisions would cause it to fall apart. This has not happened because India has had a stable democracy that capitalizes on its plurality and diversity. No one is certain what India will look like in the coming decades but we are sure that the next time we take a turn at the kaleidoscope, we will see a different set of images of this colorful, vibrant, emerging Asian power of the 21st century.

PART X

Same Metaphor, Different Meanings

Sometimes seemingly the same cultural metaphor is interpreted in a radically different manner in different nations. For example, the symbols connected with the Spanish bullfight are radically different from those related to the Portuguese bullfight. In the Spanish bullfight the bull is killed, whereas it is only harnessed or stopped physically in the Portuguese bullfight. Thus we need to be sure that the metaphor we choose to study is one for which there is no ambiguity or confusion. However, one small area in Portugal now permits the killing of bulls and we examine the reactions of various Portuguese citizens to this practice. How and why these two cultures and their respective cultural metaphors differ is the subject of this part.

CHAPTER 30

The Spanish Bullfight

The killing of the bull is either loved or hated. There is no middle way.

—"Manolo Chopera" (2002, p. 82)

It is a cultural component that comes from times that we no longer even remember. . . . It is something that is ours, something intrinsic. It is a world that once you enter, you like it more and more, and you want to be more a part of it.

—Victoria Caceres, 24, who has been attending
bullfights regularly since childhood
(quoted in Wilkinson, 2007b, p. A8)

The characteristic that on the surface best portrays Spaniards is their contagious vitality, their love of life. Therefore it may seem paradoxical that the metaphor used in this chapter to describe Spanish culture is the bullfight, a confrontation with death. Yet this feature touches on other, less obvious qualities pervasive among Spaniards: Underlying their outer joy, or *alegría*, Spaniards tend to host a deeply ingrained sensitivity to the tragic and an equally strong emotional pull toward the heroic. Perhaps these characteristics, more than their love of life, have made the bullfight *la fiesta nacional* for hundreds of years; in combination these characteristics are manifested in a unique form of individualism emphasizing self-reliance and pride. The bullfight combines a passionate celebration of life with an elaborate system of rituals, a grandiose and artistic spectacle with blood, violence, and an all-too-real danger to the valiant performers. It exemplifies Spanish pride, individualism, emotionalism, and many other characteristics of Spanish people and culture.

As might be expected, flamenco and the bullfight are closely intertwined, and both activities highlight the proud and self-sufficient flavor of individualism unique to Spain. Whatever Spaniards do, they want it to be distinctively individualistic and a personal expression of individuality. For example, in Germany a symphony may advertise that its program will be the same on both Friday and Saturday night, and there will be few if any differences in the performances. This is not true in Spain, for each bullfight and each flamenco dance is or should be unique.

However, many Spaniards are not enthusiastic about bullfighting and they regard the killing of the bull as barbaric. The European Union (EU) formally requested that Spain abolish bullfighting and it stopped subsidizing Spanish farms that raise bulls for this express purpose. The state-run Spanish television, which started live coverage of bullfighting in 1948, stopped doing so in 2007 during prime time because of the alleged barbarism and now shows only delayed coverage late at night. Still, regional state channels and Pay TV do provide live coverage. Although attendance has been declining, 65 million Spaniards still attend bullfights each year, many of them more than once. In addition, periodically a particularly brave matador who takes great risks and demonstrates unusual skills captivates the public, as in the case of both Jose Tomas (Wilkinson, 2007b) and Sebastian Castella (Johnson, 2006), and attendance and coverage markedly increase.

The ritual of the classic bullfight, too, may be losing its appeal in contemporary Spanish society because of the popularity of soccer matches and other forms of Sunday afternoon entertainment. In more than half of Spain bullfighting virtually does not exist. To stem this decline, some bullfighters have turned the bullfight into stunts. Jesulin de Ubrique, for example, has performed on TV in front of live all-female crowds numbering in the thousands, and throwing bras and other accessories into the ring is common. Traditionalists regard such displays as vulgar. Although bullfighting employs more than 200,000 Spaniards or 1% of the labor force, financial rewards tend to flow disproportionately to a small number of matadors (about 170). Furthermore, of more than 300 bullrings in Spain, fewer than 10 are deemed first class. The newer forms of bullfighting, including the use of female matadors, have helped to increase bullfighting's popularity in specific segments of the population. There are about 20,000 bullfights annually in Spain.

It is important to stress that even soccer does not capture the essence of Spanish culture in the way that the bullfight does, probably because soccer is too universalistic a sport. Although soccer draws the same kinds of excited crowds as the bullfight, soccer fans have no role in the game comparable to audience participation in the bullfight—they judge the matador. Whereas soccer players form highly cooperative teams, bullfighters' groups are much looser. The latter more accurately reflect Spanish collective behavior. Finally, in contrast to the fact that the bullfight is an elaborate ritual, soccer is a sport, albeit a popular one. Therefore, there is little chance that soccer will ever completely replace the bullfight in Spanish culture.

The following discussion will relate different aspects of Spain's culture to the many facets of the bullfight. But first, the sequence of the event itself is provided to establish the language and image of the bullfight.

The Bullfight Begins

The bullfight, or *corrida de toros*, meaning "running of the bulls" (also, *la corrida* or *los toros* for short), takes place on Sunday afternoons or on holidays, timed precisely against the setting of the sun. The *corrida* begins with the white handkerchief signal of the officiating president and the sounding of a trumpet. The commencement of the bullfight is celebrated with the colorful opening *paseillo* or parade of participants, who march in hierarchical order to the upbeat tune of the *paso doble*. Once the parade is over, the president again uses his white handkerchief to signal the entrance of the first bull, as he does to signal the beginning of each stage of the event.

There are three separate groups (*cuadrillas*) of principal participants in the *corrida*. Each group is made up of a matador and his assistants. The *cuadrillas* fight and kill two bulls each, for a total of six bullfights during the event. When the first bull is released into the ring, it is confronted by three of the matador's assistants, the *banderilleros*. These men maneuver the bull with their magenta and gold capes, revealing to the matador the bull's peculiarities of movement; sometimes the matador participates in this stage as well.

A few minutes later, the matador enters the ring and demonstrates his bravery with his more artistic cape passes, including the classic *verónica*. The matador's valor or lack thereof often evokes emotional responses among the spectators. They applaud, cheer, and shout *Olé!* when impressed by the matador's bravery, or boo and whistle when they are displeased with his performance. In this way the public interacts with the matador, prompting him to perform more and more dangerous and aesthetically beautiful moves. These in turn further arouse the audience.

After the initial demonstrations with the bull, the two *picadores* enter the ring on horseback. One stands by as a reserve, while the other provokes the bull into charging his own horse. This allows him to get close enough to the bull to jab it with his long, spiked pole. The *picador*'s task is to weaken the bull enough to decrease his speed, thus allowing the bullfighter to show his abilities.

Moving Toward the Kill

The second stage of the bullfight allows the bull to perk up from its difficult ordeal with the picador. In this more exciting part of the fight, the *banderilleros*, on foot, each place two *banderillas* (colorfully decorated wooden sticks with barbed tips) into the bull's neck, where they should remain. At this point the bull is rested and more ferocious than ever and is ready to challenge the matador in the final stage of the fight.

In the last phase, the matador has 15 minutes to display his skill and build up the drama of the "moment of truth," performing dangerous passes with the much smaller red cape (*muleta*), which is draped over his sword. Before beginning, he salutes the president and asks permission to kill the bull if it is his first bull of the afternoon, and he may dedicate the kill to a friend, a dignitary, or the public. Once the kill is complete, the response of members of the audience determines to a large extent the level of the reward the matador is to receive. Ultimately the president decides on the level of the reward. A brave bull, too, is applauded as it is dragged out of the arena.

The same three stages are repeated for each of the five remaining bulls in the *corrida*, alternating *cuadrillas* with each new bull. The bullfight officially ends when the president leaves his seat.

The bullfight has long been a popular feature of Spanish culture. Its appearance in Spain has been attributed to a number of possible sources, including festivities of the ancient Minoan culture on the island of Crete 2,000 years before Christ, Roman circus combats, and the Moors. Whichever explanation may be true, the roots of the Spanish bullfight lay buried under layers of different peoples who invaded the Iberian Peninsula (now Spain and Portugal), bringing their respective cultures with them.

The same waves of invasion that somewhere along the way gave birth to the bullfight also bred the uniquely Spanish blend of races. People have inhabited what is now Spain for more than 100,000 years. Traces of the peninsula's prehistoric inhabitants include the famous cave paintings of Altamira, which are at least 13,000 years old. Recorded history picks up with the migration of the Iberian people from northern Africa into the southern two thirds of the peninsula around 3000 BCE. A series of other peoples subsequently came to Spain to conquer, trade, or settle there. The Phoenicians (a Semitic race) came first, followed by Celts (a Nordic race), Greeks, Carthaginians (from Carthage in northern Africa), Romans, Jews, Germanic tribes, and Moors. Gypsies are the most recent racial group to have taken their place in Spanish society, arriving from Egypt in the Middle Ages.

There are several distinctive features of the bullfight and its use as a metaphor for Spain. These include *Cuadrillas*, *Sol y Sombra*, the pompous entrance parade, audience involvement, and the ritual of the bullfight. Each feature is treated below.

Cuadrillas

The successive waves of invasion and immigration can be likened to the repeated attacks of the fighting bull and, even more appropriately, to the series of six bulls that are released into the ring during the course of the event. In terms of this metaphor, foreign invasions were so common throughout Spain's early history that it would take two consecutive *corridas* for the defending *toreros* to experience as many attacking bulls as there were invading peoples.

A Varied People

The physical characteristics resulting from the coming together of these various racial groups on the Iberian Peninsula are short, dark, Semitic types (predominant in the south) and taller, fair descendents of Celtic and Germanic ancestors (most common in the north). Combinations of these traits are also frequently seen, such as the large number of people with dark hair and blue eyes. Another prevalent trait among Spaniards is the classic Roman nose. In sum, there is no single image that would accurately portray a "typical" Spaniard, although certain combinations of features are notably Spanish, especially if one looks separately at the different regions of Spain.

In addition, certain personality traits cross tribal boundaries. Wheaton (1990) reports, "The Greek philosopher Stabo found common traits among all the isolated bands living on the peninsula: hospitality, grand manners, arrogance; indifference to privation, and hatred of outside interference in community affairs" (p. 21). These are characteristics that are still frequently used to describe the Spanish temperament.

The various invading peoples are also responsible for importing a number of prominent cultural attributes into Spain. The Romans (who ruled Spain for 400 years beginning around 200 BCE) brought their "vulgar Latin" language with them, which was to evolve into today's Castilian Spanish. The Spanish word for Spain, *España*, is a slightly modified version of the Latin name *Hispania*, which the Romans gave to the country (Spain had previously been called Iberia by its Iberian inhabitants and Hesperia by the Greeks). The Romans also constructed aqueducts, bridges, roads, and walls. The famous two-level aqueduct of Segovia stands as an impressive relic of the Roman rule. Finally, the Romans were responsible for the spread of Christianity into Spain.

Attributable to the more than 700-year presence of the Muslim Arabs, or Moors, are over 4,000 words used in modern Spanish. Words for many agricultural products the Moors brought with them derive from this source, such as the Spanish word *arroz* from the Arabic *al ruzz* (rice), *aceite* from *al zait* (olive oil), and *naranja* from *naranj* (orange), foods now an integral part of the Spanish diet (Graham, 1984), as well as many words beginning with the letters *al*. The Arabic influence in Spain is also responsible for advances in the fields of irrigation, mathematics, medicine, and, most notably, architecture, among others. The light, airy, and colorful *mezquita* or mosque in Córdoba and the impressive palaces, the Alhambra and the Generalife, in Granada are present-day monuments of Moorish architectural feats.

The *cuadrillas* in the bullfight clearly show Spaniards' attitude toward teams and collectivism. In a country with strong individualism but a shared need for defense, association with others is sometimes necessary. *Cuadrilla*, translated, does not mean team; the men in the *cuadrilla* do not work closely together. Rather, *cuadrilla* means "group, bank, or gang," reflecting the fact that there is safety in numbers and the toreros combine their distinct, individual efforts to achieve the death of the bull.

Spanish history is full of heroic events when strong and proud individuals gathered to achieve some common temporary and difficult goal. For example, the French under Napoleon lost as many soldiers to the Spanish guerillas as at the famous battle and retreat from Russia. Napoleon called this approach the "Spanish cancer" that slowly ate away the enemy's determination. In fact, the term *guerrilla warfare* was first used to describe the seemingly unorganized but systematic attacks that the Spanish used to defeat Napoleon.

The repeated invasions of the Iberian Peninsula by foreigners produced a tendency for the inhabitants to cluster together for a sense of security and control over the environment. This effect, and the lack of abundant water, is reflected in the fact that Spanish peasants on the steppes of central Spain chose to build their houses near one another, forming many small hamlets, villages, and towns. Their fear of loneliness and desire for security were so strong that they were, and still are, willing to travel great distances to work their fields just so that they may live near others (Crow, 1985).

Honoring the Collective

The tendency for Spaniards to live close together has generalized to urban housing arrangements, which take the form of apartment living rather than isolated home owning. In Spain, as in much of Europe, it is desirable to live as close as possible to the center of the city, where the action is. Marvin (1988) writes, "It is felt that as one moves further from the centre, so the quality of life gradually diminishes, because at the far limits of the town there are few bars, shops and *plazas* [squares] where people gather" (p. 129). The effect of this sentiment can be compared to housing patterns in the United States. The middle-class American's ideal house is in the suburbs with a nice size yard, or for the wealthy, it is a mansion in the midst of many acres of land. In Spain, the wealthiest Spaniards choose to make their urban apartments their permanent dwellings. They may, however, enjoy a country home as a secondary weekend getaway. Describing this logic, Marvin (1988) writes:

> Nowhere is this subordination and "culturizing" of nature, in Andalusian terms, more dramatically demonstrated than in the *corrida*. It is an urban event intimately linked with the country, which brings together, in the centre of human habitation, an uncontrolled wild bull, an item from the realm of nature, and a man who represents the epitome of culture in that, more than any ordinary man, he is able to exercise control over his "natural" fear (the fear felt by the human animal), an essential prerequisite if he is to control the wild animal. . . . The *corrida* is constructed in such a way that the imposition of human will is extremely uncertain because of the difficult circumstances; a situation which in turn generates tension, emotion, and dramatic interest. (pp. 130–131)

However, Spaniards' true loyalty does not extend to the larger collective; it is generally directed toward a smaller unit: family, friends, town, or region. Like the *cuadrilla*, the family is a large and affectionate clan whose members ferociously defend each others' honor, rights, jobs, and so on. The worst insult one can give a man is to derogate his mother, and the worst thing one can say to a woman is to insult her child (Wheaton, 1990). The protective spirit of the family also extends to the upbringing of the children. Parents dress their children well and pamper them. They bring them along wherever they go and allow them to stay up until all hours of the night. This produces children who are good-natured and outgoing and who have a strong sense of self and of their own importance (Wheaton, 1990).

The backbone of the tribal system during Spain's history of battles and invasions was the women. Tough and resourceful, they held the group together and catered to the needs of family members (Crow, 1985). However, historically the woman's role is more nurturing than authoritative. Thus, it is the husband, not the wife, whose role can be compared to that of the matador, as the person who has the last word. Like the matador who adoringly dedicates the kill to his wife or sweetheart, the Spanish husband accords his wife her due respect as stronghold of the family; he thinks of her as if she were a saint. And he expects her to behave like one. Until recently, the laws supported this expectation. However, there have been many social changes in recent years, as discussed later in this chapter.

General Franco's 36-year dictatorship, from the Spanish Civil War until his death in 1975, led to more backward policies for women than any other European country. A wife had to have her husband's written permission to do many things outside the home, such as opening a bank account, obtaining a passport, or traveling abroad. Civil divorce was illegal until 1981. Abortion, homosexuality, and adultery, too, were punishable by law. Extramarital affairs were considered unconditionally criminal for women; men were at fault only if they tainted the family's honor by failing to keep their affairs clandestine (Wheaton, 1990). The macho Don Juanism is so ingrained in Spanish culture that it was considered perfectly acceptable for a husband to keep a mistress on the side. However, although the famous, nonfictional Don Juan conquered numerous women's hearts, he usually pursued them circumspectly to keep their reputations and status intact.

All this has been rapidly changing in the past 30 years. Now, several hundred thousand mothers are unmarried, and the use of contraceptives among married women is widespread. Kissing and embracing in public are common. University attendance is split equally between the sexes. However, few women in Spanish society attain high-level jobs, and Spanish heritage is, of course, still patrilineal. Children go by both their father's and mother's last names. In addition, when a woman gets married, she keeps her name but adds her husband's name to her own, linking the two by the possessive word *de* or "of." For example, if a woman named Maria Alberti married a man named Miguel Sanchez, her name would officially become Maria Alberti de Sanchez. If her husband then died, she would become Maria Alberti Viuda de Sanchez—literally, Maria Alberti Widow of Sanchez.

The family is only one side of the Spanish tendency to cluster into loose groups like the *cuadrilla*. Historically, when the people of Spain started grouping together, they did not unite under a national flag; the geographical boundaries that sliced up the country topographically and climatically also kept invaders and inhabitants within isolated regions of the peninsula. Foreign influences and native loyalties were therefore determined by region, resulting in vastly differing cultures and distinct personal appearances. Four different languages and many dialects are spoken in Spain: Castilian, known outside of Spain as Spanish; Catalonian, Galician, and Basque. Despite the similarities, the cultural differences between these regions remain incredibly large.

The greatest geographical barriers in Spain are its mountain ranges. Although foreigners normally think of Spain's sunny beaches and flat, coastal plains, it is actually one of the most mountainous countries in Europe, second only to Switzerland. Also, Spain is a large nation and has a population of 43 million. Although Spaniards in the city live close to one another, the population density is only 81 per kilometer versus 230 for Germany. Therefore distance separates people and allows for great variation in culture and the look of the land.

Sol y Sombra

Seating at the bullfight is divided into three sections: sun (*sol*), shade (*sombra*), and a combination of sun and shade (*sol y sombra*). Tickets for seats in the shade are the most expensive. They cost about double the price of seats in the sun, which are the cheapest. Because bullfight tickets are relatively expensive to begin with, the price differential in seating separates the members of the audience into wealthier and poorer groups.

With a few exceptions, Spain itself can be divided into the relatively cool, wet, more prosperous north and the hot, dry, lethargic southern regions. Spain also has a large, central plateau area (the *meseta*, or tableland), which undergoes extremes of both heat and cold and poverty and wealth, making it comparable to the *sol y sombra* sections. A more detailed description of some of the better known Spanish regions will help illustrate the depth of regional differences in geography, climate, people, and culture.

There are 17 regional states in Spain, and in recent years there has been a devolution of power from the national government to them. As *The Economist* points out ("Spanish Centrifuge," 2007), this devolution of power has not yet resulted in the loose federation found in Belgium and elsewhere where the federal government is responsible for only a few functions such as national defense, but there is a definite movement in this direction. Given their unique languages and cultures, the Basque provinces and Catalonia are of particular interest in the debate about devolution, as noted below.

In the northwestern corner of Spain just above Portugal lies Galicia. Its frequent rain and drizzle, lush vegetation, and wide *rías* (fjords) simulate the northern European landscape of the natives' Nordic ancestors who settled in the region. The Gallegans also share the fair skin, light eyes, and bagpipe music of their Celtic cousins, the Scots. Working as fishermen, shepherds, and farmers, the people in rural areas of Galicia speak their own language, a mixture of Portuguese and Castilian Spanish.

Moving eastward along the northern coastline of Spain, one eventually encounters the Basque provinces. The region is one of Spain's most prosperous. Bilbao, the nation's industrial capital, is situated on the Basque coastline. The Basque countryside is adorned with beautiful Alpine vistas of rolling hills and Swiss-style farmhouses and is home to the hard-working, rugged descendants of the Iberian Peninsula's earliest settlers, the Iberians. The Basque language, Euskara, bears no demonstrable relation to any known language in the world. It is thought to date back to the prehistoric Bronze Ages, thousands of years before Christ. The Basques have one of the strongest separatist movements of all the Spanish regions. Their notorious radical, separatist organization, the ETA (an acronym for Basque Nation and Liberty) is responsible for widespread terrorist activities against the Spanish Civil Guard, Spanish civilians, and Basque Autonomous Police. However, there is only limited support for this activity.

In Spain's northeastern corner, sandwiched between the Pyrenees mountains and the Mediterranean sea, is the other region that harbors resentment against the control of Madrid. Supported primarily by its large textile and tourist industries, Catalonia shares the prosperity of the Basque region. Its culture is more European than the rest of Spain. The Catalan language is highly related to the southern French Provençal and is taught in school and spoken by most Catalans. Barcelona, the capital of Catalonia, rivals Madrid in population and is playing an increasingly central role in European life and international affairs, for example, hosting the Summer Olympics in 1992. Catalans, often characterized by their "canny commonsense," or *seny*, also demonstrate artistic capability. The region has produced a number of Spain's master artists, such as Salvador Dalí and Joan Miró, as well as architect Antoni Gaudí, and it contributed to the artistic development of Pablo Picasso, who was born in southern Spain. In recent years the movement to become a separate nation has accelerated in Catalonia and the federal government has granted many of its institutions independence or great latitude; for example, classes in schools are taught in Catalan, not Spanish.

In the center of Spain, on its sparsely populated plateau or *meseta* at an altitude of 2,000 to 3,000 feet, lies Castile, which exerts a dominant influence on Spanish culture. It is the birthplace of Spain's official national language, known outside of Spain as Spanish and within it as Castilian. Castile is politically dominant, too, and the federal government resides in Spain's capital, Madrid. The landscape of this region consists of vast expanses of arid steppes on which 10,000 castles were built during the battles of the Middle Ages (hence, the name *Castilla*). The dryness of the Castilian air allows for extremes of temperature, making the region akin to the *sol y sombra* seating at the bullfight. A Spanish proverb says, *Nueve meses de invierno y tres de infierno* (Nine months of winter and three of hell), and it accurately describes the Castilian climate. In Madrid, for example, even on the most treacherously hot days of August, the temperature can be deliciously cool in the shade, and one quite often needs a sweater in the evening.

Finally, Andalusia, in southern Spain, is home of the bullfight, Gypsies, and stereotypically dark Spaniards. The climate is relentlessly hot and dry. The peasants' parched faces mirror the parched face of the land, teaching the Andalusians how cruel yet beautiful life can be (Wheaton, 1990). Strongly influenced by the long Moorish presence in Andalusia, the natives express their intense feelings of joy and tragedy through their flamenco dancing and their lamenting "deep song," the *cante jondo*. Andalusia's scattered towns of whitewashed houses and red-tiled roofs reveal its Mediterranean essence. The majority of Andalusian land, however, belongs to a small number of large estates where acres and acres of irrigated, rolling hills and valleys are used to cultivate olive groves and other agricultural products.

The above descriptions only hint at the incredible variety of Spain's regions and at the harshness of life in the heat of the Spanish sun. Although Spain lies roughly between the same parallels as Cape Cod, Massachusetts, and Cape Hatteras, North Carolina, in the United States, and Madrid is at the same latitude as New York City, the mountains around Spain's periphery keep the Atlantic and Mediterranean moisture out. The central plateau bakes like an oven. The resulting austere living conditions throughout much of Spain are what engendered that underlying Spanish sense of sadness and tragedy, which seems to relate more closely to the Spanish love of the bullfight than does the Spaniards' outward gaiety. It is perhaps for this reason that Spaniards don't smile quietly and peacefully but laugh loudly and cry deeply in a land of extreme change and passion.

The Pompous Entrance Parade

A more joyful side of Spanish culture is represented by the *paseillo* at the beginning of the bullfight. The *corrida* opens with the proud march of all those individuals who appear in the arena during the course of the afternoon accompanied by the sound of the happy *paso doble* music. Lavishly costumed, everyone from the matador to the *areneros*, who tidy up the sand, struts into the ring in descending hierarchical order according to the importance of the participant's role. Every last one proudly shows himself off to the public eye. The aristocratic opening stems from the bullfight's aristocratic past; until the late 1600s, only noblemen *(hidalgos)* fought bulls, to prove their manhood to their ladies, peers, and underlings (Crow, 1985).

The values permeating both the *paseillo* and Spanish culture are honor, dignity, and pride, as well as a contempt for manual labor. These aristocratic values seeped into Spanish culture during its centuries of battles and heroes. The swelling of the noble class in Spain at this time was so great that it has been estimated that about 50% of the Spanish population has some claim to a title (Wheaton, 1990). Together, these are the values that make up the most important of all overriding Spanish characteristics—its unique type of individualism. Analogously, the pompous entrance parade highlights the individualistic, proud, and fun-loving aspects of Spanish culture. Crow (1985) eloquently describes this aspect of the Spanish mind-set in the following passage:

The Spaniard, thus, does not feel that he is born to realize any social end, but that he is born primarily to realize himself. His sense of personal dignity is admirable at times, exasperating at others; selfhood is the center of his gravity. His individual person has a value that is sacred and irreplaceable. In the universe he may be nothing, but to himself he is everything. (p. 11)

Proud Spanish individualism produces a number of outcomes for the Spanish culture. It helps to explain the Spanish people's lack of effective collective efforts. In fact, the anarchist movement was particularly popular in Spain at the turn of the 19th century, as we might expect in a nation valuing proud and self-sufficient individualism so highly. Because Spaniards refuse to subordinate their personal beliefs to a collective goal, and because everyone has to have a say in everything, there is a tendency for nothing ever to get done. Many Spaniards who can afford it opt to hire a *gestor*, a person who will wait in the necessary lines and wade through the bureaucracy in their place (Wheaton, 1990). Crow (1985) observes:

Spanish individualism is still anarchic and inorganic. The race is not cohesive except when it is unified "against someone or something." If Spaniards could only work as hard for as they do against things, their country would be one of the most dynamic and most progressive in Western Europe, perhaps in the world. (p. 361)

Since 1975, due to globalization and the greater legal freedoms afforded to Spaniards, this type of individualism seems to be changing, at least partially. Still, Spain's proud and self-sufficient individualism is in marked contrast to U.S. competitive individualism, Danish interdependent individualism, and the other specific types of individualism and collectivism described in this book.

On top of their difficulties in cooperating, part of the problem Spaniards have had in being productive is that their culture does not teach them to view work in the same positive light in which U.S. Americans see it. Work is seen as a means to an end—survival—not an end in itself. For example, while U.S. teenagers are often encouraged or forced by their parents to take summer jobs or part-time jobs during the school year even though the parents could easily support them, in Spain, families prefer to give their children educational and cultural experiences, such as sending them to France or England for the summer. The independence gained through work experience is not considered as important. In fact, one Spanish king in the 18th century issued a royal document telling his people that work is indeed honorable.

Further complicating the situation, jobs in some parts of Spain are often difficult to obtain and not very lucrative. As a result, Spaniards may be extremely hardworking when necessary to make ends meet (moonlighting is common in Spain), but in many regions the economic constraints on getting ahead curb their ambition. The Spanish word for business reveals the culture's disdain for commercial and administrative activities. The term they use is *negocio*, translated as "the negation of leisure" (Graham, 1984).

Many Spaniards do not mind if they accomplish little because they value self-expression more than material success. This is why the Spaniard has to be neither a Picasso nor a successful businessperson to be self-satisfied; recall that even participants in the mundane aspects of the bullfight parade proudly around the ring. Crow (1985) quotes an Englishman he met in Spain as having said, "In this country every blessed beggar acts like a king!" (p. 362). Rather than through material goods, Spanish individuals tend to be content to express their self-sufficiency through "wit, grandiloquent phrases, appearances, courtesy, generosity, and pride" (Wheaton, 1990, p. 68). Extremely articulate and enjoying the sound of their own voices, Spaniards will often spend hours conversing in the street cafés or bars where they meet. The odd result of this manifestation of Spanish individualism is that "often the most arrogant person in a group is the most charming" (Wheaton, 1990, p. 68).

However, globalization has influenced these patterns of behavior in many ways. For example, many businesspeople have stopped taking a long midday siesta, a practice originally motivated in large part by the lack of air conditioning. Because workplaces are typically air conditioned, many managers and workers prefer a regular 9-to-5 workday rather than one ending at 8 p.m. or later so that they can spend more time at home (Woolis, 2005).

Like the bullfighters with their elaborate costumes, Spaniards tend to spend a large chunk of their monthly salary on fashionable clothes to enhance the effect of their performance. The expense is not wasted, however, because they spend much of their time under the scrutiny of the public eye, getting together in public places rather than in their homes, a situation that is partly due to their rather cramped living space.

The proud marchers of the *paseillo* adhere to a rigid hierarchical order as they file around the ring. Spaniards readily accept a certain amount of inequality in the power structure of their society. But in Hofstede's (2001) study of 53 different countries' cultural values, Spain actually clustered with those nations deemphasizing power distance, ranking 31st of 53. Similarly it ranked 20th out of 53 nations on individualism. These are predicable results, given the balance that Spaniards seek between hierarchical order and individualistic expression, and they help to confirm the concept of proud but self-sufficient individualism found in Spain.

Finally, the pride and acceptance Spaniards feel about everyday life have contributed to the current of realism portrayed in Spanish art. Wheaton (1990) writes, "An invitation to contemplation, careful observation of life as it is, and directness of expression are qualities that come to mind when one asks what it is that has remained consistently 'Spanish' about Spanish art" (p. 290). From ordinary scenes like a shepherd feeding his dog to images of Christ with streams of blood pouring from underneath his crown of thorns and Picasso's horrifying cubist depiction of the bombing of Guernica, Spanish artists concerned themselves with documenting the world around them as each artist would uniquely see it. The same tradition has allowed Spaniards over the years to face violence and death with dignity.

Audience Involvement

The excited cheers and shouts of *Olé!* as the bull's horns whisk past the matador's gracefully leaning body highlight the strong Spanish emotionalism. The shouts and cheers are a form of self-expression and also audience participation, affecting the matador's behavior and further performance. Furthermore, once the matador makes his kill, the public displays its judgment of his performance. The audience's judgmental response in turn indicates the appropriate salute for the matador to make. If the members of the audience are displeased, they boo and whistle, and the matador does not reenter the ring. If they are pleased, they cheer, and the matador will step back into the ring to acknowledge the applause. If the positive public response is prolonged, the matador will take a lap of honor around the ring with his *banderilleros.* If the matador's performance was very good, the members of the public wave handkerchiefs to request that he be honored with an ear of the bull. The president must comply if most people in the audience wave their handkerchiefs. It is then up to the president to decide whether or not to award two ears for an exceptional performance, or in rare cases, even the tail. After any of these rewards, the matador takes a lap of honor around the ring, during which the audience throws objects (wineskins, hats, articles of clothing, and so on) to the matador, which are then thrown back, much like autographs. Other objects are meant as gifts to the matador, such as flowers and cigars.

Likewise, the emotional nature of Spanish expression is not only a one-sided outpouring but also an interactive form of communication. Therefore, while Spaniards' emotionalism might on the one hand reflect their proud individualism, on the other hand its participative and empathetic qualities surpass this notion. They convey the Spanish culture's emphasis on relationships, generosity, hospitality, and recognition of social needs.

These attributes relate back to the social behavior introduced in the previous section. The Spaniards' gregarious and fun-loving nature is magnified by their uninhibited emotionalism. Spanish men are not too shy to compliment women, whether acquaintances or strangers. There is even a special name for these pleasant remarks—*piropos.* Spaniards are not afraid, either, to allow the conversation to move onto a personal level, which helps avoid the small talk of the U.S. cocktail party and shop talk between business acquaintances. Thus Spaniards come across as genuine, interesting individuals rather than as superficial socialites or narrow-minded career people.

In allowing themselves to act on their emotions, Spaniards are incredibly generous and hospitable. Just as they enthusiastically wave their handkerchiefs in the air if they are moved by a matador's performance to reward him with a trophy from the bull, the

Spaniards are moved by their relationships with other people to want to give them whatever they can. This takes the form of sharing food, cigarettes, or general kindness, which contributes to the upbeat social atmosphere.

Spanish social life, for example, is often accompanied by the communal enjoyment of food and drink. To accommodate this taste, Spanish bars serve a wide variety of appetizer-sized portions of food called *tapas*, which groups of friends order to share. Overall, the loud talking, odor of food, and joviality produced from the beer and wine create a sense of vitality and life that further intensify the emotional nature and gaiety of Spanish social interactions, much like the interactive involvement of an excited audience with a bullfighter.

Spaniards also share their day-to-day lives outside of social gatherings. If a Spaniard brings a package of food to eat on the train, the first thing he or she does after opening the packet is to offer its contents to everyone nearby. Travelers in Spain should not feel reluctant to accept a cookie or a piece of sausage (chorizo) in such a situation and should avoid hoarding their own food.

Spanish hospitality is primarily an outgrowth of the Arabic culture. Phrases and customs expressing courtesy, such as *esta es su casa* (this is your home) and *buen provecho* (enjoy your meal), are traces of Moorish domination (Wheaton, 1990). Spanish hospitality is not to be mistaken, however, for the U.S. custom of inviting virtual strangers into one's home. In Spain (as in other Latin countries), one can know people for years without being invited into their home for a meal; if a dinner invitation is called for, it is much more likely to take place in a restaurant.

In addition, like the Arabs, Spaniards are generally relationship oriented. So, as with their propensity toward material sharing, Spaniards' concept of time focuses on interpersonal interactions. Thus, for example, completing the natural cycle of a conversation or other human interaction takes on more importance for the Spaniards than punctuality. In Hall and Hall's (1990) terms, Spaniards have a polychronic orientation toward time, doing several activities simultaneously while being highly involved in interactions.

Spanish emotionalism and generosity are also related to Spain's national social policies. A strong social security program, almost totally subsidized universities, and a mandatory month's paid vacation for all workers reflect Spaniards' responsiveness to people's needs. Spain is less materialistic and aggressive than many other nations, such as Japan and the United States: In Hofstede's study (2001) of 53 nations, Spain clustered with those nations that were more feminine, as we would expect.

The Ritual of the Bullfight

The logical outcome of all the overarching features of Spanish culture described above—regionalism, a sense of tragedy, proud individualism, and emotional interaction—is an attraction to the ritual, which provides an arena for all of these characteristics. The ultimate ritual in Spanish society is the bullfight. From its ritualistic handkerchief signals to its precisely regulated sequence and strictly defined roles of the matador, *banderilleros*, and *picadores*, and from the official role of the president to the sacrifice of the wild bull in the moment of truth, the *corrida* is permeated with rituals. Rituals define the bullfight; they make it what it is. Foreigners who mistake it for some sort of cruel sport often miss this point. It is not a sport. The ritual is a way of combining festivity with solemnity and life with death. These are some of the features that draw Spaniards to the bullfight, and these are the same features that draw Spaniards to a more encompassing ritual in Spanish society: religion.

Christianity entered Spain during Roman rule in the first century after Christ. Although the Romans resisted it, Spaniards took up Christianity with an unparalleled fervor. The tales of torture, martyrdom, and sainthood of the early Spanish Christians appealed highly to their sense of tragedy; the Spaniards' many hardships made them appreciate the ascetic Catholic values. At certain times in history the Spanish zeal for Christianity went out of control, to the detriment (and death) of many. In particular, at the end of the 15th century under the reign of the Catholic monarchs, Ferdinand and Isabella, religion sowed the seeds for the persecution and expulsion of hundreds of thousands of Jews under the auspices of the Holy Inquisition. At the same time, a large missionary effort was launched to convert the "heathen" native peoples of the recently discovered New World. Whole cultures were consequently disrupted.

In addition to its fervor, Spanish Catholicism differed from that of other Catholic countries in another way. The Spanish people preserved their pagan traditions right alongside their new Christian beliefs. Therefore, to the Spaniard, God is not an intangible, elusive concept but instead an almost human, concrete presence. The Spaniard's relationship with God is thus a personal one. Spaniards see God as patient and forgiving of human weaknesses and are not afraid to have conversations with Him, asking for personal favors.

The personal aspect of Spanish religion was enhanced by the Moors as well as by Spain's pagan past. The Moors, too, enjoyed an all-encompassing relationship with God (Allah) and left evidence of this fact in the Spanish language. The expression *si Dios quiere* (if God wills it) is widely used, as is the phrase *ojalá*, meaning "I hope so," derived from the Arabic *wa shá' allah* (may God will it) (Wheaton, 1990).

However, Spanish people do not always deal directly with God. Because they like religion to have a personal flavor, it is at least as common for Spaniards to pray to the Virgin Mary and the saints. Thus one often finds small candles lit in front of images of these figureheads in churches, with a charitable contribution box nearby to compensate for the requested favor. However, favors are not asked randomly of any saint. When you lose something, you pray to St. Anthony; when you travel, you may wear a protective medallion with the image of St. Christopher. Many Spaniards also make pilgrimages to places where the Holy Virgin appeared. Finally, these religious references surface in everyday life, such as the common female name *Maria del Pilar*, usually Pilar for short, after the Virgin of the Pillar in Zaragoza, and the equally common name Conchita, a nickname for *Maria de la Conceptión* (Mary of the Immaculate Conception).

A wilder aspect of paganism that is wound into Spanish religion is the popular celebration, or fiesta. The fiesta is a ritual like the bullfight (the bullfight is actually part of many fiestas, exemplifying their pagan nature). Each region of Spain has its most famous fiesta, like the fiesta of San Fermín in Pamplona, Las Fallas in Valencia, El Roco in Andalusia, and Holy Week celebrations throughout the country. These celebrations involve extravagant costuming, floats, processions, and drinking, elements that would seem to stray far from piety. As a case in point, does the scene of dozens of men running in front of rushing bulls sound more Christian or more pagan? The answer is obvious, yet this ritual is observed as part of the celebration of a Christian saint in Pamplona. To the Spaniard, religion simply does not occupy a separate compartment in life; it can be incorporated into life's most mundane and most dramatic moments. The Spanish fiesta shares many similarities with the Mexican fiesta (see Chapter 23), but there are also many points of difference.

The pervasiveness of religion in Spanish day-to-day life would seem to work in favor of the church, but after Franco's death in 1975 there was an anticlerical period during which a negative reaction to conservative Catholicism occurred. Still, almost half of the population attends church regularly, and 90% of the population define themselves as Catholic, suggesting that Catholicism as a cultural influence is strong. In recent years there has been a renewed interest in Catholicism, as manifested in popular recordings and films. For many years movie directors had difficulty persuading actors to play priests and nuns in film, but this is no longer true.

A U.S. American's experience as an exchange student living with a Spanish family helps to illustrate the importance of bullfighting in Spanish culture. One Sunday the father turned on the televised bullfight and said to his houseguest: "I hate bullfighting, but if you want to understand the Spanish, you had better learn about it."

As already indicated, however, Spain is undergoing rapid change, particularly since Franco's death in 1975. It is now a member of the EU. After a long period of conservative Catholic behavior there has been a strong movement toward freedom and creativity that has attracted worldwide attention. The *movida* (cultural movement) has been propelled forward by the activities of such artists as Barcelo (painting), Almodovar (cinema), Dominguez (clothes designer), and Mecano (pop music). Relaxed customs can now be found in Spanish society, and some have argued that Spain is being too influenced by the United States. Although such change is real, it is difficult to imagine Spain without the rituals of the bullfight and the underlying Spanish culture that the bullfight expresses.

In summary, a passage by Crow (1985) concisely recapitulates some of the points made in this chapter:

Spain today is a composite of all that has gone before. Her taproot reaches into the bottomless past. On several successive occasions in history she has flowered in beauty, shedding her glory over the civilization of Europe. On many other occasions she has grimly closed her door on the outside world, and retired into the gloom of fixed memories. In spite of her perennially poor government, her vitality is ever present, and appears inexhaustible. The Spanish people are among the most generous, the most noble human beings on earth. Their spontaneous art places them in a unique category among the nations of Europe, both for its quantity and for its incomparable beauty. With one foot in the present and the other in the past, Spain today stands straddling the unfathomable abyss. (p. 356)

As the people of Spain pursue an accelerated course of modernization, they are evolving into the more cosmopolitan embodiment as envisioned by the founders of the EU—but without, we hope, losing the special love of life that makes Spain so unique and wonderful. Whatever the eventual outcome, the bullfight will remain as a major, if not the major, metaphor for understanding deep-seated cultural values and attitudes among the Spanish people.

CHAPTER 31

The Portuguese Bullfight

Bullfighting is like a ballet in which the bullfighter and the bull dance a perfect pas de deux.

—Pedrito de Portugal, the most famous matador
in Portugal (quoted in Bilefsky, 2007b)

Superficially Spain and Portugal are similar in many ways. As inhabitants of the Iberian peninsula, they have experienced hundreds of years of successive invasions. Both were sea-faring nations that became world powers during the era of the conquistadores. In the GLOBE study (House et al., 2004) of cross-cultural differences, both Spain and Portugal score in the high collectivism group and are placed in the Latin Europe cluster. Portugal scored higher than Spain on gender egalitarianism in practice and they both had similar scores on power distance (high).

Comparing Portugal and Spain on the indices in *Pocket World in Figures* (The Economist, 2007) yields many more similarities than differences. Spain, of course, is more than five times larger in size than Portugal, and it is wealthier ($25,300 gross domestic product per person versus $16,610), and purchasing power parity (the United States equals 100) is 63 for Spain versus 49.5 for Portugal. Portugal's population is about 10 million while Spain's is 41 million, and the languages are different. Today, Portugal is one of the poorest countries in Western Europe and its recent history has been marred by corruption scandals, a far cry from the days of the great 15th-century explorer Vasco da Gama, who along with other explorers helped build a vast empire that included the country of Brazil and parts of Africa and Asia.

By and large, however, the two Iberian nations are remarkably similar in many ways, at least superficially. In fact, for 60 years Portugal was a part of Spain until a revolt in the 17th century, the anniversary of which is annually celebrated. It is instructive to compare the cultural metaphors for these two nations: the bullfight. Both in Spain and Portugal, the bullfight celebrates emotionalism, love of life, and heroism. But the Spanish bullfight emphasizes tragedy and is a ritual of death, and there are more expressions of individualism in the Spanish version than in the Portuguese version, although such expressions are present in Portugal. Studying these two types of bullfights helps to sensitize us to subtle cultural differences that are often overlooked, such as assuming that all Asian nations have the same culture.

As we will see, there are indeed substantive cultural differences between Spain and Portugal. These differences are also reflected in their different treatment of their colonies: The Spanish conquistadores usually conquered and destroyed in South America. In Brazil, by contrast, the Portuguese took the natives, and even later their black slaves from Africa, as concubines, thus legitimizing the resulting children.

As in Spain, bullfighting in Portugal was once the sport of noblemen. In the 18th century Marquees de Pombal, the Portuguese prime minister, prohibited the killing of bulls after the son of the Duke of Arcos was killed in a bullfight. This prohibition profoundly changed the nature of Portuguese bullfighting and over time led to expressions of cultural values that are, to an outsider, startling and unique. In 2007, the celebrated Portuguese bullfighter Pedrito de Portugal was fined 100,000 euros for the crime of killing a bull, and this gave rise to a national debate between supporters and critics (Bilefsky, 2007b). Supporters felt that a death-free struggle was sacrilegious whereas critics wanted Portugal to continue its traditions of civility and humanity toward animals. In this chapter, we describe the Portuguese bullfight in terms of four cultural characteristics: Pride in traditions, stratification amid unity or collectivism, artistry and human gore, and profitless bravery.

Pride in Traditions

There are six separate rounds in the Portuguese bullfight, each with a specific *cavaleiro* (mounted horseman) and *forcados* (footmen). The bullfight spectacle begins with the grand entrance into the bullring by all the participants. The *cavaleiros* enter the ring first, wearing the traditional costumes of the 17th century. Their embroidered vests and feathered hats clearly emphasize the cultural heritage of the Portuguese. The *forcados*, or footmen, are attired in more humble yet traditional brown pants and small vests. The *toreiros*, or cape men, stand out because of their embroidered pants and vests, which are similar to those worn by the Spanish matadors. Metaphorically the *cavaleiros* stand for the noble class with its elegance and style, while the *forcados* represent the common laborer of Portugal. These *forcados*, much like their ancestors, are not flashy and don't mind getting dirty while performing their task. They understand and accept their role in society and are proud to contribute in any way possible.

The Bullfight Proceeds

The bullfight begins by pitting a fast bull against a graceful horse and rider. The *cavaleiro* must first demonstrate his control of the horse and his ability to avoid the charging bull, after which he faces the bull from across the ring and rides furiously toward him. At the last split second the *cavaleiro* swerves to avoid the bull but simultaneously drives a wooden spear into his neck. The *cavaleiro* repeats this activity six or seven times, and each encounter is more dangerous than the previous one. Then he exits the ring and the *forcados* enter.

Metaphorically the *cavaleiro* is the authority figure in the bullfight. The *toreiros*, men with the capes, are used only to support them by distracting the bull long enough so that the *cavaleiro* can transition to another horse or grab another spear. (Of course, the *torero* is the most important person in the ring in Spain.)

The *forcados*, an eight-man team, line up as if playing leapfrog. The leader hollers *O toro!* continuously until the bull charges. Fortunately the horns of the bull are covered in a leather sheath. When the bull plows into the leader, he doesn't get gored. Instead, the bull easily picks up the entire eight-man team, and the action becomes furious. Finally the group brings the bull to a standstill and one man takes hold of the tail. The intended outcome is a successful *pega*, that is, the seven descend from the bull while the eighth man hangs onto the tail and stays upright behind the speeding beast until the bull comes to a halt.

The entire process is repeated with five additional bulls. After each of the six rounds, the *cavaleiros* and *forcados* meet in the center of the ring and take a victory lap around it while the crowd shows its appreciation by throwing flowers, hats, and other accessories. Then the *cavaleiros* and *forcados* throw these items back to the crowd as a sign of respect and appreciation for the crowd's support.

This description is a poor substitute for actually seeing a Portuguese bullfight. When the bull picks up one *forcado* or several *forcados* and hurls them into the air, the sensation is stunning. It is even more stunning to see a seemingly foolhardy *forcado* break loose of the eight-person team in his attempt to stop the bull all by himself. Sometimes he is actually successful, but it is more likely that he will sail high into the air over the bull's back when the bull rushes into him, picks him up, and treats him with disdain. But is the *forcado* really being foolhardy? Only an understanding of Portuguese culture will help us to answer this question.

Community and Church

The soul of Portugal is reflected in the warmth of the people, who are spiritually connected to the land. As noted, the *forcados* represent the typical Portuguese laborers: simple and honest people who work all day and then come home at dusk to a warm supper and a carafe of homemade wine. After dinner they tend to go to the local café to drink an espresso and talk to their friends about the local soccer team. These people are repeating the pattern of simple lifestyles that their ancestors established. They realize that they are separate from the upper class (represented by the *cavaleiros*), but they are proud of their role in the bullfight and in society.

It is not surprising that religion is important in the culture; Portugal is 97% Catholic. Similarly the *cavaleiros* bless themselves, asking God for protection while in the ring with the bull. Metaphorically each of the participants in the bullfight seeks truth. For example, the *cavaleiros* test their control of horse and bull, each time being more daring and trying riskier moves. The *forcados* also search for truth and test their courage by facing the bull head on. For the most part the Portuguese seek inner truth and strength, and they believe that actions speak louder than words.

Portugal is still a Catholic nation heavily influenced by the clergy, and it did not suffer from the anticlericalism unleashed in Spain after Franco's death in 1975. As we might expect in a culture that values the search for truth, regular church attendance is widespread. As Catholics, the Portuguese believe all of us are sinners in some way, for which we must atone. Hence many people go to confession on a weekly basis and find this activity as important as actually attending Sunday Mass.

Clergymen are considered experts, and their opinions are highly regarded. Many Portuguese use the clergy in the same way that Americans use psychologists. The Catholic Church and its teachings are an important source of stability for its members.

People's role in the social structure of both the family and extended family gives them a sense of security to life. The ordinary citizens respect anyone who is a symbol of authority, whether it is a clergyman, a doctor, or a manager.

Authority is rarely challenged in this hierarchical culture although the value systems are in transition from the old authoritarian systems to the more democratic approaches of today, both in the political and economic arenas. Portugal was governed for almost half of the 20th century by a dictator, Antonio de Oliveira Salazar, who hated modernity and treated his people like children; nevertheless, he created a degree of stability and kept Portugal out of World War II. In 2007, there was quite an uproar when Salazar was voted as the "greatest Portuguese that ever lived," beating out the great Portuguese explorer, Vasco da Gama. Some people argued that the vote reflected disillusionment with the present state of affairs and a longing for the past (Bilefsky, 2007c).

Roles for Men and Women

Roles are clearly defined. While the *toreiros,* the cape men who help distract the bull for the *cavaleiro,* are men, they play a supportive role similar to that of women in Portuguese society. In everyday life roles of men and women are clearly defined. The man is seen as the conqueror and hunter who provides for the family while the woman is the supporter who does all the cooking and cleaning. Even though more and more women are entering the workforce, they are normally not considered the breadwinners of the household. Working women are still expected to come home and take care of their womanly responsibilities such as cooking, cleaning, and taking care of the children. Based on this pattern, one would expect that Portugal is a masculine country, but actually it ranks 45 of 53 nations on the masculine-feminine dimension (Hofstede, 1991) and high on gender egalitarianism in the GLOBE (House et al., 2004) study, presumably because the Portuguese place great value on the land, the environment, and relationships. Hence it is acceptable for a man to be nurturing.

Just as the *cavaleiros* and *toreiros* are loyal and demonstrate mutual respect for each other, so too do men and women. Divorce is uncommon in Portugal and considered a disgrace. Actually, when a husband cheats on a wife, or vice versa, they are said to *meter os cornos*, which literally means to be "gored by the bull." When a man is gored, as with the *pega*, he is expected to try to remarry. If he is unsuccessful, family members may well experience a sense of shame and derisively call him a quitter.

Given the significance of tradition, long-term trust is important in business. Metaphorically the bull represents an outsider trying to conduct business; the Portuguese are suspicious and afraid of getting "gored by the bull." Paradoxically, then, Portuguese people are friendly but simultaneously suspicious. Relationship building, especially outside of work, is important in gaining the needed trust, and establishing a strong rapport through personal contact is key to negotiating effectively in Portugal.

Although there are few women in management roles, the Portuguese will conduct business with women if they feel that they are figures of authority. Although gender is important, authority outweighs it. Thus, at least in business, it is more important to be a person of authority first and a woman second.

Timeliness is important and a person is viewed as disrespectful if not punctual. It is uncommon to be late for work or a meeting. Unlike the Spanish, who are known for being tardy, the Portuguese see timeliness as extremely important and reflective of character and inner strength.

Finally, in light of the emphasis on tradition and following rules, it is easy to understand why change is slow and innovation deemphasized. Deviating from the norm is rarely accepted. In a recent bullfight, for example, one of the *forcados* decided to do a 360-degree spin while the bull was charging, but this act was not viewed favorably by the public. While it was innovative and more exciting due to the increased risk, the traditional *pega* is still preferred by a majority of the crowd. This lack of innovation has retarded Portugal's development. However, one surprising innovation that emerged from the great Lisbon earthquake of 1755 was the invention by military engineers of Europe's first earthquake-proof buildings under the leadership of the Marquis de Pombal, one of Portugal's great modernizers (Schrady, 2008).

Stratification Amid Unity

One of the interesting things about Portugal is that it is a collectivist society. Although it is important to define rules clearly, the most critical feature of this culture is that the group is successful. This is clearly demonstrated in the bullfight: When a *cavaleiro* enters the ring, for example, the other five *cavaleiros* shake his hand and wish him luck in the bullfight, thus signifying that it is more important to see oneself as part of a group than to win. Likewise, before the *pega* it is customary that the lead *cavaleiro* dedicates his efforts to a spouse, teacher, or even the entire crowd. In many instances he will give his hat to the individual to whom he is dedicating the event and collect it once his "job" has been completed. This tradition dates back to the time when soldiers would go out to battle, leaving behind a token for their loved ones to remember them by. What the *cavaleiro* asks for in return is permission to proceed, which is usually signaled by a wave, hug, or applause.

After each bullfight *cavaleiros* and *forcados* meet at the center of the ring and shake hands, which symbolizes the coming together of social classes. They explicitly recognize that it is not the nobleman or the common citizen who achieves success independently but rather all parties working together.

While Spain and Portugal are similar on objective measures, as noted previously, the Portuguese have developed a decidedly different character from that of the people of Spain. International travelers who mingle with the Portuguese usually return home filled with special warmth for the people. The Portuguese are easily approachable and helpful. In Portugal, the family is perhaps the single most important institution in society. Many of the people have no close friendships outside the family unit. Although the people in the cities have become increasingly sophisticated, the more rural areas are still characterized by working families who manifest this pattern of behavior. As a result, it is common to see people marrying their neighbors. And newlyweds tend to live close to their parents, friends, and place of birth.

Similarly it is common for the entire family to gather for lunch at home. Lunch is by far the most important meal of the day, and everything in the country shuts down between noon and two in the afternoon so that people can eat. At lunchtime people do not discuss work, as there is a clear separation between work and family life that is rarely crossed. Food is important to the Portuguese, and they usually make elaborate meals at lunchtime, while dinner is much lighter and less formal. Beer and wine are consumed during lunch and constitute an important part of the culture. Many families grow their own grapes and make enough wine to last the family an entire year.

After dinner, everyone usually gathers at a local café to have an espresso and to socialize. The family will go as a group but on arrival each member will join natural subgroups. Men will sit with their friends and play cards, women will discuss the daily occurrences for an hour or two and return home to start preparing the meal for the next day, and the children will play a game of *fusebol* or watch TV with their friends. This is really the only time during the week that people socialize outside the family and spend time with their friends.

It is important to understand the stratified collectivist norms that the Portuguese emphasize. To conduct business in Portugal, it is critical that a relationship is developed. One of the best ways to establish such relationships is over lunch. One frequently repeated aphorism is that the Portuguese need 2 hours for lunch owing to the fact that 1 hour is for eating while the other hour is for talking. Many business deals are finalized during lunch, and presumably the better the lunch, the better the deal. Bribery is commonly used to get things accomplished and is an accepted way of life. If a businessperson is trying to get paperwork signed or to close a deal, he or she may want to host lunch at the most expensive restaurant in town, and by the time the dessert is served, there will probably be a successful resolution of the issue.

Artistry and Human Gore

The beauty of the Lusitano horse, the elaborate costumes worn by the *cavaleiros*, and the elegance of the opening ceremony demonstrate the artistry that is seen throughout the country. In Portugal, the performance of the horse in the bullring is perhaps one of the most important factors in the breeding and selection process of the Lusitano horse. The horses are trained by their riders and learn many things, including prancing, moving sideways, and standing on their hind legs. These skills are "shown off" during the bullfight and are needed to avoid the bull. The movements of the horse are admired by all, and the *cavaleiro* is applauded for his efforts in training the animal. It is extremely important that the *cavaleiro* demonstrate control of the bull and horse, which is done by repeatedly spearing the bull.

This artistry is also seen in music, arts and crafts, and the architecture of the country. There are *fado* houses throughout Portugal where the female *fadista*, clutching a traditional black shawl, sings the sad and romantic songs of Portugal. A rough translation of *fado* is fate, from the Latin *fatuum*, meaning prophecy. *Fado* usually tells of unrequited love, jealousy, and a longing for days gone by. *Fado* became famous in the 19th century when Maria Severa, the beautiful daughter of a gypsy, took Portugal by storm, singing her way into the hearts of the people, especially the Count of Vimioso, the most renowned bullfighter of his day.

Hand-painted ceramics called *azuleijos* constitute an important craft in Portugal. Because of their decorative design these tiles have been adopted by builders and are seen in almost every house. Other hand-painted pottery is also seen throughout the country, especially in the northern region. The architecture of Portugal is visible in the cathedrals, churches, and castles spread across the country. For the most part, Portuguese architecture imitates Italian and Spanish designs. There is little if any uniqueness to the architecture, but its beauty is undeniable.

What we see at the bullfight is not only the artistry but also human gore. The *forcados* are often injured while trying to wrestle the bull, and it is common to see blood on their faces and clothing. In the Spanish bullfight the gore comes from killing of the bull, but the Portuguese bullfight is said to be the "revenge of the bull." The first time that people watch the *pega* they are amazed at the violence and the pain inflicted by the bull on the *forcados*. In an attempt to stop the charging bull, the *forcados* are often trampled, thrown, and injured. The gore, however, is experienced only by the working class (*forcados*) and not the upper class (*cavaleiros*), which is symbolic of what has occurred during wartime. The *forcados* are similar to the front line of the army while the *cavaleiros* are the generals. Still, a *forcado* is rarely killed (less than one every 10 years or so). The Portuguese see the bullfight as an art form and not as human gore, although international visitors frequently focus on the latter.

In business, much as in the bullfight, the Portuguese are perfectionists. If something goes wrong, they will do it over and over again until they get it right. The *cavaleiro*, for example, will sometimes change horses three or four times to find the one that will help him put on the best show. The Portuguese are also detail oriented. If the bull clips the horse or the *cavaleiro* drops one of his spears, he will go back and do it again. This level of detail is seen in business negotiations involving contracts. Unlike the Spanish, who finalize many business transactions orally, the Portuguese require that everything be in writing.

Duplication, however, is more important than innovation. The Portuguese believe in the status quo and are suspicious of those who try to do things differently. Their belief is that if they have done it a certain way successfully many times in the past, what need is there to change? This thinking is different from the individualistic ideas that we see in the United States, which rewards creativity or "thinking outside the box." In Portugal this type of thinking is seen as breaking tradition, so it is probably wise for foreigners to follow the maxim "When in Portugal do as the Portuguese."

Profitless Bravery

When U.S. Americans attend the Portuguese bullfight, they tend to focus on the *pega*, probably because they see only the violence, human triumph, and misfortune (see Chapter 16, "American Football"). But whereas U.S. Americans may equate the physical violence of the *pega* to that of football, the Portuguese see the *pega* as a symbol of bravery.

In U.S. football, stars are role models for children because of their athletic ability and the money they earn. A major criticism of today's football players is that they lack the love for the game that previous generations had. Some think the players are more interested in the money than in the game itself. The average salary for a professional football player is more than $1 million a year, and many players make far more than the average. On sports talk shows and in letters to the editors, fans continually complain that players have no loyalty to their teams and cities, especially since the advent of free agency. Many fans have difficulty identifying with the football players who supposedly represent their teams and cities.

Conversely, the *forcados* are amateurs and don't make any money at all, whereas the *cavaleiros* are paid. How can this be? In many cases the *forcados* are battered and bruised by the bull, and no one in his right mind would do this without getting paid, one might think. But the Portuguese bullfight is not about making money, being a hero, or having your own sports. Rather, it is a joyous celebration of tradition and bravery. These *forcados* do it for the love of the ritual and the lessons that it has taught them. The bullfight forces a man to look deep within himself for a sense of pride, and it is a reminder of the hardships faced by previous generations.

It is typical during the *pega* that a *forcado* acting as team captain is not successful in holding the bull the first time. As opposed to quitting or giving way to another *forcado*, he will go out again. Sometimes three and four attempts must be made to grab the bull, but it is always the same brave lead *forcado* who is involved in the *pega*. Broken ribs and legs are common in the sport but are ignored until the task of capturing the bull is completed.

Saving face is a concept that is important to the Portuguese. Regardless of how badly the *forcado* is injured, nothing can be worse than injury to the ego. The shame that a person would bring to the family by quitting an important task before completion is unthinkable. Playing with pain is a part of the game with which U.S. Americans are familiar, but most have never witnessed it to the extent that is visible in the bullfight. Portuguese are extremely proud of the fact that the *pega* is performed only in Portugal and that it symbolizes the courage of the people.

One business implication of this pattern of behavior is that the Portuguese are stubborn but persistent. The *forcado* can be knocked down repeatedly but will get back up again no matter how badly beaten he is. Although persistence is an excellent characteristic, some people see this stubbornness as illogical beyond a reasonable point. Whether in a bullring or business environment, it is important to prepare for dealing with this extreme level of persistence.

Persistence typically leads to completing tasks. Hence a contract made with the Portuguese is generally ironclad. However, they are hard negotiators. Because they are so persistent and stubborn, negotiation is typically difficult. The U.S. saying when something is going to take a long time is "you better bring your lunch." In Portugal it should be "you better go out for lunch" because unless a relationship can be built outside a business setting, success will not occur. Portuguese do not consider themselves to be slaves to the clock. The process tends to be long and drawn out and will normally require many lunches.

As this chapter has demonstrated, the Portuguese and Spanish bullfights are as different as the people themselves are. The people of Portugal are less aggressive than those of Spain, and killing the bull is repulsive to them. As discussed earlier, the characters of the two peoples are reflected in the way that they administered their colonies in their glory days as colonial empires: the Spanish pillaged and the Portuguese intermarried. Sociologists argue that bullfighting in Portugal is less of a blood sport than in Spain because the Portuguese are "soft machos" (Bilefsky, 2007b).

One of the main characteristics of the bullfight is audience involvement. The Spanish will applaud an excellent performance by the matador or *torero* but will not hesitate to whistle and boo if displeased with the performance. If they boo, the *torero* will not reenter the ring. However, prolonged applause and the waving of handkerchiefs are a great honor, and the *torero* is rewarded with an ear and in some cases the tail of the bull; he is also well paid. In the Portuguese bullfight, on the other hand, rarely does the audience boo a performance. Regardless of how the *cavaleiro* and *forcado* performed, they are treated with respect. This pattern reflects the origin of the Portuguese bullfight, which is a celebration of bravery and life, not death.

The bullfight is symbolic of the battles that Portugal and its people faced during the invasion of the Moors. If people were to boo and whistle, then metaphorically it was a sign of disrespect for the soldiers who served the country. Thus it is important that, at the end of each of the six stages or rounds, the participants gather at the center of the ring so that the audience can acknowledge and applaud their efforts.

There is also a great deal of difference in the role of hierarchy in both nations. In Spain, the participants enter the ring in descending order of importance while in Portugal all the participants are viewed as of equal importance, even though the *cavaleiros* represent the noble class while the *forcados* represent the working class. In Spain the president determines the award given to the matador, but in Portugal the people decide on the reward by applauding. In the bullfight and throughout Spain the people like to show off by dressing nicely, whether rich or poor. This is clearly not the case in Portugal as evidenced by the humble clothing worn by the *forcados*.

The biggest difference between the two bullfights, however, is deeply rooted in the meaning of the bullfight for each country. In Spain, the bullfight is about control and conquest through death. The matadors are expected to control the bull and, once this control is demonstrated, they kill it. In the Portuguese version of the bullfight, death is not a part of the ritual. The true meaning for all Portuguese is the celebration of bravery demonstrated by the *cavaleiro* and *forcado*. According to a full-length documentary film, *Taking the Face: The Portuguese Bullfight,* which was released in 2008, Barrancos, a border town near Spain, is the only place in Portugal where the bull is killed in defiance of Portuguese law.

Still, similar to other nations facing globalization, Portugal is understandably concerned that its culture is not significantly altered. Today, Portugal is trying to position itself as the "West Coast of Europe," a destination for tourists as well as investors with its natural beauty and history and strength as an environmentally friendly energy producer, third in Europe behind Sweden and Austria (Pfanner, 2007). It is a member of the European Union (EU), which provides subsidies to its poorer members (including Portugal) with the expectation that higher economic growth will occur so that eventually these nations will in turn help to subsidize other activities and nations in the confederation. In the case of Ireland this approach has worked remarkably well, but clearly Irish culture has been affected (see Chapter 13, "Irish Conversations").

Some Portuguese are even concerned about the symbolism found on the new European coins in which Portugal looks as if it were a part of Spain (see Trueheart, 1998), and in 2003, 40 prominent Portuguese economists and business leaders published an open letter in which they warned against the rising danger of Portugal's corporations being purchased and controlled by other nations, particularly Spain. However, cultural patterns that have taken hundreds of years to develop are difficult to destroy, and we can expect that the Portuguese bullfight will continue to reflect the underlying values that the Portuguese hold dear, even if changes are introduced into this beloved activity.

PART XI

Perspectives on Continents

As one moves from the ethnic group to the nation to a group of nations to a continent, it is still possible but increasingly difficult to employ cultural metaphors as effective devices for profiling the underlying cultural values and assumptions. This is true even for the one nation, Australia, which is also a continent. But it is even more true for other continents. Europe, for example, has numerous cultural and ethnic groupings. However, the globalized economy requires that we at least attempt to profile the underlying values and assumptions of continents if at all possible. Some citizens of various nations on the European continent, for example, now identify with the European Union (EU) as much as, or more than, their respective nations. However, surveys indicate that only one in ten people think of themselves as European. Respondents still identify themselves as citizens of specific nations. Of course, the complexity of this issue increases in the World Trade Organization, which includes a far larger number of nations.

In this part of the book we profile two continents, Australia and Africa, although the cultural metaphor for Africa, the bush taxi, relates primarily to sub-Saharan Africa. We focus on Africa because almost all of the cross-cultural research to date, particularly among management and cross-cultural researchers, has disregarded this continent. We can only speculate on the reasons for such disregard, as we do in the chapter itself, but we maintain that it is essential for Africa to become better integrated into the world economy, and such integration is facilitated by an enriched cross-cultural comprehension of its strengths and problems.

CHAPTER 32

Australian Outdoor Recreational Activities

Australians and U.S. Americans tend to get along well with one another. Both tend to be highly informal in interactions and to downplay pomposity. Speaking the same language also facilitates comfortable communications. The two nationalities share historical experiences, as the British sent prisoners to populate the U.S. colonies until the American Revolution, after which the British shipped their prisoners to the "godforsaken" continent of Australia. Both nations are large, and Australia is about equal in size to the 48 contiguous states in the United States.

But there are vast differences between these two friendly nations. In this chapter, we discuss how and why Australia is so captivating to the imagination, not only of Americans but also of people throughout the world; we examine new realities that go beyond stereotypes; and we look at the phenomenon of equality matching among the tall poppies—cutting down to size those who manifest superior ability or high levels of motivation—especially if they are seen as pompous and self-aggrandizing.

In fact, there are so many fascinating aspects of Australian culture that it is difficult to select one of them as a cultural metaphor, and the changes in the past 50 years have been profound, thus compounding this difficulty. What seems to unify Australians, regardless of ethnic background, is outdoor recreational activities, including sports of all types, sailing and surfing, and frequenting outdoor restaurants and coffeehouses. We employ such activities as our cultural metaphor.

Capturing the Imagination

Few if any nations capture the imagination of visitors as effectively as Australia. There is quite a geographical distance separating Australia from the United States and

Europe, and even from Asia, although less so. In recent years Australia's relationship with Asia has been strengthened, as noted below. Australia is simultaneously an island, a nation, and the driest and most fragile of the seven continents. About a third of Australia is desert, but there are also mountains and rain forests.

Australia is an urbanized country with more than half of its population settled in just five cities and almost 85% living in towns usually close to the coasts. Nevertheless, the sparsely populated outback fascinates in a manner similar to America's Old West. For instance, Ned Kelly of outback fame is a real bandit whose death by hanging made him mythic, somewhat like Billy the Kid in the United States, and Australians know that he looked the judge straight in the eye before being hanged and said, "I will see you there where I go."

Even archetypal films from these two nations are similar. For example, arguably the greatest Western of all is *High Noon*, starring Gary Cooper as the marshal and Grace Kelly as his newly married wife who opposes violence because of her Quaker beliefs; the townspeople fail to support him when a gang of outlaws whom he sent to prison returns to exact revenge, and even his wife is initially noncommittal and even negative about his involvement in the coming fight. Still, he fights the gang against long odds and ultimately emerges as the winner, with some critical help from his wife at the last moment. The Australian counterpart film is *Mad Max*, in which Mel Gibson plays a lawman in the outback seeking justice because his wife and young child are brutally murdered. This film was so popular in the United States that it led to two sequels. Both films idealize the individual standing up against the unjust group and the environment surrounding him.

Robert Basham, CEO and one of the three founders of the very successful Outback Steakhouse restaurant chain in the United States, is among the many U.S. business executives capitalizing on the fascination surrounding Australia. After many years in the industry, he and his two partners came to believe that successful restaurants frequently fail because the most competent cooks and waiters quit due to overwork resulting from having to serve both lunch and dinner. Hence a good part of their initial business strategy revolved around serving only dinner. Just as important, however, was branding the image of the new restaurant in the minds of potential customers. They did so by naming the chain "Outback Steakhouse" and marketing and advertising extensively with this theme, even though they had never visited Australia until sometime after the chain was founded. Later they expanded the simple menu to include a taste of New Orleans cuisine, thus differentiating Outback Steakhouse from competitors, but they have successfully continued to emphasize the Australian theme.

Australia is fascinating for many other reasons, including the fact that it is home to a host of unusual creatures and animals, including wombats, koala bears, and kangaroos. It is also host to some dangerous snakes against which it is difficult to protect. In the outback, campers routinely construct fences around their sleeping areas to ensure that wandering crocodiles do not enter.

Widely known landmarks include the 1,200-mile Great Barrier Reef, not to mention the ground-breaking architecture of the Sydney Opera House, and the monumental Ayers Rock in the interior. Australians are keenly aware that this cornucopia of unusual and attractive features sparks the imagination of long-term and short-term visitors and joke among themselves about promoting "The Rock, the Reef, and Ropera." It is not surprising that tourism generates more than 10% of Australia's export revenues with new sources opening up every year, especially from the Chinese mainland.

Australian English is also a great source of fascination, as it is a mixture of British English, aboriginal sayings, and locally derived terms that the settlers used in everyday life. The Australian Tourist Commission publishes a quaintly titled short pamphlet, *A Fair Dinkum Aussie Dictionary*, which is useful for the visitor seeking to understand the Australian mind-set. For example, a *hoon* is a lout; *hard yakka* is hard work; and a *dill* is a simpleton. The two different but overlapping types of English used in the United States and Australia help to facilitate a common bond, as both are replete with colorful words and slang that enrich conversations and experiences, sometimes humorously. One Australian couple driving in Arizona in the winter was stopped by a policeman for speeding, but he forgave them once he ascertained their citizenship, with the admonition that they should watch out for "snowbirds." They literally and diligently were on the lookout for snowbirds for 2 days before they understood what snow birds were, that is, retirees escaping from the colder parts of the United States during the winter months.

In some ways Australians are much more colorful and direct than their U.S. counterparts when it comes to language and insults. Former prime ministers are on record in Parliament as calling their political opponents scumbags, "perfumed gigolos," and "brain-damaged looney crims." This tendency to debate publicly and intermittently to spice up the debate is clearly traceable to Australia's British heritage. Still, what is acceptable speech in Australia might well constitute a basis for legal action in the United States.

Given the climate, which in some parts of the nation is warm 12 months of the year, and the large number of enjoyable activities that the outdoors offers, it is little wonder that Australians emphasize outdoor recreational activities. Australians are sports-mad—perhaps the most sports-mad nation in the world, although this is difficult to believe if one is a U.S. American—and engage in and watch a large variety of sports, including those with English origins (cricket, tennis, rugby, and lawn tennis), soccer or football, sailing, and horse racing. Australian football is unique and difficult to describe, other than by saying that it is a cross between soccer and rugby and simultaneously a version of U.S. football without protective outfits. Hiking and camping are also popular.

Still, many and perhaps most Australians do not exercise regularly or engage in outdoor sports. They do, however, watch them intensively and at great length. In 2007, *BBC News* online reported that 80% of the adult population of Australia gambles, making it first among nations in gambling, and it is a safe bet that a large portion of the gambling involves sports. This is a country which has the 53rd-largest population but the most gambling machines. According to the report, one gambler joked that Australians would gamble on two flies climbing a wall. Sydney, Brisbane, and Melbourne have some of the largest gambling casinos in Australia. The one in Sydney is reported to be the size of seven football fields and the Crown Entertainment Complex in Melbourne attracts 12 million visitors per year.

New Realities: Beyond Stereotypes

Because Australia is so vast and varied, there are major differences between various parts of the nation. Still, given its urban flavor, it is difficult to reconcile the stereotypes about Australia with modern realities. For example, Australians tend to be amused by the popularity of the *Crocodile Dundee* films, whose theme is that someone expert in living in the bush experiences all sorts of difficulties and adventures when visiting cities such as New York and Los Angeles. Most Australians are far removed from such bush experiences.

Australia is a nation whose majority still claims British-derived heritage. But it has become the most immigrant-influenced nation in the world. About 24% of the estimated population of 21.1 million people was born in another nation, and about an equal percentage of citizens come from families in which at least one parent was born elsewhere. Australians now come from 165 nationalities.

The relationship with Britain, in many ways, has weakened considerably, perhaps irretrievably. It was strong until World War I, during which Britain at the urging of Winston Churchill pursued a controversial and ultimately senseless attack on Turkey at Gallipoli; thousands of soldiers died, and a large percentage of these were Australians. During World War II Churchill made the decision not to use English troops to protect Australia against the invading Japanese; this role was assumed by U.S. troops, which helps to account for the good feelings between the two nationalities. Despite the unpopularity of the Iraq policy of President Bush among the Australian people, when Prime Minister John Howard was defeated in 2007 and replaced by Kevin Rudd, a member of the opposition party, the new leader planned to follow a U.S.-centric foreign policy.

Gradually Australians realized that they would be economically vulnerable if they continued to bolster their relationship with Britain at the expense of other relationships. Former Prime Minister Paul Keating in the early 1990s publicly proclaimed that Australia had to cease being a "branch office of empire," to become a republic, and to aim for "enmeshment" in Asia. In a national election in 1999, voters narrowly defeated a proposal to replace the monarchy with a republic; had the proposal passed, the Queen of England would have been removed as the titular head of state.

As in the United States, there has been organized opposition to immigrants, which was legally incorporated into the "White Australia" policy from 1901 until 1973 and specifically directed against Asians. The aborigines, the original settlers in Australia and currently 2% of the population, also suffered greatly for decades, and they were treated in about the same inhumane manner that settlers in the United States treated Native Americans, for example, by subjecting them to deliberate massacres and the involuntary removal of aboriginal children from their families (the Stolen Generations). This is portrayed very movingly in the award-winning 2002 film, *The Rabbit Proof Fence,* which tells the true story of Molly Craig, who escapes with her younger sister and cousin from a government camp where they were kept after being forcibly removed from their mother. Molly leads the other girls on a journey of 1,500 miles in search of the rabbit-proof fence that bisects the continent and will take them home.

Following the abolition of the White Australia policy in 1973, several initiatives were launched to build racial harmony based on a policy of multiculturalism. Australia is still coming to terms with its past as it welcomes waves of immigrants every year. In 2008, Prime Minister Rudd led the parliamentary apology to members of the Stolen Generations of aborigines who were forcibly taken from their families and communities. The apology comes 11 years after a report found that during the period 1910 to 1970, between one in three and one in ten aboriginal children were taken from their families.

Asian immigration from several nations has increased significantly since 1990. As might be expected, there was a backlash against the recent wave of immigrants, and a small but vocal opposition developed. Pauline Hanson, a member of Parliament, was the leader of this opposition. In her maiden speech in Parliament in 1996 she indicated that she would represent all Australians "except aborigines and Torres Strait islanders." She went on to say that "we are in danger of being swamped by Asians. They have their own culture and religion, form ghettoes and do not assimilate." Both Mrs. Hanson and her party have now become a footnote in history, which speaks well of the Australians' receptivity to immigration.

The wisdom of expanding Australia's focus beyond Britain has generally been confirmed. The gross domestic product (GDP) per person is $32,320. Life expectancy is comparable to other industrialized nations at 78.5 years for men and 83.4 years for women. Australia has 1% of total world exports, ranking 25th among nations. Similarly, it ranks 6th among nations on global competitiveness and 3rd among nations with its score of 95.5 on the Human Development Index, a measure combining GDP per person, adult literacy, and life expectancy (only Iceland and Norway are higher at numbers 1 and 2 respectively). According to the 2006 census, the median age of Australians was 37 years, the proportion of the population age 65 years and older has increased to 13.3%, and the proportion of children age 0 to 14 years has decreased to 19.8% when compared to 1996 figures. Females outnumbered males in 2006 with 97 males for every 100 females.

Perhaps the best way to capture the relationship between these profound changes and our cultural metaphor, outdoor recreational activities, is to explore the evolution of the barbecue, which supposedly originated in Australia and is common in many nations today. Originally a family held a barbecue at home on Sunday after church and invited only family members and perhaps a few close friends. Over time, however, the barbecue grew into a larger affair, with many families and friends participating at someone's home. Everyone would prepare and bring food, and there was generally a clear separation of sexes, with the men hanging out by the barbecue pit, playing sports or cards, or talking and eating.

Conversations at barbecues tend to be casual, involving sports, children, and the weather. Most Australians do not attend church regularly, and religion is seldom discussed at barbecues; the subject of sex is not completely taboo but it is handled gingerly if at all in mixed or polite company; and political discussions tend to focus on "poli bashing" or the awful behavior and low intelligence of politicians. Soon the barbecue was brought to the beach and included even more individuals and organized sports such as beach volleyball and sailing races.

Because of the diversity of Australia's current population, the barbecue is not important among some ethnic groups. They have replaced it by eating out at ethnic restaurants or popular coffeehouses, especially in the eight largest cities, each with more than a million inhabitants. Two of these cities, Sydney and Melbourne, rank 9th and 17th worldwide among cities on the quality of city life index.

Equality Matching Among the Tall Poppies

Until recently, Australia was ashamed of its criminal origins. This embarrassment found expression in such phrases as "the cultural cringe," meaning that Australia looked to its supposed superior, Britain, for guidance and help; and "down under," signifying that Australia was far away from the centers of power in Britain and the United States. Even the most important national holiday, ANZAC Day, celebrates the defeat of the Australian army at Gallipoli, whereas national holidays elsewhere usually celebrate victories. Unlike U.S. Americans, who love winners and ignore losers, Australians hold the "Aussie battler" in high esteem, especially if he tries valiantly but loses. The popular song, "Waltzing Matilda," connotes a despairing outlook on life.

Such historical experiences led to the well-known phenomenon, "cutting down the tall poppies." Australia is an equality matching nation, as its scores on the Hofstede (2001) dimensions and the more recent GLOBE (House et al., 2004) study dimensions confirm: It is both highly individualistic and low on power distance simultaneously. But it is unlike other nations that fit into this generic category such as Sweden and Norway, which tend to emphasize individual development or related personal characteristics. Australia historically reinforced a negative approach, being neutral or negative about achievement and cutting down anyone striving to be "better than others," regardless of motivation, effort, or reason for doing so. Australians especially disdain pomposity in any form. Richard Brislin (1993), a prominent cross-cultural psychologist, advised colleagues on lecture tours in Australia to be aware that at least a third of any Australian audience would openly question their credentials.

Similarly another cross-cultural psychologist, N. T. Feather (1994), developed a scale to measure the tendency to cut down tall poppies and compared Japanese and Australian respondents using it. He confirmed that Australians have a significant tendency to cut down the tall poppies, but he also discovered that this tendency was higher among his Japanese respondents, a result he attributed to their collectivistic orientation. Also, Feather found that Australian respondents did not begrudge others their success as long as it was earned; they were suspicious of the validity of such success, especially if it was self-promoted as is commonly done in the United States. In recent times, however, the tall poppy syndrome is beginning to be viewed as an obstacle to success, wealth creation, and excellence, with former Prime Minister John Howard arguing that "if there's one thing that we need to get rid of in this country it is our tall poppy syndrome" (Bowring, 2007).

Given modern Australia's international prominence in many areas such as sports, films, and theater, it seems clear that the "cultural cringe" is losing its potency and appeal. Still, there is ambivalence about the type of success most commonly associated with market pricing cultures such as the United States and Britain. For several years Martin Gannon lectured to and interacted with about 45 Australian students per year, during which time they completed a 34-item scale, each of whose items require that the respondent select only one of four responses tailored to the four generic cultures highlighted in this book: community sharing, authority ranking, equality matching, and market pricing (see Figure 1.2; Gelfand & Holcombe, 1998). Overwhelmingly the results indicate that these Australian students, regardless of ethnic background, are equality matching in orientation.

However, when he then presented a scenario requiring the person either to go immediately to a business meeting at which a contract will be signed and a million-dollar bonus for the person will be the result or to help out a long-term friend who needs immediate help (no other responses are allowed), most but not all of the students opt for the million dollars (Gannon, 2003). One student even opined that a million dollars would buy a lot of friends, which evoked an immediate roar of disapproval. The student papers also revealed this ambivalence, as they tended to classify Australia as an equality matching culture that was quickly taking on many of the characteristics of a market pricing culture.

Government policy favors equality matching, and Australia does not confront some of the social problems found elsewhere, for example, the 47 million U.S. citizens without health insurance, its large number of homeless people, and so forth. The Australian concept of equality matching seems to have evolved from the concept of "mate," that is, helping out one another in the bush when the environment is harsh and dangerous and resources are scarce. Australians historically have been individualistic, and this is also borne out by more recent studies, such as the GLOBE study (House et al., 2004), in a remarkably consistent manner. Once mates, however, they are close to one another emotionally and psychologically. There is a clear distinction between members of the in-group and out-group, and becoming a member of the in-group requires time, patience, and the appropriate behaviors.

Although they are friendly, Australians historically have been passive with those not in their intimate circle. The Australian pub, so prominent in films from and about Australia, reflects this passivity and in-group, out-group distinction. Most probably this tendency is even stronger among the newer immigrants, the majority of whom come from authority ranking cultures.

The ambivalence that Australians have about the conflict between different outcomes associated with market pricing and equality matching tends to manifest itself in various ways. Australians strongly support the welfare system and individuals as long as they make a "fair go." It is not necessary to be outstanding to receive help, support, and even respect. However, Australians expect individuals to be "Aussie battlers" and speak disparagingly about "dole bludgers," people who are unfairly taking advantage of the welfare system. In one instance there was a public debate centering on a young man who refused to take work if he had to cut his long hair. Fortunately a hairdresser saved the day by offering him a position without this requirement, which satisfied most people that he was not really a "dole bludger."

In 1999 there was an outcry and public debate over adding the concept of *mate* to the constitution. The proposed and offending passage originally read:

> Australians are free to be proud of their country and heritage, free to realize themselves as individuals and free to pursue their hopes and ideals. We value excellence as well as fairness, independence as dearly as mateship.

Given the multicultural diversity in modern Australia and the country's ineluctable movement toward becoming a key player in the World Trade Organization and a globalized economy, such an outcry is understandable. Australian culture is changing, and so is its economy; their interlocked nature will lead to additional and major changes in the future. A significant challenge facing Australia in the future is its inability (thanks to a poor ecology) to sustain large population growth. Some experts suggest that the country cannot maintain the same quality of life and rank high on the Human Development Index if the population increases much beyond 24 million, which it is expected to reach by the middle of the 21st century. Overseas migration plays an important role in increasing the aging population with the majority coming from the United Kingdom and New Zealand and increasingly from China and India. The country also experienced several years of drought in recent years, and it is feared that climate change will pose the biggest threat to Australian agriculture in the coming years. The government is cognizant of the policy choices that it will have to make to sustain the GDP growth of the last several years in the face of demographic and climate challenges.

Still, the fascination that the world has about Australia, while it may diminish, will still be strong. It is one of the most dynamic economies in the world, a country where mateship coexists with individualism, with strong ties to both Europe and the Americas and to Asia. As a native put it, "it has all that the U.S. has to offer with very few of its negatives." Similarly it is difficult if not impossible to imagine Australia without an amazing array of outdoor recreational activities. And, while tall poppies will stand, we can also expect that equality matching will not disappear, as it is fundamental to the Australian mind-set and culture.

CHAPTER 33

The Sub-Saharan African Bush Taxi

Early in my tenure as Country Director in Niger, I became convinced that the greatest threat to the safety and security of our Peace Corps Volunteers is not terrorism (as one might be led to conclude from the amount of attention and resources devoted to it) but bush taxis.

—J. R. Bullington (2004)

Before discussing the African bush taxi as a cultural metaphor for the whole of sub-Saharan Africa, it is necessary to offer an opening explanation. How can one apply a single metaphor to most of a continent, especially a continent of such diversity? After all, five of the world's six major divisions of humanity can be found in Africa (Diamond, 1999, p. 377)—blacks, whites (in South Africa), Asians (in Madagascar), Khosians (the Bushmen of Southern Africa), and Pygmies (in Central Africa)—and three of those originated as natives in Africa (blacks, Khosians, and Pygmies). One quarter of the world's languages are spoken only in Africa (Diamond, 1999, p. 377). By contrast, applying a single metaphor to the European subcontinent would be difficult, if not impossible, and it has only a fraction of the diversity of Africa.

Africa is not a monolithic entity, but notwithstanding its cultural diversity, writers such as Edoho (2001) and Choudhury (1986) have recognized that "certain indigenous trends of thought, cultural influence, and value orientation are commonly shared by the majority of people in Africa" (Edoho, 2001, p. 79). The danger is that these similarities will blind us to the underlying differences and the richness of cultures and attitudes found in Africa.

We must also add a second explanation. Our metaphor applies only to sub-Saharan Africa. The Sahara has always been a huge physical and hence cultural barrier in

Africa. Of course, there has always been trade and exchange of ideas and peoples across the Sahara, but for the most part, the regions north and south of the Sahara are more dissimilar than alike. The peoples and cultures of North Africa are probably closer to those of the European Mediterranean and the Middle East than to those in sub-Saharan Africa. Hence, for convenience, when we refer to Africa, we mean Africa south of the Sahara and not the entire geographical continent.

This is an area that has a population of more than 743 million and includes 47 countries, the largest being Nigeria with more than 130 million and the smallest being Seychelles with less than 0.1 million. South Africa has the largest gross domestic product (GDP; $161 billion) and Guinea Bissau has the smallest GDP ($213 million). In recent years, the region's public stock markets have been attracting investors and it may well be that what billions in aid failed to achieve in terms of economic progress will be achieved by investors interested in generating good returns from some of these resource-rich nations. In 2007 alone, the investing world's "final frontier" attracted $2.6 billion in private equity (Farzad, 2007). Countries such as China and India have also come courting with several partnership opportunities.

In most parts of Africa, ownership of private vehicles is limited due to economic conditions. Travel is difficult, there are few paved roads, and paved and dirt roads alike are poorly maintained. Few buses travel African roads and the old colonial railroads are, for the most part, in poor repair. The main form of transport in many parts of Africa is the bush taxi. These vehicles go by many different names: bush taxi in Anglophone countries, *taxi brousse* in most Francophone countries, and *matatus* in Kenya. Bush taxis take many forms, for example, they may be cars (usually station wagons) or vans. In many of the former French colonies, they are often Peugeot station wagons, called *sept places* (seven-seaters) although they seldom conform to their name. In some cases, they are converted pick-up trucks with bench seating installed, and in other cases large trucks fill the role.

Whatever they are called, they share many similarities: They are always privately owned; they seldom, if ever, operate on a fixed schedule; they are usually overcrowded; and they are almost always in poor condition. Riszard Kapuscinski (2001), a Polish journalist and author, describes a bush taxi trip he took from Dakar in Senegal to Abdallah Wallo, a small village on the Senegalese side of the Senegal River, marking the frontier of that country and Mauritania:

> Finally a small Toyota van pulled up. These vehicles are built to seat twelve, but here they accommodate more than thirty passengers. It is difficult to describe the number and nature of all the additions, the welded-on extensions, the little benches inside such a bus. When it is full, for one passenger to either get out or get in, all the others must do so as well; the intricate calibration of the internal arrangements is as tight and precise as the workings of a Swiss watch, and who-ever occupies a place must take into account the fact that for the next several hours, for all intents and purposes, he will not be able to move so much as a toe. (p. 214)

Kapuscinski's description of the bush taxi captures many of the paradoxes of Africa to an outside observer. Behind the seeming chaos of Africa there is an ordered structure, and the workings of an African village, or an African business, may be as tight and precise as the workings of a Swiss watch. As we will see, African culture is bound by conventions and rules, many not immediately obvious to an outsider.

Basic Operations

Bush taxis are usually owned by an individual who has managed to save the money to make an investment in a vehicle. If all goes well, this local entrepreneur may acquire several such vehicles, hiring drivers to operate them. They fill an important role in a community, often serving as the only regular link (apart from walking or cycling) between a village and the outside world. Often the driver will carry letters or verbal messages from one village to another. The bush taxis transport both people and goods. Africans seldom travel light, and people returning to their villages by bush taxi may bring pots, pans, sacks of produce or grain, chickens or goats, bags of cement, and furniture, all competing for space with the other passengers.

The importance of bush taxis varies from one country to another depending on the economic conditions of the country. In Benin, a small country in West Africa, bush taxis are the predominant form of transport, even between the main towns. By contrast, in neighboring Burkina Faso, efficient private bus companies service the main towns, but bush taxis are the main transport in rural areas (where 80% of Burkinabé live). In Nigeria, large buses ply the crumbling freeway system mainly at night (with armed guards on board to deter bandits), while by day the Peugeot seven-seaters take over, dodging giant potholes and oncoming traffic at speeds usually reserved for German autobahns.

Safety is not a concern of most bush taxi drivers. For instance, two tourists in Nigeria, having seen cars driving the wrong way on Nigerian freeways, were concerned about safety in taking a night bus. They were reassured by the owner of the bus that "at night the cars stay off the road, the buses are so big and powerful they crush any cars in their path." Author Paul Theroux (2003) describes a bush taxi trip to Nairobi, the Kenyan capital:

> Our overstuffed Peugeot was doing eighty. I said to Kamali, "Mind asking the driver to slow down?" "Pole-pole, bwana" [slow-slow boss], Kamali said, leaning forward. Insulted by this suggestion, the driver went faster and more recklessly. Because so many of the roads were better than before, people drove faster and there were more fatal accidents. (p. 174)

A Short History of Africa

Mankind was born in Africa, with human history beginning there about 7 million years ago, when a population of African apes gave rise to the ancestor of modern humans

(Diamond, 1999, p. 36). These early hominids developed the ability to walk upright. About 2.5 million years ago, a more modern precursor, *Homo habilis*, the toolmaker, arose in the African river valley (present-day Ethiopia, Kenya, and Tanzania). An even closer ancestor, *Homo erectus,* arose 1.5 million years ago and is believed to have spread from Africa to populate the rest of the globe (BBC World Service, 2003). *Homo erectus* looked a lot like modern man with a larger brain than the earlier hominids. *Homo erectus* gradually evolved into modern man: *Homo sapiens*.

The first African civilization was that of Egypt. The Sahara desert was fertile 8,000 years ago. As the climate changed and the deserts advanced, the Nile river valley gained importance, and the people living there assumed great power and influence. These people gave rise to the great Egyptian civilization, which from about 3,000 BCE to the birth of Christ was the most powerful the world had ever seen.

Until 2,000 or 3,000 years ago, Africa south of the Sahara was populated by scattered groups of hunter-gatherers. Then the face of the continent changed when the Bantu peoples left the rainforests of West Africa (present-day Cameroon) and gradually dispersed south and eastward to cover most of equatorial and subequatorial Africa (Davidson, 1984, pp. 47–60). These peoples possessed metallurgy and agriculture (based on yams), which was enough of a technological advantage to allow them to displace the original non-Bantu inhabitants.

Agriculture and metallurgy allowed the creation of kingdoms. The non-Bantu peoples of West Africa, with a millet and sorghum-based agriculture, created several powerful kingdoms from about 200 BCE. The ancient kingdom of Ghana (centered around present-day Mali) was renowned in the Arab world for its power and learning, beginning in the 5th century. Meanwhile, ancient kingdoms also flourished in Zimbabwe (Davidson, 1984, p. 120), and in East Africa the Swahili civilization was founded by seafaring traders about 700 CE. The Swahili civilization and many of the West African empires were predominantly Muslim.

The Impact of Slavery

As in most parts of the world, slavery was always practiced in Africa. However, the arrival of Europeans about 1500 led to an explosion in the practice. Slaves, mainly from West Africa, were transported from the Gold and Slave Coasts[1] to the plantations of the Americas. By the late 19th century when slavery was finally abolished legally in the Western world, an estimated 15 million Africans had been sold into slavery, most of them the strongest and the most productive members of society. A smaller trade to Arab countries was centered in the Swahili regions of East Africa.

The European slave trade in Africa slowly transitioned into a colonial invasion, preceded by Christian missionaries. Davidson (1984) quotes an African saying: "First they had the Bible and we had the land. After a while we found things had changed round. Now they have the land and we have the Bible" (p. 167). While Europeans had always maintained trading posts along the African coasts, the so-called "scramble for Africa" did not begin in earnest until around 1880. Within 30 years five countries—Britain, France, Germany, Portugal, and Italy—and one individual, King Leopold of Belgium,[2] controlled virtually the whole of Africa, apart from Ethiopia and Liberia (Pakenham, 1997, p. xxii). The end of World War II gave rise to decolonization, and between 1959 (Ghana was the first African nation to achieve independence) and 1973—when the Portuguese were finally driven out of Angola, Mozambique, and Guinea Bissau—47 independent nations were born. Many would see the final death knell of colonization as the end of the apartheid regime in South Africa in 1991.

Modern Africa is best known in the West for its problems rather than its successes. Famine, civil wars, military coups, poverty, AIDS, Ebola, and genocide seem to be the only headlines to come from Africa. Few African countries have been able to overcome the colonial legacy of artificial frontiers, absence of infrastructure, lack of civil institutions, and the scarcity of economic and educational opportunities for their people. The impact of HIV/AIDS, currently affecting 24.5 million in sub-Saharan Africa (UNAIDS, 2007), has wiped out past advances in life expectancy and health care, leaving the continent facing an ever more uncertain future.

Foreign direct investment in Africa is minuscule, as uncertainty keeps Western investors away. There is little industry, and economies are dependent on export of raw commodities, such as oil, diamonds, gold, cotton, cocoa, and coffee, which have little value added in-country. Subsidies by Western governments to their agricultural producers mean that African producers cannot leverage their competitive advantages in low labor and land costs. Crude oil comprises more than half of Africa's exports and in 2005, 60.5% of net foreign direct investment in sub-Saharan Africa went to oil-exporting countries. According to the *Doing Business* (2008) Web site, Mauritius (24), South Africa (32), Botswana (38), and Namibia (51) were ranked among the top third of 181 countries in ease of doing business. The average rank of African countries, however, lingered at a dismal 136 among 178 countries. Botswana is a wonderful success story as the world's largest producer of diamonds and one of Africa's most stable countries. In two thirds of sub-Saharan Africa, one or two products account for more than 60% of the exports.

Yet, despite all the problems, there is an optimism and a vitality about Africa and Africans that is unmatched by any other continent. Anybody who has visited the continent will readily attest to the energy and warmth found there. African culture is unique, and intrinsic African values such as communalism and respect for elders should eventually allow Africa to solve its own social and economic problems in its own unique way and to fully exploit its natural resources for the benefit of its people.

African Time Orientation and Fatalism

African bush taxis do not operate on any fixed schedule. It is normal for people to arrive early in the morning for a departure and then to spend the entire day waiting for their transport to leave. Africans like to joke with U.S. Americans that in the United States, "time is money." Well, in Africa, time most certainly is not money. Gas is money, leaving with unoccupied space for one more fare is money, but waiting all day just to fill the vehicle is not money. To Westerners, time is a commodity to be bought and sold but not to Africans. People will wait patiently, mostly unconcerned about any schedule they might have, because they know they will eventually get to where they want to go. Impatient expatriates or even "modern" Africans, with more money than sense, will be allowed to purchase extra seats to fill the bush taxi, but otherwise all will wait patiently for the last passenger to be crammed in. Once that happens the scene will suddenly change, the calm will be shattered, and there will be a frantic haste to get away as people scramble for their seats and the taxi pulls away, often with several passengers running after the moving vehicle.

The Sense of Fatalism

Closely related to the African attitude about time is the sense of fatalism that many Africans possess. There is a sense that what will be will be, and there is little an individual can do about it. This sense of fatalism helps people to wait patiently or react calmly when their tire blows out (with no spare) in the middle of nowhere. However, bush taxi riders in West Africa are known to protect themselves against the "demons of the road" by obtaining talismans because they are never sure if they will arrive safely at their destinations without these. In some ways the sense of fatalism could be considered an aspect of Islam, where believers put their trust in God, with everything that happens depending on God's willingness to let it happen.

Islam is the predominant religion in West Africa and in parts of East Africa, and Africans have readily adapted to the Muslim idea of *Inshallah* or God willing. So, if a desire to meet a friend for a drink the next day elicits only an *Inshallah*, it does not mean that your friend is reluctant to meet but only that the decision to meet is out of his hands; it literally rests with God. However, African fatalism goes deeper than Islam, which is a relatively recent arrival to Africa. African fatalism has its roots in ancient African culture and the African environment. Dorothy Pennington (1985) discusses the African concept of time in terms of the extent to which people feel in control of their destiny or feel subject to the control of external forces: "To what extent does a group rationalize its fate by postulating mysterious forces and beings in nature and mysterious powers among their fellows or higher forces?" (p. 123).

U.S. Americans feel that they can, to a large extent, control their destiny. Witness the difficulty people have in accepting an incurable illness or a death; there is a sense of being cheated or being let down by medical science. Africans, on the other hand, find it easier to accept the hand fate has dealt them. Although the expatriate will become upset at sitting around all day waiting for the bush taxi to leave, for the African, it's not such a big deal. Where does this fatalism come from?

It is common in Burkina Faso, when drinking a beverage (especially an alcoholic one), to pour a small amount onto the ground as an offering to the ancestors. This is done in an unself-conscious manner, even by Western-educated Africans, Christians, or Muslims, as if it were a perfectly normal thing to do. Expatriates have adopted this custom (although, truth be told, probably more because it doubles as an opportunity to rinse their glass before drinking). Davidson (1984, p. 47) notes the same practice throughout West Africa.

For an African, including Christians or Muslims, the ancestors are an everyday presence. In African religious ontology there are five divisions ranked in order of importance (Mbiti, 1970, p. 20): God, spirits (ancestors), man, animals and plants, and inanimate objects. God is the supreme being, but the ancestors and the past define man. Man is at the center of the universe and has responsibilities to maintain balance and order in the universe. Living in harmony with nature and fulfilling obligations are important concepts. When people die, time does not come to an end; they live on as spirits.

A story illustrates this concept. Ouidah, in present-day Benin, was one of the main Portuguese slaving ports in the 18th and 19th centuries. On the route the slaves took from the town to the beach where the slave ships waited, there was a sacred tree. It was believed that if the slaves rounded the tree a certain number of times, their souls would return to Africa as spirits when they died. Deeply upsetting as a life of slavery was, even more upsetting to the slaves was the idea of their spirits being lost in a foreign land, far away from their native place and their descendants. The slavers used this to control their charges because denial of this ritual was truly a fate worse than death.

Ideas About Time

Africans also have little concept of future time, as Mbiti (1970) argues:

> The linear concept of time in Western thought, with an indefinite past, present and infinite future is practically foreign to African thinking. The future is virtually absent because events which lie in it have not taken place, they have not been realized, and cannot, therefore, constitute time. (p. 20)

Future events are simply potential events, but actual events are what has already taken place and are more important. People tend to look backward rather than forward, and thus there is not the same belief in progress and improvement that characterizes Western thought. There is a living past going on all around, and the African sense of time is more circular or cyclical than linear (Lessem & Palsule, 1997, p. 60). People who see time as a linear progression are likely to be more time conscious than those who see time as a repeating cycle. Africans do not perceive time in terms of a clock, but rather in terms of events, or how time is used. Sitting around talking all day with friends is not a waste of time; rather it is an event, making a connection with people and worth spending time on (Upkabi, 1990, p. 20). Thus for an African, social relations in the workplace are important. Time spent chatting is not a waste of time but part of relationship building and essential for group harmony.

To an African, fate lies in the hands of a living and omnipotent God. This is regardless of whether the person is nominally Christian, Muslim, or a follower of traditional religion. Most Africans are tolerant of religious freedom but cannot understand atheism. An educated Catholic African recounted how he met a European tourist who professed to be an atheist. To the African this was just absurd. He could understand somebody being Muslim, Christian, Hindu, or Buddhist, but to be an atheist, this European must have been crazy, he thought. The idea of atheism may be deeply upsetting to many Africans. It is not uncommon for Africans to change religions, and they tend to be quite flexible in their beliefs.

A Peace Corps volunteer teacher in Burkina Faso noticed one of his students, a Muslim, wearing a crucifix one day. When questioned, the student said that she had decided to become a Catholic and her parents didn't care what religion she held. A voodoo fetish priest in Lome, the capital of Togo, reported that some of his best clients were Catholic (although he said some of the new Protestant converts were deeply hostile to his practice). The living omnipresent God intervenes regularly in human affairs, as do the spirits, according to this way of thinking. Because humans are not the ultimate arbiters of their fate, there is little point in planning for the future. People who believe in an all-powerful God are more likely to be less time conscious than people who believe in their own power to influence events.

That is not to say that Africans listlessly drift through life, passively awaiting the hand of God and the ancestors to decide events. First, we are speaking of tendencies, not strict causal relationships. Second, as argued by Pennington (1985) and Albert (1970), fatalism is not complete futility; rather, it means dignity in the face of adversity and humility in the event of prosperity. The distinction between fatalists and doing-oriented people is one of the degree to which people control their lives: "No matter how conscientiously a man may perform his duties, he may fail or suffer, whereas some men who are worthless or evil may prosper" (Albert, 1970, p. 106).

Because events are unpredictable, planning and regularity become less important. Also, as Africa becomes more and more Westernized with urbanization, modern education, and exposure to Western media, Africans subscribe more and more to Western concepts of time and doing. Many buses now operate on regular schedules, whereas the bush taxis only leave when they are full.

Of course, religion is only one of the contributors to an African sense of time and fatalism. A second interpretation depends on the nature of the activities in which a society is engaged. In a technological, commercial society, time is a valuable commodity to be bought and sold, whereas in an agrarian society, time is controlled by the cycle of nature. In traditional African society, life is unhurried, travel and the pace of life are slow, and so strict timekeeping is not important. In a business setting this cuts both ways. Africans may sit around talking for several hours during the workday and then work late (without overtime) to complete their daily tasks. An expatriate manager, accustomed to doing a day's work and then leaving work to socialize, would consider this bizarre, but to an African this is perfectly normal behavior. And agrarian societies are not indolent. For instance, African farmers, when the rains come, will work feverishly to plant their crops because that is what is needed. Then, when the planting is done, an unhurried pace will reassert itself.

Africans then have a more relaxed view of time than U.S. Americans or Europeans. Time is a complex affair and Africans tend to be polychronic, often working on several tasks at the same time. There are many consequences to this African sense of time (Edoho, 2001, p. 58). For instance, it may take many meetings to arrange a deal, and patience in taking the time to get to know partners is important. Deadlines may be flexible, and no offense should be taken if people arrive late for meetings. Important people are expected to arrive late for meetings or events; the more important the personage, the later he or she turns up.

Communalism and Community Sharing

An African bush taxi is an intimate experience. Westerners who are used to having a fairly broad personal space sometimes find the intimacy upsetting. People are crammed into the bush taxi, and in the mainly tropical regions of Africa maintaining personal hygiene is difficult. Food is readily shared among passengers. As a bush taxi passes through a town, there are frequent stops—for traffic, police checkpoints, breakdowns, and so on. Each stop is an opportunity for an army of vendors—women selling peanuts or fruit, men selling grilled chicken, boys selling plastic bags filled with water or juice, young men selling the latest in women's fashions—to descend on and mob a bush taxi or bus. Business is conducted through the open windows, with the vendors often forced to run to obtain payment from passengers as the bush taxi suddenly speeds off.

Food and drink are shared among passengers; it is impolite not to offer to share your meal with your fellow passengers and only slightly less impolite to refuse an offer. It is not just food and drink that are shared. For example, a U.S. Peace Corps volunteer in Senegal ended up holding a stranger's baby for the duration of a trip; it is normal for friends and neighbors or acquaintances to care for others' children, and this extends to the communal bush taxi.

The bush taxi is an important community service, serving as a link between a community and the outside world. Although all passengers are expected to pay a fare, exceptions may be made. Normally, seated passengers pay a regular fare, whereas those who lack the means are allowed to cling to the outside of the bush taxi or ride on the roof. Departure of the bush taxi is accompanied by a scramble of young men who jump on the fender or climb on the roof to make a precarious and dangerous journey. This is more common in remote rural areas where transport options are limited and the driver/owner is more likely to know the people concerned.

Sometimes a seat may be given to a non-fare-paying passenger, although he may be asked to vacate the place if a paying passenger is picked up. These casual passengers are always male; a woman's movements are more restricted and a woman would never travel (or be allowed to travel) without the means of paying.

The Trait of Communalism

Communalism is a strong trait in African culture. Land is often held in common with relatives or an extended family. Communalism starts with the family and extends outward. The African family is an extended one covering several generations:

> [The African family] extends outwards to a great distance and backward to many generations, and may even include unborn children. So, whereas Westerners generally have little family awareness beyond, say, first cousins, and to map out a family tree would be no more than amusing diversion, to an African such a mental map of his family is the focus and center of his identity. (McCarthy, 1994, p. 14)

Again we see the importance of ancestors (see Chapter 26, "The Chinese Family Altar"). The strength of the African family was seen in Ghana in 1983 when 1.3 million Ghanaians (10% of the total population of Ghana) were deported suddenly from Nigeria. The expected refugee crisis never materialized, however, for the deportees were quickly reassimilated by their families (Siegel, 1992, p. 175). The concept of family sharing has been seen by some as an impediment to entrepreneurship and enterprise. Many educated Africans, living in cities, bemoan the fact that they are expected to take care of several distant relatives. Educated, salaried Africans, in addition to their immediate families, may have several nieces, nephews, and cousins living with them, for whom they are paying school and living expenses. In addition, those who are relatively well off may send money home to their villages for medical expenses or a tin roof on a house. The more successful the person is, the more he or she is expected to give back. Often, one son will be selected from a family to go to school and be educated, at the expense of the whole extended family, in the hope that this person will obtain a salaried position that will allow him to provide for the family in the future. However, this may serve as a disincentive to educated Africans to work to improve their position.

The African sense of community extends beyond the family to the village, which is seen almost as an extended family. When Africans are asked where they are from, they will typically give the name of their village. Even if they were born in a city, and even if their parents before them were born in a city, they will still name their parents' or grandparents' village. For example, an urban-educated Burkinabé took a detour from a business trip to visit his village. He himself had been born in a city several hundred kilometers to the south but still identified himself with the village. He was deeply embarrassed and upset when he got lost in the bush looking for the village, finding it only after seeking directions. In urban settings Africans will form village committees to plan social events, to network for jobs, to help recent arrivals, and in general to retain their village identity. While this network weakens over time, it is still a very real presence in urban Africa.

Associated with communalism is a different sense of personal space held by Africans. The people crammed together in the bush taxi have little problem with the intimacy this imposes. By contrast, a local politician in Fairfax, Virginia, opposing the proposed installation of bench-style seating in the Washington, D.C., Metro, was aghast that people might be forced to establish eye contact with their fellow passengers. Africans are comfortable with physical contact (although between the sexes certain public contact is inappropriate). Handshakes are long, often maintained for minutes on end, and accompanied by backslapping. In common with many Muslim countries, it is common for men in many parts of Africa to hold hands in public to signify a close friendship. Privacy is not well understood, doors are left open, and friends and neighbors often arrive uninvited to pass an evening. Expatriates are often uncomfortable with this and often pretend to be out to avoid unwelcome guests. Kapuscinski (2001) describes this sense of space and openness as he walks with his host in a Senegalese village greeting his neighbors:

> As we walked the village it became clear that in the tradition of its inhabitants, and even in their imagination, the concept of divided, differentiated, segmented space does not exist. There are no fences, boards, or wires, no hedges, nets, ditches or demarcation lines anywhere. The space is single, communal, open— transparent even; there are no screens in it, no erected barriers, dams and walls; it puts up no limits, offers no resistance. (p. 216)

Communalism or collectivism in contrast to individualism is one of the dimensions Hofstede (2001) used to differentiate national cultures. Individualism implies a loose social structure with people taking care of themselves and their immediate families only. Collectivism means people belong to tightly woven in-groups. African cultures, while varying from country to country (or between ethnic groups), are collectivistic. The Individualism Index values of Hofstede were low for East and West Africa and high for the United States (Nnadozie, 2001, p. 55). West Africa scored a 20 on individualism and East Africa scored a 27. By contrast the United States scored a 91. In writing about the sub-Saharan cluster in the more recent GLOBE study, Gupta and Hanges (2004) suggest that a distinctive philosophical concept in sub-Saharan Africa is *Ubuntu* or the humaneness that individuals and groups display for one another, a concept that is summed up in the phrase *umntu ngumntu ngabanye* (a person is a person through others). The countries included in sub-Saharan Africa in their study are Namibia, Zambia, Zimbabwe, Nigeria, and South Africa (black sample).

Business Consequences

There are many important consequences of communalism for managers in Africa (Blunt & Jones, 1992, p. 191). Because people often owe their first loyalty to their extended family, village, or clan, communalism may complicate managerial and supervisory relationships. Personal relationships between individuals may be more important than the business context. Rules and regulations may be circumvented due to the reciprocal relationships that exist between people. Loyalty to family and financial demands on employees by the family may lead to corrupt practices. It is not surprising that sub-Saharan Africa scores high on humane orientation in the GLOBE study (House et al., 2004).

Often the sense of communalism extends to the corporation. Managers are expected to take care of their employees. Layoffs are serious in any society but perhaps more so in Africa, where there is little paid employment and an individual may be supporting a large number of people. Managers must remember this when attention turns to layoffs or downsizing. Blunt and Jones (1992, p. 194) have applied the low individualistic orientation of African society and research on corresponding cultural orientations (unemotionality and subordination) to suggest that African organizations will have associated outcomes of excessive caution and lack of decisiveness and creativity in problem solving. However, communalism can have benefits if managers can extend the communalistic orientation to the organization.

Hierarchy in African Society: Seating Arrangements in the Bush Taxi

Although bush taxis are uncomfortable and impossibly crowded, that is not to say that there is no method to the seating arrangements. Given the speed at which bush taxis drive and the absence of seat belts, non-Africans might not think of the front seat as the prime seat. But it is the seat of honor and is normally reserved for elders of status or expatriate visitors. This illustrates the African concept of hospitality. A visitor is a guest to a country and will be treated as such. Visitors to Africa rarely need maps or timetables, for often all that is needed is to ask directions of a passerby. Invariably visitors will be swept away and brought to the point of departure of the bush taxi they seek, installed in a good seat with a negotiated fare, and given advice as to whom to trust and whom not to trust, the name of a relative at their destination who will take care of them, and perhaps a gift of peanuts or a drink. Because it is normal to reciprocate, and those that have wealth are expected to share with the less well off (and expatriates are invariably better off then Africans), a gift from visitors, whether monetary or not, will not give offense, even if it is not expected.

Associated with hierarchy, African culture can be characterized by a high power distance. As defined by Hofstede (2001), power distance is the willingness of people to accept unequal power distributions. East Africa and West Africa scored 64 and 77, respectively, for power distance, higher than the U.S. score of 40. In Africa, however, power does not necessarily equate to material wealth; age is also respected. It is common in a bush taxi for a passenger to be demoted to a less desirable seat when a more important passenger gets on. Similarly, a U.S. American visiting his daughter, a Peace Corps volunteer in Mali, tells of how a bush taxi he was riding in broke down. A young boy passed by on his bicycle. The driver commandeered the boy's bicycle to ride into town in search of a mechanic. The boy was clearly upset but could do nothing about it for he was expected to obey the instructions of any adult or elder.

This tendency is reflected in many aspects of organizational culture. Promotion is often based on seniority, not ability. Teachers, for instance, are held to be infallible sources of knowledge, and students do not appreciate teachers who profess ignorance or uncertainty. Managers and supervisors are expected to direct and command, and there is strong centralization and little initiative taking. This can lead to many problems, with junior staff unwilling to make any decision without first clearing it with their supervisors and criticism of managers (even positive criticism) being frowned on (Blunt & Jones, 1992, p. 190). There may be a tendency among African managers not to delegate or engage in participatory forms of management. Managers may be reluctant to share information with their subordinates.

Expatriates, as outsiders, are in some ways removed from the African hierarchy. So a young expatriate manager may not have the same status problems as an African manager of the same age. Nevertheless, an awareness of status and the role it plays is vital. Formality is important. An expatriate manager in Ouagadougou (the capital of Burkina Faso) complained that even though he lived only a 5-minute walk from his office, he was expected to be driven to work by his chauffeur in the morning. If he walked, his standing would have dropped among his staff. Leaders are expected to act, dress, and behave in a manner befitting their position. This can give rise to a "big man" syndrome with a preference for powerful autocratic rule.

For example, an expatriate watched the film *Mobuto: Roi de Zaire*, a documentary about the late dictator Mobuto Sese Seko of Zaire (Democratic Republic of Congo), in a movie theater in Burkina Faso. As the film displayed the opulent, lavish, and decadent lifestyle of the dictator, there were gasps of both awe and admiration from the audience. Although all were shocked by the atrocities he committed, there was also a sense of admiration for the absolute power he wielded and the lifestyle he enjoyed. Again, this is only a tendency. In the same country the most revered political leader is Thomas Sankara, president from 1983 to 1987, when he was overthrown and murdered in a coup. Sankara was a man of the common people who rejected the trappings of power, selling the government Mercedes in exchange for Renaults and hitching rides on the airplanes of other African leaders. Similarly Nelson Mandela is revered throughout Africa for the simplicity of his lifestyle.

Politeness and Respect

A great deal of politeness is expected in daily interactions. In learning Mooré, the language spoken by the Mossi, the major ethnic group of Burkina Faso, the first thing one learns are the ritual greetings that seem to go on and on endlessly (Hello, how are you, and your family, and your health, and your sleep, and your wife?—and so on). To a Westerner, accustomed to a passing nod to acquaintances, this endless repetition may seem amusing and childish, but for Africans this is an important ritual, establishing status and position and reinforcing communal ties. Although in a business setting the greetings are certainly shortened, it is vital to greet coworkers verbally and shake hands. To pass a colleague without a greeting can cause offense and lead the offended party to wonder what he did to receive such rude treatment.

Along with long formal greetings, it is also important to use titles. A secondary school teacher in Francophone Africa is expected to greet his school principal with Monsieur Le Proviseur, never the principal's first name or even family name. It is invariably better to use a person's title (doctor, chairman, chief, and so on) rather than his or her name. Note that we are again speaking about tendencies, and the notion of hierarchy varies between different ethnic groups.

As an example, Burkina Faso, a landlocked Sahelian country in West Africa, has more than 60 different ethnic groups. The main ethnic group, the Mossi, comprises more than 40% of the population and had a highly evolved, rigid hierarchical system before colonization by the French in the 1890s. The Mossi had an overall king (Naba) with absolute power. The Naba had a standing army and exacted tribute from regional kings and chiefs, and so the system continued down to the village level. The French retained this feudal system as a means of control, and today Mossi villages typically have a powerful chief. By contrast, in the southwest of the country in the border areas with Ghana, the Lobi people live in separate extended family compounds, loosely joined to form a village, with a titular chief, but real power resides in a group of elders. Lobi family compounds resemble small fortresses and were built out of arrow range of their neighbors, such was the suspicion they held about their out-group clan members. Today, Lobi villages are characterized by a more democratic if sometimes chaotic approach to decision making.

It is more difficult to apply the bush taxi metaphor to other aspects of African culture, specifically Hofstede's (2001) concepts of masculinity and uncertainty avoidance. There is some ambiguity about how Africa measures on these dimensions. Hofstede measures both East and West Africa as moderately masculine, with a score lower than the United States. Yet in Africa, gender roles are well defined, and males have far more opportunities than females. Gender imbalances are far greater than in the United States. Polygamy is common in some parts of Africa and is tolerated under both Islam and many traditional religions. Female excision is also widely practiced. Multilateral aid from institutions like the World Bank is often specifically linked to the participation rates of young girls in education. Certainly one would never meet a female bush taxi driver.

The Role of Women

There is a great deal of variation within sub-Saharan Africa. For instance, according to the World Bank in 2007, in Seychelles, 92% of women are literate but in Niger and Chad, the figures are 15% and 13% respectively. In Sierra Leone, two women die for every 100 live births but in Mauritius, 24 die for every 100,000 live births. Burundi has the highest proportion of women in its labor force (90.5% in 2005) and Sudan has the lowest (22.5%). In the GLOBE study (House et al., 2004), the sub-Saharan Africa cluster had a mean score of 3.29 in gender egalitarianism, very similar to the Southern Asian cluster (3.28) but lower than the Eastern Europe cluster, which scored 3.84, and higher than the Middle Eastern cluster (2.95).

One would expect African cultural groups to score high on uncertainty avoidance or risk acceptance. As a traditional society, without a strong belief in modernization and continuous improvement, but rather tending to look toward the past for inspiration, one would expect Africans to be less willing to accommodate change and uncertainty. This is borne out by the findings of the GLOBE study (House et al., 2004) in which the sub-Saharan Africa cluster was among the highest in uncertainty avoidance together with Southern Asia, the Middle East, Eastern Europe, and Latin America.

Triandis (1984) has produced a useful summary of African cultural values compared to those of the Western developed nations. As we have already discussed, Africans tend to believe in external causation and in their limited potential to change events. They are past oriented, more focused on being than doing, and more passive than active. Success in endeavors is judged more on the process or on the well-being of the group, rather than guided by absolute principles and measured by results. The influence of contextual factors may negate Western principles of management (Gray, 2001, p. 271).

Summing up, we can say that the African bush taxi captures much of the essence of Africa and its culture. These beat-up and bedraggled cars and vans that furiously dash across the savannahs, jungles, deserts, and cityscapes of Africa summarize the energy and ingenuity of Africa. They have been repaired and added to so many times that they scarcely resemble the vehicle that was originally made. They are held together with rope, twine, rubber bands, nails, and pieces of clothing. People wait patiently for the eventual departure of their bush taxi, rarely complaining, for what is the point of complaining? People keep their good spirits, even when 30 are crammed into a vehicle built for 12. They share both their food and their jokes with their neighbors, and they watch out for each other's children, for this is the way in the African community. They give up their seats for an elder, for age is to be respected. They give up their seats for foreigners because they are guests and should be treated with respect.

The bush taxi is Africa, in the creativity and ingenuity involved in getting them to go at all and in the cheerfulness and vitality of the people who ride them. Too often African values of family, communalism, hierarchy, and respect for age are cited as obstacles to good management and organization in Africa. Yet cultures with many similar values, such as South Korea, Taiwan, and Japan, have shown that individualism and a doing orientation are not prerequisites to economic development and good management (Ugwuegbu, 2001, p. 36). Good managers, whether African or expatriate, should be able to exploit rather than overcome African cultural values to build strong organizations.

When Africa does "develop," when it finally can offer its people decent health care, a decent education, and economic opportunities, it will be based on African cultural values such as communalism, teamwork, concern for others, respect for age, and a strong sense of place and tradition. Perhaps then the bush taxis will be able to afford to leave on time, full or not.

Notes

1. Approximately corresponding to the Gulf of Guinea area from Côte D'Ivoire to Cameroon, although the slave trade extended as far south as Angola.

2. The Belgian Congo (present-day Democratic Republic of Congo or Congo-Kinshasa) was originally the personal fiefdom of King Leopold of the Belgians. Eventually he had so mismanaged the colony that he was forced, in 1908, to hand it over to the Belgian state (Pakenham, 1997, p. 662).

PART XII

What We've Learned

In this part of the book we emphasize what we have learned from this in-depth exploration of 29 national cultures, two continents, diversity as represented by India but implicitly by at least 90% of the world's nations, and the extension of cultures across national borders as represented by the example of the Chinese. We integrate this part, which consists of only one chapter, through the use of the frameworks described in Chapter 1. In this fashion we can provide an integrative summary that takes into account the great amount of material presented throughout the book.

CHAPTER 34

An Integrative Summary

In each chapter of this book we have presented a large amount of material focusing on the specific factors used to derive each cultural metaphor such as religion, language, and history (see the Constructing Cultural Metaphors section in Chapter 1). We then described each national culture in depth in terms of the distinctive features of each metaphor. This method has allowed us to standardize the format across chapters. In this chapter we present an integrative summary in terms of these specific factors. Our discussion will incorporate these factors within the following categories: Perspectives on culture; institutions and culture; additional factors; and beyond ethnic and national cultures.

Perspectives on Culture

Some standard textbooks suggest that cultures change very slowly. However, the pace of change has accelerated dramatically since 1970 when globalization began to become prominent. Also, some parts of a culture can change very quickly while other parts change very slowly. An example of a quick change is lifetime employment in Japan. Given the collectivistic emphasis in Japan, it is understandable why this concept has been so engrained in Japanese thinking. However, the onslaught of globalization, including the acquisitions of Japanese companies by non-Japanese, has decreased significantly the guarantee of lifetime employment. Similarly, China has changed quickly and has become more individualistic largely as a result of three influences: the largest transition in history from agrarian to urban living; the one-child-per-family law; and the move to a market economy integrated with other national economies through the World Trade Organization and other international agencies. Still, both in Japan and China the emphasis on collectivism is apparent. Witness, for example, the unbelievably clogged transportation systems in China during the Lunar New Year as people try to return to the villages and cities of their birth, in accordance with Confucian traditions.

Some of the confusion as to whether cultures change quickly or slowly relates to the definition of culture. Many of us shy away from defining a murky area such as culture. There are, literally, thousands of definitions of culture. However, there is a consensus among most if not all experts that any definition must include a culture's values, norms, and behaviors. What the majority of people value in a culture tends to become enshrined in norms or expected standards of behavior. At times it is possible to obtain some initial insight into the linkages among a culture's values, norms, and expected behaviors, for example, by looking at whether greetings are predominantly expressed in terms of kissing or bowing or shaking hands. There is, however, an unfortunate tendency to emphasize behaviors that are disconnected from core values and norms. By way of example, many U.S. Americans encounter non-Western people dressing in Western attire and speaking English. These U.S. Americans automatically assume similarity of values and norms when dissimilarity is prevalent.

One key value, and perhaps the key value, is whether a culture values individualism over collectivism or vice versa, particularly whether an individual will make a decision in terms of his or her own desires or subordinate such desires to the demands and norms of the group. In all or most chapters we have described the type of individualism or collectivism found within each national culture, for example, Scandinavian individualism, U.S. competitive individualism, *kata*-based Japanese collectivism, and Hindu-based and familial Indian collectivism. We have also provided a summary of many key values by describing the various scales that Hofstede (2001) and the GLOBE researchers (House et al., 2004) derived and used the national rankings on these scales throughout the book. See Chapter 1.

Throughout history both conquerors and insiders have frequently attempted to change the core values of a national culture, for example, the British occupation of southern Ireland during which the speaking of Gaelic was forbidden. A related example is Mao Tse Tung's program to rid China of traditional Confucian values stressing the family, relationships, and education. Frequently these attempts lead to unfortunate results, such as the extreme poverty and feelings of powerlessness that the Native Americans in North and Latin America have experienced at the hands of European conquerors and the deaths of an estimated 50 million Chinese due largely to Mao's programs. Many if not most of the key values of a culture reemerge in some form. China, for instance, has seen an explosion of interest in Confucian thought, as confirmed by recent best-selling books in that nation emphasizing this topic. Some wealthy Chinese parents are sending their children to educational programs stressing Confucian thought and the obligations to both relationships and the state that these programs promote.

Although many cultural analysts employ the nation as the basic unit of analysis, as in the case of the Hofstede and GLOBE studies, less than 10% of the 220 nations in the world are monocultural. Even nations such as India, which is more than 80% Hindu, experience significant multicultural and multireligious differences. To complicate matters, because of war and their political solutions, national boundaries are sometimes changed abruptly and influence the relationships among subcultures in each nation. For example, the British mandated that four major tribes, including at least 300 distinct subtribes within them, form the nation of Nigeria. One unfortunate outcome of this mandate is the periodic and bloody confrontations between Muslim-dominant tribes and Christian-dominant tribes over value-laden issues such as whether Islamic law should be used and whether the state should be theocratic.

The concept of national culture is subject to intense scrutiny and debate in many nations as subcultures within them seek national sovereignty or at least regionalization, which would allow each region within a nation to operate independently on most issues. Such regionalization and fractionalization of the nation has occurred in Belgium, Spain, and elsewhere. This tendency has increased significantly since the fall of the Berlin Wall and the breakup of the Soviet Union in the early 1990s. In 1990 there were about 190 nations; today there are 220.

In studying culture, we believe it is essential to use multiple approaches and we have done so in this book. Although we accept the use of the most popular cross-cultural management perspective, as represented by the Hofstede and the GLOBE study, we are aware of the inherent limitations of such a survey-based approach. In particular, there is a tendency to emphasize the rankings of national cultures on a small and limited number of cross-cultural dimensions derived from questionnaire data. This occurs at the expense of an in-depth treatment and understanding of each national culture. Cultural metaphors are designed to minimize this problem while incorporating the rankings of national cultures within each description of each national culture.

Although the use of cultural metaphors possesses many advantages, there are also inherent limitations associated with this method. Understanding each cultural metaphor is but a first step in understanding a host culture. Clearly the most effective method of cross-cultural understanding is total immersion in a culture over an extended period of time. Ideally, a thorough understanding of the culture provided by the multiple approaches currently available should supplement total immersion.

Furthermore, as Hofstede (2001) demonstrated, there is a movement from collectivism to individualism as the gross domestic product (GDP) of a nation increases. His study extended over just four years, but most of the emerging behaviors we have witnessed since 1970 seem to support his point of view. However, there have been reactions to this movement toward individualism, even when GDP per capita increases. For example, Russia has moved back toward a more collectivistic orientation since 2000 and many Muslims would prefer to stop individualism and globalization completely.

There is a tendency to believe that the United States is fundamentally different from other national cultures because of its unqualified acceptance of competitive individualism. As discussed in Chapter 16, "American Football," the United States is distinct from other developed nations because of the widespread acceptance of both traditional values and competitive individualism. As noted above, there are different forms of individualism, for example, egalitarian individualism in Scandinavian nations. The different types of individualism and collectivism and the values associated with them frequently express themselves in terms of the degree to which individuals are protected in the culture. As we might expect, there are greater welfare protections in Scandinavia than in the United States. These protections are supported by taxing Scandinavian citizens at a higher rate than in the United States.

Furthermore, it is important to go beyond national boundaries and emphasize clusters of nations on some issues. Nations do cluster together in terms of both basic values expressed in religious affinities (e.g., Buddhist nations) and geographical locations. For many good reasons we tend to emphasize such clusters, for instance, South Asian nations, Latin American nations, northern European nations, southern European nations, and so on. Harry Triandis, in the introduction to the GLOBE study (House et al., 2004), points out that different independent researchers have identified about 10 such clusters. We have described the 10 clusters that the GLOBE study derived, and they overlap significantly with the clusters identified by Samuel Huntington (1996) and other researchers.

Finally, it is critical to understand why we titled the major parts of this book as we did. Clearly there is overlap among the various parts of the book. We have described national cultures in each part as illustrative of the part title, such as a torn culture or one torn from its roots, as in the case of Mexico and Russia. However, both of these nations could also fit into other parts of the book. Cultures are complicated and it is not appropriate to pigeonhole them into one category. But the categories do exist and all of us need to understand them as we attempt to comprehend the changes occurring both within each culture and across cultures.

Cultures and Institutions

There is an ongoing debate about the relationship between cultures and their institutions. Specifically, do cultural values determine the value-laden institutions that dictate norms and behavior or can institutions influence the culture? There is no definitive answer to this debate. Frequently cultural values determine the legal system, the educational system, the economic system, the political governance system, and the dominant family system, as we would expect. At other times key leaders can change the culture itself to bring it into conformity with the institutions they champion. Perhaps the most pertinent example of such a leader is Ataturk, who changed Turkey's culture critically in the early 1920s by mandating the use of the Roman alphabet and separating church and state in a nation that is overwhelmingly Muslim. However, even in the case of Ataturk, it was his prestige as war hero and the support of the Turkish military that allowed these key changes to occur.

Our view is systemic: Culture influences institutions and vice versa. It is clear that, once an educational system is established, it tends to reinforce the values that its leaders and documents espouse, making change to the system difficult. However, changes do occur because of globalization and related influences. For example, there is a movement in Europe to standardize university courses across nations and to make

European educational practices consistent with those of other nations such as the United States. Historically European universities employed a grading system that recognized only one of two grades in each course: pass or fail. Sometimes the pass grade would signify only that the student had taken a course. These universities primarily emphasized success on a grueling final written examination followed by an oral examination of each student that involved at least two professors as examiners. Some of these universities are now giving letter grades in courses and deemphasizing these examinations.

Legal systems also influence cultures. For the sake of simplicity, we can divide legal systems into those that apply universal rules to specific cases deductively and those that inductively require the justification of a particular action by citation of legal precedents and cases. Frequently the deductive and inductive approaches overlap, although one tends to dominate in a nation. For example, Japan historically stressed the use of a deductive system, which many major international but non-Japanese companies operating there found inappropriate to meet the challenges posed by globalization. To complicate matters, there was no jury in most business cases in Japan and the judge made the final decision independently. For years international companies attempted to avoid going to court in Japan because of the high probability that they would lose their cases, regardless of their merits. In more recent years Japan has taken case law into greater consideration, a position more consistent with the British and U.S. traditions emphasizing prior cases and legal precedents.

However, cultural values also influence legal systems. Perhaps the outstanding example of this influence is found in the use of day fines in egalitarian Scandinavian nations. If a person is found guilty in court and a financial penalty is imposed, it is adjusted to a person's income level. Thus one person may receive a fine of several thousand dollars while another person committing the same offense may receive a fine of $50 or $100.

Sometimes commentators treat economic systems as if they are divorced from cultural influences. However, this is an incorrect assumption. Economic systems do reflect the values in a culture. U.S. capitalism is clearly consistent with the cultural values found in the United States. Immigrants coming to the United States frequently do so because of the economic opportunities that such a system fosters. However, there is a downside to this system, as the government limits the amount of financial support it gives to unemployed workers as compared to what they would receive in many European nations such as France and Sweden. Resolving this dilemma of providing too little or too much support without harming the dynamics of the marketplace is difficult. Still, some nations are actively trying to do so. Perhaps the outstanding example cited in this book is Denmark's emphasis on *flexicurity*, which requires the active and ongoing collaboration of the government, corporations, and trade unions (see Chapter 11).

At other times the dynamics of the economic system influence cultural values. For example, the U.S. government has been emphasizing an unregulated marketplace since at least 1970. However, the subprime mortgage crisis and the actions of banks underwriting mortgages—immortalized in *Fortune* magazine's recent title of its lead article, "What Were They Smoking?"—has led to initiatives to regulate the market-place more strictly.

Culture also influences the political governance system. Israel, for example, has multiple political parties, each representing different interests, and it is difficult for one party to govern without the support of at least some of the minor political parties. Because of the historical treatment of the Jews at the hands of dictators such as Hitler and the importance of avoiding such dictatorial leadership, Israelis have constructed a political governance system in which all of these political parties receive representation. This has led to a situation in which there is a great deal of vote swapping, hidden agreements, open hostility on the floor of the Knesset between individuals representing diametrically opposed political parties, and periodic expo-sures of corruption. Israel represents but one example, and a quick perusal of other chapters in this book will easily confirm the fact that a nation's culture critically influences the political governance system.

As we might expect, the political governance system can influence culture. For instance, when Mao ruled China, he attacked Confucian values by breaking up families, having children report on the activities of their parents, and having members of the Red Guard humiliate and punish their elders, who occupy a revered place in Confucian thought. Similarly the Soviet Union's adoption of communism at the turn of the 20th century led to a greater emphasis on collectivism and the subordination of the individual to the group or collective.

Religions represent another major institution influencing culture. It is possible to visualize the regions of the world in terms of the major religions in each of them, such as Buddhism in Southeast Asia, Christianity in Western Europe, and Hinduism in India. Religions, whether major or minor, express cultural values and norms, and we can increase cross-cultural understanding by analyzing how and why these religions evolved as well as their major creeds, dogmas, and points of view. In various chapters of this book we have briefly explained most of these major religions from these perspectives. These religions include Christianity, Islam, Judaism, Buddhism, Hinduism, Confucianism, Taoism, and Shintoism. While a persuasive argument can be made that religion is losing its place of prominence in some nations because of the low percentage of citizens actively participating in it, such as is the case in Western Europe, there are examples of nations in which religious participation has increased. Even in the European nations, the cultural importance of religion continues to be strong and expressive of the values and norms that have guided them for centuries.

Arguably the most important institution influencing a culture is family. How a culture views this institution is critical. In many if not most nations the family and the extended kinship system are viewed as more important than the individuals who are their members, which has given rise to the key distinction between individualism and collectivism. In the extended Chinese kinship network, it is common to make fine distinctions between relationships, such as that between a fifth and a sixth cousin. Such distinctions reinforce the traditional hierarchical nature of Chinese families and their kinship networks. Also, there is a good amount of evidence to suggest that individuals who feel that they will not receive fair treatment in courts of law tend to rely on kinship and familial groups rather than the legal system. Such a perceived unfair system encourages lying for friends, particularly family members, even when under oath, and holding back critical information that might help to resolve an impasse.

Cultural values also help to explain socialization practices in families, such as the manner in which many collectivistic Asians strap the baby to a mother's body for the first three years of life so that the bond between mother and child is the focus of importance. This is in contrast to individualistic cultures in which the baby is allowed great latitude in exploring his or her environment a few months after birth.

In sum, institutional systems—particularly legal, educational, economic, religious, familial, and political governance systems—are critically related to culture. These institutional systems and culture mutually influence one another over time and frequently historical events have occurred at critical times because of the relative influence of one over the other.

Additional Factors

Some additional critical factors also influence culture: its history, geographical location, language, the use of space, unique or distinctive concepts within each culture, and demographics. If a national culture has experienced decades and even centuries of subordination at the hands of another nation, its outlook can be decisively influenced. Hindu nationalists, for example, still express deep resentment against Britain because of the oppressive nature of British rule until 1948. In the Balkan Wars of the late 1990s, the national press and TV programs in Orthodox Catholic Serbia emphasized the fact that the Catholics had been defeated and treated—from their perspective—very unfairly by their Muslim victors during a battle occurring in the 14th century. Such examples bring into prominence the role that history plays in the development of cultural values, many of which are retained for centuries.

Similarly, geography helps to reinforce cultural identity. Japan, for example, is an island nation and this fact has influenced its emphasis on being Japanese. While it is fashionable to talk about the death of distance occurring because of modern transportation and communication systems, geography is still critical. For example, although 156 out of the world's 220 nations are members of the World Trade Organization, regional economic blocs of nations have emerged in recent years and become powerful, for example, the European Union (EU) and NAFTA (the United States and its two largest trading partners, Canada and Mexico). Becoming a member of such a bloc is determined largely by geographical nearness. As these economic blocs become solidified, we can expect that citizens will begin to identify not only with their national cultures but also with their respective blocs. Such a pattern is clearly emerging in several European nations in the EU. For example, Ireland was largely a poor, agrarian nation until 1970 but its inclusion in the EU has allowed it to advance economically in a very significant manner, and many of the Irish tend to see their futures and even their identities as inseparable from the EU.

Geography is also associated with language. Linguists, at the broadest level, distinguish between languages developed in cold and warm climates. Cold climates tend to be associated with short words and direct language. By contrast, warm climates tend to be associated with more flowery language that embellishes a message and statement. Generally speaking, languages in warm climates tend to involve more emotionally charged and longer words containing more vowels than consonants. Even within one small nation, there can be hundreds of languages that reflect regional geographical preferences. However, because of increased communication through TV and other methods, there is a tendency to use one dominant language in a nation and to stress uniform pronunciation of its words. There is strong evidence indicating that languages are dying at an alarming rate, from an estimated 15,000 100 years ago to 7,000 today (Gannon, 2008), at least partially due to the death of distance facilitated by modern communication and transportation systems. Still, the evidence also suggests that the major language groups such as English and Chinese continue to be robust.

Language is of key importance in reinforcing cultural values. There has been much conflict involving ethnic groups speaking different languages in a nation, as we have seen in the cases of Belgium, Canada, and Spain. In Malaysia, the Chinese tend to send their children to Chinese-language schools, thus reinforcing their Chinese identities at the expense of the national culture, which has three main groups: the Chinese, the native Malays, and the Indians. Sometimes those partial to one language in a nation will not respond to those speaking another language, even when they are fluent in both languages. In India there are 16 official languages, reflective of the cultural diversity found in that nation.

An extreme emphasis on the importance of language is found in linguist Benjamin Whorf's belief that language must precede culture and is, in effect, equivalent to culture. It is not necessary, however, to accept this emphasis uncritically. Anthropologist Edward T. Hall, for example, argues that knowing the language of a culture is similar to understanding how to read a musical score, and without such knowledge it is difficult if not impossible to play a musical piece. He also views each culture as existing along a continuum going from low-context to high-context communication. Although it is undoubtedly true that the deepest understanding of a culture involves knowing its language, other factors can be as important or more important, particularly knowing and being sensitive to the values, norms, and behaviors of a culture.

The use of space is another factor that distinguishes cultures. Some collectivistic cultures emphasize a family system in which several related families live within the same protected compound. In individualistic cultures, there is a tendency for nuclear families to live separately from related families, even when they get together for holidays and parties. Similarly, in northern Europe there is a tendency to keep office doors closed, whereas in collectivistic Japan workers sit facing one another, with the supervisor walking among them to help small groups or individuals working alone. Even the acceptable amount of distance between two individuals engaged in a conversation varies by culture.

Furthermore, there are idiosyncratic interpretations of the same words in different cultures. *Privacy* in Saudi Arabia and *I* in such collectivistic cultures as Japan have negative connotations associated with them, whereas in individualistic nations such as the United States and Great Britain such words are perceived positively.

There are also words and phrases in a particular culture that express distinctive values. Frequently it is difficult if not impossible to translate these words and phrases accurately into English, as we have seen with the Thai phrase *Mai Pen Rai* and the Finnish word *sisu*. Similarly there are proverbs in a culture that convey a unique value and even worldview. Such words, phrases, and proverbs serve as vehicles for understanding each national culture and tend to supplement the insights provided by cultural metaphors.

One unique concept found in Asia is that of face. As we have seen, Westerners and Asians agree on the first type, saving face. This type of face is equivalent to the Western concept of saving one's honor. But Asians are also concerned about giving face, that is, protecting not only themselves but others in a transaction. In this sense face represents the unwritten set of rules or norms governing the behaviors of those involved in a transaction so that everyone's honor remains intact.

Culture is also influenced by demographic changes. For example, several of the world's developed nations have aging populations whereas in many developing nations more than half of the population is less than 20 years old. When the population of a developing nation is disproportionately young, there is a greater tendency toward radical actions. Similarly, immigration can change a culture. In the United States, for example, White Anglo Saxon Americans will be in a minority by the year 2050 or even earlier while Hispanic Americans could represent more than 25% of the population. As minority groups gain power and prominence in a nation, they will attempt to change the institutions to make them more compatible with their native cultures and, in the process, change the national culture at least in some major ways.

One cultural change that is becoming more prominent in recent years is the emphasis on gender equality. In Finland, for example, women constitute more than half of the legislature. The prime ministers in several nations are now female. Even in conservative Saudi Arabia, there will be a Saudi Arabian University of Science and Technology in which males and females attend integrated classes for the first time. As more and more women enter the workforce, they will demand gender equality, which will tend to increase it both at work and outside of it.

Beyond Ethnic and National Cultures

Culture is the key concept in this book. As indicated, we have described each national culture in depth through an analysis of numerous factors (see Chapter 1) summarized in this final chapter. However, it is possible to go beyond ethnic and national cultures. For example, Gannon (2008) has developed 93 cross-cultural paradoxes that have emerged because of globalization. A paradox is a statement containing inconsistent or contradictory subparts that does not seem to be true but is in fact true. Some of the paradoxes are implied above, for example, geography has become less important as modern communication, transportation, and financial systems have been able to integrate nations, whereas regional economic blocs based on geographical nearness have become more important.

Using the lens of cross-cultural paradoxes, we can analyze a large number of issues that go beyond ethnic and national cultures. These include such behavioral issues as leadership, motivation, and group behavior across cultures; communicating across cultures, especially in terms of the relative advantages of face-to-face versus mediated communication such as that occurring on the worldwide Web; conflicts generated by cross-cultural ethical differences; and cross-cultural negotiations. These issues also include broad concerns such as multiethnicity, religious differences, immigration, and geography. Understanding cross-cultural paradoxes helps us to move beyond culture perceived only in terms of ethnic and national boundaries.

Globalization has sensitized most of us to perceive the world in new and different ways. It represents the increasing interdependence among individuals, national governments, nonprofit organizations, and business firms. Three mechanisms facilitating globalization are the free movement of goods, services, talents, ideas, knowledge, capital, and communications across national boundaries; the creation of new technologies such as the Internet and highly efficient airplanes expediting such movement; and the lowering of tariffs and other impediments to this movement through the actions of such organizations as the World Trade Organization. As globalization proceeds, we can expect to see some increased cross-cultural conflicts. However, globalization's success may lead to a world in which individuals from different cultures not only increase their understanding of other cultures but also begin to change their own values, norms, and behaviors, making them more consistent across cultures.

There are many good reasons to justify globalization, including the dramatic reduction of poverty in Asia from 76% in the 1970s to 15% in 1998, in large part because of the increased encouragement of globalization among Asian nations (Sala-i-Martin, 2002). However, there are many problems that have emerged. Environmental issues such as air pollution and access to clean water have become problematic. Although globalization has clearly led to prosperity in nations participating significantly in the global economy, economic inequality within nations has also risen.

On the other hand, it is possible to envision alternative scenarios in which those opposing globalization will ultimately triumph. At this point in time such a probability is low unless there are unforeseen events such as a Third World War or a global financial collapse. The World Bank has identified 64 financial and banking crises between 1970 and 1998, any one of which could have triggered a chain reaction leading to such a global financial collapse (Goldin & Reinert, 2006, p. 109). Even given such crises, it is our belief that globalization will proceed and, as it does, cultures will need to adapt to one another. There is clear evidence that such adaptation is occurring, for example, the movement to standardize education within Europe to make it consistent with other educational systems worldwide and the movement to standardize accounting systems in a similar manner. As such global changes increase, we can expect adaptation to occur, including changes in values and norms. The U.S. Supreme Court, for instance, recently discussed the justifications for the death penalty and the justices included information about the lack of a death penalty in many European nations in formulating their own points of view. Possibly moving away from the death penalty in the United States is but one example of how cultural values and norms may change. Throughout the book we have included many similar examples.

In sum, it is important to examine the specific factors used to derive cultural metaphors. It is also important to understand how these factors influence culture and how culture in turn influences them. Such factors include various types of institutions. There are also several additional factors that we have described in this chapter. However, it is possible to go beyond national and ethnic cultures, especially in our era of globalization. Although we have emphasized cross-cultural paradoxes, we believe that other perspectives will help us to gain insight into ethnic and national cultures as we learn to adapt to and to change aspects of our globalizing world for mutual benefit. And, although citizens tend to identify with their own nations and to some extent with regional economic blocs, there may come a time when in fact we will have a world government and all of us will become citizens of the world.

References

Accents: We want to talk proper. (2002, December 7). *The Economist,* p. 55.

Adler, N., with Gundersen, A. (2007). *International dimensions of organizational behavior.* Cincinnati, OH: South-Western.

Albert, E. (1970). Conceptual systems in Africa. In J. N. Paden (Ed.), *The African experience.* Chicago: Northwestern University Press.

Alotaibi, M. (1989). *Bedouin: The nomads of the desert.* Vero Beach, FL: Rourke.

Al-Zahrani, S., & Kaplowitz, S. (1993). Attributional biases in individualistic and collectivist cultures: A comparison of Americans and Saudis. *Social Psychology Quarterly, 56,* 223–233.

American football: The search for supermen. (2006, September 30). *The Economist,* p. 95.

Andersen, P. (1994). Explaining intercultural differences in nonverbal communication. In L. Samovar & R. Porter (Eds.), *Intercultural communication: A reader* (7th ed., pp. 4–24). Belmont, CA: Wadsworth.

Andrews, J. (1999, July 24). Special report, Canada: Holding its own. *The Economist,* pp. 1–18.

Arana, M. (2002, February 24). Edna O'Brien, doyenne of Irish letters. *The Washington Post* [Book World], p. 10.

Arden, N. (1990, May). Searching for India along the great trunk road. *National Geographic,* pp. 177–185.

Aronson, D. R. (1978). *The city is our farm.* Cambridge, MA: Schenkman.

The ascent of British man. (2002, December 7). *The Economist,* p. 53.

Ashkanasy, N., Gupta, V., Mayfield, M. S., & Trevor-Roberts, E. (2004). Future orientation. In R. J. House, P. J. Hanges, M. Javidan, P. W. Dorfman, & V. Gupta (Eds.), *Culture, leadership, and organizations: The GLOBE study of 62 societies* (pp. 282–300). Thousand Oaks, CA: Sage.

Banerji, P. (1983). *Erotica in Indian dance.* Atlantic Highlands, NJ: Humanities Press.

Barnlund, D. (1989). *Public and private self in Japan and the United States.* Yarmouth, ME: Intercultural Press.

Barzini, L. (1964). *The Italians.* New York: Atheneum.

Barzini, L. (1983). *The Europeans.* New York: Simon & Schuster.

BBC World Service. (2003). Origins of the human race. *The story of Africa.* Retrieved December 5, 2008, from http://www.bbc.co.uk/worldservice/specials/1624_story_of_africa/index.shtml

Beckett, J. D. (1986). *A short history of Ireland.* London: Cresset Library.

Belgium. (2009). *New York Times.* Retrieved February 13, 2009, from http://topics.nytimes.com/top/news/international/countriesandterritories/belgium/index.html

Belgium: Fading away. (1992, October 31). *The Economist,* p. 52.

Bellman, E., & Range, J. (2008, May 1). Shortage of laborers plagues India. *The Wall Street Journal,* p. A1.

Benet, S. (1951). *Song, dance, and customs of peasant Poland.* London: Dennis Dobson.

Benoit, B. (2007). Different divide: Why the German economic schism is now north-south. *Financial Times*, p. 9.

Benton, W. (Ed.). (1970). *Encyclopedia Britannica*. Chicago: Encyclopedia Britannica.

Berry, M. (1990). The limits of community: Finland as a symbol of frontier-infused liberalism. In M. Berry, G. Maude, & J. Schuchalter (Eds.), *Frontiers of American political experience* (pp. 31–86). Turku, Finland: Turku University Press.

Berry, M. (2008a). Epilogue. In M. Berry (Ed.), *"That's not me": Learning to cope with sensitive cultural issues* (ISBN 978-952-92-2540-8, 76-78). Finland: Author.

Berry, M. (2008b). Finnish intercultural business overlaps with student experience. In M. Berry (Ed.). *"That's not me": Learning to cope with sensitive cultural issues* (ISBN 978-952-92-2540-8, 86-88). Finland: Author.

Berry, M., Carbaugh, D., & Nurmikari-Berry, M. (2004). Communicating Finnish quietude: A pedagogical process for discovering implicit cultural meanings in languages. *Language and Intercultural Communication, 4,* 261–280.

Berry, M., Carbaugh, D., & Nurmikari-Berry, M. (2008). "Discomfort with silence" in a culture "comfortable with quietness." In M. Berry (Ed.), *"That's not me": Learning to cope with sensitive cultural issues* (ISBN 978-952-92-2540-8, 39-60). Finland: Author.

Bettleheim, B. (1969). *The children of the dream*. New York: Macmillan.

Bilefsky, D. (2007a, April 9). Belgians hail the Middle Ages (well, not the plague part). *New York Times*. Retrieved November 8, 2008, from http://www.nytimes.com/2007/04/09/world/europe/09medieval.html?_r=1&scp=13&sq=dan+bilefsky+and+belgium&st=nyt&oref=slogin

Bilefsky, D. (2007b, August 12). Matador wins. Bull dies. The end? Not in Portugal. *New York Times*. Retrieved December 20, 2008, from http://www.nytimes.com/2007/08/12/world/europe/12portugal.html?_r=1&pagewanted=print

Bilefsky, D. (2007c, July 25). A TV contest sets off a furor over Portugal's ex-dictator. *The New York Times*. Retrieved December 20, 2008, from http://www.nytimes.com/2007/07/25/world/europe/25portugal.html?pagewanted=print

Bisbee, S. (1951). *The new Turks*. Philadelphia: University of Pennsylvania Press.

Blunt, P., & Jones, M. (1992). *Managing organizations in Africa*. Berlin: de Gruyter.

Boerner, S., & Frieherr von Streit, C. (2005). Transformational leadership and group climate: Empirical results from a symphony orchestra. *Journal of Leadership and Organizational Studies, 12,* 31–41.

Bonavia, D. (1989). *The Chinese*. London: Penguin.

Bond, M. (1986). *The psychology of the Chinese people*. New York: Oxford University Press.

Bond, M., & Smith, P. (1998). *Social psychology across cultures* (2nd ed.). London: Blackwell.

Boorstin, D. (1965). *The Americans: The national experience*. New York: Random House.

Boudreaux, R. (2007, December 27). Struggle rooted in land. *Los Angeles Times*, pp. A1, A10.

Bowring, P. (2007, February 26). Tall poppies flourish Down Under. *The International Herald Tribune*. Retrieved November 8, 2008, from http://www.iht.com/bin/print.php?id=4725833

Braganti, N., & Devine, E. (1992). *European customs and manners* (2nd ed.). Minneapolis, MN: Meadowbrook.

Brett, J. (2001). *Negotiating globally*. San Francisco: Jossey-Bass.

Briefing American power: the hobbled hegemon. (2007, June 30). *The Economist,* pp. 29–32.

Brint, S. (1989, July-August). Italy observed. *Society,* pp. 71–76.

Brislin, R. (1993). *Understanding culture's influence on behavior*. New York: Harcourt Brace.

Brown, D. L. (2001, July 15). The boundaries of ignorance: Canadian TV show satirizes Americans. *The Washington Post*, pp. C1, C7.

Brzezinski, Z. (2000, January 7). War and football. *The Washington Post*, p. A23.

Bullington, J. R. (2004, November). Letter from Niger. Retrieved November 8, 2008, from http://www.unc.edu/depts/diplomat/archives_roll/2004_10-12/bullington_0411/bullington_0411.html

Burns, A. C. (1963). *History of Nigeria.* London: Allen & Unwin.

Burr, A. (1917). *Russell H. Connell and his work.* Philadelphia, PA: Winston.

But did they buy their own furniture? (2006, August 12). *The Economist,* p. 46.

By hook or by crook: Italy and the Italian Church. (2007, June 2). *The Economist,* pp. 49–50.

Cagaptay, S. (2006, August 18–20). Islamists in charge. *The Wall Street Journal,* p. 13.

Cahill, T. (1995). *How the Irish saved civilization.* New York: Doubleday, Anchor Books.

Campbell, J. (1962). *The masks of god: Oriental mythology.* New York: Penguin.

Cannadine, D. (1998). *The rise and fall of class in Britain.* New York: Columbia University Press.

Carroll, R. (1987). *Cultural misunderstandings: The French American experience.* Chicago: University of Chicago Press.

Carroll, S., & Gannon, M. (1997). *Ethical dimensions of international management.* Thousand Oaks, CA: Sage.

Chen, M. (2001). *Inside Chinese business: A guide for managers worldwide.* Boston: Harvard Business School Press.

Chhokar, J. S. (2008). India: Diversity and complexity in action. In J. S. Chhokar, F. C. Brodbeck, & R. J. House (Eds.), *Culture and leadership around the world: The GLOBE book of in-depth studies of 25 societies* (pp. 971–1011). Thousand Oaks, CA: Sage.

Chinese Culture Connection. (1987). Chinese values and the search for culture-free dimensions of culture. *Journal of Cross-Cultural Psychology, 18*(2), 143–164.

Choi, Hee Seop. (2005, Oct. 18). Hee Seop Choi quotations. *Baseball almanac.* http://www.baseball-almanac.com/quotes/hee_seop_choiquotes.shtml.

Choi, H. W. (2001, June 26). After massive layoff, Daewoo's CEO begs a thousand pardons. *The Wall Street Journal,* pp. A1, A14.

Choi, H. W. (2002, January 9). Korean club scene: How guys and gals use cupid for hire. *The Wall Street Journal,* pp. A1, A9.

Choudhury, A. (1986). The community concept of business: A critique. *International Studies of Management and Organization, 16*(2), 79–95.

Chu, H. (2008, March 30). From low caste to high office in India. *Los Angeles Times,* p. A9.

Chua, A. (2003). *World on fire.* New York: Doubleday.

Clarke, M., & Crisp, M. (1976). *Understanding ballet.* New York: Harmony.

Clausewitz, C. von. (1989). *On war* (M. Howard & P. Paret, Trans.). Princeton, NJ: Princeton University Press. (Original work published 1832)

Clayre, A. (1985). *The heart of the dragon.* Boston: Houghton Mifflin.

Colvin, J. (2002, July 22). Sick of scandal? Blame football. *Fortune,* p. 44.

Colvin, J. (2003, February 17). For touchdown, see rule 3, section 38. *Fortune,* pp. 17, 34.

Coming apart? It is time to start worrying about Turkey—and looking for ways to revive its European hopes. (2006, May 6). *The Economist,* p. 15.

Conty, M. (1999, Fall). Belle of the ball. *Hermes* (Columbia Business School), pp. 42–44.

Coomaraswamy, A. (1969). *The dance of Shiva.* New York: Sunwise Turn. (Original work published 1924)

Copland, A. (1957). *What to listen for in music.* New York: McGraw-Hill. (Original work published 1939)

Cottrell, J. (1986). *Library of nations: Germany.* Alexandria, VA: Time-Life Books.

Could Flanders be reinvested? (1997, September 20). *The Economist,* p. 54.

Craig, J. (1979). *Culture shock: What not to do in Malaysia and Singapore.* Singapore: Times Book International.

Crook, C. (1991, May 4). A survey of India. *The Economist,* pp. 1–18.

Crow, J. (1985). *Spain: The root and the flower* (3rd ed.). Berkeley: University of California Press.

Crozier, M. (1964). *The bureaucratic phenomenon.* Chicago: University of Chicago Press.

Culturegrams: The nations around us (Vol. 2). (1991). Garrett Park, MD: Garrett Park Press.

Daniszewski, J. (2005a, April 17). Catholicism losing ground in Ireland. *Los Angeles Times*, pp. A1, A8, and A9.

Daniszewski, J. (2005b, April 24). A twist for an ancient tongue trying to survive. *Los Angeles Times,* pp. A1, A8, and A9.

Daun, A. (1991). Individualism and collectivity among Swedes. *Ethnos, 56,* 165–172.

Davidson, B. (1984). *The story of Africa.* London: Mitchell Beazley.

De Gramont, S. (1969). *The French.* New York: Putnam.

de Jorio, A. (2000). *Gesture in Naples and gesture in classical antiquity* (A. Kendon, Trans.). Bloomington: Indiana University Press.

Delany, M. (1974). *Of Irish ways.* Minneapolis, MN: Dillion.

De Mente, B. (1990). *The kata factor.* Phoenix, AZ: Phoenix Books.

Demick, B. (2006, May 21). Koreans' kimchi adulation, with a side of skepticism. *Los Angeles Times,* p. A30.

de Soto, H. (2000). *The mystery of capital: Why capitalism succeeds in the West and fails everywhere else.* New York: Basic Books.

Diamond, J. (1999). *Guns, germs, and steel: The fates of human societies.* New York: Norton.

Dindi, H., & Gazur, M. (1989). *Turkish culture for Americans.* Boulder, CO: International Concepts.

Doing Business. (2008). Economy rankings. Retrieved December 17, 2008, from http://www.doingbusiness.org/EconomyRankings/

Dougherty, C. (2007, December 26). Denmark feels the pinch as young workers flee to lands of lower taxes. *The New York Times.* Retrieved December 20, 2008, from http://www.nytimes.com/2007/12/26/business/worldbusiness/26labor.html?pagewanted=print

The Economist (Ed.). (2007). *Pocket world in figures.* New York: John Wiley.

Edoho, F. (2001). Management in Africa: The quest for a philosophical framework. In F. Edoho (Ed.), *Management challenges for Africa in the twenty-first century*. New York: Praeger.

An eerie lull in the violent delta. (2007, November 8). *Economist.com.*

Eine kleine samba. (1995, November 4). *The Economist,* p. 49.

Elon, A. (1971). *The Israelis.* New York: Penguin.

Emsden, C. (2008, January 26). Euro-Zone growth may rest with Germany. *The Wall Street Journal,* p. A2.

England's shame. (1998, June 20). *The Economist,* p. 49.

Erez, M. (1986). The congruence of goal-setting strategies with sociocultural values and its effects on performance. *Journal of Management, 12,* 83–90.

Esterl, M. (2008, March 4). In Germany, scandals tarnish business elite. *The Wall Street Journal,* p. A1.

Fackler, M. (2007, August 6). Career women in Japan find a blocked path. *The New York Times.* Retrieved December 20, 2008, from http://search.conduit.com/Results.aspx?q=fackler+%22career+women%22&ctid=CT1798526&octid=CT1798526

Fackler, M. (2008, January 2). Japan's new education model: India. *International Herald Tribune.* Retrieved December 20, 2008, from http://www.iht.com/articles/2008/01/02/business/yen.php

Fahri, P. (2002, November 20). Popping the question. *The Washington Post,* pp. C1, C9.

Faiola, A. (1997, October 2). Priest tries to rap Brazilians back into fold. *The Washington Post,* pp. A1, A18.

Fan, M. (2007, December 26). Children of China's elite get lessons in family values. *The San Diego Union-Tribune,* p. A16.

Fang, T. (1999). *Chinese business negotiating style.* Thousand Oaks, CA: Sage.

Farzad, R. (2007, December 10). Can greed save Africa? *Business Week*, pp. 46–54.

Feather, N. (1994). Attitudes toward high achievers and reactions to their fall: Theory and research concerning tall poppies. *Advances in Experimental Social Psychology, 21,* 697–715.

Ferguson, N. (2002). *Empire: The rise and demise of the British world order and the lessons for global power.* New York: Basic Books.

Ferguson, N. (2006, February 27). The crash of civilizations. *Los Angeles Times,* p. B13.

Fieg, J. (1976). *A common core: Thais and Americans.* Yarmouth, ME: Intercultural Press.

Fieg, J., & Mortlock, E. (1989). *A common core: Thais and Americans* (Rev. ed.). Yarmouth, ME: Intercultural Press.

Fisher, G. (1988). *Mindsets.* Yarmouth, ME: Intercultural Press.

Fiske, A. (1991a). The four elementary forms of sociality: Frameworks for a unified theory of social relations. *Psychological Review, 99,* 689–723.

Fiske, A. (1991b). *Structures of social life.* New York: Free Press.

Fitzgerald, M. (2007, March 25). How to improve it? Ask those who use it. *The New York Times.* Retrieved December 20, 2008, from http://www.nytimes.com/2007/03/25/business/yourmoney/25Proto.html

Fleckenstein, B. (1999). Germany: Forerunner of a post-national military? In C. C. Moskos, J. A. Williams, & D. R. Segal (Eds.), *The postmodern military* (pp. 31–53). New York: Oxford University Press.

Fleishman, J. (2007, January 21). Looking west for work. *Los Angeles Times,* pp. C1, C10.

Fleming, C., & Lavin, D. (1997, November 5). How many protests can the French take? About 10,000 a year. *The Wall Street Journal Europe,* p. 1.

The flowers of Kobe. (1995, January 21). *The Economist,* pp. 35–36.

Friedman, T. (1989). *From Beirut to Jerusalem.* Garden City, NY: Doubleday.

Friedman, T. (2005). *The world is flat.* New York: Farrar, Straus, & Giroux.

Frost, E. (1987). *For richer, for poorer.* New York: Council on Foreign Relations.

Furness, N., & Tilton, T. (1979). *The case for the welfare state.* Bloomington: Indiana University Press.

Gabrielidis, C., Stephan, W., Ybarra, O., Dos Santos-Pearson, V., & Villareal, L. (1997). Preferred styles of conflict resolution. *Journal of Cross-Cultural Psychology, 28,* 661–677.

Galbraith, J. (1984). *The affluent society* (4th ed.). Boston: Houghton Mifflin.

Gamerman, E. (2008, Feb. 29). What makes Finnish kids so smart? *The Wall Street Journal,* pp. W1, W10.

Gannon, M. (2003). Cultural metaphors: Their use in management practice and as a method for understanding cultures. In W. J. Lonner, D. L. Dinnel, S. A. Hayes, & D. N. Satler (Eds.), *OnLine readings in psychology and culture.* Western Washington University, Department of Psychology, Center for Cross-Cultural Research. Website: http://www. wwu.edu/~culture.

Gannon, M. (2008). *Paradoxes of culture and globalization.* Thousand Oaks, CA: Sage.

Gannon, M. J., Gupta, A., Audia, P., & Kristof-Brown, A. (2005–2006). Cultural metaphors as frames of reference for nations: A six-country study. *International Studies of Management and Organization, 35*(4), 37–47.

Gardner, A. (2003). The new calculus of Bedouin pastoralism in the Kingdom of Saudi Arabia. *Human Organization,* pp. 1–15.

Geertz, C. (1973). *The interpretation of cultures.* New York: Basic Books.

Gelfand, M., & Holcombe, K. (1998). Behavioral patterns of horizontal and vertical individualism and collectivism. In T. Singelis (Ed.), *Teaching about culture, ethnicity, & diversity* (pp. 121–132). Thousand Oaks, CA: Sage.

George, P. (1987). *University teaching across cultures.* Bangkok, Thailand: U.S. Information Service.

Gilbert, T. (2007). *A Danish Christmas Carol.* Denmark official Web site. Retrieved November 7, 2008 from http://www.denmark.dk/en/servicemenu/News/FocusOn/Archives2007/ADanishChristmasCarol.htm

Gledhill, R. (2007, February 15). Catholics set to pass Anglicans as leading UK church. *Times online*. Retrieved November, 10, 2008, from http://www.timesonline.co.uk/tol/news/article1386939.ece.

Glyn, A. (1970). *The British: Portrait of a people*. New York: Putnam.

Goldin, I., & Reinert, K. (2006). *Globalization for development*. New York: Palgrave Macmillan.

Goldstein, P. (2004, September 12). The game of life. *Los Angeles Times*, pp. E25, E26.

Gopal, R., & Dadachanji, S. (1951). *Indian dancing*. London: Phoenix House.

Gotchenour, T. (1977). The albatross. In D. Batchelder & E. Warner (Eds.), *Beyond experience* (pp. 131–136). Brattleboro, VT: The Experiment Press.

Graham, R. (1984). *Spain: Change of a nation*. London: Michael Joseph.

Gray, K. R. (2001). Small business management in Africa: Prospects for future development. In F. Edoho (Ed.), *Management challenges for Africa in the twenty-first century*. New York: Praeger.

Griffith, S. B. (Trans.). (1963). *Sun Tzu and the art of war*. London: Oxford University Press.

Gross compensation: New CEO pay figures make top brass look positively piggy [Editorial]. (1996, March 18). *Business Week*, pp. 32–34.

Gupta, V., & Hanges, P. J. (2004). Regional and climate clustering of societal cultures. In R. J. House, P. J., Hanges, M. Javidan, P. W. Dorfman, & V. Gupta (Eds), *Culture, leadership, and organizations: The GLOBE study of 62 societies* (pp. 178–215). Thousand Oaks, CA: Sage.

Haire, M., Ghiselli, E., & Porter, L. (1966). *Managerial thinking: An international study*. New York: John Wiley.

Hall, E. (1966). *The hidden dimension*. Garden City, NY: Doubleday.

Hall, E. (1983). *The dance of life*. Garden City, NY: Doubleday.

Hall, E., & Hall, M. (1990). *Understanding cultural differences*. Yarmouth, ME: Intercultural Press.

Halonen, T. (2007, February 16). Opening address by President of the Republic of Finland, Tarja Halonen, at the Innovation Seminar in Sydney, Australia. Retrieved from www.finland.com.au/news_innovation_seminar.html

Hand of God, hand of Italian man. (1998, May 16). *The Economist*, p. 56.

Handley, P. (2006). *The king never smiles*. New Haven, CT: Yale University Press.

Hann, C. (1985). *A village without solidarity: Polish peasants in years of crisis*. New Haven, CT: Yale University Press.

Harden, B. (2008, May 6). Japan steadily becoming a land of few children. *The Washington Post*, p. A10.

Harris, E. (2008, March 2). Nigerians' English goes its own way. *USA Today*. Retrieved February 14, 2009, from http://www.usatoday.com/news/world/2008-03-02-2756073788_x.htm

Haskell, A. (1963). *The Russian genius in ballet*. Elmsford, NY: Pergamon.

Haskell, A. (1968). *Balletomania*. New York: AMS Press.

Haycraft, J. (1985). *Italian labyrinth*. New York: Penguin.

Heclo, H., & Madsen, H. (1987). *Policy and politics in Sweden*. Philadelphia: Temple University Press.

Heinrich, B. (1990). A matter of degrees. In F. R. Thomas (Ed.), *Americans in Denmark* (pp. 35–53). Carbondale: University of Southern Illinois Press.

Hess, D. (1995). *The Brazilian puzzle: Culture on the borderlands of the Western world*. New York: Columbia University Press.

História do Brasil: Curso moderno. (1971). São Paulo, Brazil: Companhia Editora Nacional.

Hitt, M., Ireland, D., & Hoskisson, R. (2007). *Strategic management: Competitiveness and globalization* (7th ed.). Cincinnati, OH: Thomson South-Western.

Hofstadter, R. (1955). *Social Darwinism in American thought*. Boston: Beacon.

Hofstede, G. (1980a). *Culture's consequences*. Beverly Hills, CA: Sage.

Hofstede, G. (1980b). Motivation, leadership, and organization: Do American theories apply abroad? *Organizational Dynamics, 9*, 42–63.

Hofstede, G. (1991). *Cultures and organizations: Software of the mind.* New York: McGraw-Hill.

Hofstede. G. (2001). *Culture's consequences* (2nd ed.). Thousand Oaks, CA: Sage.

Hogg, C. (2006). Japan bows to code of respect. *BBC news online.* Retrieved from http://news.bbc.co.uk/go/pr/fr/-/hi/asia-pacific/4594782.stm

Hollinger, C. (1977). *Mai pen rai means never mind.* Tokyo: John Weatherhill. (Original work published 1967)

House, R., Hanges, P. J., Javidan, M., Dorfman, P., & Gupta, V. (2004). *Culture, leadership, and organizations: The GLOBE study of 62 societies.* Thousand Oaks, CA: Sage.

How the elite has changed. (2002, December 7). *The Economist,* p. 15.

Huntington, S. (1996). *The clash of civilizations.* New York: Simon & Schuster.

Huxley, A. (1951). *Antic hay.* New York: Modern Library.

Ibison, D. (2007, February 5). Sweden moves toward privatizing companies. *Los Angeles Times*, p. C4.

In a league of its own. (2006, April 29). *The Economist*, pp. 63–64.

Isiramen, C. O. (2003, November 1). Women in Nigeria: Religion, culture, and AIDS. *International Humanist and Ethical Union.* Retrieved from http://www.iheu.org/node/979

Iwao, S. (1990). Recent changes in Japanese attitudes. In A. Romberg & T. Yamahoto (Eds.), *Same bed, different dreams* (pp. 55–73). New York: Council on Foreign Relations.

Iyer, V. R. K. (2000, October 22). Personality par excellence. *The Hindu.* Retrieved from http://www.hinduonnet.com//2000/10/22/stories/1322078e.htm

Jacobkson, M. (1987). *Finland: Myth or reality?* Helsinki: Otava.

Jago, A., Maczynski, J., Reber, G., & Bőhnisch, W. (1996). Evolving leadership styles?: A comparison of Polish managers before and after market economy reforms. *Polish Psychological Bulletin, 27*(2), 107–115.

Jago, A., Reber, G., Bőhnisch, W., Maczynski, J., Zavrel, J., & Dudorkin, J. (1993). Culture's consequences? A seven-nation study of participation. In D. F. Rogers & A. S. Raturi (Eds.), *Proceedings of the 24th Annual Meeting of the Decision Sciences Institute* (pp. 451–454). Washington, DC: Decision Sciences Institute.

Jarvenpaa, S., Knoll, K., & Leidner, D. (1998). Is anybody out there? Antecedents of trust in global virtual teams. *Journal of Management Information Systems, 14*(4), 29–36.

Jenkins, D. (1968). *Sweden and the price of progress.* New York: Coward, McCann & Geohegan.

Joe's rant. (2000). Retrieved August 16, 2003, from http://lilith 69.tripod.com/Canadian.html

Johnson, H. (1985). *How to enjoy wine.* New York: Simon & Schuster.

Johnson, K. (2006, July 24). Bullfighting, besieged in Spain, has found young French savior. *The Wall Street Journal,* pp. A1, A6.

Joyce, J. (1964). *Portrait of the artist as a young man.* New York: Viking. (Original work published 1914–15)

Kagitcibasi, C. (1990). Family and home-based intervention. In R. Brislin (Ed.), *Applied cross cultural psychology* (pp. 121–141). Newbury Park, CA: Sage.

Kahn, M. (2007). You can always get greener: Even the cleanest countries have serious environmental problems. *Mőkkikunta* Web site. Retrieved from mokkikunta.blogspot.com/2007/10/ranking-of-best-and-worst-countries-to.html

Kahn, G., & Di Leo, L. (2007, June 28). In Italian crackdown, tax cheats get the boot. *The Wall Street Journal,* pp. A1, A10.

Kaiser, R., & Perkins, L. (2005, June 3). The Finnish sauna (Finland Diary). *The Washington Post.* Retrieved November 11, 2008, from http://blog.washingtonpost.com/finlanddiary/2005/06/the_finnish_sau.html

Kakar, S. (1978). *The inner world.* New York: Oxford University Press.

Kalita, M. S. (2005, December 27). Hope and toil at India's call centers. *The Washington Post,* p. A01.

Kanter, R. (1979). Power failures in management circuits. *Harvard Business Review, 57*(4), 65–75.

Kapuscinski, R. (2001). *The shadow of the sun.* New York: Knopf.

Karp, J., & Williams, M. (1998, April 22). Leave it to Vishnu: Gods of Indian TV are Hindu deities. *The Wall Street Journal,* pp. A1, A10.

Kashiwagi, A. (2002, January 13). Unlikely TV show hooks recession-weary Japanese: Themes of devotion, teamwork boosts spirits of millions. *The Washington Post,* p. A23.

Kaufman, J. (1999, October 22). Why doesn't business, like baseball, create improbable heroes? *The Wall Street Journal,* pp. A1, A8.

Keefe, E. (1977). *Area handbook for Italy.* Washington, DC: American University.

Keirsey, D. (1998). *Please understand me II: Character and temperament types.* Del Mar, CA: Prometheus Nemesis Book.

Kesselman, M., Krieger, J., Allen, C., DeBardeleben, J., Hellman, S., Pontrisson, J., & Ost, D. (1987). *European politics in transition.* Lexington, MA: D.C. Heath.

Kettenacker, L. (1997). *Germany since 1945.* Oxford, UK: Oxford University Press.

Khalid, M. (1979). The sociocultural determinants of Arab diplomacy. In G. Atiyeh (Ed.), *Arab and American cultures* (pp. 123–142). Washington, DC: American Enterprise Institute for Public Policy Research.

Khalil, A. (2007, December 7). In Israel, it's temple vs. state over farming. *Los Angeles Times,* p. A15.

Kightly, C. (1986). *The customs and ceremonies of Britain.* London: Thames & Hudson.

Kluckholn, F., & Strodtbeck, F. (1961). *Variations in value orientations.* Evanston, IL: Row, Peterson.

Knight, C. (2007, July 25). The grandstand erected by Italy. *Los Angeles Times,* pp. E1 and E5.

Kong, L. (2007). *Singapore hawker centres: People, places, food.* National Environment Agency. Retrieved from http://app.nea.gov.sg/cms/htdocs/article.asp?pid=2986

Kotkin, J. (1993). *Tribes.* New York: Random House.

Kraar, L. (1994, October 31). The overseas Chinese. *Fortune,* pp. 91–114.

Kras, E. S. (1989). *Management in two cultures.* Yarmouth, ME: Intercultural Press.

Kresge, N. (2008, February 17). Aging guest workers are fresh poser for Germany. Reuters. Retrieved December 20, 2008, from http://www.reuters.com/article/worldNews/idUSL0521752720080218

Krich, J. (1993). *Why is this country dancing?* New York: Simon & Schuster.

Kripalani, M. (2008, May 19). In India, death to global business. *Business Week,* pp. 42–47.

Kuttner, R. (2008). The Copenhagen consensus: Reading Adam Smith in Denmark. *Foreign Affairs, 87*(2), 78–94.

Laaksonen, P. (1999). Sauna. In O. Alho, H. Hawkins, & P. Vallisaari (Eds.), *Finland: a cultural encyclopedia* (pp. 278–279). Helsinki: Finnish Literature Society.

Lakoff, G., & Johnson, M. (1980). *Metaphors we live by.* Chicago: University of Chicago Press.

Lakshmi, R. (2008, May 11). U.S. legal work blooms in India. *The Washington Post,* p. A20.

Lancaster, J. (1996, November 14). Curtains in Riyadh. *The Washington Post,* pp. A1, A26.

Lannoy, R. (1971). *The speaking tree: A study of Indian culture and society.* New York: Oxford University Press.

Laurent, A. (1983). The cultural diversity of Western concepts of management. *International Studies of Management and Organization, 13*(1–2), 75–96.

Leggett, K. (2005, May 26). Pay-as-you go kibbutzim. *The Wall Street Journal,* pp. B1, B2.

Lessem, R. (1987). *The global business.* London: Prentice-Hall International.

Lessem, R., & Palsule, S. (1997). *Managing in four worlds: From competition to co-creation.* London: Blackwell Business.

Levine, I. (1963). *Main street, Italy.* Garden City, NY: Doubleday.

Lewis, B. (1995). *The Middle East: Brief history of the last 2,000 years.* New York: Simon & Schuster.

Lewis, M. (2006). *The blind side.* New York: Norton.

Lichfield, G. (2004, May 22). Special report: Having it both ways: A survey of Russia. *The Economist,* pp. 1–16.

Lipsyte, R. (2007, January 30). The church of football. *The Nation.* Retrieved December 7, 2008, from http://www.thenation.com/doc/20070212/lipsyte/single

Living with a superpower: Special report, American values. (2003, January 4). *The Economist,* pp. 18–20.

Lucas, E. (2006, May 13). Cheer up: A survey of Poland. *The Economist,* pp. 1–12.

Luce, E. (2008). *In spite of the gods: The rise of modern India.* New York: Anchor Books.

Mackey, S. (1987). *The Saudis: Inside the desert kingdom.* Boston: Houghton Mifflin.

Mackey, S. (1992). *Passion and politics: The turbulent world of the Arabs.* New York: Penguin.

Magnier, M. (2006, April 15). China's honor code. *Los Angeles Times,* pp. A1, A17.

Magnier, M. (2007, February 12). Dying large in China. *Los Angeles Times,* pp. A1, A5.

Major, J. (1989). *The land and the people of China.* New York: J. B. Lippincott.

Manolo Chopera. (2002, September 14). *The Economist,* p. 82.

Marvin, G. (1988). *Bullfight.* New York: Basil Blackwell.

Mbiti, J. (1970). *African religions and philosophy.* New York: Doubleday.

McBride, E. (2002, March 2). A new order: A survey of Thailand. *The Economist,* pp. 1–16.

McBride, S. (2003, August 7). In corporate Asia, a looming crisis over succession. *The Wall Street Journal,* pp. A1, A6.

McCarthy, S. (1994). *Africa: The challenge of transformation.* London: Tauris.

McClave, D. (1996). The society and its environment. In E. Solsten (Ed.), *Germany: A country study* (pp. 125–156). Washington, DC: Library of Congress.

McCourt, F. (1996). *Angela's ashes.* New York: Random House.

McDating. (2002, February 2). *The Economist,* p. 30.

McFaul, M., & Stoner-Weiss, K. (2008). The myth of the authoritarian model: How Putin's crackdown holds Russia back. *Foreign Affairs, 87*(1), 68–84.

McNeilly, M. (1996). *Sun Tzu and the art of business.* New York: Oxford University Press.

McPherson, M., Smith-Lovin, L., & Brashears, M. (2006). Social isolation in America: Changes in core discussion networks over two decades. *American Sociological Review, 71*(3), 353–375. Retrieved February 14, 2009, from http://www.asanet.org/galleries/default-file/June06ASRFeature.pdf

Md. Zabid, A. R., Anantharaman, R. N., & Raveendran, J. (1997). Corporate cultures and work values in dominant ethnic organizations in Malaysia. *Journal of Transnational Management Development, 2*(4), 51–65.

Mead, W. R. (1989). Perceptions of Finland. In M. Engman & D. Kirby (Eds.), *Finland: People, nation, and state* (pp. 1–15). London: C. Hurst.

Meritocracy in America (Special report). (2005, January 1). *The Economist,* pp. 22–24.

Mexico, haunted by new ghosts. (1999, November 6). *The Economist,* p. 36.

Meyer, L. (1982). *Israel now: Portrait of a troubled land.* New York: Delacorte.

Michon, J. (1992, December 5). *Crown and crisis: The time machine.* New York: Arts & Entertainment (television network).

Milbank, D. (1993, March 17). We make a bit more of St. Patrick's Day than the Irish do. *The Wall Street Journal,* pp. A1, A8.

Miller, L., & Hustedde, R. (1987). Group approaches. In D. E. Johnson, L. R. Miller, & G. F. Sommers (Eds.), *Needs assessment: Theory and methods* (pp. 105–131). Ames: Iowa State University Press.

Milner, H. (1989). *Sweden: Social democracy in practice.* New York: Oxford University Press.

Mooij, M. de. (2005). *Global marketing and advertising: Understanding cultural paradoxes* (2nd ed.). Thousand Oaks, CA: Sage.

Moore, T. (1859). *The poetical works of Thomas Moore.* Boston: Philips, Sampson.

More state pupils in universities. (2007, July 19). *BBC news online.* Retrieved November 10, 2008, from http://news.bbc.co.uk/2/hi/uk_news/education/6905288.stm

Moriguchi, C., & Ono, H. (2006). Japanese lifetime employment: A century's perspective. In M. Blomstrom & S. La Croix (Eds.), *Institutional change in Japan: Why it happens, why it doesn't* (pp. 152–176). New York: Routledge.

Morin, R. (1998, September 6). Shy nations. *The Washington Post*, p. C5.

Morrison, T., & Conaway, W. A. (2006). *Kiss, bow, or shake hands* (2nd ed.). Avon, MA: Adams Media.

Morrison, T., Conaway, W., & Douress, J. (1995). *Dun & Bradstreet's guide to doing business around the world.* Englewood Cliffs, NJ: Prentice Hall.

Mouawad, J. (2008, January 20). The construction site called Saudi Arabia. *The New York Times.* Retrieved December 20, 2008, from http://www.nytimes.com/2008/01/20/business/worldbusiness/20saudi.html?n=Top/Reference/Times%20Topics/People/M/Mouawad,%20Jad

Munshi, K. (1965). *Indian inheritance* (Vol. 1). Bombay: Bharatiya Vidya Bhavan.

Musashi, M. (1974). *A book of five rings: The classic guide to strategy* (H. Victor, Trans.). Woodstock, NY: The Overlook Press.

Nakane, C. (1973). *Japanese society.* London: Penguin.

Narayana, G., & Kantner, J. (1992). *Doing the needful.* Boulder, CO: Westview.

Needham, J. (1954). *Science and civilization in China.* Cambridge, UK: Cambridge University Press.

Newman, P. (1987, December 7). A national contempt for the law. *MacLean's*, p. 40.

Ni, C. (2007, May 7). She makes Confucius cool again. *Los Angeles Times*, pp. A1, A8.

Nnadozie, E. (2001). Managing African business culture. In F. Edoho (Ed.), *Management challenges for Africa in the twenty-first century* (pp. 51–62). New York: Praeger.

No black and white matter. (2006, July 15). *The Economist*, p. 38.

Nydell, M. (1987). *Understanding Arabs: A guide for Westerners.* Yarmouth, ME: Intercultural Press.

O'Brien, F. (1961). At Swim-Two-Birds. In U. Mercier & D. Greene (Eds.), *1000 years of Irish prose.* New York: Gross & Dunlap.

O'Brien, F. (1974). *The poor mouth: A bad story about the hard life.* New York: Seaver. (Original work published 1940)

Olson, M. (1982). *The rise and decline of nations.* New Haven, CT: Yale University Press.

Onishi, N. (2008, June 13). Japan, seeking trim waists, measures millions. *The New York Times.* Retrieved December 20, 2008, from http://www.nytimes.com/2008/06/13/world/asia/13fat.html?pagewanted=print

Ortony, A. (1975). Why metaphors are necessary and not just nice. *Educational Theory, 25*(1), 45–53.

Ouchi, W. (1981). *Theory Z.* Reading, MA: Addison-Wesley.

Paglia, C. (1997, September 17). Gridiron feminism. *The Wall Street Journal*, p. A22.

Pakenham, T. (1997). *The scramble for Africa 1876–1912.* London: Abacus.

Parker, J. (2003, November 8). A nation apart: A survey of America. *The Economist*, pp. 1–20.

Paul, A. (2007, January 19). The year of the pigskin. *The Wall Street Journal*, pp. W1, W6.

Paz, O. (1961). *The labyrinth of solitude: Life and thought in Mexico.* New York: Grove.

Pearson, L. (1990). *Children of glasnost: Growing up soviet.* Seattle: University of Washington Press.

Pearson, V., & Stephens, W. (1998). Preferences for styles of negotiation: A comparison of Brazil and the U.S. *International Journal of Intercultural Relations, 22,* 80–99.

Peet, J. (2004, October 16). The luck of the Irish: A survey of Ireland. *The Economist*, pp. 1–12.

Peet, J. (2005, Nov. 26). Addio, dolce vita: A survey of Italy. *The Economist*, pp. 1–16.

Pennington, D. (1985). The rhythms of unity: Time in African culture. In M. K. Asante (Ed.), *African culture* (pp. 123–139). Westport, CT: Greenwood.

Peters, S. (1980). *Bedouin.* Cambridge, MA: Harvard University Press.

Pfanner, E. (2007, December 20). Portugal polishes its image as the "West Coast of Europe." *The New York Times.* Retrieved December 20, 2008, from http://www.nytimes.com/2007/12/20/business/media/20adco.html?pagewanted=print

Pierson, D. (2006, Jan. 3). Cantonese is losing its voice. *Los Angeles Times,* pp. A1, A6.

Pomfret, J. (2003, April 27). My place in the procession of tradition. *The Washington Post,* p. B2.

Pomfret, J. (2008, July 27). A long wait at the gate to greatness. *The Washington Post,* pp. B1, B6.

Pope, H. (1997, March 14). The new middle: Turks add their voices to contest of generals and fundamentalists. *The Wall Street Journal,* pp. A1, A13.

Punctured football. (1993, January 9). *The Economist,* p. 83.

Putnam, R. (1991). *Making democracy work: Civic traditions in modern Italy.* Princeton, NJ: Princeton University Press.

Race in Malaysia: Failing to spread the wealth. (2005, August 27). *The Economist,* p. 49.

Rachman, G. (1999, December 18). The globe in a glass: Christmas survey: Wine. *The Economist,* pp. 91–105.

Radzinsky, E. (2006, July 10). What is Russian civilization? *The Wall Street Journal,* p. A10.

Ramstad, E. (2007, April 25). Why Korea makes the world's best women golfers. *The Wall Street Journal,* pp. A1, A11.

Reid, M. (2006, November 18). Time to wake up: A survey of Mexico. *The Economist,* pp. 1–16.

Reid, T. (2000, January 14). The Yanks have landed. *The Washington Post,* pp. A24, A26.

Reischauer, E. (1988). *The Japanese today: Change and continuity.* Cambridge, MA: Belknap.

Richard, C. (1995). *The new Italians.* New York: Penguin.

Richmond, Y. (1992). *From nyet to da.* Yarmouth, ME: Intercultural Press.

Ricks, D. (2006). *Blunders in international business* (4th ed.). Malden, MA: Blackwell.

Ries, N. (1994). Mystical poverty and the rewards of less: Russian culture and conversation during *perestroika. Dissertation Abstracts International,* 54(09A), University Microfilms AA (No. 9406121).

Robinson, E. (1995, December 10). Over the Brazilian rainbow. *The Washington Post,* pp. C1, C8.

Robinson, E. (1997, September 28). Sound treks, musical adventures in India, Brazil and Italy. *The Washington Post,* pp. 12, 27.

Robinson, E. (1999). *Coal to cream.* New York: Free Press.

Rodriguez, G. (2007, March 11). Is success killing South Koreans? *Los Angeles Times,* p. M4.

Ross, A. (1969). *What are U?* London: Andre Deutch.

Ross, H.C. (1979). The beauty of Bedouin jewelry. *Saudi Aramco World, 30*(2), 4–9.

Rugman, A. (2005). *The regional multinationals.* Cambridge, UK: Cambridge University Press.

Russia under Putin: The making of a neo-KGB state. (2007, August 25). *The Economist,* pp. 25–28.

Russian exceptionalism: Is Russia different? (1996, June 15). *The Economist,* pp. 19–21.

Ruth, A. (1984). The second new nation: The mythology of modern Sweden. *Daedalus, 113,* 53–96.

Sajavaara, K., & Lehtonon, J. (1997). The silent Finn revisited. In A. Jaworski (Ed.), *Silence: Interdisciplinary perspectives* (pp. 263–283). Berlin: Mouton de Gruyter.

Sala-i-Martin, X. (2002, May). *The world distribution of incomes estimated from individual country distributions* (Working paper No. 8933). Cambridge, MA: National Bureau of Economic Research.

Samovar, L., & Porter, R. (1994). *International communication: A reader* (7th ed.). Belmont, CA: Wadsworth.

Sanford, C. (1961). *The quest for paradise.* Urbana: University of Illinois Press.

Schaap, J. (2008, February 8). We will catch excellence. *Parade,* pp. 8–9.

Schein, E. H. (1985). *Organizational culture and leadership.* San Francisco: Jossey-Bass.

Schneider, R. (1996). *Brazil, culture and politics in a new industrial powerhouse.* Boulder, CO: Westview.

Schrady, N. (2008). *The last day: Wrath, ruin and reason in the great Lisbon earthquake of 1755.* New York: Viking Press.

Schrank, D. (2008, January 30). Belgians limp along, hobbled by old language barriers. *The Washington Post,* p. A10.

Schrotenboer, B. (2006). Legal in state since December, mixed martial arts sells out arenas. *San Diego Union Tribune,* pp. A1, A14.

Schulze, H. (1938). *The story of musical instruments: From shepherd's pipe to symphony.* Elkhart, IN: Pan-American Band Instruments.

Schulze, H. (1998). *Germany: A new history* (D. Schneider, Trans). Cambridge, MA: Harvard University Press.

Schwartz, S. (1994). Beyond individualism and collectivism: New cultural dimensions of work values. In U. Kim, H. Triandis, C. Kagitcibasi, S. Choi, & G. Yoon (Eds.), *Individualism and collectivism: Theory, method, and applications* (pp. 19–40). Thousand Oaks, CA: Sage.

Sendut, H., Madsen, J., & Thong, G. (1989). *Managing in a plural society.* Singapore: Longman.

Shamberg, M. (Producer), & Crichton, C. (Director). (1988). *A fish called Wanda.* Hollywood, CA: Metro-Goldwyn-Mayer Pictures.

Shiflett, D. (2000, September 1). Praise the lord and pass the football. *The Wall Street Journal,* p. A10.

Siegel, B. (1992). Family and kinship. In A. A. Gordon (Ed.), *Understanding contemporary Africa* (pp. 175–200). Boulder, CO: Rienner.

Siegele, L. (2006, February 11). Waiting for a wunder: A survey of Germany. *The Economist,* pp. 1–16.

Smith, H. (1958). *The religions of man.* New York: Harper & Row.

Smith, H. (1991). *The world's religions.* San Francisco: HarperCollins.

Smith, K., Grimm, C., & Gannon, M. (1992). *Dynamics of competitive strategy.* Newbury Park, CA: Sage.

Smith, L. (1990, February 26). Fear and loathing of Japan. *Fortune,* pp. 50–60.

Smith-Spark, L. (2006, June 1). Finland's trailblazing path for women. BBC News. Retrieved December 3, 2008, from http://news.bbc.co.uk/1/hi/world/europe/5036602.stm

Solomon, J. (2001, August 17). Stinky national dish seeks smell of success and a global market. *The Wall Street Journal,* pp. A1, A4.

Solomon, M. (1996). *Consumer behavior.* Englewood Cliffs, NJ: Prentice Hall.

Solsten, E. (1996). Introduction. In E. Solsten (Ed.), *Germany: A country study.* Washington, DC: Library of Congress.

The Spanish centrifuge. (2007, March 3). *The Economist,* pp. 56–57.

Srinivas, M. (1980). *India: Social structure.* Delhi, India: Hindustan.

Standage, T. (2007, December 1). Going hybrid: A special report about business in Japan. *The Economist,* pp. 1–18.

Starr, J. (1991). *Kissing through glass.* Chicago: Contemporary Books.

Stecklow, L. (2001, January 20). Helsinki on wheels: Fast Finns find fines fit their finances. *The Wall Street Journal,* pp. A1, A4.

Stewart, E., & Bennett, M. (1991). *American cultural patterns: A cross cultural perspective* (2nd ed.). Yarmouth, ME: Intercultural Press.

Stigler, J. (2002). *Globalism and its discontents.* New York: Norton.

Stoller, R. (1975). *Perversion: The erotic form of hatred.* New York: Pantheon.

Sugawara, S. (1998, August 21). From debt to desperation in Japan. *The Washington Post,* p. G3.

Sullivan, K. (2002, December 26). The union boss is the only man to see. *The Washington Post,* pp. A1, A18.

Sundberg, G. (1910). *Det Svenska folklynnet.* Stockholm, Sweden: Norstedt & Soners.

Swardson, A. (1996, November 30). Surly Parisians taught to grin and bare their friendly side. *The Washington Post,* pp. A12.

Szalay, L. (1993). *The subjective worlds of Russians and Americans: A guide for mutual understanding.* Chevy Chase, MD: Institute of Comparative Social and Cultural Studies.

Tan, A. (1991). *The kitchen god's wife.* New York: Ballantine.

Tasker, P. (1987). *The Japanese: Portrait of a nation.* New York: Meridian.

Tavernise, S. (2008, February 9). Turkey's parliament votes to lift head scarf ban. *International Herald Tribune.* Retrieved December 20, 2008, from http://www.iht.com/articles/2008/02/09/europe/10turkey.php

Taylor, S. (1990). *Culture shock: France.* Portland, OR: Graphic Arts Center.

Thais tickled pink by unlikely fashion trendsetter: king. (2007, December 1). *San Diego Union Tribune,* p. A2.

Tharoor, S. (2007). *The elephant, the tiger, and the cell phone: Reflections on India, the emerging 21st-century power.* New York: Arcade.

Theroux, P. (2003). *Dark star safari.* New York: Houghton Mifflin.

Thomas, F. (Ed.). (1990). *Americans in Denmark.* Carbondale: University of Southern Illinois Press.

Tolan, S. (2006, May 26). The incredible shrinking Palestine. *Los Angeles Times,* p. M3.

Tornblom, K., Jonsson, D., & Foa, U. (1985). National resource, class, and preferences among three allocation rules: Sweden vs. USA. *International Journal of Intercultural Relations, 9,* 51–77.

Triandis, H. C. (1984). Toward a psychological theory of economic growth. *International Journal of Psychology, 19,* 79–95.

Triandis, H. (2002). Generic individualism and collectivism. In M. Gannon & K. Newman (Eds.), *Handbook of cross-cultural management* (pp. 16–51). London: Blackwell.

Triandis, H., Brislin, R., & Hui, C. (1988). Cross-cultural training across the individualism-collectivism divide. *International Journal of Intercultural Relations, 12,* 269–289.

Triandis, H., & Gelfand, M. (1998). Convergent measurement of horizontal and vertical individualism and collectivism. *Journal of Personality and Social Psychology, 74,* 118–128.

Trofimov, Y. (2008, March 10). Vote rocks Malaysian government. *The Wall Street Journal,* p. A3.

Trompenaars, F., & Hampden-Turner, C. (1998). *Riding the waves of culture* (2nd ed.). New York: McGraw-Hill.

Trueheart, C. (1998, January 30). Can Portugal survive the Euro? *The Washington Post,* pp. A25, A28.

Tuchman, B. (1962). *The guns of August.* New York: Macmillan.

Twitchin, J. (n.d.). *Training notes for the video, crosstalk at work: Cross cultural communication in the workplace.* London: BBC Training Videos.

Ugwuegbu, D. (2001). *The psychology of management in African organizations.* New York: Quorum.

UNAIDS. (2007). *Estimated number of people living with HIV by country, 1990–2007.* Retrieved December 17, 2008, from http://data.unaids.org/pub/GlobalReport/2008/20080818_gr08_plwh_1990_2007_en.xls

Unger, B. (2007, April 14). Dreaming of glory: A special report on Brazil. *The Economist,* pp. 1–16.

Upkabi, C. (1990). *Doing business in Africa.* Amsterdam: Royal Tropical Institute.

Vedel, A. (Ed.). (1986). *The Hachette guide to French wines.* New York: Knopf.

Venezia, E. (1997). *Cross-cultural management: UN solo management nella differenziata realta Italiana?* Milan: Universita Boccon, Tesi di laurea.

Verleyen, F. (1987). *Flanders today.* Tielt, Belgium: Lannoo Editions.

Vincent, J. 2003. *Culture and customs in Brazil.* Westport, CT: Greenwood Press.

Vogel, E. (1979). *Japan as number one: Lessons for America.* Cambridge, MA: Harvard University Press.

Wadhwa, V. (2008, July 23). What the U.S. can learn from Indian R & D. *Business Week.* Retrieved December 3, 2008, from http://www.businessweek.com/technology/content/jul2008/tc20080722_958899.htm

Wadhwa, V., de Vitton, U. K., & Gereffi, G. (2008, July 23). How the disciple became the guru. Abstract retrieved February 13, 2009, from http://ssrn.com/abstract=1170049

Wakin, D. J. (2006, May 5). An organ recital for the very, very patient. *International Herald Tribune.* Retrieved December 20, 2008, from http://www.iht.com/articles/2006/05/05/europe/web.0505cage.php

Waldman, P. (2007, June 26). As Israel prospers, some fear its defenses may grow soft. *The Wall Street Journal,* pp. A1, A12.

Wallace, P. (1999). *The psychology of the Internet.* Cambridge, UK: Cambridge University Press.

Wang, G. (1962). *Latar belakang kebudayaan cina: Basis of Chinese culture.* Kuala Lumpur, Malaysia: Dewan Bahasa dan Pustaka.

Waters, M. (1984). *The comic Irishman.* Albany: State University of New York Press.

Wax, E. (2008, April 19). Redskins cheerleaders shake up cricket in modest India. *The Washington Post,* p. A01.

Waxman, S. (1993, June 9). Brussels, capital of confusion. *The Washington Post,* pp. D1, D6.

Weber, M. (1947). *Theory of social and economic organization* (A. Henderson & T. Parsons, Trans.). New York: Free Press.

What game? (2008, February 1). *The Wall Street Journal,* pp. W1, W10.

Wheaton, K. (Ed.). (1990). *Insight guide: Spain.* Singapore: APA.

When good men turn racist. (2002, October 19). *The Economist,* p. 43.

Whitlock, C. (2008, February 2). Germany rebuffs U.S. on troops in Afghanistan. *The Washington Post,* p. A10.

Whitlock, C., & Smiley, S. (2008, March 14). Non-European PhDs in Germany find use of 'Doktor' verboten. *The Washington Post,* p. A01.

Wilkinson, T. (2005, August 24). Wiretaps unfold Italian tycoons' dirty laundry. *Los Angeles Times,* pp. A1, A8.

Wilkinson, T. (2007a, Feb. 16). Italy's rich pay to look richer. *Los Angeles Times,* p. A5.

Wilkinson, T. (2007b, Oct. 16). Rebirth in the arena. *Los Angeles Times,* pp. A1, A8.

Wilkinson, T. (2007c, January 9). Taking the "honor" out of killing women. *Los Angeles Times,* p. A4.

Willey, D. (1984). *Italians.* London: British Broadcasting Corporation.

Williams, C. (1998, February 23). A tough new course in Moscow schools: Manners. *The Washington Post,* p. A13.

Wilson, J. Q. (2007, April 20). In defense of guns. *Los Angeles Times,* p. A31.

Wilson, P., & Graham, D. (1994). *Saudi Arabia: The coming storm.* Armonk, NY: Sharpe.

Women in the workforce: The importance of sex. (2006, April 12). *The Economist.* Retrieved December 3, 2008, from http://www.economist.com/opinion/displaystory.cfm?story_id=6800723

Woolis, D. (2005, December 29). Spaniards' long lunch losing its allure. *The San Diego Union Tribune,* p. C4.

Yew, L. K. (2000). *From third world to first: The Singapore story: 1965–2000.* New York: HarperCollins.

Zachary, G. (1999, March 17). Barring entry: Ireland faces a shortage of pubs, and the blame falls on old rules. *The Wall Street Journal,* pp. A1, A10.

Zaman, A. (1999, December 2). Spreading faith through fashion: Turkish chain promotes Islamic clothing. *The Washington Post,* p. A32.

Zetterberg, H. (1984). The rational humanitarians. *Daedalus, 113,* 72–79.

Index

About the Authors

Martin J. Gannon (Ph.D., Columbia University) is Professor of International Management and Strategy, College of Business Administration, California State University San Marcos. He is also Professor Emeritus, Smith School of Business, University of Maryland at College Park. At Maryland he held several administrative positions, including the Associate Deanship for Academic Affairs and the Founding Directorship of the Center for Global Business, and received the University's International Landmark Award.

Professor Gannon has authored, co-authored, or co-edited 80 articles and 17 books, including *Paradoxes of Culture and Globalization* (2008), *Handbook of Cross-Cultural Management* (2001), *Dynamics of Competitive Strategy* (1992), *Managing without Traditional Methods: International Innovations in Human Resource Management* (1996), and *Ethical Dimensions of International Management* (1997). Professor Gannon has been the Senior Research Fulbright Professor at the Center for the Study of Work and Higher Education in Germany and the John F. Kennedy/Fulbright Professor at Thammasat University in Bangkok, and he has served as a visiting professor at several Asian and European universities. He has also been a consultant to many companies and government agencies. Professor Gannon has lived and worked in more than 25 nations for various periods of time as a visiting professor, consultant, and trainer. For additional information on Professor Gannon, please visit his home page at California State University San Marcos: www2.csusm.edu/mgannon/.

Rajnandini "Raj" Pillai (Ph.D., State University of New York at Buffalo, 1994) is a Professor of Management at the College of Business, California State University San Marcos (CSUSM). She is also Executive Director and founding member of the Center for Leadership Innovation and Mentorship Building (CLIMB) at the university. Her areas of research interest are leadership and cross-cultural management. She has published her work on charismatic and transformational leadership, leadership and voting behavior, and cross-cultural differences in organizational justice in some of the leading journals in her field such as *The Leadership Quarterly, Journal of Management,* and the *Journal of International Business Studies*. She has also co-edited two books, *Teaching Leadership: Innovative Approaches for the 21st Century* (2003) and *Follower*

Perspectives on Leadership (2007). She serves on the editorial board of *The Leadership Quarterly*. Professor Pillai has held mid-level management positions in the banking industry in India, consulted with U.S. organizations on leadership effectiveness, and conducted workshops on leadership and global issues. She has received awards for excellence in teaching and research including the College of Business Outstanding Professor Award, the Western Academy of Management Ascendant Scholar Award, and the CSUSM President's Award for Scholarship and Creative Activity.

Supporting researchers for more than 40 years

Research methods have always been at the core of SAGE's publishing program. Founder Sara Miller McCune published SAGE's first methods book, *Public Policy Evaluation*, in 1970. Soon after, she launched the *Quantitative Applications in the Social Sciences* series—affectionately known as the "little green books."

Always at the forefront of developing and supporting new approaches in methods, SAGE published early groundbreaking texts and journals in the fields of qualitative methods and evaluation.

Today, more than 40 years and two million little green books later, SAGE continues to push the boundaries with a growing list of more than 1,200 research methods books, journals, and reference works across the social, behavioral, and health sciences. Its imprints—Pine Forge Press, home of innovative textbooks in sociology, and Corwin, publisher of PreK–12 resources for teachers and administrators—broaden SAGE's range of offerings in methods. SAGE further extended its impact in 2008 when it acquired CQ Press and its best-selling and highly respected political science research methods list.

From qualitative, quantitative, and mixed methods to evaluation, SAGE is the essential resource for academics and practitioners looking for the latest methods by leading scholars.

For more information, visit **www.sagepub.com**.